The Communicative Cardiac Cell

ANNALS OF THE NEW YORK ACADEMY OF SCIENCES
Volume 1047

The Communicative Cardiac Cell

Edited by
Samuel Sideman, Rafael Beyar, and Amir Landesberg

The New York Academy of Sciences
New York, New York
2005

Library of Congress Cataloging-in-Publication Data

Larry and Horti Fairberg Cardiac Workshop (3rd : 2005 : Sintra, Portugal)
The communicative cardiac cell / edited by Samuel Sideman, Rafael Beyar, and Amir Landesberg.
p. ; cm. — (Annals of the New York Academy of Sciences ; v. 1047)
Result of the 3rd Larry & Horti Fairberg Cardiac Workshop held Jan. 15–19, 2005 in Sintra, Portugal.
Includes bibliographical references and index.
ISBN 1-57331-547-8 (cloth : alk. paper) — ISBN 1-57331-548-6 (pbk. : alk. paper)
1. Heart cells — Congresses. 2. Cell interaction — Congresses. 3. Cellular signal transduction — Congresses. 4. Calcium channels — Congresses.
[DNLM: 1. Cell Communication — physiology — Congresses. 2. Myocytes, Cardiac — physiology — Congresses. 3. Myocytes, Cardiac — pathology — Congresses. WG 280 L3344c 2005] I. Sideman, S. II. Beyar, Rafael. III. Landesberg, Amir. IV. Title. V. Series.
Q11.N5 vol. 1047
[QP114.C44]
500 s — dc22
[612.1 2005014802

Asterisk/CCP
Printed in the United States of America
ISBN 1-57331-547-8 (cloth)
ISBN 1-57331-548-6 (paper)
ISSN 0077-8923

ANNALS OF THE NEW YORK ACADEMY OF SCIENCES

Volume 1047
June 2005

The Communicative Cardiac Cell

Editors
SAMUEL SIDEMAN, RAFAEL BEYAR, AND AMIR LANDESBERG

Conference Chairs
SAMUEL SIDEMAN AND RAFAEL BEYAR

International Scientific Advisory Committee
JAMES B. BASSINGTHWAIGHTE, RAFAEL BEYAR, GIANLUIGI CONDORELLI, AMIR LANDESBERG, ANDREW D. MCCULLOCH, MARTIN MORAD, JAMES POTTER, YORAM RUDY, SAMUEL SIDEMAN, AND HEINRICH TAEGTMEYER

International Organizing Committee
SAMUEL SIDEMAN, RAFAEL BEYAR, AND AMIR LANDESBERG

This volume is the result of the 3rd Larry & Horti Fairberg Cardiac Workshop entitled *The Communicative Cardiac Cell*, held January 15–19, 2005 in Sintra, Portugal, and organized by the Technion, Israel Institute of Technology, Haifa, Israel and the Portuguese Ministry of Science & Technology, Portugal.

CONTENTS

Part I. Prologue

Part II. Cardiac Cell Development

Part III. Calcium Signaling

Part IV. Intracellular Signaling

Part V. Energetics and Transport

Part VI. Electromechanical Interactions and Modeling

Part VII. Novel Therapeutics for the Failing and Arrhythmogenic Cardiac Cell

Part VIII. Epilogue: A View into the Future

Financial assistance was received from:

- TECHNION, ISRAEL INSTITUTE OF TECHNOLOGY, HAIFA, ISRAEL
- DEPARTMENT OF BIOMEDICAL ENGINEERING
- THE BRUCE RAPPAPORT FACULTY OF MEDICINE
- AMERICAN TECHNION SOCIETY, NEW YORK CITY, USA
- THE LARRY AND HORTI FAIRBERG FUND FOR CARDIAC WORKSHOPS
- THE PORTUGUESE MINISTRY OF SCIENCE & TECHNOLOGY, LISBON, PORTUGAL

In Memory of
Horti and Larry Fairberg

Preface

The Cellular Communications Maze

SAMUEL SIDEMAN

Faculty of Biomedical Engineering, Technion, Israel Institute of Technology, Haifa, Israel

ABSTRACT: Inter- and intracellular ionic and molecular communications are indispensable to the preservation of life and all organic functions. The cell constantly responds to a myriad of extracellular ionic and molecular signals, and cell behavior in single-cell or multicellular organisms is coordinated by these signals moving into, inside, or between the cells. The signals pass through the cell's phospholipidic plasma membrane by diffusion and, mostly, via protein transporters: gates, receptors, and/or ion pumps imbedded in the membrane. Each signaling pathway is a complex cause and effect chain of events involving intricate networks of interactions. For example, the extracellular signals, which are monitored by the cell's membrane cognate receptors, form ligand–receptor complexes. These are typically amplified by interaction with a coupling protein, mostly a G protein, and are diversified via intracellular signal transductions, either directly or via the activation of intracellular second messengers. The present volume focuses on signal-induced interactions that trigger specific responses of the effector system. Highlighting the cellular communication framework and the major parameters encountered in these complex interactive phenomena may help to familiarize the uninitiated with the complex phenomena involved in sustaining life.

KEYWORDS: ligands; membranes; protein gates; receptors; signal pathways

INTRODUCTION

The Goldberg/Fairberg Cardiac Workshops

The Fairberg Cardiac Workshops continue the tradition of the Henry Goldberg Cardiac Workshops,[1–12] initiated in 1984, of manifesting the crucial need for scientific interaction and cooperation in the development of cardiac science. Goldberg's workshops brought together leading scientists and practitioners of the cardiac art, introduced quantitative and analytical procedures so as to better understand the cardiac physiology and the practical clinical picture, and explored the interactions between the major parameters affecting cardiac performance. Thus, the history of the Goldberg/Fairberg Workshops started with an organ-based, integrative approach to the cardiac system, with the focus on cardiac organ function in health and disease. The first

Address for correspondence: Prof. Sam Sideman, D.Sc., D.Sc.Hon., Faculty of Biomedical Engineering, Technion, IIT, Haifa, Israel 32000. Fax: 972 4 8294599.
sam@bm.technion.ac.il

Ann. N.Y. Acad. Sci. 1047: xi–xxiv (2005). © 2005 New York Academy of Sciences.
doi: 10.1196/annals.1341.000

Goldberg Workshop,[1] held in Haifa in 1984, explored the interactions of measurable organ scale mechanics, electrical activation, perfusion, and metabolism, and emphasized imaging in the clinical environment. However, it became very clear that a molecular and cellular approach must be implemented in order to understand the basic cardiac phenomena, even from a mechanistic viewpoint. The concept of the Cardiome was thus established as a multilevel, multiparameter description of the cardiac system.[13] The twelfth workshop, the second Fairberg Workshop[12] held in Sicily in 2003, endeavored to decipher the Cardiome by combining an analytical systems approach and engineering principles with molecular and/or genetic factors and macroscale interactions, and pursuing in depth the characteristics of the control mechanisms affecting the various interactions and functions.

The present third Fairberg Workshop continues the quest for the mechanisms that affect cardiac function by exploring, analytically and experimentally, the physiological and clinical manifestations of the inter- and intracellular signaling and communication. As such, this workshop pursues the goals of the past cardiac workshops: (1) to foster interdisciplinary interaction, so as to catalyze new research frontiers; (2) to relate basic microscale, molecular, and subcellular phenomena to the global clinically manifested cardiac performance; (3) to apply conceptual modeling and quantitative analysis to better explore, describe, and understand cardiac physiology; (4) to interpret available clinical data and design new revealing experiments; and (5) to enhance international cooperation in the endless search for better understanding of the cardiac phenomena and their implication in cardiac pathophysiology. Obviously, this international gathering of outstanding cardiac scientists exemplifies human ingenuity and demonstrates the interdisciplinary cooperation in the pursuit of new discoveries and practical medical benefits.

CELLULAR COMMUNICATIONS

The last century brought great insight into the molecular basis of life, establishing that life depends on the existence of molecular networks as integral elements within individual cells. The cell matrix formation of a multicellular human blastocyst starts with differentiation into three primary germ layers: the ectoderm, mesoderm, and endoderm, forming small collections of multipotent cells in the embryo. These proceed by continuous cell division and specialization into some 200 cell types to produce an entire community of cooperating pluripotent cells, each with its own role in the life of the organism. The pluripotent cells fated to differentiate into heart muscle cells, the cardiomyocytes, are directed through a signaling pathway (genes Nkx2.5, Tcx5, etc.) that contains both activating and inhibitory signals. Thus, the signals allow the embryonic mesoderm to differentiate into the cardiac progenitor cells and then to the various cell types that construct the functional heart. Multipotent cells need to communicate with their environment, their neighboring cells, as well as between their various intracellular compartments, in order to survive. Due to cell cooperation, the organism becomes much more than the sum of its component parts.

Cellular communications involve a complex network of interactions between proteins and other biomolecules. Most of these networks are composed of protein-protein interactions that need to be tightly regulated for proper cell life. The cellular and intracellular bilayered membranes are embedded with protein gates, receptors, or pumps that play a critical role in facilitating and controlling the signaling pathways. These plasma membrane protein receptors and transporters are particularly interesting in pathogenesis as failures in intercellular/intracellular communication networks can lead to many diseases. An illuminating review of signaling phenomena, receptors, coupling proteins, and second messengers is given by Arnold M. Katz.[14] A detailed survey of signaling pathways and protein complexes that participate in cell communication is given by Dennis Bray.[15]

Much is still to be learned by deciphering the complex cellular communication systems, including transcription schemes, metabolic networks, and signaling pathways. The latter is in effect the particular goal of the present workshop. Each signaling pathway is a complex cause and effect chain of interactions, and the present volume represents an attempt to elucidate some of the issues encountered in the study of these complex phenomena.

CELLULAR CHARACTERISTICS

From a chemical point of view, a cell, which has the ability to reproduce itself, is a special collection of molecules enclosed by a thin membrane. This reproductive process occurs by cell growth, by ingesting certain simple molecules from its surroundings, followed by cell division. Cells can thus be seen as a self-replicating network of catalytic macromolecules engaged in a carefully balanced series of energy conversions that drive biosynthesis and cell movement. But energy alone is not enough to make self-reproduction possible, and the cell must contain detailed instructions that dictate exactly how that energy is to be used. These instructions are stored and transferred by the cell's DNA, RNA, and protein molecules (e.g., enzymes and hormones).

From a physical point of view, the cell consists of four basic elements: (1) the bounding membrane, (2) the cytoplasm (i.e., the gel-like fluid filling the cell) (3) the cytoskeleton—the fibrous network that keeps its shape, and (4) the various functional organelles.

The Membrane

Each cell is enclosed by a thin, 5-nm-thick, lipid bilayer membrane—embedded with functional protein elements—that forms a selective barrier allowing ions and molecules to enter and waste products to leave. The lipid bilayer is permeable to water and small uncharged lipophilic molecules, including O_2, NO, and CO_2 or H_2CO_3, but it is impermeable to most essential molecules and ions, both cations (K^+, Na^+, Ca^{2+}) and anions (Cl^-, HCO_3^-), as well as hydrophilic molecules like glucose and macromolecules like proteins and RNA. The chemical structure of the plasma membrane makes it remarkably flexible and adjustable to rapidly growing and dividing cells.

Like the cell's plasma membrane, the membranes of intracellular organelles incorporate transport proteins, or permeases, that allow chemical communication between the organelles via the cytoplasm. Thus, protein gates, receptors, and ion pumps, which are imbedded in the membrane, control the influx and outflux of vital ions (Ca^{2+}, Na^+, K^+) and molecules so as to ensure a steady function of the intracellular elements. Permeases in the lysosomal membrane, for example, allow amino acids generated inside the lysosome to cross into the cytoplasm, where they can be used for the synthesis of new proteins.

The Cytoplasm

The cytoplasm, a gel-like fluid that contains the organelles and accounts for about half of the cell's total volume, is the locus for protein synthesis. The cytoplasm is enclosed by the "insulating" cell membrane and contains organelles and large, water-soluble, highly charged molecules, such as proteins, nucleic acids, carbohydrates, and various substances involved in cellular metabolism.

The Cytoskeleton

The cytoskeleton, a fibrous network formed by different types of long protein filaments, is distributed throughout the cytoplasm of eukaryotic cells and creates a framework that organizes the cell constituents and maintains the shape of the cell. Cytoskeletons are constantly evolving shapes. The main component parts of the cytoskeleton are actin microfilaments, intermediate stretch resilient microfilaments, and microtubules, which form the tracks for the transport of molecules within the cell.

The Organelles

The cell's interior includes a number of specialized compartments, or organelles, each surrounded by a separate membrane. The intracellular membranes are specific to these "compartments" and have very complex structures and shapes. The cardiac muscle cell contains *mitochondria*, the chemical factory that produces ATP; the *sarcomere*, the mechanical engine of muscle contraction and motion; and the *sarcoplasmic reticulum* (SR), a dense membranous network that is the dynamic Ca^{2+} storeroom and the regulator of intracellular calcium cycling. These three organelles play the key role in the cardiac excitation–contraction coupling. To ensure cardiac contraction under different external loads, organelles are in constant communication via the transport of free calcium ions (Ca^{2+}) and ATP and its oxidation products ADP and phosphate. Other important cellular organelles include the following:

The *nucleus* is the information center of the cell in all higher organisms. It is separated from the cytoplasm by an enveloping lipid membrane, and it houses the genome—the double-stranded, spiral-shaped deoxyribonucleic acid (DNA) molecules that contain the genetic information necessary for the cell to retain its unique character as it grows and divides (i.e., the genetic information necessary for cell growth and reproduction).

The *endoplasmic reticulum* (ER), a tubular, 3-D cobweb-like network covering most of the intracellular space—either as a meshwork of fine tubular membrane vesicles (smooth ER) or as a series of connected vesicles and flattened sacs (rough ER)—is a crucial element in the information chain within the cell. The ER plays a central role in the synthesis and export of proteins and glycoproteins and in the synthesis of phospholipids and cholesterol—major components of the plasma and internal membranes.

The *Golgi complex*, a most important communication element, is the site for the modification, completion, and export of secretory proteins and glycoproteins. This organelle, first described by the Italian cytologist Camillo Golgi in 1898, acts as post office to vesicles arriving from the ER and departing to pre-assigned destinations. Proteins move from the cis to the trans face of the Golgi in transport vesicles that bud and fuse with other membranes, in much the same manner as the transport from the ER to the Golgi. This trip across the Golgi often acts to concentrate the secretory product.

Not all proteins synthesized on the ER are destined for export: many remain inside the cell, and some become anchored in the membranes of internal organelles or in the plasma membrane. *Autocrine signaling* involves cells that respond to self-released substances and trigger receptors within their membrane. Such signals often initiate differentiation in developmental processes. It is presumed that each protein has some type of marker that fits a specific location in the cell.

INTERCELLULAR COMMUNICATIONS

Ionic and Molecular Signaling

The extracellular space outside the cells is filled with a composite material, which is composed of a gel with suspended fibrous proteins. The gel consists of large polysaccharide molecules in a water solution of inorganic salts, nutrients, and waste.

Cells in a multicellular organism communicate with each other via chemical signals. The survival of multicellular organisms depends on elaborate intercellular communications that coordinate the growth, differentiation, apoptosis, and metabolism in single cells of various tissues. This intercellular communication is realized either directly by coupling the cells within a tissue/organ—mainly through the gap junctions between adjacent cells—or indirectly by extracellular signaling molecules, for example, hormones, growth factors, and neurotransmitters that can diffuse or be transported in the blood to reach the appropriate cell(s).

Gap Junctions

Apart from a few terminally differentiated cells (e.g., skeletal muscle, red blood cells, lymphocytes), most body cells communicate via gap junctions. Gap junctions are intercellular channels, approximately 1.5 nm in diameter, composed of 2 integral proteins—connexons—one in each membrane, connected in mirror symmetry. The connexon is an oligomer of six intrinsic phosphoproteins named connexins that form the channel that makes up the gap junction. The major physiological role of gap junctions is to synchronize metabolic or electronic signals between cells in a tissue. The gap junctions form continuous aqueous pathways that provide free passage of signal-carrying ions or small (<1,000 da) molecules between the adjacent cells and allow the communicating cells to equilibrate their regulatory ions and molecules (e.g., Na^+, K^+, Ca^{2+}, cAMP, glutathione, amino acids sugars, nucleotides). The adheres junctions that provide mechanical strength and hold the cardiac cells tightly during muscle contraction/relaxation are, like the gap junctions, constructed of transmembrane proteins.

The intercellular ion flow allows the propagation of the membrane electrical action potential along the adjacent cells throughout the tissue. The strength of the action potential is constant, being an inherent cellular characteristic, and is independent of the strength of the stimulation, which does, however, affect the frequency of formation of the action potential. Increased intracellular Ca^{2+} and lower pH decrease the gap junction permeability.

Extracellular Signals

Extracellular signals must be recognized by membrane-bound receptors in the plasma membrane on the target cell. Some signal molecules are able to cross the plasma membrane into the cytosol, where they bind intracellular receptors. Also, certain membrane-bound proteins in one cell can directly signal an adjacent cell. Clearly, the cell signaling processes ensure that crucial activities occur in the right cells at the right time and in synchrony with the other cells.

In 1912, Otto Loewi was first to demonstrate that nerve stimulation could generate chemical *neurotransmitters*, which can modify the cardiac cellular function. Acetylcholine, a chemical mediator parasympathetic neurotransmitter generated by vagal stimulation, does not "generate" a second messenger and, via a group of inhibitory coupling proteins, reduces cAMP production, thus blunting response to the cAMP sympathetic stimulation. Other chemical mediators released by sympathetic stimulation include cardioaccelerators like catecholamine and epinephrine (adrenalin). Note that the extracellular messengers and neurotransmitters include dopamine, histamine, serotonin, glucagons, glutamate as well as peptides (e.g., angiotensin II, bradykinin and vasopressin, which must bind to specific plasma membrane receptors in order to modify the cardiac function). These specific substances produce a specific response only in target cells that have receptors for these very signaling molecules. Thus, specific responses to the signals of neurotransmitters are determined by the structure of the neurotransmitters as well as by the nature of the receptor recognizing the particular signal. This involves detection and recognition of the arriving signal and its chemical binding to the receptor protein molecules, each specialized for different functions.

In addition to signaling ions (H^+, Na^+, K^+, Ca^{2+}) and neurotransmitters, the cells secrete various soluble intercellular signaling molecules (e.g., growth factors and cytokines) as well as *extracellular membrane vesicles* composed of different kinds of proteins and lipids. These bead-like vesicles also represent a common form of chemical signaling via secreted molecules with inscribed information that move through the extracellular space. This discovery won Günther Blobel of Rockefeller University in New York the Nobel Prize for Medicine in 1999.

Signaling by secreted molecules can be classified by two basic mechanisms:

(1) *Paracrine signaling* (e.g., growth regulators). The signaling molecules affect only nearby target cells. The influence of the signal molecule is short lived and is removed from the environment. Many cells can simultaneously receive and respond to chemical signals produced by a single cell in their vicinity. Note that molecules on the surfaces of adjacent cells can directly contact each other, and specific surface molecules on plasma membranes can serve as signals. Cell recognition markers are important in embryo development.

(2) *Endocrine signaling*. The signaling molecules travel through the organ and act on target cells located at a distance from their origin of synthesis. Cells of endocrine organs secrete hormones into the circulatory vessels where they travel to the target cells in other parts of the body.

TRANSMEMBRANE TRANSPORT

Nature has devised many ways for the transfer of information between the hydrophilic extracellular and intracellular environments across the hydrophobic lipid plasma membrane, mostly by utilizing various types of membrane-bound protein receptors.

External cell signals can use various mechanisms to pass the "message" into the cell's interior, either directly or indirectly via an ion-channel-linked receptor, a G protein–linked receptor, or an enzyme-linked receptor (see below). *Direct* transfer includes signal molecules as, for example, hydrophobic molecules (e.g., steroid hormones) or nitric oxide (NO). NO often binds to the enzyme that converts GTP into cyclic GMP, an internal cell signal molecule. *Indirect* transfer occurs when the signaling ions or molecules (e.g., ligands, neurotransmitters, or vesicles) bind to the receptor protein and trigger some changes that initiate the process of transduction. This brings about a change, often a conformational change, in the receptor molecule, thus effectively translating the signal into a form that the target cell recognizes and that can bring about a specific cellular response. This transduction may be a single step, wherein the activated protein complex affects the target molecule directly, or a relay pathway of chemical reactions involving the creation of *second messengers* (see below) within the cell. This cascade of intracellular signals via the secondary messengers ultimately alters the behavior of the target cell.

Transport Modes across the Cell Membranes

Transport across cell membranes involves a number of major mechanisms:

(1) *Passive diffusion*. Molecules and ions move spontaneously down their concentration gradient (e.g., lipid soluble substances diffuse across the membrane bilayer), including ligands that dissolve in the membrane bilayer before reaching the protein receptor.

(2) *Integral protein-mediated transport*
   a. Passive movement of water through aquaporin channels (water only)
   b. Passive facilitated transport mediated by binding to a specific transporter protein
   c. Passive ionic diffusion along an electrochemical gradient through time- and voltage-dependent ion selective channels
   d. Ionic exchangers including co- and countertransporters, depending on the electrochemical gradients
   e. Active, energy-dependent, transport against the concentration or potential gradient, mainly of ions, by the ion pumps. Some active transporters bind ATP directly and use the energy of its hydrolysis to drive the ions or molecules against a concentration gradient. For example, almost one-third of all the energy generated by the mitochondria is used just to run the Na^+/K^+ ATPase pump.

(3) *Vesicular facilitated transport*. Information-carrying molecules may be transported into and inside the cell while encapsulated in vesicles. An important example is the transport of free fatty acid molecules, encapsulated in the lipolytic chylomicron triacylglycol vesicles in the extracellular fluid/capillary lumen, across the cell membrane to the intracellular sites of triacylglycerol activation, esterification, lipolysis, and fatty acid oxidation.

Transmembrane Transporters

Transmembrane transporters are protein receptors that facilitate ionic or molecular transport across the membrane and include *chemically induced ligand gates, mechanical gates*, and *voltage gates*.

Ligand gates are activated by a ligand, a signaling ion or molecule, that opens/closes the gated channels by binding extracellularly (external ligands) or intracellularly (internal ligands) to the channel.

Mechanically gated ion channels operate by responding to deformation of the cells, the membranes, and the stretch receptors. Stress-induced opening of ion channels controls the Na^+ H^+ cytoplasm concentrations and the cell's pH. Specifically, the protein transporter NHE1, which is present in membranes of nearly all cells, is regulated by the stretch and pull of the membrane as the cell changes volume. NHE1 changes its shape rapidly: turning "off" as cells expand and "on" as cell volume is reduced. This allows extracellular Na^+ to come in and the protons H^+ to leave the cell, thus lowering the cell's acidity. The Na^+ H^+ concentration control is critical as cells become very acidic in ischemia when the blood supply is cut off. On the other hand, too much Na^+ is a major mechanism of cell death. Blocking, or inhibiting, the NHE1 transporter in ischemia seems to prevent cell death.

Voltage-gated ion channels in "excitable" cells open or close in response to the electric charge across the plasma membrane. The channel is either fully open or fully closed and some 7,000 Na^+ ions/ms pass through each open channel. It is interesting to note that inherited ion-channel diseases are related to K^+ (heartbeat defects), Na^+ (muscle spasms, Liddle's syndrome: elevated osmotic pressure and hypertension), and Cl^- (cystic fibrosis, kidney stones).

Receptors, G Proteins, and Second Messengers

Receptors are phosphoproteins made of peptide chains with seven loops that transverse the bilayered membrane and expose extracellular and intracellular domains. Extracellular signals must be recognized by the membrane-bound receptors in the plasma membrane on the target cell. When the signaling ions or molecules (e.g., ligands, neurotransmitters, or vesicles) bind to the receptor protein, it triggers some changes that initiate the process of transduction.

There are basically two general types of membrane receptors:

(1) *Non-G protein–coupled receptors*. The intracellular domain is activated when a ligand (e.g., a hormone) binds to the extracellular domain. Consequently, the intracellular portion either demonstrates an intrinsic enzyme activity itself or directly activates other enzymes inside the cell. These enzymes are generally involved in the phosphorylation of proteins (the so-called kinase activity).

(2) *G protein–coupled receptors (GPCRs).* The formation of a ligand-receptor complex in the extracellular plasma membrane domain does not necessarily modify the cardiac function and often requires the intervention of "second" intracellular messengers to transfer the desired information to the targeted effectors. This involves the activation of coupling proteins in the intracellular domain, commonly G proteins, to form GPCRs that affect other membrane phospholipids or proteins (e.g., membrane-bound enzymes) to produce intracellular second messengers that transmit the signal elsewhere within the cell. GPCRs represent the majority of the membrane receptors, corresponding to some 1,000 unique GPCR genes and over 100 different GPCRs, many of which are orphans, without currently defined endogenous agonists.[16] The signals can be either stimulatory or inhibitory (e.g., they can cause an increased level of enzyme activity or a decreased level of activity in the second messenger systems).

G proteins are protein receptors that bind the guanine nucleotides GDP and GTP. They consist of three different subunits and undergo allosteric changes when a hormone, or an otherwise suitable ligand, binds to them. This triggers a replacement of GDP by GTP, which in turn activates an effector molecule; for example, adenylyl cyclase, an enzyme located inside the plasma membrane that catalyzes the conversion of ATP into the second messenger cAMP. Thus, the response to α-adrenergic receptor agonists is mediated by phospholipids-derived diacylglycerol 1,4,5 triphosphate. The response to β-adrenergic receptor yields adenosine 3′5′-cyclic monophosphate (cAMP) produced from ATP.

GPCRs are a major mechanism for converting an external signal into an intracellular message. For example, hormone binding can use a G protein intermediate to produce changes in the concentrations of intracellular messengers. Also, G proteins cause K^+ channels to open in response to neurotransmitter binding in the heart muscle. In general, the guanine nucleotides have the important effects of modifying the affinity of the receptor as well as the affinity of the G protein for the effector. Changes in the affinity of the receptor for the ligand following the G protein activation help end the receptor-mediated stimuli by dissociating the ligand from the receptor. It is noteworthy that the G protein signal transduction plays a critical role in modifying the performance of the heart in response to autonomic and other stimuli.

cAMP, which is produced from ATP by adenylate cyclase, is the most important intracellular messenger, second to Ca^{2+}, in the regulation of cardiac function. It is involved in enhancing Ca^{2+} entry to the cytosol and in promoting relaxation. Adrenaline, glucagons, and adrenocorticotropic hormones can change the cAMP concentration within the target cells. Increased concentrations of cAMP affect cardiac arrhythmogenesis as well as the activation of a kinase that leads to an increase in the rate of glucose release from glycogen. The cAMP signal is controlled by adenylyl cyclase, which synthesizes it, and phosphodiesterases, which degrades it. Ca^{2+} plays a dual role in inhibiting adenylyl cyclase and stimulating the phosphodiesterases, thus demonstrating the complex interactions between cAMP and Ca^{2+} in modulating cardiac function.

Vasopressin, thrombin, and acetylcholine hormones, for example, use the inositol phosphate pathway that is associated with the release of Ca^{2+} from the SR. Tyrosine kinases activate the GTP binding protein, Ras, which leads to a cascade of activations of kinases and eventual changes in gene regulatory proteins. Gene transcription and thus gene expression are changed, causing changes in cell proliferation and cell differentiation. Note that about 30% of human cancers have mutations in the Ras genes.

Finally, we note the existence of *signaling microdomains*. As recently reported by Paul Insell *et al.*,[17] multiple-signal transduction events occur via plasma membrane receptors located in multiple subcellular GPCR-mediated signaling microdomains. Several GPCRs, their cognate heterotrimeric G proteins and effectors, localize to lipid rafts/caveolae. Lipid rafts, enriched in cholesterol and sphingolipids, form such a microdomain along with a subset of lipid rafts/caveolae, that are 50–100 nm invaginations of the plasma membrane, enriched in the protein caveolin-3. Though discovered some 50 years ago, the precise functional role of the caveolae is still not completely understood.

INTRACELLULAR SIGNAL TRANSFER

Some signaling molecules may remain on the cell surfaces, while other information-carrying molecules are transported across the membrane. Some signaling molecules are mobilized in the cell by molecular motors, kinesins, via the elongated tubular structures that serve as railway tracks and carry the molecules towards their destination. Clearly, biophysical constraints imposed by macromolecular crowding of the cell have a significant influence on the evolution of the cell-signaling pathways as diffusion partakes in controlling the number and locations of the signaling complexes.

Intercompartmental information transfer involves different elements in direction selection and in the actual transmission. An important integrated intracellular event involves Ca^{2+} signaling in the cardiac excitation–contraction coupling phenomenon. Ca^{2+} enters the cytoplasm through the cell's membrane voltage-gated channels during repolarization. The Ca^{2+} in the cell is stored in the SR. The membrane of the SR contains active ion pumps for the uptake of Ca^{2+} from the cytoplasm, thus keeping the Ca^{2+} cytoplasm at very low concentration (about 10^{-7} M). The Ca^{2+} is released from the SR into the cytoplasm at subsequent contractions. Unlike the intracellular Ca^{2+} in the skeletal muscle, which is mobilized from the SR by the action potential and its concentration is independent of the extracellular Ca^{2+}, the intracellular Ca^{2+} concentration in the cardiac muscle depends on the extracellular Ca^{2+} concentration, although some 85% of the Ca^{2+} is supplied by the SR in each cycle. The rates and extent of myocardial relaxation and contraction are correspondingly determined by the rate and extent of Ca^{2+} uptake from the cytoplasm into the SR and from the SR into the cytoplasm, where it is available for binding to the troponin complex in the sarcomere. Clearly, the control of Ca^{2+} transport across the SR membrane represents a tuning mechanism for regulating muscle relaxation and the heart's response to increased workloads. This characteristic of "calcium-induced calcium release" phenomenon places Ca^{2+} in the second messengers category.

The Ubiquitin–Protein Death Signal

The maintenance of life depends on well-behaved, well-established, characteristic processes that are expressed by, and centered in, the cells. Selective degradation of proteins in eukaryotes is primarily conducted by the ubiquitin proteasome system. Ciechanover, Hershko, and Rose received the 2004 Nobel Prize for their discovery of the "ubiquitin-mediated protein degradation." Accordingly, the cells attach a "marker," a protein molecule named ubiquitin, onto proteins that need be destroyed. The ubiquitin marked protein complex moves to the proteasomes, and the bound ubiquitin acts to activate the transfer of the doomed protein into the molecular disassembly complex. The ubiquitin is released and the doomed protein is squeezed into the proteasome, the cell's waste disposer, where it is broken down, so that its constituent subunits can be reused in the cell. Examples of processes governed by ubiquitin-mediated protein degradation are cell division, DNA repair, quality control of newly produced proteins, and parts of the immune defense system. Disruptions in this "doom targeted" process can lead to diseases, such as cystic fibrosis and cancer.

DISCUSSION

Addressing the task of deciphering the highly complex system of cell communication brings to mind the difficult, intractable, and insolvable challenge of unraveling the Gordian knot. As the story goes, Midas erected a shrine and placed his wagon in the center of it, yoked to a pole with a large knot, with no ends exposed. Hundreds of tightly interwoven thongs of cornel bark made the knot an impressive centerpiece and an important symbol for the Phrygians' shrine. It was eventually moved and housed near the temple of Zeus Basileus in the ancient city Gordium, ruled by Midas' father Gordius. The lore grew that whoever opened the Gordian knot would become lord over the whole world. Many attempted to unravel it. Alexander the Great finally cut it open with his sword and went on to conquer and rule all of Asia. Legend has it that he considered this victory over the Gordian knot the most decisive battle he ever fought.

Considering the complex cellular communications system with it numerous interacting parameters, it is evident that the task of deciphering all its secrets is much more demanding than unraveling the Gordian knot, which challenged the Great Alexander. It seems that no one man, even one endowed with exceptional genius and equipped with a very sharp virtual sword, can in one lifetime unveil all the cellular communications pathways and identify the potential risks of their malfunctions, and the means to control them. Being somewhat less blessed with fortitude, and missing the sharp sword of Alexander, we must address these challenges carefully with finely honed tweezers. Praying for inspiration and being supported by the cooperation of our fellow scientists, we should eventually succeed in unraveling the secrets of life hidden in that little cell.

The brief overview of the cell communication modes—with their multiple transduction pathways and their numerous parameters presented here—is by necessity lacking in details as well as in subject matter. For example, we have omitted the highly important genetic expressions, electrical sinus and nodal ventricular activation, as well as most neural related communications, as these have gained much attention in scientific literature for many years. Thus, we have concentrated on major parameters and intelligible phenomena in the cardiac system that are involved in the initiation and activation of chemically induced signal transduction pathways. The noticeable omission of genetic signaling pathways is partly compensated for in the following presentations. There is a large amount of published information dealing with genetic signaling pathways that control the gene expression and cellular proliferation, including an integrated Signaling Pathway Database (SPAD) for genetic information. (SPAD was recently developed in Kyushu University, Japan, to explain the extracellular signaling transduction associated with growth factor, cytokine, and hormones that initiate intercellular signaling pathways. Also, SPAD highlights information on protein–protein interactions and protein–DNA interactions.)

No attempt was made in this presentation to integrate and explain the interactions between the various signaling pathways and communication highways. To paraphrase the quip from actress Bette Davis that "getting old is not for sissies," the obvious justification for this omission is that "getting wise is not easy" and a bold imaginative approach based on integrated system biology and bioengineering analysis may bring us to this goal. On a philosophical level, we can argue that the cell is a very complicated system, and large complicated systems are by necessity robust: able to self-correct, to self-adjust, and to compensate for unexpected changes and deviations from the normal steady state conditions. Clearly, the well-known interplay between the sympathetic and the parasympathetic systems that keeps our system balanced suggests that this notion has merits. The obvious question that comes to mind is whether the minute 10-micron cell with its uncountable nanoscale elements operating in a picosecond timescale can indeed be considered as a "large," robust, self-sufficient system. The available know-how may not yield a definitive answer, but while the jury is still out, intuition says that the cell is a very complex, multiparameter, robust system that needs sophisticated analytical tools to be deciphered. The following are some specific questions that come to mind. Why do cells use so many protein complexes? Is there more than one molecular signal that, like the ubiquitin, determines the fate and function of the addressed recipients of the information-carrying signal? Why do many receptor complexes form when active and then dissociate when inactive? These and other basic questions are open for discussion in the following presentations.

SUMMARY

The cardiac cellular communication and signaling pathways form a monumentally complex system of interactions involving information transfer. This brief overview addresses some of the most important parameters involved and highlights some of the better-known pathways. Clearer insight and better understanding of this important system, which has a singular responsibility to maintain life and generate disease, may eventually be achieved by an all-encompassing effort, by integrating know-how, and by developing analytical tools and bio-analytical models for anticipating pathological phenomena.

ACKNOWLEDGMENTS

We gratefully thank all those who helped transform this workshop from vision to reality. First and foremost are the workshop participants whose enthusiasm and expertise made the meeting an exciting affair. Next, we thank the International Program

Committee and the members of the Organizing Committee for endless suggestions and continuous vigilance. Special thanks go to our local organizer Dona Maria Fernanda Afonso, of the Portuguese Ministry of Science, who brought us to the beautiful city of Sintra and its surrounding mountains in Portugal. She vigorously helped make the workshop a success. Thanks are also due to our secretary, Mrs. Betty Kazin, for devotion and care. Special personal thanks go to Mrs. and Mr. Jack George and Mr. Martin Kellner of Los Angeles, California, for their support and trust that we are climbing on the right mountain, and that we may eventually reach the top. Finally, we thank our sponsors, the Technion, Israel Institute of Technology, and the Bruce Rappaport Family Medical School. Most sincerely, we gratefully remember our departed friends, Horti and Larry Fairberg of the American Technion Society, New York, whose endowment facilitated this important international event.

REFERENCES

1. SIDEMAN, S. & R. BEYAR, Eds. 1985. Simulation and Imaging of the Cardiac System—State of the Heart. Proc. 1st Henry Goldberg Workshop, Technion, Haifa, Israel. 1984. Martinus Nijhoff. Dordrecht; Boston.
2. SIDEMAN, S. & R. BEYAR, Eds. 1987. Simulation and Control of the Cardiac System. 3 vols. Proc. 2nd Henry Goldberg Workshop, Haifa, Israel. 1985. CRC Press. Boca Raton, FL.
3. SIDEMAN, S. & R. BEYAR, Eds. 1987. Activation, Metabolism and Perfusion of the Heart. Simulation and Experimental Models. Proc. 3rd Henry Goldberg Workshop, Rutgers University, New Brunswick, NJ. 1986. Martinus Nijhoff. Dordrecht; Boston.
4. SIDEMAN, S. & R. BEYAR, Eds. 1989. Analysis and Simulation of the Cardiac System—Ischemia. 3 vols. Proc. 4th Henry Goldberg Workshop, Tiberias, Israel. May 1987. CRC Press. Boca Raton, FL.
5. SIDEMAN, S. & R. BEYAR, Eds. 1990. Imaging, Analysis and Simulation of the Cardiac System. Proc. 5th Henry Goldberg Workshop. Cambridge, U.K., 1988. Freund. London.
6. SIDEMAN, S., R. BEYAR & A. KLEBER, Eds. 1991. Cardiac Electrophysiology, Circulation and Transport. Proc. 7th Henry Goldberg Workshop. Gwatt, Switzerland. 1991. Kluwer. New York.
7. SIDEMAN, S. & R. BEYAR, Eds. 1991. Imaging, Measurements and Analysis of the Heart. Proc. 6th Henry Goldberg Workshop, Eilat, Israel. Dec. 1989. Hemisphere. New York.
8. SIDEMAN, S. & R. BEYAR, Eds. 1993. Interactive Phenomena in the Cardiac System. Proc. 8th Henry Goldberg Workshop, Bethesda, MD. 1992. Plenum Press. New York.
9. SIDEMAN, S. & R. BEYAR, Eds. 1995. Molecular and Subcellular Cardiology: Effects on Structure and Function. Proc. 9th Henry Goldberg Workshop, Haifa, Israel. 1994. Plenum Press. New York.
10. SIDEMAN, S. & R. BEYAR, Eds. 1997. Analytical and Quantitative Cardiology: From Genetics to Function. Proc. 10th Henry Goldberg Workshop, Haifa, Israel. 1996. Plenum Press. New York.
11. SIDEMAN, S. & A. LANDESBERG, Eds. 2002. Visualization and Imaging in Transport Phenomena. Proc. 1st Larry and Horti Fairberg Workshop, Antalya, Turkey. 2002. Ann. N.Y. Acad. Sci. **972:** 1–344.
12. SIDEMAN, S. & R. BEYAR, Eds. 2004. Cardiac Engineering: From Genes and Cells to Structure and Function. Proc. 2nd Larry and Horti Fairberg Workshop, Erice, Sicily. 2003. Ann. N.Y. Acad. Sci. **1015**.
13. BASSINGWAIGHTE, J.B. 1995. Toward modeling the human physionome. *In* Molecular and Subcellular Cadiology. S. Sideman & R. Beyar, Eds.: 331–339. Plenum Press. New York.
14. KATZ, A.M. 1992. Receptors, coupling proteins and second messengers. *In* Physiology of the Heart. 2nd ed.: 274–302. Raven Press. New York.

15. BRAY, D. 1998. Signaling complexes: Biophysical constraints on intracellular communication. Annu. Rev. Biophys. Biomol. Struct. **27:** 59–75.
16. TANG, C.M. & P.A. INSEL. 2004. GPCR expression in the heart; "new" receptors in myocytes and fibroblasts. Trends Cardiovasc. Med. **14:** 94–99.
17. INSEL, P.A., B.P. HEAD, R.S. OSTROM, *et al*. 2005. Caveolae and lipid rafts: G Protein–coupled receptor signaling microdomains in cardiac myocytes. Ann. N.Y. Acad. Sci. **1047:** 166–172. This volume.

From Organ to Molecules: Steps and Consequences

RAFAEL BEYAR

Faculties of Medicine and Biomedical Engineering, Technion, Israel Institute of Technology, Haifa, Israel

ABSTRACT: The classic cardiac research programs revolved around measurable properties such as pressures, work done, vascular blood flow, electrical propagation, and other such parameters that defined the global heart functions in health and decease. Consistently, the first Henry Goldberg Workshop, held in Haifa in 1984, focused on the interactions between cardiac mechanics, electrical activation, perfusion, and metabolism of the whole heart. Questions focused on the macroscale cardiac function and performance. These studies involved engineering science, simulation, and modeling tools that were essential for the understanding of the complex interactions within the cardiac system. Three-dimensional imaging, ventricular structure, fiber mechanics, circulation and cardiovascular flow, electrical propagation, and blood pumping were all major foci of research at that time. However, it was soon obvious that in order to understand organ level characteristics, one must explore the complex cellular and intracellular control mechanisms; these became the foci of our subsequent workshops. Better understanding of organ level performance required integrated studies of organ and tissue structure and function with genetic, molecular, and cellular characteristics, including cellular communication and ionic and molecular signaling. Analysis of the cardiac system thus depends on continuous probing of the heart system with modern measurement techniques and on integrating data and acquired knowledge with analytical models, constantly evolving to match reality.

KEYWORDS: cells and molecules; 3-D imaging; interacting parameters; macroscale

INROADS TO CARDIAC RESEARCH

Organ level measurements and imaging are essential to the understanding of the interactions between structure and function of the heart, and new imaging modalities are critical to our understanding of therapeutic modalities and complex surgical procedures in the heart muscle. Parameters such as heart torsion and deformation are directly linked to structure and function of individual elements within the myocardium. However, the evolution of research at the end of the last century dictated the

Address for correspondence: Prof. Rafael Beyar, M.D., D.Sc., Division of Invasive Cardiology, Rappaport Faculty of Medicine and Faculty of Biomedical Engineering, Technion, Haifa 32000, Israel. Voice: 972-4-8542181; fax: 972-4-8543451.
rafael@tx.technion.il

Ann. N.Y. Acad. Sci. 1047: 1–12 (2005).
doi: 10.1196/annals.1341.001

need to dive into the cellular and subcellular dimensions for a better understanding of cardiac functions.

Molecular and subcellular aspects of the cardiac system were the main focus of the Henry Goldberg meeting in Haifa in 1994. Understanding of evolution and signaling was critical to the fields of cardiac physiology and pathology. The link to genetics and to genomic and proteomic interactions were emphasized in Haifa in 1996. Developmental aspects of the cardiac cells and the understanding of proteomics within the heart became a critical aspect in our meeting and research, which involved complex cellular- and ion-channel modeling, molecular structure analysis, and development. To understand the complex control mechanism at the molecular level, models of muscle contractions needed further sophistication. Developmental aspects of cardiac myocytes were highlighted by the progress obtained in stem cell research with the groundbreaking work from Dr. Gepstein's laboratory at Technion on differentiation and integration of human embryonic stem cells into cardiac structures. Finally, the Nobel Prize in Chemistry 2004 was awarded to Avram Hershko and Aaron Ciechanover, from the Technion-Faculty of Medicine, and Irwin Rose, from the United States, for highlighting the process of ubiquitin-mediated protein degradation as a major control mechanism for life processes. The applicability of this discovery to the heart is obvious. An example of its importance is the cell cycle, tightly controlled by the ubiquitin protein degradation system, affecting cell death, apoptosis, and regeneration—topics that are the highlight of cardiac research.

THE GOLDBERG/FAIRBERG WORKSHOPS: TWO DECADES OF ORGAN-TO-MOLECULE APPROACH

The titles of the Goldberg and Fairberg Workshop series on the cardiovascular system are presented in TABLE 1. Each meeting focused on the leading cardiac science at the time. The early workshops centered on the global cardiovascular system in relation to transfiber mechanics. Integration of mechanics, electrical activity, metabolism, and coronary circulation were intensely discussed. Three-D imaging was used as an important tool to understand and study heart performance in health and disease. Molecular and genetic aspects of cellular function started to appear as a major thrust of the Haifa meeting a decade later, in 1994. Intracellular control mechanisms and subcellular analysis of cardiac performance became an important part of subsequent meetings. Complex studies involving genomics and proteomics of different cell functions led to sophisticated modeling attempts of the cardiac cell and the entire heart. Research of the molecules, the gene, the protein, cell performance, and the overall organ behaviors became a continuum, as is clearly reflected in this meeting in Sintra, Portugal.

ORGAN LEVEL STUDIES: GLOBAL AND MYOCARDIAL MECHANICS

The role of the heart as a pump under a wide variety of loading conditions continues to be a major topic for investigations and analyses. Overall, models of heart mechanics and circulation were highlighted in the first proceedings,[1] where topics of transmural distribution of myocardial stress, oxygen demand, and coronary flow

TABLE 1. The Goldberg/Fairberg Workshops: two decades of organ-to-molecule approach

Haifa, Israel, 1984[1]	Interactions between mechanics, electrical activation, perfusion, and metabolism
Haifa, Israel, 1985[2]	Cardiovascular control
Rutgers, NJ, USA, 1986[3]	Transformation of microscale activation phenomena to macroscale performance, electrical, metabolism and mechanics
Tiberia, Israel, 1987[4]	Parameters affecting cardiac performance, ischemia
Cambridge, U.K., 1988[5]	Nonhomogeneity of the cardiac muscle structure and performance
Eilat, Israel, 1989[6]	New imaging techniques for analysis of local and global cardiac performance
Gwatt, Switzerland, 1991[7]	Basic microlevel phenomena, with particular emphasis on electrocardiology
Bethesda, MD, 1992[8]	Interactions of basic phenomena and their application to clinical practice
Haifa, Israel, 1994[9]	Molecular and subcellular aspects of the cardiac system
Haifa, Israel, 1996[10]	Analytical and quantitative analyses from basic genetics to cardiac function. The interface between structure and function
Antalya, Turkey, 2002[11]	Modern imaging and visualization techniques
Sicily, Italy, 2003[12]	The Cardiome, combining analytical system approach and engineering
Sintra, Portugal, 2005	The communicative cardiac cell

were presented.[13] The role of mathematical modeling in assessment of wall stress and myocardial function was addressed by Mirsky;[14] models of elastance were addressed by Sagawa *et al.*;[15] energetics and coronary flow were presented by world leaders; models of electrical propagation were presented by Rudy;[16] the metabolic and neural control of coronary circulation was addressed by Berne[17] and Feigl;[18] and the approach to modeling and analysis of metabolic events, and the approach to complex modeling were highlighted by Bassingthwaighte.[19]

A particular aspect related to modeling and analysis, which was highlighted in subsequent meetings, was the effect of ventricular torsion, or twist, on transmyocardial fiber mechanics.[20,21] A series of modeling and experimental work[20–25] have shown that transmyocardial mechanics is inherently a distributed parameter that follows a typical pattern. The transmyocardial distribution of parameters is offset by the presence of torsion or twist. Furthermore, twist was also shown to be a dynamic event[25] with a considerable amount of twist-recoil during isovolumic relaxation, as demonstrated in FIGURE 1.[26] This observation is confirmed by additional studies,[27–31] which have demonstrated that twist and torsion have direct implication on systolic and diastolic heart function and the transmural distribution of cardiac mechanics

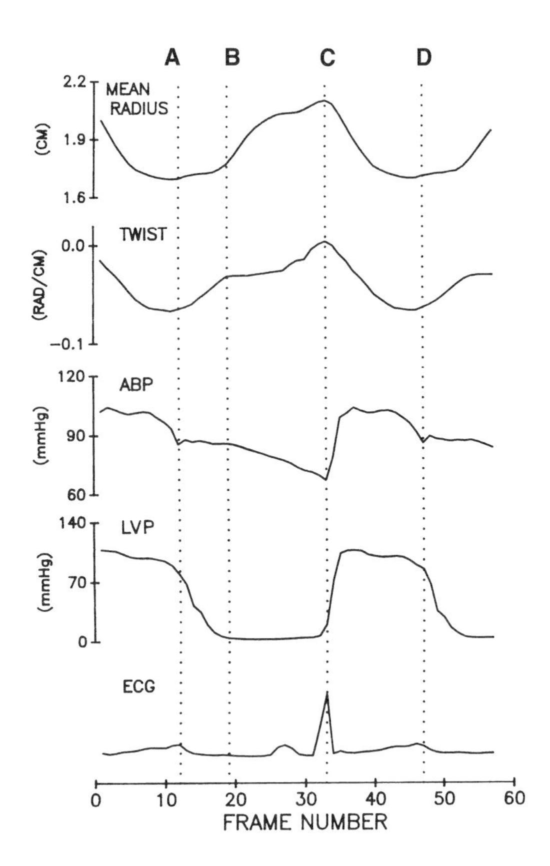

FIGURE 1. **A–B** is the isovolumic relaxation phase and **C–D** is the ejection phase. Note that during ejection, twist and shortening (change in mean radius) are linear; during isovolumic relaxation, rapid twist recoil happens before lengthening. ABP: aortic blood pressure; LVP: left ventricular pressure. Reproduced with permission from Beyar *et al.*[26]

(FIG. 2). An elegant study of patients by Knudtson *et al.*[32] has shown the acute effects of ischemia on twist mechanics during balloon angioplasty.[32]

The problem of transmural mechanics and torsion has triggered numerous publications in this area. An important study, which links these mechanical events to molecular and genetic data, was presented by Davis *et al.* in 2001.[33] Their study was first to suggest how the gradient of phosphorylation can facilitate an overall pattern of cardiac contraction, as viewed with MRI imaging. Based on previous modeling and experimental observations, these authors have suggested that evolution has incorporated a variety of successful strategies for myocardial motion, by specifically adding the efficiency of torsion to compression, for the generation of hemodynamic work. They demonstrate using a spatial gradient of myosin light chain phosphorylation across the heart wall and argue that this gradient facilitates torsion by inversely altering tension production (Epi>Endo) and the stretch activation response (Endo>Epi). In order to demonstrate the importance of cardiac light chain phosphor-

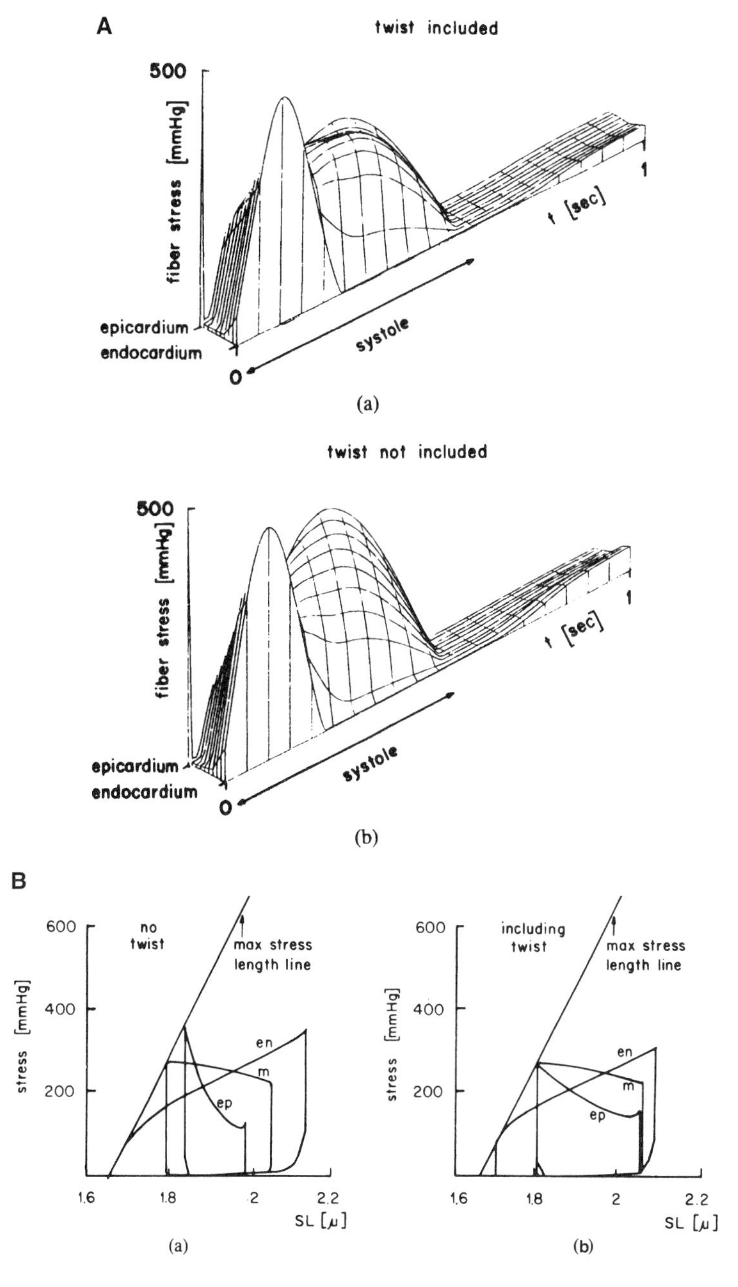

FIGURE 2. (**A**) LV model prediction of fiber stress distributed from endocardium to epicardium. Note that the inhomogeneity in stressed fiber is reduced when twist is included in the model. (**B**) Sarcomere stress-length loops with and without twist in an LV model. Note that twist reduced transmyocardial stress and work distribution across the wall. Reproduced with permission from Beyar and Sideman.[22] en, endocardium; m, middle layer; ep, epicardium.

ylation on the development of structure and function, the authors cloned myosin light chain kinase from a human heart, and were able to study the effect of that mutation in a genetically modified mouse. They have also identified a gain-in-function mutation in two individuals with a specific type of cardiac hypertrophy.

The Goldberg Workshops in 1991, 1992, and 1994 have focused considerable attention on the molecular and genetic aspects of myocardial diseases and the function of the myocardial fibers. Imaging and molecular cardiology became extremely important tools for enhancing our understanding of the cell and the organ. An NIH meeting, "Form and Function: New Views on Development, Diseases and Therapies for the Heart," held in April 2002 in Bethesda, MD, highlighted these studies and provided the following key recommendations for future research strategies:[34]

- Increase our understanding of normal and abnormal shape and fiber mechanics
- Elucidate signaling pathways by which stress and strain regulate gene expression, myofiber pattern formation, and structural remodeling in vivo
- Investigate alterations in diastolic deformation that may explain changes in diastolic filling patterns
- Understand the synchronization of electrical depolarization and repolarization, and the coordination of myofiber contraction and relaxation
- Advance cardiac imaging
- Advance therapeutics through surgery and devices

Note that these recommendations highlight the need for cellular signaling, gene expression regulation, and myofiber pattern, and urge the need for computational models that will allow the integration of complex interactions between different experimental observations. The typical interdisciplinary approach taken by the Goldberg/Fairberg scientists for two decades is elegantly reflected by the NIH panel recommendations.

CELL AND MOLECULE LEVEL STUDIES: CARDIAC CELL DEVELOPMENT, COMMUNICATIONS AND PROTEOMICS

Experimental and modeling endeavors have made it clear that accounting for cellular and molecular aspects is critical for research development. Two such attempts, which are also discussed in the present workshop, are particularly noteworthy.

Cell developmental aspects are critical for the understanding of cardiac structure and function. Human Embryonic Stem Cell research, pioneered by the work of Thompson and Itskovitz-Eldor,[35] has opened up a new era in cardiac stem cell research. The pioneering work of Kehat *et al.*[36] from Gepstein's Laboratory have shown for the first time the differentiation of human embryonic stem cells to cardiomyocytes (FIG. 3) and have characterized their individual properties and their intercommunication pathways.[37]

Using an elegant model of multielectrode array, the *cell–cell communications via the spread of the electrical signal* were shown to be inherent developmental aspects of the individual myocytes. In addition, the ability to develop cellular communications among cells from different organisms and different species were beautifully demonstrated *in vitro* (FIG. 4) and *in vivo* in swine heart models.[38]

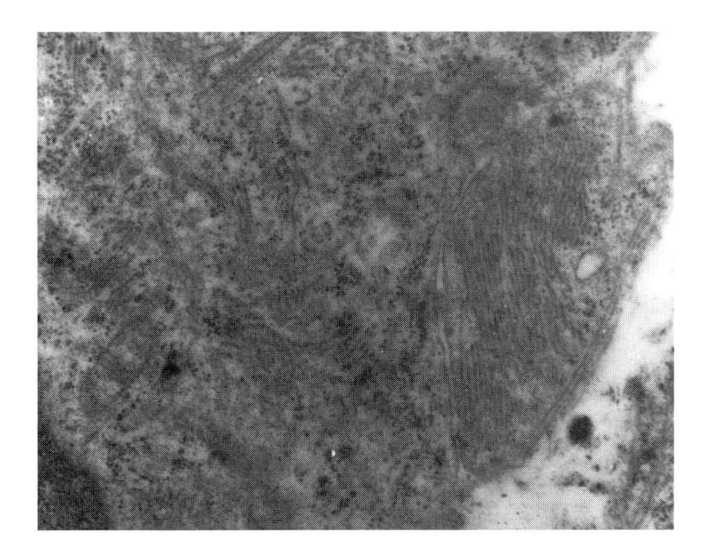

FIGURE 3. Two neighboring 13-day-old human embryonic cells differentiating into cardiac myocytes (counting from the day of plating). In one cell (*right*) the sarcomere is more ordered. In the other (*left*) myofibrils are more dispersed and are arranged as stack of hay. Provided by I. Kehat and L. Gepstein.

Many aspects of the different mechanisms of contraction and its development with time as well as to the complex signaling pathways within the cells and in between the different cells remain open. Furthermore, the question as to how do these communications lead to modifications of global function is critical to the understanding of the global organ function.

Signaling and proteomics are extremely important in understanding cell and intercellular function and communications. The 2004 Nobel Prize in chemistry, awarded to two Technion faculty members, Hershko and Ciechanover, together with Irwin Rose from the United States, for their work on the ubiquitin protein degradation system, highlights the importance of proteomics in cell protein structure and signaling.[39–40] The ubiquitin–proteasome system is the major pathway of nonlysosomal degradation of intracellular proteins (FIG. 5). A review discussing the involvement of the ubiquitin system in cardiovascular diseases and atherosclerosis was recently published.[41] This system is multifunctional and plays a major role in the elimination of damaged and unneeded proteins as well as in the regulation of cellular mediators. Among the pathways that are tightly controlled by the ubiquitin system are those associated with inflammation, cell proliferation, and apoptosis. There is ample evidence regarding the potential involvement of the ubiquitin–proteasome system in the initiation, progression, and complication stages of atherogenesis. In view of its major role in controlling intracellular proteins and its association with physiological and pathological cell proliferation, it is expected that the ubiquitin protein degradation system will have great potential for use in cardiovascular therapy.

Signaling and communications involve many molecules and various modalities. The different proteins that carry information from genetic material and cellular organs

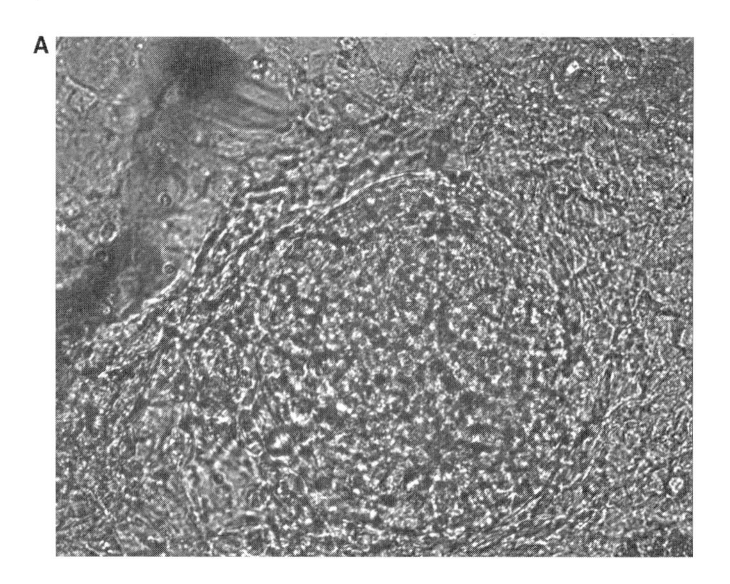

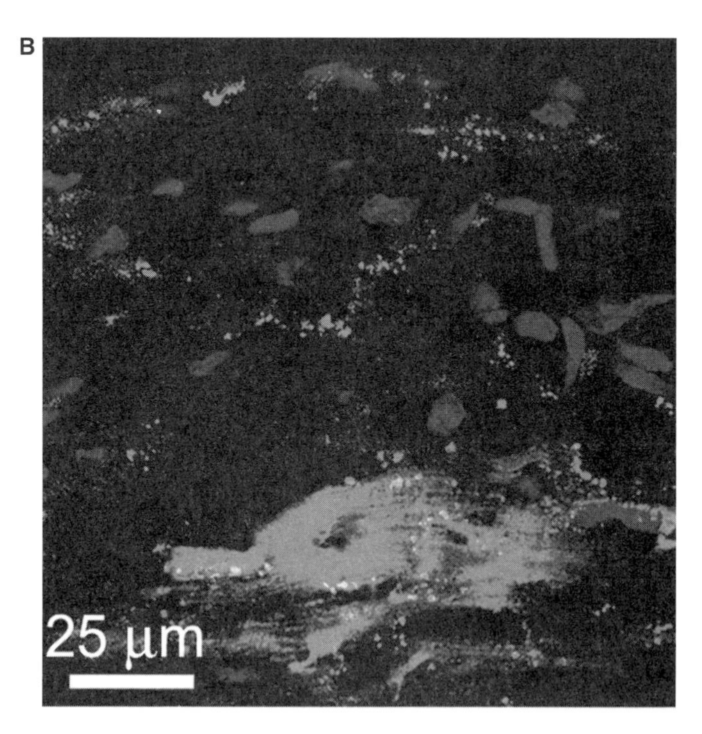

FIGURE 4. Cell–cell communications: (**A**) *In vitro* integration of human embryonic stem (ES) cell colony on top of mouse embryonic fibroblast layer (*arrows* indicate colony borders). (**B**) Structural integration of human ES cell–derived cardiomyocytes stained with the cell tracker CM-DiI (*red*) to the normal myocardium of a pig as demonstrated by Connexin 43 staining (*green*); nuclear staining (To-Pro) is coded *blue* [in color in Annals Online]. Provided by L. Gepstein, I. Kehat, and O. Caspi.

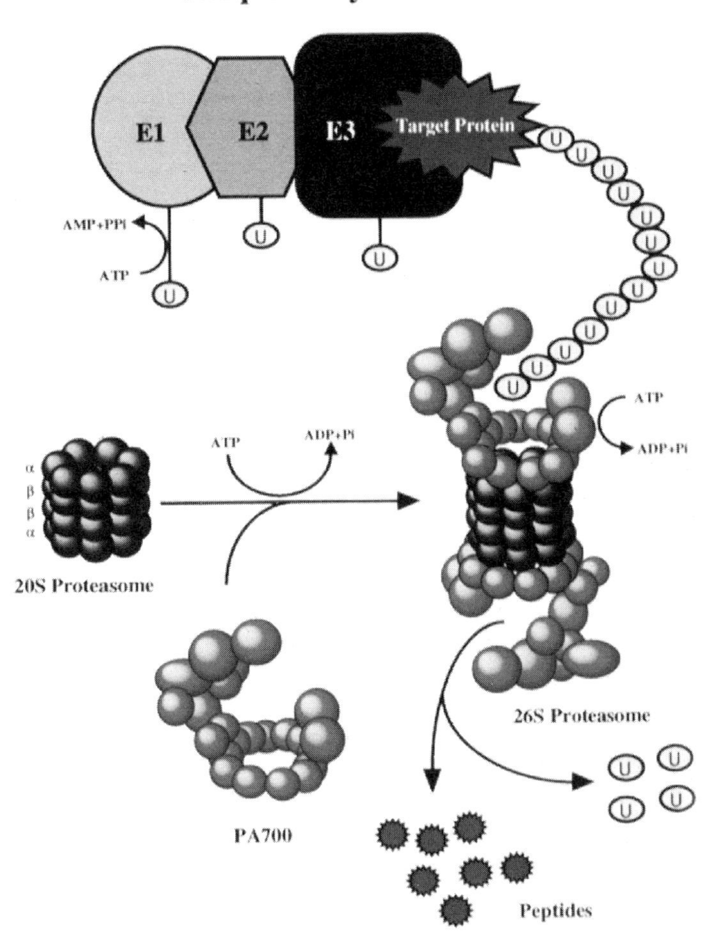

FIGURE 5. The ubiquitin protein degradation pathway. The target protein is polyubiquinated by the series of enzymes E1, E2, and E3, and is then degraded by the 26S proteasome. The ubiquitin molecule is released and becomes available for further tagging. (Reproduced by permission of Ciechanover and Taneka, personal communications).

are associated with ionic molecules as well as ion currents that generate the membrane potential that is essential for maintaining contraction. The role of calcium and all the different molecular mechanisms in calcium transport and signaling have been mathematically quantified to yield a comprehensive understanding of cardiac excitation–contraction coupling. This problem is tightly coupled to the mechanism of force generation, and the various feedbacks and controls of calcium binding and the troponin myosin interaction. A beautiful modeling approach to that problem has been provided by Landesberg and Sideman (FIG. 6) in studies following their 1994 modeling paper of the cardiac contractile unit.[42]

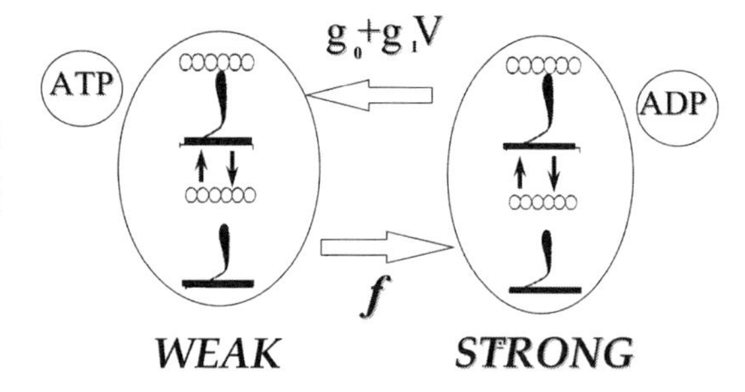

FIGURE 6. A four-stage model of the cardiac contractile unit. Based on Landesberg and Sideman.[42,43]

CONCLUSION

A major part of this meeting will be dedicated to electromechanical coupling and the mechanism of contraction. It is exciting to see how our overall view of the organ function has highlighted molecular cellular mechanisms, which are now used in an integrative way to build our understanding of the heart by complex modeling and integrative approaches.

REFERENCES

1. SIDEMAN, S. & R. BEYAR, Eds. 1985. Simulation and Imaging of the Cardiac System. State of the Heart. Martinus Nijhoff. The Hague.
2. SIDEMAN, S. & R. BEYAR, Eds. 1987. Simulation and Control of the Cardiac System. 3 vols. Proc. 2nd Henry Goldberg Workshop, Haifa, Israel. 1985. CRC Press. Boca Raton, FL.
3. SIDEMAN, S. & R. BEYAR, Eds. 1987. Activation Metabolism and Perfusion of the Heart. Martinus Nijhoff. The Hague.
4. SIDEMAN, S. & R. BEYAR, Eds. 1989. Analysis and Simulation of the Cardiac System—Ischemia. CRC Press. Boca Raton, FL.
5. SIDEMAN, S. & R. BEYAR, Eds. 1990. Imaging, Analysis and Simulation of the Cardiac System. Freund. London.
6. SIDEMAN, S. & R. BEYAR, Eds. 1991. Imaging, Measurement and Analysis of the Heart. Hemisphere. New York.
7. SIDEMAN, S., A. KLEBER & R. BEYAR, Eds. 1991. Cardiac Electrophysiology, Circulation and Transport. Kluwer. Boston.
8. SIDEMAN, S. & R. BEYAR, Eds. 1993. Interactive Phenomena in the Cardiac System. Plenum Press, New York.
9. SIDEMAN, S. & R. BEYAR, Eds. 1995. Molecular and Subcellular Cardiology: Effects of Structure and Function. Plenum Press. New York.
10. SIDEMAN, S. & R. BEYAR, Eds. 1997. Analytical and Quantitative Cardiology: From Genetics to Function. Plenum Press. New York.
11. SIDEMAN, S. & A. LANDESBERG. Eds. 2002. Visualization and Imaging in Transport Phenomena. Ann. N.Y. Acad. Sci. **972:** 1–344.
12. SIDEMAN, S. & R. BEYAR. 2004. Cardiac Engineering: From Genes and Cells to Structure and Function. Ann. N.Y. Acad. Sci. **1015:** 1–404.
13. BEYAR, R. & S. SIDEMAN. 1984. The interrelationship between the left ventricular contraction, transmural blood perfusion, and spatial energy balance. A new model of the cardiac system. *In* Simulation and Imaging of the Cardiac System. S. Sideman & R. Beyar, Eds.: 331–357. Martinus Nijhoff. The Hague.
14. MIRSKY, I. 1984. The role of myocardial models in the assessment of myocardial function. *In* Simulation and Imaging of the Cardiovascular System. S. Sideman & R. Beyar, Eds.: 35–63. Martinus Nijhoff. The Hague.

15. SAGAWA, K., L. MAUGHAN & K. SUNAGAWA. 1984. A cardiac model convenint for vascular load coupling. *In* Simulation and Imaging of the Cardiovascular System. S. Sideman & R. Beyar, Eds.: 90–102. Martinus Nijhoff. The Hague.
16. RUDY, Y. 1984. Critical aspects of forward and inverse problems in electrocardiography. *In* Simulation and Imaging of the Cardiovascular System. S. Sideman & R. Beyar, Eds.: 279–298. Martinus Nijhoff. The Hague.
17. BERNE, R.M. 1984. Metabolic control of coronary blood flow. *In* Simulation and Imaging of the Cardiovascular System. S. Sideman & R. Beyar, Eds.: 309–322. Martinus Nijhoff. The Hague.
18. FEIGL, E.O. 1984. Neural control of coronary blood flow. *In* Simulation and Imaging of the Cardiovascular System. S. Sideman & R. Beyar, Eds.: 323–330. Martinus Nijhoff. The Hague.
19. BASSINGTHWAIGHTE, J. 1984. Constraints in the interpretation of dynamic images of metabolic events. *In* Simulation and Imaging of the Cardiovascular System. S. Sideman & R. Beyar, Eds.: 378–390. Martinus Nijhoff. The Hague.
20. BEYAR, R. & S. SIDEMAN. 1984. Computer study of the left ventricular performance based on fiber structure, sarcomere dynamics, and transmural electrical activation propagation velocity. Circ. Res. **55:** 358–375.
21. ARTS, T., S. MEERBAUM, R.S. RENEMAN & E. CORDAY. 1984. Torsion of the left ventricle during the ejection phase in the intact dog. Cardiovasc. Res. **18:** 183–193.
22. BEYAR, R. & S. SIDEMAN. 1985. Effect of twisting motion on the nonuniformities of transmyocardial fiber mechanics and energy demand—a theoretical study. Special issue on modeling and simulations. IEEE Trans. Biomed. Eng. **32:** 764–769.
23. BEYAR, R. & S. SIDEMAN. 1986. Left ventricular mechanics related to the distribution of oxygen demand throughout the wall. Circ. Res. **58:** 664–677.
24. BEYAR, R. & S. SIDEMAN. 1986. Spatial energy balance within a structural model of the left ventricle. Ann. Biomed. Eng. **14:** 467–487.
25. BEYAR, R. & S. SIDEMAN. 1986. The dynamic twisting of the LV: a computer study. Ann. Biomed. Eng. **14:** 547–562.
26. BEYAR, R., F. YIN, M. HAUSKNECHT, *et al.* 1989. Dependence of left ventricular twist-radial shortening relations on cardiac cycle phase. Am. J. Physiol. **257:** H1119–H1126.
27. GIBBONS KROEKER, C.A., H.E. TER KEURS, M.L. KNUDTSON, *et al.* 1993. An optical device to measure the dynamics of apex rotation of the left ventricle. Am. J. Physiol. **265:** H1444–H1449.
28. BUCHALTER, M.B., J.L. WEISS, W.J. ROGERS, *et al.* 1990. Noninvasive quantification of left ventricular twist and torsion in normal humans using magnetic resonance myocardial tagging. Circulation **81:** 1236–1244.
29. GIBBONS KROEKER, C.A., J.V. TYBERG & R. BEYAR. 1995. The effects of load manipulations, heart rate and contractility on apical rotation: an experimental study in anesthetized dogs. Circulation **92:** 130–141.
30. KROEKER, C.A., J.V. TYBERG & R. BEYAR. 1995. Effects of ischemia on left ventricular apex rotation. An experimental study in anaesthetized dogs. Circulation **92:** 3539–3548.
31. YUN, K.L., M.A. NICZYPORUK, G.T. DAUGHTERS, *et al.* 1991. Alterations in left-ventricular diastolic twist mechanics during acute human cardiac allograft-rejection. Circulation **83:** 962–973.
32. KNUDTSON, M.L., P.D. GALBRAITH, K.L. HILDEBRAND, *et al.* 1997. The dynamics of left ventricular apex-rotation during angioplasty: a sensitive index of ischemic dysfunction. Circulation **96:** 801–808.
33. DAVIS, S.L., S. HASSANZADEH, S. WINITSKY, *et al.* 2001. The overall pattern of cardiac contraction depends on a spatial gradient of myosin regulatory light chain phosphorylation. Cell **107:** 631–641.
34. BUCKBERG, G., M. WEISFELDT, M. BALLESTER, *et al.* 2004. Left ventricular form and function: scientific priorities and strategic planning for development of new views of disease. Circulation **110:** E333–E336.
35. THOMPSON, J.A., J. ITSKOVITZ-ELDOR, S.S. SHAPIRO, *et al.* 1998. Embryonic stem cell lines derived from human blastocysts. Science **282:** 1145–1147.
36. KEHAT, I., D. KENYAGIN-KARSENTI, M. DRUCKMANN, *et al.* 2001. Human embryonic stem cells can differentiate into myocytes portraying cardiomyocytic structural and functional properties. J. Clin. Invest. **108:** 407–414.

37. Kehat, I., A. Gepstein, A. Spira, *et al.* 2002. High-resolution electrophysiological assessment of human embryonic stem cell–derived cardiomyocytes: a novel in vitro model for the study of conduction. Circ. Res. **91:** 659–661.
38. Kehat, I., L. Khimovich, O. Caspi, *et al.* 2004. Electromechanical integration of cardiomyocytes derived from human embryonic stem cells. Nat. Biotechnol. **22:** 1282–1289.
39. Ciechanover, A., Y. Hod & A. Hershko. 1978. A heat-stable polypeptide component of an ATP-dependent proteolytic system from reticulocytes. Biochem. Biophys. Res. Commun. **81:** 1100–1105.
40. Hershko, A., A. Ciechanover & A. Rose. 1979. Resolution of an ATP-dependent proteolytic system from reticulocytes: a component that interacts with ATP. Proc. Natl. Acad. Sci. U.S.A. **76:** 3107–3110.
41. Herrmann, J., A. Ciechanover, L.O. Lerman & A. Lerman. 2004. The ubiquitin–proteasome system in cardiovascular diseases: a hypothesis extended. Cardiovasc. Res. **61:** 11–21.
42. Landesberg, A. & S. Sideman. 1994. Mechanical regulation of cardiac muscle by coupling calcium kinetics with cross-bridge cycling: a dynamic model. Am. J. Physiol. **267:** H779–H795.
43. Landesberg, A. & S. Sideman. 1994. Coupling calcium binding to troponin C and cross-bridge cycling in skinned cardiac cells. Am. J. Physiol. **266:** H1260–H1271.

Evolution of the Heart from Bacteria to Man

NANETTE H. BISHOPRIC

Department of Molecular and Cellular Pharmacology, University of Miami, Miami, Florida 33101, USA

ABSTRACT: This review provides an overview of the evolutionary path to the mammalian heart from the beginnings of life (about four billion years ago) to the present. Essential tools for cellular homeostasis and for extracting and burning energy are still in use and essentially unchanged since the appearance of the eukaryotes. The primitive coelom, characteristic of early multicellular organisms (~800 million years ago), is lined by endoderm and is a passive receptacle for gas exchange, feeding, and sexual reproduction. The cells around this structure express genes homologous to NKX2.5/tinman, and gradual specialization of this "gastroderm" results in the appearance of mesoderm in the phylum Bilateria, which will produce the first primitive cardiac myocytes. Investment of the coelom by these mesodermal cells forms a "gastrovascular" structure. Further evolution of this structure in the bilaterian branches Ecdysoa (Drosophila) and Deuterostoma (amphioxus) culminate in a peristaltic tubular heart, without valves, without blood vessels or blood, but featuring a single layer of contracting mesoderm. The appearance of Chordata and subsequently the vertebrates is accompanied by a rapid structural diversification of this primitive linear heart: looping, unidirectional circulation, an enclosed vasculature, and the conduction system. A later innovation is the parallel circulation to the lungs, followed by the appearance of septa and the four-chambered heart in reptiles, birds, and mammals. With differentiation of the cardiac chambers, regional specialization of the proteins in the cardiac myocyte can be detected in the teleost fish and amphibians. In mammals, growth constraints are placed on the heart, presumably to accommodate the constraints of the body plan and the thoracic cavity, and adult cardiac myocytes lose the ability to re-enter the cell cycle on demand. Mammalian cardiac myocyte innervation betrays the ancient link between the heart, the gut, and reproduction: the vagus nerve controlling heart rate emanates from centers in the central nervous system regulating feeding and affective behavior.

KEYWORDS: apoptosis; conduction system; evolution; heart development; mammals; metabolism; phylogeny; stem cells

FROM METEORITES TO METABOLISM (4.5–3.5 BILLION YEARS AGO)

At the root of the tree of life are the prokaryotes, which appeared approximately 3.8 billion years ago (BYA). The earth was emerging from the Hadean period, an extremely hot and volatile era characterized by incessant collisions, explosions, and

Address for correspondence: Prof. Nanette H. Bishopric, M.D., Prof. Pharmacology, Medicine, and Pediatrics, Dept. of Molecular and Cellular Pharmacology, University of Miami, P.O. Box 016189 (R-189), Miami, FL 33101, USA. Voice: 305-243-6775; fax: 305-243-6082.
nhb@chroma.med.Miami.edu

Ann. N.Y. Acad. Sci. 1047: 13–29 (2005). © 2005 New York Academy of Sciences.
doi: 10.1196/annals.1341.002

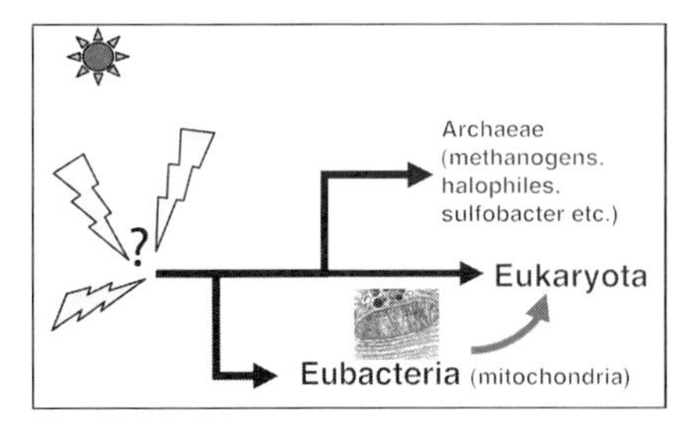

FIGURE 1. In the beginning. The creation of the Earth was followed rather quickly by the appearance of life, c. 4 billion years ago. Ingestion of one prokaryote by another leads to the first eukaryotic life form [in color in Annals Online].

volcanic eruptions (FIG. 1). Consequently, life is likely to have first appeared under conditions vastly different from those existing today. Research into modern-day microbial populations at extremes of heat and pressure may lead to a better picture of the properties of the first living organism. However, it is likely that this ancestor was a single cell, bounded by a protein–lipid membrane externally but without defined internal compartments segregating metabolic and reproductive functions.[1] This cell would have been highly heat tolerant and able to obtain energy and carbon from inorganic substrates. It would have been anaerobic, for there was no oxygen in the Hadean atmosphere. Based on analyses of organisms harvested from deep ocean volcanic vents, which are thought to replicate features of the Hadean environment, molecular hydrogen and sulfur may have served as electron donors and oxidized sulfur and CO_2 as electron acceptors for energy generation. From this early prokaryote branched the eubacteria and the so-called archaea, including the halophiles and methanogens. Comparative analysis of 16S rRNA sequences suggests that the archaeae may have evolved relatively little since then.

The appearance of enzymes, or catalytic proteins, was a critical and probably essential first step in the formation of living organisms. In the absence of oxygen, primitive monists utilized enzymes capable of accelerating chemical reactions that could generate energy through the breakdown of carbohydrates. These glycolytic enzymes have been highly conserved throughout subsequent evolution, as have been other eubacterial features, including HMG CoA reductases, required for cell membrane components, and ribosomes. Although well-developed cytoskeletons do not appear until later in evolution, membrane- and DNA-associated forms of actin and homologues of other cytoskeletal proteins, including tubulin, have been identified in the eubacteria.[2–4]

The appearance of photosynthetic cyanobacteria ~3.5 BYA eventually caused the appearance of oxygen in the atmosphere beginning about 2.8 BYA. This profound environmental change precipitated a number of new phenomena, including rust, oxidative stress, and the development of a capacity for oxygen metabolism in certain bacteria. The enzymes in this pathway were probably derived from similar enzymes used in the oxidation of sulfur, but with a considerably greater relative yield of energy per reducing equivalent.[5] This increase in bioenergetic efficiency provided the footing for a quantum expansion of biological complexity.

EUKARYOTES: THE ACCIDENTAL SYMBIONTS (2.1–1.8 BILLION YEARS AGO)

Just as today, the early prokaryotes lived in close proximity and frequently used each other as food. It is now generally believed that one such ingestion resulted not in the breakdown of the eaten, but in the establishment of an endosymbiosis in which the meal, most likely a member of the alpha-proteobacteria, remained metabolically active within the organism that ate it: thus appeared the first mitochondria.[6] This hypothesis is supported by evidence from *Reclinomonas americana*, a single-celled organism whose mitochondria contain an almost intact alpha-proteobacterial genome. The alpha-proteobacteria are related to the rickettsia, and many species exhibit obligate or opportunistic intracellular symbiotic lifestyles.[7] Similarly, an endosymbiosis involving cyanobacteria in the ancestors of green plants likely resulted in the development of the chloroplast. The appearance of other bilamellate organelles, such as the lysosomes, may have come about in the same way.

The presence of multiple autonomously replicating mitochondria in this new life-form meant a tremendous increase in energy generating capacity. Moreover, each of these mitochondria could realize a 20-fold increase in the energy yield from carbohydrates simply by facilitating a move from glycolytic to oxidative metabolism when oxygen was available. In what could be thought of as a downside to the relationship, the mitochondria-dependent apoptosis pathway could have developed from residual components of an ancient prokaryotic defense system.[8–10]

Even the eukaryotic nucleus may have an endosymbiotic origin, although this is more controversial.[11–15] Molecular evidence suggests that the eukaryotes resemble both the eubacteria and archaeae, as if they arose from a symbiosis between these two prokaryotes.[16] DNA- and RNA-processing functions appear to have come from the archaeae: both archaea and eukaryotes have genes for histones, but the eubacteria do not.[17] On the other hand, several genes involved in metabolism appear to have a eubacterial rather than archaeal origin.[18] One view is that archaeae may have taken up residence within bacteria and gradually lost the need for independent metabolic activity. An even more contentious suggestion is that viruses present in the primordial soup were responsible for the transfer to prokaryotes of genetic material within an envelope, the nucleus then being a sort of chronic viral infection.[19] Whatever its origin, the nucleus provides a separation between transcription (nuclear) and translation (cytosol), and hence an opportunity for greater RNA editing before protein translation is initiated.

PLANTS AND ANIMALS: A COMMON ORIGIN FOR MULTICELLULAR ORGANISMS?

Eukaryotes branched into more than 60 separate lines, including the green plants, red algae, diatoms, and protists. However, resolving the kinships among fungi, plants, and metazoans has been difficult[20] (FIG. 2). Molecular phylogeny of a number of ancient genes, including elongation factor 1a, support a sister relationship between the fungi and the metazoans, with a more distant kinship to the green plants. On this basis a new eukaryotic superfamily has been proposed, the Opisthokonts (meaning "flagellum in the rear"). This kingdom includes the fungi, collar flagellates (including *Giardia lamblia*) and our direct ancestors, the metazoans. In this model, our earliest common eukaryotic ancestor would have been capable of directed movement, facilitated not only by a new abundance of energy but also by the development of a cytoskeleton, from which the flagellum likely derived. Evolution

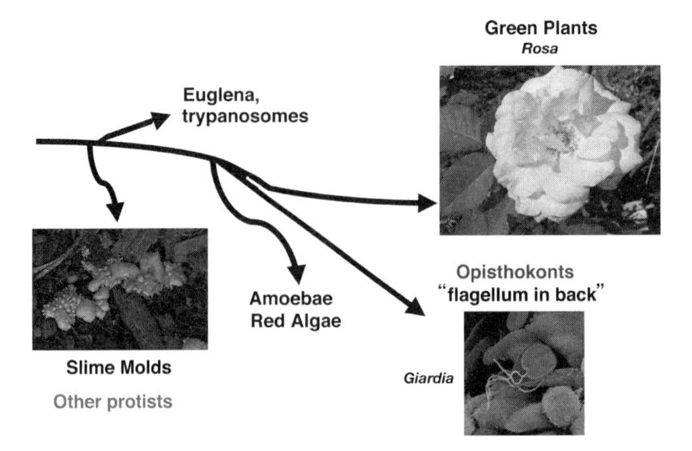

FIGURE 2. Divergence of the eukaryotes. Eukaryotes, still single-celled, split into more than sixty separate lines. The timing of plant divergence remains conjectural [in color in Annals Online].

of the cytoskeleton, and of a more pliable outer membrane, may have been driven by the need to eat other organisms.[21] The earliest cytoskeletal structures have been expanded into the cilia, mitotic spindles, and components of cytokinesis. The actomyosin filaments that permit cellular contraction are similar to those in budding yeast (*Saccharomyces cerevisiae*), which form an actomyosin-containing constriction ring during cell division.[22] This model suggests that green plants took a radically different evolutionary path in which movement was a low priority; meanwhile, traits we share with plants, such as multicellularity and tissue differentiation, are the result of convergent evolution rather than a common inheritance.

However, other lines of evidence, including analysis of a different set of genes, suggest a closer relationship between plants and metazoans that excludes the fungi. For example, plants and metazoans express homologues of p300/CBP acetyltransferases, but fungi do not.[23,24] Atrial natriuretic peptides have been identified in both metazoans and green plants, and serve to regulate solute flow through tissues in both groups.[25,26] GATA transcription factors, similar to those in metazoans, are present in plants and regulate transcription in response to light and nitrates.[27] MADS-box transcription factors are also found in plants, where they regulate aspects of development including flower formation.[28] The entire suite of G1-regulating proteins, including E2F and Rb, appear to be exclusive to the metazoans and green plants, as are the signals regulating the designation of totipotent stem cells.[29] These findings will need to be resolved in future models of the plant–animal–fungus trifurcation.

METAZOANS AND MULTICELLULARITY (800–700 MILLION YEARS AGO)

Some protists (algae, kelp) and many fungi exist as complexes of cells rather than as free-living organisms. However, multicellularity is a hallmark of metazoan life

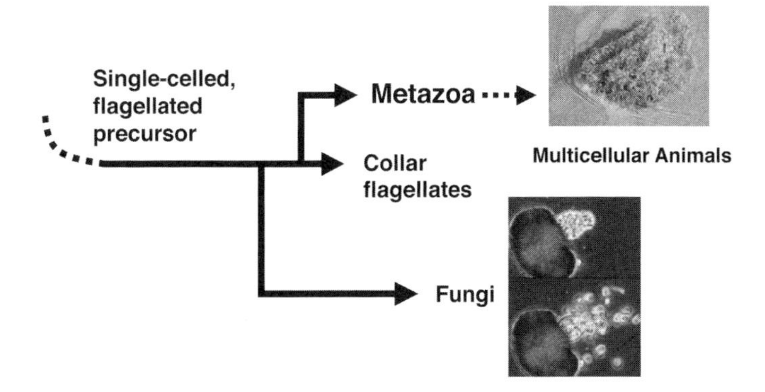

FIGURE 3. The opisthokonts, ancestors of multicellular life. A representative metazoan (a sponge) is shown [in color in Annals Online].

(FIG. 3). The most primitive extant metazoan phylum, Poriphera (sponges), as well as members of the sister phylum Cnidaria (corals) form motile, flagellated blastulae that migrate and eventually attach to substrate to form new multicellular organisms with two basic cell layers separated by a gelatinous matrix. New insights into the molecular origins of multicellular life, and the basal position of Poriphera in the animal kingdom, have been recently reviewed.[30,31] Most of the important fundamental features of multicellular life are present in Poriphera and simply undergo further specialization and refinement as evolution proceeds. Several of the more important innovations will be mentioned here.

Separation of Cell Layers

Sponges, like corals and jellyfish, are diploblastic: that is, they have only two embryonic cell layers, endoderm and ectoderm. They have a single central body cavity, or coelom, in which eating, gas exchange, and reproduction occur. The endodermal lining of this cavity is involved in three major functions: circulation of seawater (through ciliary action), nutrient absorption, and reproduction, through capture of sperm that may filter into the coelom. Subsequent evolutionary events will gradually separate these functions in higher metazoans, but this observation from Poriphera suggests that a single endodermal cell type was responsible for all three in the last common ancestor of all animals.

Matrix Molecules

An extracellular matrix is required to hold cells together, and collagen and beta-crystallin first appear in this context.[32] Collagen will go on to become the most abundant protein in the metazoan kingdom. In sponges, collagen synthesis has been shown to be regulated by a homologue of myotrophin, evidence of an early autocrine/paracrine loop directing cell growth, and analogous to the induction of cardiac hypertrophy by mammalian myotrophin.[33]

Cell–Cell Communication and Recognition

Disaggregated sponge cells can re-associate in a process that depends on a proteoglycan aggregation factor and its receptor, a membrane protein with immunoglobulin and scavenger receptor homologies.[34] The sponge possesses the ability to recognize and reject allografts from other sponges, and the grafted cells undergo programmed death with upregulation of a death domain–containing receptor homologous to Fas.[35] Related innovations thought to have arisen in Urmetazoan ancestors include cell–cell and cell-matrix adhesion molecules, epithelial tight junctions, an early immune system, and the allocation of pluripotent stem cells with surface markers resembling those on mammalian precursor cells.[36,37] Proteins with homology to Bcl-2 can be found in sponges,[38–41] showing that both immune recognition and apoptosis are phylogenetically ancient. Apoptotic signaling pathways are also found in plants,[42] strengthening the argument that all multicellular organisms are related from a common ancestor (see above).

Body Patterning Molecules

The body plans of the earliest multicellular animals, including Poriphera, do not include a head or a tail or indeed any axis of symmetry. Radial or axial body plan symmetry are likely to have appeared in an ancestor common to both the Cnidaria (e.g., corals, jellyfish, sea anemones, hydra) and Bilateria, from which mammals are descended (FIG. 4). It is this hydra-like ancestor that may have taken the first steps toward mesoderm as well.[43,44] As noted earlier, jellyfish are diploblastic, with outer and inner cell layers separated by a jellylike substance, the mesohyl; however, during

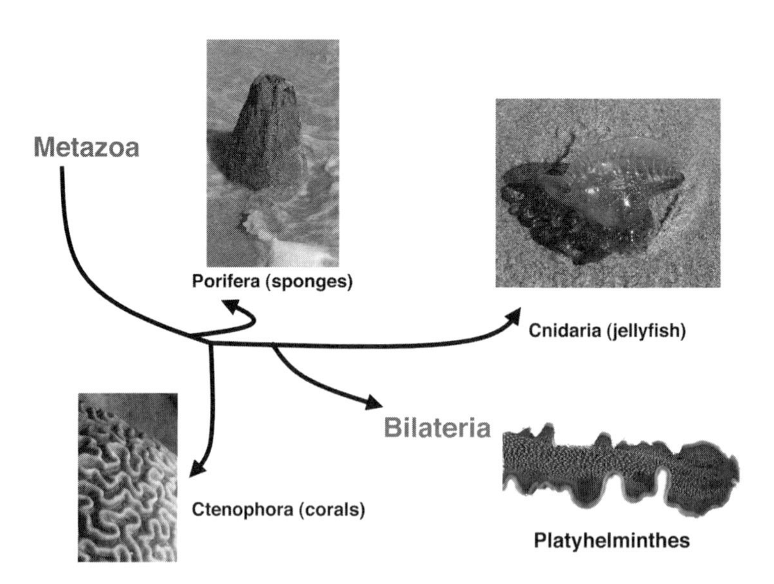

FIGURE 4. The first animals. Early branching of the metazoan family, of which Porifera are considered basal members [in color in Annals Online].

the medusa, or sexually reproductive, stage of the jellyfish life cycle, a number of specialized non-myoepithelial cells are found, including striated muscle cells, and sensory and nerve cells.[45] The striated muscle cells derive from a third germ layer that separates from the ectoderm early in medusa development,[46] possibly analogous to the middle germ layer of Bilateria, and express muscle-specific forms of troponin and myosin heavy chain. Further evidence that this third germ layer is a prototype of mesoderm is that it expresses the mesoderm-patterning genes Twist and Brachyury, as well as the muscle regulatory proteins MEF-2 and Snail.[45]

THE SYMMETRICAL BODY PLAN: BILATERIA (700–600 MILLION YEARS AGO)

All Bilateria are unequivocally triploblastic. At the root of this group are the flatworms, with later branches including the Ecdysoa, ancestors of the insects; the Lophotrochozoa, ancestors of annelids and mollusks; and the Deuterostomes. Our bilaterian ancestors were probably wormlike organisms with a recognizable front and back end (including a one-way intestine with anus), but lacking a recognizable head or eyes.[47] Patterning of the sagittal axis, and the development of a head with a separate opening to the environment, occurred with the radiation of the Bilateria and apparently coincident with the expansion of the Hox gene cluster.[48] Evidence suggests that a single prototype of these pattern-dictating genes underwent one or more rounds of tandem duplication around the time of the great Cambrian explosion, a time when a large increase in the number of life-forms appears in the archaeological record.[49] The Hox gene cluster dictates segmentation, headness, and tailness in the Bilateria,[50,51] and increasing copy number and divergence of these genes has been accompanied by increasing body plan complexity. In addition to the anterior–posterior axis, a clear dorsoventral axis becomes established sometime during bilaterian evolution.

The Tubular Heart

A critical step, the appearance of a single or paired heart primordium, occurred in a bilaterian ancestor—most likely prior to the divergence of Deuterostomes and Ecdysoa. Based on commonalities among the major branches of Bilateria (FIG. 5), this structure was most likely a tubular, pulsatile structure that lacked an enclosed vascular system but instead served to force fluid through pericellular interstices. Lacking chambers, septa, and valves, these early heart tubes probably did not drive unidirectional blood flow. During embryogenesis in insects, mollusks and annelids, this heart tube begins as an invagination from the gut, and continuity between the heart tube(s) and gut persists into the adult form. The myocytes investing the heart tube may have features of myoepithelium, vertebrate cardiomyocytes, and/or striated muscle, depending on the organism, and may have self-renewal properties.[52] The Drosophila heart contains additional features such as a cardioaortic valve and pericardial cells.

Whether the insect and vertebrate hearts evolved independently has been argued for nearly 200 years, largely on anatomic grounds: the insect heart tube is dorsal, while the vertebrate heart is ventral. Although it is possible that the emergence of

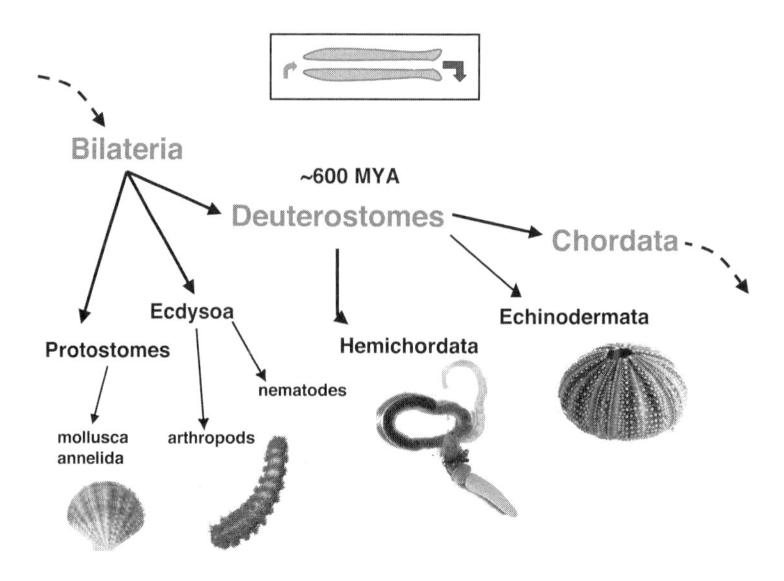

FIGURE 5. The two-ended body plan. Major branches of the Deuterostomes are shown. Two important model organisms, Drosophila and *C. elegans*, belong to the arthropods and nematodes, respectively. The diagram depicts the novel two-stoma body plan [in color in Annals Online].

insect and vertebrate hearts from non-cardiac mesoderm were separate events after the divergence of Deuterostomes and Ecdysoa, another point of view is that a simple longitudinal heart tube was present in the earliest bilaterians and subsequently underwent independent morphological and gene regulatory modifications. Functional and genetic, if not anatomic, evidence supports a close relationship between this primordial structure and the mammalian, chambered heart. The Drosophila heart tube is induced by expression of a homeotic gene, tinman, which was later found to have a mammalian homologue with the same function (Nkx2.5). Nkx homologues play similar roles in amphioxus and tunicates and have been cloned from octopus and cuttlefish.[53] Drosophila heart cells also express homologues of MEF2 and Hand, two genes involved in cardiogenic differentiation in vertebrates.[54–56] Orthologous gene regulatory patterns dictating nervous system development and anteroposterior patterning are also shared between Drosophila and vertebrates. These similarities can be explained by body plan inversion or structural relocations in the descendents of a common ancestor that already possessed a heart tube, notochord, and anteroposterior axis.[57] Interestingly, the single *Caenorhabditis elegans* MEF-2 homologue, *ceMEF2*, is dispensible for myogenesis, indicating that nematodes may have evolved alternative functions for this protein.[58]

THE TWO-ENDED ANIMAL: DEUTEROSTOMES (~600 MILLION YEARS AGO)

The Deuterostomes include the Echinodermata (e.g., sea urchins, starfish), Hemichordates (acorn worms) and Chordates. The Deuterostome body plan has a through-gut, with two openings at the head and rear of the bilaterally symmetric body plan, in place of the single gastrovascular–reproductive opening of earlier metazoans. Creation of the new opening, the mouth, involves specialization of head structures that require expression of the patterning genes Otx, Forkhead, Brachyury, and Goosecoid, among others. These gene products dictate both the appearance of the anterior gut and of a dorsal cluster of nerve cells that replaces a diffuse net of nerves.[43,59]

With these promising innovations, the original Deuterostome is still thought to have been a headless, eyeless, worm-like creature similar to the modern acorn worm. This latter organism is an enteropneust ("gut-breather") suggesting that our common ancestor still lacked segregation of gas exchange and digestive functions. Appearance of the mouth marks the beginning of the separation of gut and vascular system and is a critical feature distinguishing Deuterostomes from the other two major branches of Bilateria: Ecdysoa (e.g., fruit flies), and Lophotrochozoa (e.g., annelid worms).

Deuterostome evolution coincides with the beginning of the multiplication and functional divergence of genes encoding components of the contractile proteins. Superimposition of molecular phylogenetic trees suggests that significant radiation of striated muscle actin, myosin, and troponin isoforms occurred during the Cambrian Explosion sometime between 570 and 540 million years ago (MYA). In particular, divergence of smooth and striated muscle isoforms of the EF-hand proteins (myosin essential light chains, cardiac myosin regulatory light chains, and tropinin C), as well as striated- and smooth-muscle forms of myosin heavy chain and actin, appear to have taken place during or before the appearance of the chordates.[60]

THE NERVOUS SYSTEM GETS ORGANIZED: CHORDATES (~550 MILLION YEARS AGO)

The Chordates are defined by the presence of a dorsal nerve cord and a muscular supporting notochord (FIG. 6). Three major chordate branches extant today are the Urochordates (tunicates), the Cephalochordates (e.g., amphioxus), and the Vertebrates, which have a segmented bony enclosure for the axial nervous system. Myotomes and skeletal muscle are well established in the chordates, although phylogenetic trees suggest that the myogenic factors MyoD and Myf arose earlier. The basal members of this family, the sea squirts, typically have a notochord only during the larval stage, which involutes after attachment and immobilization. Other new features in animals from this group include development of a filter-feeding pharynx or gill structures,[61] the first appearance of an organized liver (amphioxus), and complete self-containment of the vascular system.

Both amphioxus and tunicates have a tubular heart that contracts rhythmically. The direction of flow may alternate in isolated tunicate heart preparations (M. Morad, personal communication). As in earlier Deuterostomes, these vessels are centered on the gut and pharynx, although no longer continuous with it. Features unique to amphioxus are the presence of both dorsal and ventral vessels, several parts of which are contractile.[62] The notion that these indeed constitute a heart equivalent is supported by the identification of an amphioxus Nkx2.5 homologue (*AmphiNk2-tin*) that is initially expressed in the ventral vessel.[53] This heart lacks chambers, valves, or endothelial lining and consists entirely of myocardial cells.[52]

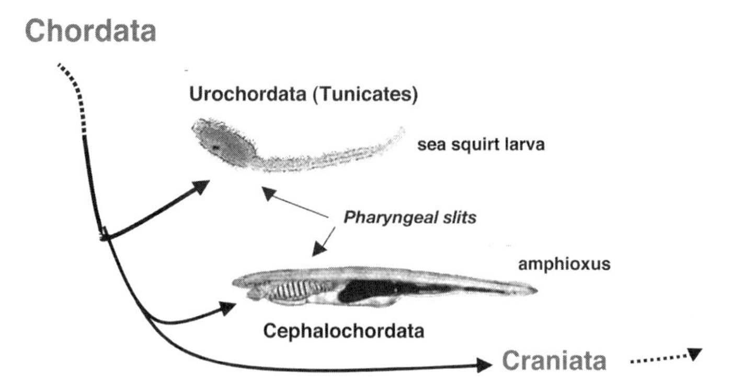

FIGURE 6. Chordates: the gut and gas exchange separate. Development of the enclosed vascular system, distinct from the gastrointestinal tract, appears first in Chordates. Tunicates (sea squirts) are basal members of the chordate family, as seen here in larval form. After anchoring to the sea floor, adult tunicates lose their nervous system, gills, and mobility, but retain a cardiovascular system [in color in Annals Online].

BRAINS AND HEARTS: THE VERTEBRATES (~500 MILLION YEARS AGO)

At the base of the Vertebrate branch are the hagfish and lampreys (FIG. 7). Hagfish (*Myxinidae*) have a partially calcified skull, but a cartilaginous vertebral column, so that they only just qualify as vertebrates. Their closed vascular system has several contractile components in series, including the portal vein heart and the systemic

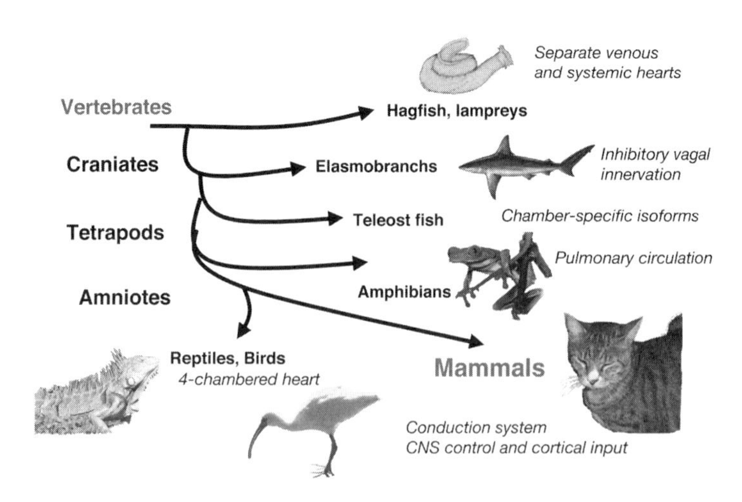

FIGURE 7. Some vertebrate innovations in cardiac morphology and function. Note the apical position of Felis domesticus [in color in Annals Online].

heart. The portal vein heart delivers blood to the gill vasculature and is roughly analogous to the right-sided cardiac chambers in birds and mammals. The Myxine portal vein heart and both atria express a natriuretic peptide.[63,64] Subsequent vertebrate evolution has produced many radical innovations in cardiac shape and function, including chambers and valves, separate respiratory and systemic circulations, endothermy, gap junctions, and the conduction system. A detailed discussion of each of these is beyond the scope of this review; the reader is referred to several excellent papers on general and specific aspects of vertebrate heart development.[65–67]

Endothermy and Energy Conservation

A recurring theme in vertebrate evolution is the development of adaptations to manage energy consumption and storage. Steps to improve bioenergetic efficiency, such as optimizing the efficiency of blood circulation, modifying metabolic and contractile protein functions under periods of stress, and providing central nervous system control over cardiac activity can be identified at various stages of vertebrate evolution. The development of endothermy in avians and mammals (and possibly therapods) conferred greater environmental adaptability and reproductive success[68] but also increased the metabolic cost of daily living.[69] This will have increased demand on the heart and cardiovascular system to deliver nutrients, as well as have increased the amount of effort needed to obtain nutrients from the environment. Many vertebrates undergo metabolic slowing during periods of limited resources, and some, including hagfish, can survive extended periods of anoxia. Similar adaptive methods, such as a switch to anaerobic glycolysis under hypoxia, can be observed in the mammalian myocardium.[70]

Isoform Specialization

Gene duplications likely facilitated many of the innovations of the vertebrate heart. The Devonian period (~420 MYA) saw a huge proliferation of fish species, a movement from marine into fresh water environments, and eventually the first scramble of vertebrates onto land (~300 MYA). The accompanying radiation of many genes encoding contractile and cytoskeletal elements, creatine kinase, and natriuretic peptides led to altered molecular properties for these proteins, and provided for functional specialization.[71–74] Cardiac actin and cardiac troponin C diverged from their skeletal counterparts prior to the frog–mammal divergence (350 MYA), while the three classes of cardiac, slow skeletal, and fast skeletal muscle isoforms diverged some time before the bird–mammal separation. Functional specialization is illustrated by trout cardiac TnC, which exhibits amino acid changes that confer increased calcium sensitivity, offsetting the desensitizing effect of cold and allowing cardiac function to be maintained at low water temperatures.[75] The acquisition of novel transcriptional regulatory properties for contractile proteins permits great plasticity in the vertebrate heart, including the ability to increase contractile protein mass rapidly.[76] Products of gene duplication may also acquire divergent regulatory properties: although cardiac and skeletal actin isoforms are tightly conserved at the protein level, their gene control regions are highly dissimilar.[77–80] Divergence at the promoter level provides for different levels of transcriptional response depending on the extracellular milieu, and for the recruitment of isoforms with specific adaptive properties during periods of increased demand.

Separation of the Pulmonary Circulation

Definitive organization of the original tubular heart into separate systemic and pulmonary circulations occurred sometime between separation of the teleost fish and amphibians (350–300 MYA).[81] Teleost fish hearts have only two chambers that serve both for oxygenation and tissue perfusion, connected by an atrioventricular valve, while the amphibian heart has two atria and a single ventricular chamber. In reptiles, a muscular ridge running from the apex to the base of the ventricle creates directed flow of oxygenated and deoxygenated blood. This ridge is small in turtles, and considerable mixing occurs, but in lizards and snakes, the ridge is larger and separation is more efficient.[82] Birds and mammals have complete atrial and ventricular septa, allowing separation between oxygenated and deoxygenated blood, and a full division of the systemic and pulmonary circulations. The importance of this divided and directional flow can be observed when the system is defective.[83]

Myocardial Regeneration (Or Not)

The ability to efficiently replace lost myocytes disappeared sometime around the appearance of warm-bloodedness. Unlike mollusks, arthropods, and amphibians,[52] mammals cannot generate significant numbers of new ventricular myocytes after birth, either through self-renewal or replacement. This loss has become a significant liability only very recently, with the appearance of *Homo sapiens sapiens* and the unique ability of Western societies to meet and exceed daily caloric requirements.[84] Recent, detailed studies have identified a pool of undifferentiated mononuclear precursor cells in human ventricular myocardium, identifiable by expression of one or more surface markers including *c-kit* and sca-1;[85–87] a second population of isl-1–positive precursor cells has recently been observed in the atria and right ventricular outflow tract.[88] The extent to which these cells resemble the various types of precursor cells identified in myocardial tissue from arthropoda to ascidians and whether myocardial regenerative capability will be re-created in humans are currently topics of intensive research.

Cell–Cell Communication

Connexins are proteins that form intracellular junctions that transmit several types of chemical and electrical signals. The appearance of connexin-based gap junctions parallels the increasing complexity and size of the vertebrate heart. A basal connexin has been reported in chordates and more than 20 separate connexin family members are found in mammals. Rapid communication through gap junctions is a critical factor allowing the vertebrate heart to contract in a rapid, synchronous manner, as distinct from the more peristaltic, wave-like contractions of invertebrate hearts.[89] Synchronous contraction contributes to biomechanical efficiency and was further promoted in vertebrate evolution by the appearance of a specialized electrical conduction system. The development of the major components of this system has been reviewed elsewhere[90,91]

Love and the Mammalian Heart

The process of gradual separation of the primitive body cavity into three distinct functional systems is nearly complete in mammals. The combined digestion-sex-respiration organ has become the gastrointestinal tract, the reproductive system, and the heart. Clues to their ancient unity can be found in their early embryonic gene expression programs and differentiation cues.[92] Interestingly, central nervous system control over these functions has evolved in tandem, re-investing the heart with a role in reproduction. In hagfish, a single, unmyelinated nerve (the vagus) sends branches to the viscera, but there is no direct innervation of the heart. In lower vertebrates, this innervation couples cardiac function to perception of danger by inducing a state of immobilization.[93,94] Higher vertebrates utilize the well-known fight-or-flight response, which inhibits vagal immobilization and stimulates heart rate through secretion of sympathoadrenal hormones into the bloodstream. A recent evolutionary feature, specific to mammals, is the development of a myelinated vagus nerve with sources in the medullary nucleus ambiguus. Changes in vagal tone are critically responsible for rapid slowing and speeding of heart rate in mammals. At the same time, efferent pathways from the mammalian frontal cortex allow vagal tone to be influenced by perceptions of safety and desire for social proximity. The individual is thus able to become calm and receptive to interaction, which might otherwise be terrifying, and invoke death-feigning or flight. This function of the vagus nerve critically enables the behavior required for courtship, copulation, and sustained pair bonding.[94] The tendency to identify our hearts as the location of affection, passion, and constancy may therefore have a sound basis in mammalian neurophysiology. Is the latest chapter in cardiac evolution to be the development of love?

ACKNOWLEDGMENTS

This work was supported by grants from the National Institutes of Health (R01-HL071094) and the Miami Heart Research Institute. I am grateful to Dr. Sam Sideman for suggesting the topic of this paper, and to Dr. Keith Webster and the late Drs. Peter Hochacha and Peter Lutz for their inspiration and support. The Tree of Life Web Project (http://tolweb.org/tree/phylogeny. html), provided much general background information for this review as well as several images: FIGURE 2, "Giardia"; FIGURE 3, "Fungi"; FIGURE 4, "platyhelminthes." The slime mold image in FIGURE 2 was provided by Dennis Curtin (www.shortcourses.com). The hagfish illustration in FIGURE 7 was drawn by Nancy Vander Velde and first appeared in the September 2003 issue of *California Diving News*.

REFERENCES

1. LAKE, J.A. 1989. Origin of the eukaryotic nucleus: eukaryotes and eocytes are genotypically related. Can. J. Microbiol. **35:** 109–118.
2. GARNER, E.C., C.S. CAMPBELL & R.D. MULLINS. 2004. Dynamic instability in a DNA-segregating prokaryotic actin homolog. Science **306:** 987–989.
3. LEWIS, P.J. 2004. Bacterial subcellular architecture: recent advances and future prospects. Mol. Microbiol. **54:** 1135–1150.
4. KRUSE, T., J. BORK-JENSEN & K. GERDES. 2005. The morphogenetic MreBCD proteins of *Escherichia coli* form an essential membrane-bound complex. Mol. Microbiol. **55:** 78–89.
5. WEBSTER, K.A. 2003. Evolution of the coordinate regulation of glycolytic enzyme genes by hypoxia. J. Exp. Biol. **206:** 2911–2922.

6. GRAY, M.W., G. BURGER & B.F. LANG. 1999. Mitochondrial evolution. Science **283:** 1476–1481.
7. BATUT, J., S.G. ANDERSSON & D. O'CALLAGHAN. 2004. The evolution of chronic infection strategies in the alpha-proteobacteria. Nat. Rev. Microbiol. **2:** 933–945.
8. BLACKSTONE, N.W. & D.R. GREEN. 1999. The evolution of a mechanism of cell suicide. BioEssays **21:** 84–88.
9. AMEISEN, J.C. 2002. On the origin, evolution, and nature of programmed cell death: a timeline of four billion years. Cell Death Differ. **9:** 367–393.
10. KOONIN, E.V. & L. ARAVIND. 2002. Origin and evolution of eukaryotic apoptosis: the bacterial connection. Cell Death Differ. **9:** 394–404.
11. MANS, B.J., V. ANANTHARAMAN, L. ARAVIND & E.V. KOONIN. 2004. Comparative genomics, evolution and origins of the nuclear envelope and nuclear pore complex. Cell Cycle **3:** 1612–1637.
12. SERAVIN, L.N. 1986. [The origin of the eukaryotic cell. IV. The general hypothesis of the autogenous origin of eukaryotes]. Tsitologiia **28:** 899–910.
13. GOFSHTEIN, L.V. 1978. [Plant histones. Relevance to the evolution of prokaryotes to eukaryotes]. Biokhimiia **43:** 947–958.
14. GUPTA, R.S., K. AITKEN, M. FALAH & B. SINGH. 1994. Cloning of *Giardia lamblia* heat shock protein HSP70 homologs: implications regarding origin of eukaryotic cells and of endoplasmic reticulum. Proc. Natl. Acad. Sci. U.S.A. **91:** 2895–2899.
15. PENNISI, E. 2004. The birth of the nucleus. Science **305:** 766–768.
16. Hartman, H. & A. Fedorov. 2002. The origin of the eukaryotic cell: a genomic investigation. Proc. Natl. Acad. Sci. U.S.A. **99:** 1420–1425.
17. BAILEY, K.A., S.L. PEREIRA, J. WIDOM & J.N. REEVE. 2000. Archaeal histone selection of nucleosome positioning sequences and the procaryotic origin of histone-dependent genome evolution. J. Mol. Biol. **303:** 25–34.
18. MARTIN, W., H. BRINKMANN, C. SAVONNA & R. CERFF. 1993. Evidence for a chimeric nature of nuclear genomes: eubacterial origin of eukaryotic glyceraldehyde-3-phosphate dehydrogenase genes. Proc. Natl. Acad. Sci. U.S.A. **90:** 8692–8696.
19. BELL, P.J. 2001. Viral eukaryogenesis: was the ancestor of the nucleus a complex DNA virus? J. Mol. Evol. **53:** 251–256.
20. LANG, B.F., C. O'KELLY, T. NERAD, *et al.* 2002. The closest unicellular relatives of animals. Curr. Biol. **12:** 1773–1778.
21. NASMYTH, K. 1995. Evolution of the cell cycle. Philos. Trans. R. Soc. Lond. B. Biol. Sci. **349:** 271–281.
22. WATTS, F.Z., D.M. MILLER & E. ORR. 1985. Identification of myosin heavy chain in Saccharomyces cerevisiae. Nature **316:** 83–85.
23. BORDOLI, L., M. NETSCH, U. LÜTHI, *et al.* 2001. Plant orthologs of p300/CBP: conservation of a core domain in metazoan p300/CBP acetyltransferase-related domains. Nucl. Acids Res. **29:** 589–597.
24. PANDEY, R., A. MULLER, C.A. NAPOLI, *et al.* 2002. Analysis of histone acetyltransferase and histone deacetylase families of *Arabidopsis thaliana* suggests functional diversification of chromatin modification among multicellular eukaryotes. Nucl. Acids Res. **30:** 5036–5055.
25. BILLINGTON, T., M. PHARMAWATI & C.A. GEHRING. 1997. Isolation and immunoaffinity purification of biologically active plant natriuretic peptide. Biochem. Biophys. Res. Commun. **235:** 722–725.
26. VESELY, D.L., W.R. GOWER & A.T. GIORDANO. 1993. Atrial natriuretic peptides are present throughout the plant kingdom and enhance solute flow in plants. Am. J. Physiol. **265:** E465–E477.
27. REYES, J.C., M.I. MURO-PASTOR & F.J. FLORENCIO. 2004. The GATA family of transcription factors in *Arabidopsis* and rice. Plant Physiol. **134:** 1718–1732.
28. NAM, J., C.W. DE PAMPHILIS, H. MA & M. NEI. 2003. Antiquity and evolution of the MADS-box gene family controlling flower development in plants. Mol. Biol. Evol. **20:** 1435–1447.
29. STILLER, J.W. 2004. Emerging genomic and proteomic evidence on relationships among the animal, plant and fungal kingdoms. Genomics Proteomics Bioinformatics **2:** 70–76.
30. KING, N. 2004. The unicellular ancestry of animal development. Dev. Cell **7:** 313–325.

31. MULLER, W.E., M. WIENS, T. ADELL, *et al.* 2004. Bauplan of urmetazoa: basis for genetic complexity of metazoa. Int. Rev. Cytol. **235:** 53–92.
32. KRASKO, A., I.M. MULLER & W.E. MULLER. 1997. Evolutionary relationships of the metazoan beta gamma-crystallins, including that from the marine sponge *Geodia cydonium*. Proc. R. Soc. Lond. B. Biol. Sci. **264:** 1077–1084.
33. SCHRODER, H.C., A. KRASKO, R. BATEL, *et al.* 2000. Stimulation of protein (collagen) synthesis in sponge cells by a cardiac myotrophin-related molecule from *Suberites domuncula*. FASEB J. **14:** 2022–2031.
34. BLUMBACH, B., Z. PANCER, B. DIEHL, *et al.* 1998. The putative sponge aggregation receptor. Isolation and characterization of a molecule composed of scavenger receptor cysteine-rich domains and short consensus repeats. J. Cell Sci. **111:** 2635–2644.
35. WIENS, M., A. KRASKO, B. BLUMBACH, *et al.* 2000. Increased expression of the potential proapoptotic molecule DD2 and increased synthesis of leukotriene B4 during allograft rejection in a marine sponge. Cell Death Differ. **7:** 461–469.
36. MULLER, W.E., M. WIENS, I.M. MULLER & H.C. SCHRODER. 2004. The chemokine networks in sponges: potential roles in morphogenesis, immunity and stem cell formation. Prog. Mol. Subcell. Biol. **34:** 103–143.
37. BOSCH, T.C.G. 2004. Control of asymmetric cell divisions: will cnidarians provide an answer? BioEssays **26:** 929–931.
38. WIENS, M., A. KRASKO, C.I. MULLER & W.E. MULLER. 2000. Molecular evolution of apoptotic pathways: cloning of key domains from sponges (Bcl-2 homology domains and death domains) and their phylogenetic relationships. J. Mol. Evol. **50:** 520–531.
39. ARAVIND, L., V.M. DIXIT & E.V. KOONIN. 1999. The domains of death: evolution of the apoptosis machinery. Trends Biochem. Sci. **24:** 47–53.
40. MITTLER, R., L. SIMON & E. LAM. 1997. Pathogen-induced programmed cell death in tobacco. J. Cell Sci. **110:** 1333–1344.
41. KAWAI, M., L. PAN, J.C. REED & H. UCHIMIYA. 1999. Evolutionally conserved plant homologue of the Bax inhibitor-1 (BI-1) gene capable of suppressing Bax-induced cell death in yeast(1). FEBS Lett. **464:** 143–147.
42. BALK, J., C.J. LEAVER & P.F. MCCABE. 1999. Translocation of cytochrome c from the mitochondria to the cytosol occurs during heat-induced programmed cell death in cucumber plants. FEBS Lett. **463:** 151–154.
43. MARCELLINI, S., U. TECHNAU, J.C. SMITH & P. LEMAIRE. 2003. Evolution of Brachyury proteins: identification of a novel regulatory domain conserved within Bilateria. Dev. Biol. **260:** 352–361.
44. LARTILLOT, N., M. LE GOUAR & A. ADOUTTE. 2002. Expression patterns of fork head and goosecoid homologues in the mollusc *Patella vulgata* supports the ancestry of the anterior mesendoderm across Bilateria. Dev. Genes Evol. **212:** 551–561.
45. SPRING, J., N. YANZE, C. JOSCH, *et al.* 2002. Conservation of Brachyury, Mef2, and Snail in the myogenic lineage of jellyfish: a connection to the mesoderm of bilateria. Dev. Biol. **244:** 372–384.
46. KUHN, A. 1910. Die Entwicklung der Geschlectsindividuen der Hydromedusen. Zool. Jahrbuch. **30:** 145–164.
47. BAGUNA, J. & M. RIUTORT. 2004. The dawn of bilaterian animals: the case of acoelomorph flatworms. BioEssays **26:** 1046–1057.
48. POPODI, E., J.C. KISSINGER, M.E. ANDREWS & R.A. RAFF. 1996. Sea urchin Hox genes: insights into the ancestral Hox cluster. Mol. Biol. Evol. **13:** 1078–1086.
49. ZHANG, J. & M. NEI. 1996. Evolution of Antennapedia-class homeobox genes. Genetics **142:** 295–303.
50. PETERSON, K.J., S.Q. IRVINE, R.A. CAMERON & E.H. DAVIDSON. 2000. Quantitative assessment of Hox complex expression in the indirect development of the polychaete annelid Chaetopterus sp. Proc. Natl. Acad. Sci. U.S.A. **97:** 4487–4492.
51. HART, C.P., A. FAINSOD & F.H. RUDDLE. 1987. Sequence analysis of the murine Hox-2.2, Hox-2.3 and Hox-2.4 homeo-boxes: evolutionary and structural comparisons. Genomics. **1:** 182–195.
52. MARTYNOVA, M.G. 2004. Proliferation and differentiation processes in the heart: muscle elements in different phylogenetic groups. Int. Rev. Cytol. **235:** 215–250.
53. HOLLAND, N.D., T.V. VENKATESH, L.Z. HOLLAND, *et al.* 2003. AmphiNk2-tin, an

amphioxus homeobox gene expressed in myocardial progenitors: insights into evolution of the vertebrate heart. Dev. Biol. **255:** 128–137.
54. Kolsch, V. & A. Paululat. 2002. The highly conserved cardiogenic bHLH factor Hand is specifically expressed in circular visceral muscle progenitor cells and in all cells of the dorsal vessel during Drosophila embryogenesis. Dev. Genes Evol. **212:** 473–485.
55. Bodmer, R. & M. Frasch. 1999. Genetic determination of Drosophila heart development. *In* Heart Development. R.P. Harvey & N. Rosenthal, Eds.: 65–90. Academic Press. San Diego.
56. Cripps, R.M., T.L. Lovato & E.N. Olson. 2004. Positive autoregulation of the Myocyte enhancer factor-2 myogenic control gene during somatic muscle development in Drosophila. Dev. Biol. **267:** 536–547.
57. Gerhart, J.C. 2000. Inversion of the chordate body axis: are there alternatives? Proc. Natl. Acad. Sci. U.S.A. **97:** 4445–4448.
58. Dichoso, D., T. Brodigan, K.Y. Chwoe, *et al.* 2000. The MADS-box factor CeMEF2 is not essential for *Caenorhabditis elegans* myogenesis and development. Dev. Biol. **223:** 431–440.
59. Tagawa, K., N. Satoh & T. Humphreys. 2001. Molecular studies of hemichordate development: a key to understanding the evolution of bilateral animals and chordates. Evol. Dev. **3:** 443–454.
60. Ota, S. & N. Saitou. 1999. Phylogenetic relationship of muscle tissues deduced from superimposition of gene trees. Mol. Biol. Evol. **16:** 856–867.
61. Dominguez, P., A.G. Jacobson & R.P.S. Jefferies. 2002. Paired gill slits in a fossil with a calcite skeleton. Nature **417:** 841–844.
62. Simoes-Costa, M.S., M. Vasconcelos, A. Sampaio, *et al.* 2005. The evolutionary origin of cardiac chambers. Dev. Biol. **277:** 1–15.
63. Reinecke, M., D. Betzler & W.G. Forssmann. 1987. Immunocytochemistry of cardiac polypeptide hormones (cardiodilatin/atrial natriuretic polypeptide) in brain and hearts of *Myxine glutinosa* (Cyclostomata). Histochemistry **86:** 233–239.
64. Endo, H., T.K. Yamada, M. Kurohmaru & Y. Hayashi. 1997. Structure of portal vein heart wall in the brown hagfish (*Paramyxine atami*). J. Morphol. **231:** 225–230.
65. Chen, J.N. & M.C. Fishman. 1996. Zebrafish tinman homolog demarcates the heart field and initiates myocardial differentiation. Development **122:** 3809–3816.
66. Fishman, M.C. & K.R. Chien. 1997. Fashioning the vertebrate heart: earliest embryonic decisions. Development **124:** 2099–2117.
67. Chen, J.N. & M.C. Fishman. 2000. Genetics of heart development. Trends Genet. **16:** 383–388.
68. Hillenius, W.J. & J.A. Ruben. 2004. The evolution of endothermy in terrestrial vertebrates: Who? When? Why? Physiol. Biochem. Zool. **77:** 1019–1042.
69. Suarez, R.K., C.J. Doll, A.E. Buie, *et al.* 1989. Turtles and rats: a biochemical comparison of anoxia-tolerant and anoxia-sensitive brains. Am. J. Physiol. **257:** R1083–R1088.
70. Webster, K.A. & N.H. Bishopric. 1992. Molecular regulation of cardiac myocyte adaptations to chronic hypoxia. J. Mol. Cell. Cardiol. **24:** 741–752.
71. Uda, K., T. Suzuki & W.R. Ellington. 2004. Elements of the major myofibrillar binding peptide motif are present in the earliest of true muscle type creatine kinases. Int. J. Biochem. Cell Biol. **36:** 785–794.
72. Kawakoshi, A., S. Hyodo, A. Yasuda & Y. Takei. 2003. A single and novel natriuretic peptide is expressed in the heart and brain of the most primitive vertebrate, the hagfish (*Eptatretus burgeri*). J. Mol. Endocrinol. **31:** 209–220.
73. Fock, U. & H. Hinssen. 2002. Nebulin is a thin filament protein of the cardiac muscle of the agnathans. J. Muscle Res. Cell Motil. **23:** 205–213.
74. Aouacheria, A., C. Cluzel, C. Lethias, *et al.* 2004. Invertebrate data predict an early emergence of vertebrate fibrillar collagen clades and an anti-incest model. J. Biol. Chem. **279:** 47711–47719.
75. Gillis, T.E. & G.F. Tibbits. 2002. Beating the cold; the functional evolution of troponin C in teleost fish. Comp. Biochem. Physiol. A. Mol. Integr. Physiol. **132:** 763–772.

76. ANDERSEN, J.B., B.C. ROURKE, V.J. CAIOZZO, *et al.* 2005. Postprandial cardiac hypertrophy in pythons. Nature **434:** 37–38.
77. BISHOPRIC, N.H. & L. KEDES. 1991. Adrenergic regulation of the skeletal alpha-actin gene promoter during myocardial cell hypertrophy. Proc. Natl. Acad. Sci. U.S.A. **88:** 2132–2136.
78. BISHOPRIC, N.H., R. GAHLMANN, R. WADE & L. KEDES. 1992. Gene expression during skeletal and cardiac muscle development. *In* The Heart and Cardiovascular System, 2nd ed., H.A. Fozzard *et al.*, Eds.: 1587–1597. Raven Press. New York.
79. SARTORELLI, V., N.A. HONG, N.H. BISHOPRIC & L. KEDES. 1992. Myocardial activation of the human cardiac alpha-actin promoter by helix-loop-helix proteins. Proc. Natl. Acad. Sci. U.S.A. **89:** 4047–4051.
80. SLEPAK T.I., K.A. WEBSTER, J. ZANG, *et al.* 2001. Control of cardiac-specific transcription by p300 through myocyte enhancer factor-2D. J. Biol. Chem. **276:** 7575–7585.
81. SIMOES-COSTA, M.S., M. VASCONCELOS, A.C. SAMPAIO, *et al.* 2005. The evolutionary origin of cardiac chambers. Dev. Biol. **277:** 1–15.
82. WANG, T., J. ALTIMIRAS & M. AXELSSON. 2002. Intracardiac flow separation in an *in situ* perfused heart from Burmese python *Python molurus*. J. Exp. Biol. **205:** 2715–2723.
83. RISHNIW, M. & B.P. CARMEL. 1999. Atrioventricular valvular insufficiency and congestive heart failure in a carpet python. Aust. Vet. J. **77:** 580–583.
84. EATON, S.B., M. KONNER & M. SHOSTAK. 1988. Stone agers in the fast lane: chronic degenerative diseases in evolutionary perspective. Am. J. Med. **84:** 739–749.
85. BELTRAMI, A.P., L. BARLUCCHI, D. TORELLA, *et al.* 2003. Adult cardiac stem cells are multipotent and support myocardial regeneration. Cell **114:** 763–776.
86. URBANEK, K., F. QUAINI, G. TASCA, *et al.* 2003. Intense myocyte formation from cardiac stem cells in human cardiac hypertrophy. Proc. Natl. Acad. Sci. U.S.A. **100:** 10440–10445.
87. OH, H., S.B. BRADFUTE, T.D. GALLARDO, *et al.* 2003. Cardiac progenitor cells from adult myocardium: homing, differentiation, and fusion after infarction. Proc. Natl. Acad. Sci. U.S.A. **100:** 12313–12318.
88. LAUGWITZ, K.L., A. MORETTI, J. LAM, *et al.* 2005. Postnatal isl1+ cardioblasts enter fully differentiated cardiomyocyte lineages. Nature **433:** 647–653.
89. BECKER, D.L., J.E. COOK, C.S. DAVIES, *et al.* 1998. Expression of major gap junction connexin types in the working myocardium of eight chordates. Cell Biol. Int. **22:** 527–543.
90. THOMPSON, R.P., M. RECKOVA, A. DE ALMEIDA, *et al.* 2003. The oldest, toughest cells in the heart. Novartis Found. Symp. **250:** 157–174; discussion 174–176, 276–279.
91. SEDMERA, D., M. Reckova, A. DE ALMEIDA, *et al.* 2003. Functional and morphological evidence for a ventricular conduction system in zebrafish and *Xenopus* hearts. Am. J. Physiol. Heart. Circ. Physiol. **284:** H1152–H1160.
92. NAKAMURA, T. & M.D. SCHNEIDER. 2003. The way to a human's heart is through the stomach: visceral endoderm-like cells drive human embryonic stem cells to a cardiac fate. Circulation **107:** 2638–2639.
93. TAYLOR, E.W., D. JORDAN & J.H. COOTE. 1999. Central control of the cardiovascular and respiratory systems and their interactions in vertebrates. Physiol. Rev. **79:** 855–916.
94. PORGES, S.W. 1998. Love: an emergent property of the mammalian autonomic nervous system. Psychoneuroendocrinology **23:** 837–861.

The Miscommunicative Cardiac Cell

When Good Proteins Go Bad

ALDRIN V. GOMES, GAYATHRI VENKATRAMAN, AND JAMES D. POTTER

Department of Molecular and Cellular Pharmacology, Leonard M. Miller School of Medicine, University of Miami, Miami, Florida 33101, USA

ABSTRACT: Troponin (Tn) is made up of three subunits, troponin T (TnT), troponin I (TnI), and troponin C (TnC). In cardiac muscle, TnI can exist as two isoforms, slow skeletal TnI (ssTnI) or cardiac TnI (cTnI), whereas TnT occurs as multiple isoforms. The predominant form of TnI in fetal cardiac muscle is ssTnI, which is derived from a different gene than cTnI. However, the predominant form of cardiac TnT (cTnT) in fetal muscle is cTnT1, which is derived from the same gene that produces the adult cTnT isoform (cTnT3). Fetal cardiac muscle is more sensitive to Ca^{2+} than adult muscle and this may be due in part to the fetal cTnT1 and ssTnI isoforms. cTnT1 and/or ssTnI by themselves cause a significant increase in Ca^{2+} sensitivity when compared to cTnT3 and/ or cTnI. Mutations in the gene for cTnT can cause hypertrophic cardiomyopathy or dilated cardiomyopathy (DCM). Investigation of DCM mutations in the fetal cTnT1 isoform showed that the cTnT isoform is an important determinant of the effect of the mutation. The TnI isoform also affects the physiological function of the cardiac muscle. The presence of both the fetal TnT isoform, containing a DCM mutation, and ssTnI results in larger changes in Ca^{2+} sensitivity than the same DCM mutant in the adult TnT isoform and in the presence of cTnI (when compared to their respective wild-type TnT controls). These recent results suggest that some mutations may have different severities in fetal and adult hearts.

KEYWORDS: calcium; dilated cardiomyopathy; hypertrophic cardiomyopathy; isoforms; troponin

INTRODUCTION

The regulation of the contractile activity in striated muscle is dependent upon the cooperative interaction between thick and thin filament proteins. The regulatory proteins troponin (Tn) and tropomyosin (Tm) form a complex on the actin filament that inhibits actomyosin ATPase and force development under resting intracellular Ca^{2+} concentrations. The Tn complex is the intracellular Ca^{2+} sensor that initiates striated muscle contraction. Tn is composed of three subunits, troponin T ([TnT], the tropomyosin binding subunit), troponin I ([TnI], the inhibitory subunit), and troponin C ([TnC], the Ca^{2+} binding subunit). When Ca^{2+} binds to the regulatory Ca^{2+} bind-

Address for correspondence: Prof. James D. Potter, Ph.D., Chairman, Dept. of Molecular and Cellular Pharmacology, Leonard M. Miller School of Medicine at the University of Miami, 1600 N.W. 10th Avenue, Miami, FL 33136, USA. Voice: 305-243-5874; fax: 305-324-6024.
jdpotter@miami.edu

Ann. N.Y. Acad. Sci. 1047: 30–37 (2005). © 2005 New York Academy of Sciences.
doi: 10.1196/annals.1341.003

ing sites of TnC, this results in changes in the tertiary structure of TnC. Conformational changes in TnC then cause changes in the structure of the other Tn subunits (TnT, TnI) and Tm, resulting in Tm moving to a position that allows a greater number of myosin heads to bind to the actin filament, leading to muscle contraction.

Mutations in Tn subunits are associated with all three main types of cardiomyopathy: hypertrophic cardiomyopathy (HCM), dilated cardiomyopathy (DCM), and restrictive cardiomyopathy (RCM). HCM is one of the most frequently occurring inherited cardiac disorders and is characterized by hypertrophy of the interventricular septum and left ventricular wall, and is suggested to be familial (that is, inherited in an autosomal fashion) in ~55% of the cases. DCM is a relatively common disorder that results in heart failure and premature death and occurs ~2.5 times more commonly than HCM.[1] HCM and DCM can be caused by mutations in any of the three troponin subunits.[2] RCM has only been found to be associated with the TnI subunit but is likely to be found in other Tn subunits in the future.[3]

TROPONIN T ISOFORMS

There are at least four different TnT isoforms expressed in the human heart at the protein level.[4,5] Alternative splicing of exon 4 and 5 generates the different isoforms (c[cardiac]TnT1→cTnT4) with varying electrophoretic mobility. TnT3 (lacking exon 5) is the dominant TnT isoform in the adult heart, whereas TnT1 (when both exons 4 and 5 are present) is the dominant TnT isoform in the fetal heart (FIG. 1). TnT isoforms have been previously found to be important for Ca^{2+} sensitivity of

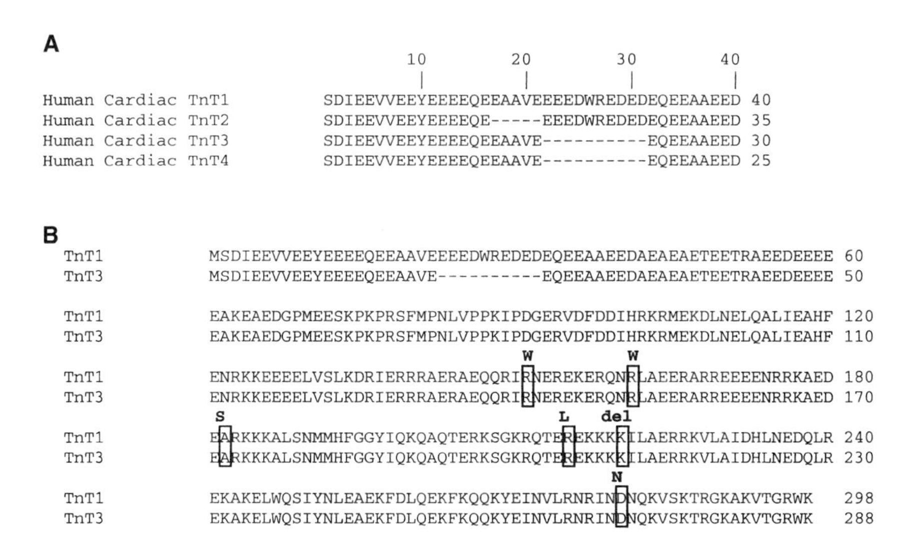

FIGURE 1. Primary structure of troponin T (TnT) isoforms. (**A**) N-terminal hypervariable region of human cardiac troponin T. (**B**) Amino acid sequence of the predominant fetal (TnT1) and adult (TnT3) human cardiac TnT isoforms showing the location of mutations associated with dilated cardiomyopathy.

force development, maximal and minimal force, and maximal and basal ATPase activity.[6–12] In the presence of cTnI or slow skeletal TnI (ssTnI), cTnT1-containing fibers showed increased Ca^{2+} sensitivity of force development (ΔpCa_{50} = +0.12) compared with cTnT3-containing fibers. Another cTnT isoform, cTnT4, which is predominantly found in fetal cardiac tissue, is also re-expressed in some failing hearts. In the presence of cTnI or ssTnI, cTnT1-containing fibers showed increased Ca^{2+} sensitivity of force development (ΔpCa_{50} = +0.17) compared with cTnT4-containing fibers. The maximal force in ssTnI reconstituted skinned fibers was significantly greater in the presence of cTnT1 when compared to cTnT3-containing fibers. Tn complexes (in the presence of cTnI or ssTnI) containing cTnT1 had a reduced ability to inhibit actomyosin ATPase activity when compared to Tn complexes containing cTnT3. Maintenance of normal myocardial contractility depends on both the proper homeostasis of Ca^{2+} in the muscle cells and on the contractile responsiveness of the myofilaments to Ca^{2+}. The change in the Ca^{2+} sensitivity of force development that occurs when different cTnT isoforms are present are likely to modulate myocardial contractility resulting in functionally distinct thin filament responsiveness to Ca^{2+}.[9] In this review, we discuss a new area of study that will help to determine if Tn mutations associated with cardiomyopathy have different functional properties in fetal and adult cardiac muscle. To emphasize the potential importance of these studies and the need for early screening of infants for mutations associated with cardiomyopathy, recent results of two TnT DCM mutations in a fetal setting are presented.

DILATED CARDIOMYOPATHY

DCM is characterized by progressive ventricular dilation and impaired systolic function.[13] Pathological manifestation of DCM includes modest hypertrophy, myocyte degeneration, and increased interstitial fibrosis. DCM can be caused by a variety of factors, including ischemia, alcohol toxicity, viral infections, as well as familial and genetic factors. Among these factors, it is estimated that 25% to 30% of all DCM cases are due to familial and genetic factors. Genetic analysis of family members with familial DCM revealed that the disease is primarily transmitted by autosomal dominant mode of inheritance followed by autosomal recessive, X-linked, and maternal modes of inheritance. Recently, mutations in genes for various cytoskeletal and sarcomeric proteins were shown to cause DCM.[14] The first sarcomeric protein mutations associated with DCM were identified by Olson and coworkers in the gene coding for the cardiac actin (missense mutations R312H and E361G).[15] Subsequently, mutations were identified in other sarcomeric proteins, including all three Tn subunits—cardiac TnT, cardiac TnI, cardiac TnC, as well as in α-Tm, β-myosin heavy chain, cardiac myosin binding protein C, and titin.

TROPONIN T MUTATIONS ASSOCIATED WITH DILATED CARDIOMYOPATHY

Kamisago *et al.*[16] reported the first TnT mutation as the cause of DCM in two unrelated families. This mutation occurs in exon 13 of cardiac TnT due to a deletion of the basic amino acid lysine at position 210. The second TnT mutation identified by

Li *et al.*[17] in a large family of 72 members was a missense mutation, R141W, which occurred in the tropomyosin binding region of TnT. Seventeen members of this family died of heart failure before the genotypic studies were carried out. Fourteen living members of the family clinically manifested DCM, predominantly in the second decade of their life. None of the individuals bearing the TnT mutations exhibited ventricular hypertrophy, which is a hallmark feature of HCM.

There are now four other mutations in TnT (R131W, A171S, R205L, and D270N) that cause DCM,[18,19] and the list will grow significantly as more genetic testing is performed on patients affected with DCM (FIG. 1). The A171S mutation is interesting as this mutation is associated with a significant gender difference in the severity of the observed phenotype, with adult males demonstrating more severe left ventricular dysfunction than adult females.[19] There are two other mutations in TnT that cause DCM—R92W and K273E. However, these mutations in affected patients initially cause HCM and the disease later progresses to DCM.[20,21] It is unclear what factors contribute to this transition from hypertrophy to progressive ventricular dilation.

PHYSIOLOGICAL EFFECTS OF DCM MUTATIONS IN ADULT TnT (TnT3)

Of the five TnT mutations currently known to cause only DCM, the R141W and the ΔK210 mutations have been studied by several different groups. Morimoto *et al.*[22] investigated the first DCM TnT mutation ΔK210 and found a decrease in the Ca^{2+} sensitivity of force generation with no change in maximal force or cooperativity in permeabilized rabbit cardiac muscle fibers. They also found an increase in the free Ca^{2+} concentrations required for the activation of ATPase activity. Morimoto *et al.*[22] suggested the Ca^{2+} desensitization of force generation as the primary cause for DCM. Investigation of the ΔK210 mutation by Robinson *et al.*[23] showed an increase in Ca^{2+} sensitivity of ATPase activation when 100% mutant proteins were used in the study. However, equimolar mixtures of wild-type and ΔK210 mutant gave a lower Ca^{2+} sensitivity compared to wild type in ATPase assays. Our group also found that the ΔK210 mutation decreased the Ca^{2+} sensitivity of force development as well as the maximal isometric force generated in skinned fiber studies.[24] We also observed a decrease in the maximal actomyosin ATPase activity. The ΔK210 mutation caused a decrease in the Ca^{2+} sensitivity of force development in contrast to most TnT HCM mutations that sensitize the muscle to Ca^{2+}. Investigation of the TnT R141W mutation showed that this mutation caused a slight but not significant decrease in Ca^{2+} sensitivity of force development.[24] At present, it is not clear if a significant decrease in the Ca^{2+} sensitivity is common to all TnT DCM mutations because the number of mutations identified and investigated so far is too few to make that determination. Based on results from HCM mutations, it is likely that some DCM mutations will cause little or no change in Ca^{2+} sensitivity of force development.[2]

PHYSIOLOGICAL EFFECTS OF DCM MUTATIONS IN FETAL TnT (TnT1)

Clinical data show that familial DCM affects infants as well as adults. The clinical data for the two TnT DCM mutations studied (R141W and ΔK210) suggest that

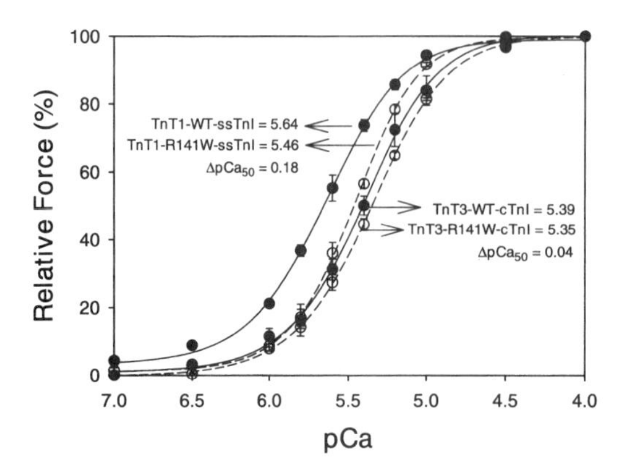

FIGURE 2. Effect of the troponin (Tn) T mutation R141W on the Ca^{2+} sensitivity of force development in the presence of adult or fetal Tn isoforms. In the presence of cardiac TnI, the TnT3 R141W mutation caused a small but not statistically significant decrease in Ca^{2+} sensitivity of force development ($\Delta pCa_{50} = 0.04$) when compared to wild-type TnT3. However, in the presence of slow skeletal TnI, the TnT1 R141W mutation caused a significant decrease in the Ca^{2+} sensitivity of force development ($\Delta pCa_{50} = 0.18$) when compared to wild-type TnT1.

these mutations may cause a worse functional phenotype in a fetal setting than in an adult setting. Patients with the TnT ΔK210 mutation show a high incidence of sudden cardiac death in infants (20% of affected individuals).[16] Clinical data suggests that the R141W mutation may also cause infantile cardiomyopathy.[17]

When TnT DCM causing mutations, R141W and ΔK210, were present in the primary fetal TnT isoform (TnT1), these mutations caused a significant decrease in the Ca^{2+} sensitivity of force development and maximal actomyosin ATPase activity in the presence of cTnI.[25] The magnitude of decrease in Ca^{2+} sensitivity in the TnT1 isoform was larger when compared to the adult TnT3 isoform for both mutations. The only difference between TnT1 and TnT3 is the presence of 10 extra amino acids in the N-terminus of the TnT molecule. The predominant TnI isoform throughout human fetal life is ssTnI, which gradually decreases during the first few months of postnatal development. Irrespective of the TnT isoform utilized, ssTnI-reconstituted fibers showed a greater Ca^{2+} sensitivity of force development than cTnI-reconstituted fibers.[9] When ssTnI (predominant fetal TnI isoform in the heart) was used instead of cTnI, both TnT1 R141W– and TnT1 ΔK210–treated fibers decreased the Ca^{2+} sensitivity compared to TnT1–treated fibers (similar to the decrease that was observed in the presence of cTnI) (FIG. 2).[25] The decrease in Ca^{2+} sensitivity was greater when the TnT1 mutations were studied in the presence of ssTnI (the fetal isoform of TnI in the heart) than when the TnT3 mutants were studied in the presence of cTnI (the adult isoform of TnI in the heart). Thus, the change in Ca^{2+} sensitivity for the DCM mutations would be greater in the fetal form of cardiac Tn than in the adult form. Although, for the two mutations so far investigated (R141W and ΔK210), the main effect on the Ca^{2+} sensitivity of force development was due to the presence of the DCM mutations in the fetal TnT isoform, it is likely that for some mutations the presence of ssTnI would have significant effects on the physiological function of these mutations.

TROPONIN T POLYMORPHISMS

A five base pair insertion/deletion polymorphism in intron 3 of TnT was recently found by Komamura *et al.*[26] The presence of this polymorphism was shown to be associated with a wider interventricular septal wall thickness and increased left ventricle mass/height ratio. The frequency of the deletion allele was significantly higher in the hypertrophy population than in the control population (healthy subjects recruited from medical checkups). It should be noted, however, that the presence of this polymorphism alone did not seem to be sufficient to cause prominent cardiac hypertrophy. These results suggest that the TnT deletion allele may be associated with a predisposition to prominent left ventricular hypertrophy.

The five base pairs (CTTCT) that were deleted may be part of an intronic polypyrimidine tract sequence that affects splicing. *In vitro* studies showed that the deletion allele significantly affected the mRNA expression pattern by skipping exon 4 during splicing. These results could explain the possible presence of the TnT4 isoform (which is similar to TnT3 but lacks exon 4) in failing hearts. Several reports have shown increased expression of a fetal cardiac isoform of TnT (TnT4) in human heart failure.[4] Other polymorphisms present in TnT may also affect N-terminal alternatively spliced region. Some of these changes may be more important in the fetal heart than in the adult heart as TnT4 has similar physiological features to TnT3 in the presence of cardiac TnI (adult TnI isoform) but has lower maximal force than TnT3 in the presence of ssTnI (fetal TnI isoform).[8,9]

PERSPECTIVES

cTnT is the first TnT isoform seen in developing embryonic and fetal skeletal muscle.[27,28] cTnT is also expressed in diseased adult human skeletal muscle. Samples from patients with Duchenne muscular dystrophy (DMD) contained approximately five-and-one-half times more cTnT than normal skeletal muscle. DMD patients contained as much as 45% of the cTnT found in cardiac muscle.[29] Mutations associated with HCM or DCM in cTnT isoforms would also be present in the fetal skeletal muscle and may cause skeletal muscle problems. Over 300 mutations are currently known to cause HCM of which Tn subunits account for >50 mutations. To date, at least 30 different mutations in the cTnT gene have been reported to cause HCM.[30] This ever-growing number of mutations in TnT makes it likely that some patients with diseases that increase the levels of cTnT in adult skeletal muscle could also have mutations in TnT. This might explain the occurrence of patients with TnT mutations together with another disease.

CONCLUSIONS

Clinical data on the effect of mutations on children less than 1 year old are extremely limited or not available. More data are needed on the treatment of younger patients as this group also has a risk of sudden death. Although the clinical phenotype of patients with cardiomyopathy represents the integrated effect of multiple interacting factors, two features found in most cardiomyopathies are altered Ca^{2+} sensitivity and altered bioenergetics. We hypothesize that the magnitude of the change in Ca^{2+} sensitivity of force development is a major determinant of the patient outcome and the risk of sudden death. Both the Ca^{2+} sensitivity and bioenergetics of the fetal heart may be altered in the presence of mutations in TnT associated with either HCM or DCM when compared to normal fetal heart.

ACKNOWLEDGMENTS

This work was supported in part by NIH Grants HL-42325 and HL-67415.

REFERENCES

1. MICHELS, V.V., P.P MOLL, F.A MILLER, *et al.* 1992. The frequency of familial dilated cardiomyopathy in a series of patients with idiopathic dilated cardiomyopathy. New Engl. J. Med. **326:** 77–82.
2. GOMES, A.V. & J.D. POTTER. 2004. Molecular and cellular aspects of troponin cardiomyopathies. Ann. N.Y. Acad. Sci. **1015:** 214–224.
3. MOGENSEN, J., T. KUBO, M. DUQUE, *et al.* 2003. Idiopathic restrictive cardiomyopathy is part of the clinical expression of cardiac troponin I mutations. J. Clin. Invest. **111:** 209–216.
4. ANDERSON, P.A., A. GREIG, T.M. MARK, *et al.* 1995. Molecular basis of human cardiac troponin T isoforms expressed in the developing, adult, and failing heart. Circ. Res. **76:** 681–686.
5. TOWNSEND, P.J., P.J. BARTON, M.H. YACOUB & H. FARZA. 1995. Molecular cloning of human cardiac troponin T isoforms: expression in developing and failing heart. J. Mol. Cell. Cardiol. **27:** 2223–2236.
6. TOBACMAN, L.S. & R. LEE. 1987. Isolation and functional comparison of bovine cardiac troponin T isoforms. J. Biol. Chem. **262:** 4059–4064.
7. TOBACMAN, L.S. 1988. Structure-function studies of the amino-terminal region of bovine cardiac troponin T. J. Biol. Chem. **263:** 2668–2672.
8. GOMES, A.V., G. GUZMAN, J. ZHAO & J.D. POTTER. 2002. Cardiac troponin T isoforms affect the Ca^{2+} sensitivity and inhibition of force development. Insights into the role of troponin T isoforms in the heart. J. Biol. Chem. **277:** 35341–35349.
9. GOMES, A.V., G. VENKATRAMAN, J.P. DAVIS, *et al.* 2004. Cardiac troponin T isoforms affect the Ca^{2+} sensitivity of force development in the presence of slow skeletal troponin I: insights into the role of troponin T isoforms in the fetal heart. J. Biol. Chem. **279:** 49579–49587.
10. REISER, P.J., M.L. GREASER & R.L. MOSS. 1992. Developmental changes in troponin T isoform expression and tension production in chicken single skeletal muscle fibres. J. Physiol. **449:** 573–588.
11. OGUT, O., H. GRANZIER & J.P. JIN. 1999. Acidic and basic troponin T isoforms in mature fast-twitch skeletal muscle and effect on contractility. Am. J. Physiol. **276:** C1162-C1170.
12. NASSAR, R., N.N. MALOUF, M.B. KELLY, *et al.* 1991. Force-pCa relation and troponin T isoforms of rabbit myocardium. Circ. Res. **69:** 1470–1475.
13. COHN, J.N., M.R. BRISTOW, K.R. CHIEN, *et al.* 1997. Report of the National Heart, Lung, and Blood Institute Special Emphasis Panel on Heart Failure Research. Circulation **95:** 766–770.
14. SEIDMAN, J.G. & C. SEIDMAN. 2001. The genetic basis for cardiomyopathy: from mutation identification to mechanistic paradigms. Cell **104:** 557–567.
15. OLSON, T.M., V.V. MICHELS, S.N. THIBODEAU, *et al.* 1998. Actin mutations in dilated cardiomyopathy, a heritable form of heart failure. Science **280:** 750–752.
16. KAMISAGO, M., S.D. SHARMA, S.R. DEPALMA, *et al.* 2000. Mutations in sarcomere protein genes as a cause of dilated cardiomyopathy. New Engl. J. Med. **343:** 1688–1696.
17. LI, D., G.Z. CZERNUSZEWICZ, O. GONZALEZ, *et al.* 2001. Novel cardiac troponin T mutation as a cause of familial dilated cardiomyopathy. Circulation **104:** 2188–2193.
18. MOGENSEN, J., R.T. MURPHY, T. SHAW, *et al.* 2004. Severe disease expression of cardiac troponin C and T mutations in patients with idiopathic dilated cardiomyopathy. J. Am. Coll. Cardiol. **44:** 2033–2040.
19. STEFANELLI, C.B., A. ROSENTHAL, A.B. BORISOV, *et al.* 2004. Novel troponin T mutation in familial dilated cardiomyopathy with gender-dependant severity. Mol. Genet. Metab. **83:** 188–196.

20. FUJINO, N., M. SHIMIZU, H. INO, *et al.* 2001. Cardiac troponin T Arg92Trp mutation and progression from hypertrophic to dilated cardiomyopathy. Clin. Cardiol. **24:** 397–402.
21. FUJINO, N., M. SHIMIZU, H. INO, *et al.* 2002. A novel mutation Lys273Glu in the cardiac troponin T gene shows high degree of penetrance and transition from hypertrophic to dilated cardiomyopathy. Am. J. Cardiol. **89:** 29–33.
22. MORIMOTO, S., Q.W. LU, K. HARADA, *et al.* 2002. Ca^{2+}-desensitizing effect of a deletion mutation Delta K210 in cardiac troponin T that causes familial dilated cardiomyopathy. Proc. Natl. Acad. Sci. U.S.A. **99:** 913–918.
23. ROBINSON, P., M. MIRZA, A. KNOTT, *et al.* 2002. Alterations in thin filament regulation induced by a human cardiac troponin T mutant that causes dilated cardiomyopathy are distinct from those induced by troponin T mutants that cause hypertrophic cardiomyopathy. J. Biol. Chem. **277:** 40710–40716.
24. VENKATRAMAN G., K. HARADA, A.V. GOMES, *et al.* 2003. Different functional properties of troponin T mutants that cause dilated cardiomyopathy. J. Biol. Chem. **278:** 41670–41676.
25. VENKATRAMAN, G., A.V. GOMES, W.G. KERRICK & J.D. POTTER. 2005. Characterization of troponin T dilated cardiomyopathy mutations in the fetal troponin isoform. J. Biol. Chem. **280:** 17584–17592.
26. KOMAMURA, K., N. IWAI, K. KOKAME, *et al.* 2004. The role of a common TNNT2 polymorphism in cardiac hypertrophy. J. Hum. Genet. **49:** 129–133.
27. SABRY, M.A. & G.K. DHOOT. 1991. Identification of and pattern of transitions of cardiac, adult slow and slow skeletal muscle-like embryonic isoforms of troponin T in developing rat and human skeletal muscles. J. Muscle Res. Cell. Motil. **12:** 262–270.
28. TOYOTA, N. & Y. SHIMADA. 1981. Differentiation of troponin in cardiac and skeletal muscles in chicken embryos as studied by immunofluorescence microscopy. J. Cell Biol. **91:** 497–504.
29. BODOR, G.S., L. SURVANT, E.M. VOSS, *et al.* 1997. Cardiac troponin T composition in normal and regenerating human skeletal muscle. Clin. Chem. **43:** 476–484.
30. GOMES, A.V., J.A. BARNES, K. HARADA & J.D. POTTER. 2004. Role of troponin T in disease. Mol. Cell. Biochem. **263:** 115–129.

Multiple Stem Cell Populations Contribute to the Formation of the Myocardium

LEONARD M. EISENBERG, RICARDO MORENO, AND ROGER R. MARKWALD

Department of Cell Biology and Anatomy, Medical University of South Carolina, Charleston, South Carolina 29425, USA

ABSTRACT: Owing to the very rapid growth of the vertebrate embryo following fertilization, an efficient circulatory system needs to be established during the initial stages of development. For that reason, the first functional organ that develops in both the bird and mammalian embryo is the heart. Until recently, the narrative of cardiac development was portrayed in a straightforward manner, with all the myocardium in the mature heart being generated from the expansion of an original pool of myocardial cells present in the early gastrula. It is now known that the story of the developing myocardium is more dynamic, as it is comprises cellular components of multiple ancestries. The *de novo* addition of myocytes to the developing heart occurs at various points during embryogenesis, as cardiac muscle takes on new members by the absorption of cells that either reside in neighboring nonmuscle tissue or come into contact with the myocardium by entering the heart upon migration or via the circulation. This article reviews what is presently known about cellular populations that contribute to the myocardium and examine reasons why the embryo utilizes multiple cellular sources for forming the cardiac muscle.

KEYWORDS: cardiac development; heart fields; myocardium; neural crest; plasticity; stem cells

INTRODUCTION

One of the great hopes in modern medicine is to develop the means to restore diseased or damaged hearts to full cardiac function. Among the most promising research today in cardiovascular biology involves using stem cells to generate new myocardial tissue. Many different stem cell populations have been studied for both their capacity to generate cardiomyocytes and their latent utility in repairing the adult myocardium. Yet, it remains controversial which stem cell populations possess myocardial phenotypic potential, and it is still unclear what are the properties of a myocardial-competent stem cell. To resolve these issues, it may be pertinent to consider the stem cell populations that contribute to the formation of the myocardium during embryogenesis.[1] Recent studies in cardiac developmental biology have changed scientists' perception of what constitutes a cardiac stem cell. In the past, it was believed that the myocardium was constituted entirely of cells originating from

Address for correspondence: Prof. Roger R. Markwald, Ph.D., Medical University of South Carolina, Department of Cell Biology and Anatomy, BSB Rm. 642, 171 Ashley Avenue, Charleston, SC 29425, USA. Voice: 843-792-5880; fax: 843-792-0664.
markwald@MUSC.edu

**Ann. N.Y. Acad. Sci. 1047: 38–49 (2005). © 2005 New York Academy of Sciences.
doi: 10.1196/annals.1341.004**

paired mesodermal fields of the early gastrula, with the subsequent and substantial growth of the cardiac muscle involving only the expansion of this pre-existing pool of cardiomyocytes. It is now known that the cellular constituents of the myocardium have multiple ancestries, and that *de novo* addition of myocytes to the developing heart occurs at various points during embryogenesis. In the following sections, we describe what is presently known about the cell populations that contribute to the developing myocardium and discuss these new findings in the context of defining what is a myocardial stem cell.

INITIATION OF CARDIOGENESIS

The formation of an operative, beating heart occurs during the early stages of avian and mammalian embryogenesis, as the survival of a rapidly growing organism requires circulating blood. Cardiogenesis is initiated at the onset of gastrulation, when the embryo is a relatively simple structure consisting of the three primary germ layers. Myocardial progenitors localize to the anterior half of the embryo within bilaterally distributed mesodermal regions,[2–4] which are referred to as the precardiac mesoderm, heart-forming fields, or the primary heart fields (FIG. 1A). As development proceeds, the lateral mesoderm separates into distinct dorsal and ventral layers, with the precardiac cells segregating to the ventral splanchnic mesodermal layer.[5–7] Thereafter, the bilateral heart-forming fields move anteriorly toward the ventral midline and begin to merge to form a three-dimensional structure—the primary heart tube.[6,8] Concurrent with these morphological events, cells within these heart fields start to reveal a cardiac phenotype. By the time the fusion of the heart fields is initiated, several contractile proteins are exhibited in a prestriated pattern by the cardiogenic cells.[9–15] Concomitant with the appearance of striated muscle by the 10-somite stage, is the first display of beating by the primitive heart. The formation of the primary heart tube is completed by embryonic-day 2 and -day 8 of the chick and mouse embryo, respectively.

FORMATION OF THE OUTFLOW TRACT

The process of heart field fusion does not stop with the emergence of a beating and functional heart, as the trailing posterior ends of the primary heart fields are still bilaterally separated. As demonstrated in the embryonic chick, the initial form of the contractile heart is constructed of just the primitive right and left ventricular segments. Subsequently, the remainder of the heart-forming fields continue to merge progressively at the posterior pole of the primary heart tube to form in sequence the primitive atrioventricular canal and atrial segments.[16] However, the final major cardiac segment—the outflow tract—is added to the primitive heart by a somewhat different scenario.[17–19] Following the integration of the other segments into the primary heart tube, the outflow tract is transmuted from mesodermal tissue located anterior to the newly formed primitive heart (FIG. 1B).

The conotruncal region of the cardiac outflow tract is demarcated from the contiguous aortic sac region by the muscular construction of its outer wall.[20] Although the primary heart field was until recently believed to provide the entirety of the

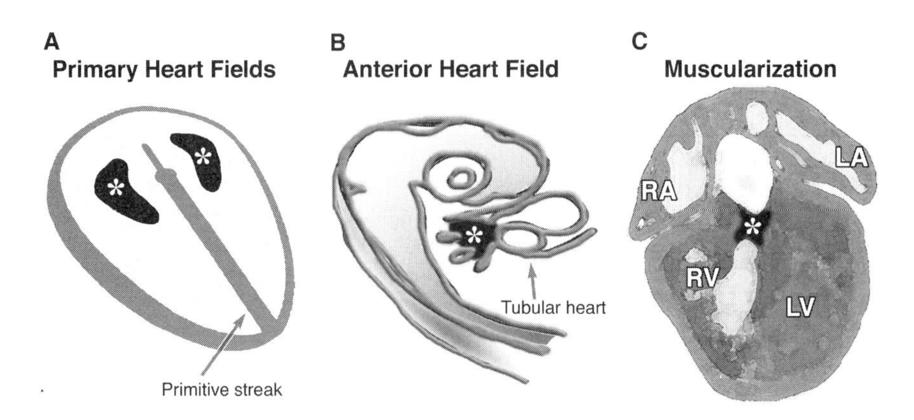

FIGURE 1. The myocardium is generated from multiple stem cell sources within the developing embryo.[2,17,27] (**A**) Drawing of an early gastrula-stage chick embryo in dorsal view showing the location of the precardiac mesoderm (*asterisks*), which is also referred to as the heart-forming or primary heart fields. (**B**) Illustration of the upper half of a tubular heart-stage chick embryo highlighting the pharyngeal mesoderm location of the anterior heart field (*asterisk*), which is the primordium of the outflow tract myocardium. (**C**) During later stages of development, several episodes of mesenchymal cell recruitment contribute to the growth and remodeling of the myocardium. Here is drawn the myocardial layer of a sectioned mid-gestational heart, with the increasingly muscularized outlet septum highlighted (*asterisk*). The cellular composition of the outlet septum changes at this point in development by outgrowth of existing myocardium and recruitment of intracardiac mesenchyme to the myocardial lineage. RA, right atrium; LA, left atrium; RV, right ventricle; LV, left ventricle.

myocardial progeny for the mature heart, it is now clear that conotruncal myocardium has a distinct origin, as it arises from mesoderm cells residing beyond the arterial pole of the tubular heart.[17–19] Nonetheless, the formation of the outflow tract does proceed coactively with the development of the primary heart tube. Between days 2 and 3.5 of chick development, mesodermal cells flanking the anterior end of the primitive ventricle both consolidate into the developing heart and differentiate to myocardial tissue. The mesodermal region that gives rise to the outflow tract is variously referred to as the anterior heart field or secondary heart field.[17,18] *In vivo* labeling experiments, with either fluorescent dye or lacZ-expressing adenovirus, have indicated that the anterior heart field in the chick consists of mesoderm located anterior to the distal outlet surrounding the aortic sac and extending into each of the pharyngeal arches. Studies in the mouse have provided evidence for a similarly located anterior heart field in mammals. For example, mice containing a *Fgf10* enhancer trap lacZ transgene display within the outflow tract high-level expression of lacZ, which prior to the formation of this final cardiac segment is most conspicuously exhibited within a domain that extends from the arterial pole of the developing heart into pharyngeal arches 1 through 6.[21] Other supportive data for the presence of the anterior heart field in mice is provided by studies investigating Islet-1, which is a LIM homeodomain transcription factor. Islet-1 has an intriguing distribution pattern, as the expression of this gene is most pronounced within the primary heart fields and throughout the pharyngeal mesoderm. Moreover, homozygous Islet-1 null mice suffer from severe cardiac abnormalities, including the mal-development of the outflow tract.[22] In addition to demonstrating that the pharyngeal mesoderm is a major cellular contributor to the developing outflow tract in both the bird and mammalian embryo, the anterior heart field data also gave the first suggestion that the location of cardiac stem cells is not unique to a single time and place in embryonic development.

INTRACARDIAC AND EXTRACARDIAC CONTRIBUTIONS TO THE MUSCULARIZATION OF THE HEART

The transformation of the anterior heart field into the final segment of the primitive heart completes the initial phase of cardiac development. Yet, this does not end the process of cell recruitment to a continuing growing myocardium. When the outflow tract becomes adjoined to the primitive ventricle, the heart is still a relatively simple tubular structure consisting of an external myocardial sheet surrounding an inner endocardial layer. Although the contractile activity of the primary heart tube is sufficient for circulating blood throughout the early embryo, the heart will require substantial growth and remodeling to meet the circulatory needs of an ever-growing embryo and adult. The acquisition of greater structural complexity, as the heart undergoes the transition from a simple tubular pump into a multichambered structure, represents the next phase of cardiac development.

The renovation of a simple tube into the mature multichambered heart involves several critical remodeling events. First, looping of the primary heart tube juxtaposes the anterior outlet and posterior inlet ends of the heart. Within a now bent tube, the primitive segments start to reposition to allow for the proper alignment of the future chambers of the mature heart.[23] Concurrently, the developing cardiac segments become subdivided by septa that initially are mesenchymal in composition.[24] Later in development, the mesenchymal septa become highly muscularized.[25–28] Moreover, at the venous pole of the heart, the mesenchymal perimeter of the caval and pulmonary veins become displaced by a muscular sheath.[25,26] These episodes of cardiac muscularization result from two distinct, but not mutually exclusive, mechanisms involving the growth and expansion of preexisting myocardium, and/or the recruitment of new cardiomyocytes from intracardiac and extracardiac mesenchyme adjacent to the expanding myocardium (FIG. 1C).[25–28] The discovery that cells are added to the cardiac muscle throughout its development by the recruitment of adjacent non-muscle cells has been of major significance, as it demonstrates the continually dynamic nature of the embryonic myocardium. That the myocardium may be more dynamic still is suggested by findings described in the next two sections.

THE MYOCARDIAL POTENTIAL OF THE NEURAL CREST

The neural crest is a cellular population contained within the leading edge of the neural folds that dissociates from the neural tube flowing closure. Neural crest cells migrate to various embryonic tissues and, depending on their precise destination within the embryo, will generate neurons, glia, pigment cells, chondrocytes, and smooth muscle cells.[29–31] In contrast, no evidence has ever been produced to suggest that this particular extracardiac cell population contributes progenitor cells to the myocardium of the bird and mammalian heart. Nonetheless, neural crest cells play a critical role in cardiac maturation, as they contribute smooth muscle cells for the developing aortic arch arteries and participate in formation of the outflow septum.[32,33] When cardiac neural crest cells are ablated or their migration is disrupted by gene perturbation, the outflow tract will not subdivide properly. Accordingly, a major consequence of interrupting neural crest interactions with the developing heart is that the aorta and pulmonary arteries fail to separate, resulting in persistent truncus arteriosis.[32]

In lower vertebrates, such as the fish, the outflow tract is not septated. Since the separation of the pulmonary and aortic circulatory systems in higher vertebrates is dependent on a neural crest-mediated subdivision of the outflow tract, it was expected that the involvement of the neural crest in heart development would be significantly less in the fish. Yet, this did not turn out to be the case, as recent studies of the zebrafish embryo found that the neural crest provides a substantial cellular contribution to the myocardium of this species.[34,35] In late stage zebrafish embryos, the percentage of cardiomyocytes of neural crest origin ranged from 13.9% to 30.4% among the various segments of the heart. In addition, fate mapping studies indicated that the neural crest region participating in cardiac development is more expansive in the zebrafish than it is for the bird and mammal (FIG. 2A).

The demonstration that the cardiac neural crest provides an even greater, but distinct cellular contribution in the zebrafish heart may appear to be paradoxical. Yet, new thinking about the plasticity of stem cells may provide the proper context for the zebrafish neural crest observations. Since neural crest cells as a whole are known

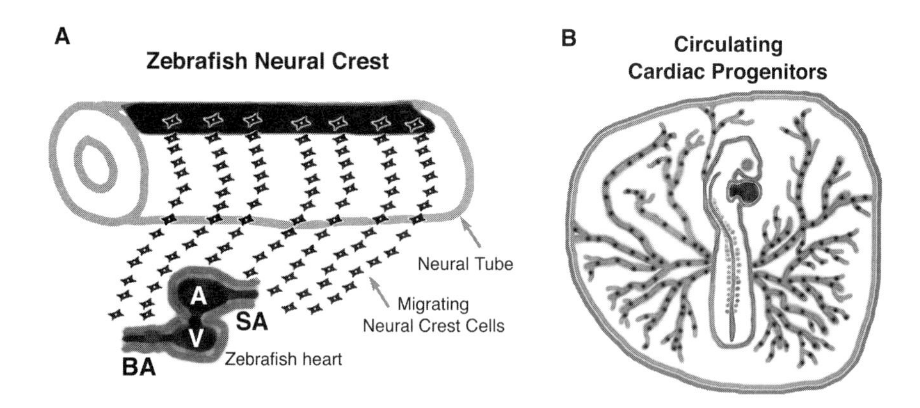

FIGURE 2. Myocardial potential of migratory cell populations. (**A**) Schematic drawing depicting the migration of neural crest cells within the zebrafish embryo. Fate mapping studies in the zebrafish have demonstrated that the neural crest provides a substantial cellular contribution to the myocardium of the outflow tract: bulbus arteriosus (BA), ventricle (V), atrium (A); and inflow region: sinus venosus (SA).[34,35] (**B**) Illustration of a tubular heart-stage chick embryo showing a simplified view of the embryonic circulation, with both the primitive heart and circulating blood cells highlighted in black. Results presented in this paper substantiate the hypothesis that circulating progenitors contribute cells to the developing myocardium.[1,51]

to diversify to a broad range of differentiated cell phenotypes, it has been debated whether this cellular population is comprised of multiple lineage restricted precursors or individual multipotent progenitor cells.[31,36] Prior to the recent published reports on the zebrafish embryo, it was not suspected that neural crest possessed myocardial potential. Although it can not be ruled out that zebrafish possess a distinct population of myocardial-competent neural crest cells not found in higher vertebrates, a more compelling explanation is that the potential to form cardiomyocytes is an attribute of all neural crest cells regardless of whether this potential is ever realized in the animal. As has been shown in stem cell studies in the adult mammalian heart,[37] the fate of zebrafish neural crest to form cardiomyocytes may be related to their direct association with the myocardium. Accordingly, the inability of the cardiac neural crest in the bird and mammalian embryo to provide myocardial progenitors might be a consequence of their distribution upon entering the heart, which is limited to the outflow tract cushions. In contrast, neural crest cells in the zebrafish migrate deeper into the developing heart and become exposed to cardiac muscle[34,35]—an interaction that may account for why this extracardiac cell population uniquely realizes its myocardial phenotypic potential in this animal. An implication of the zebrafish neural crest story is the overriding importance of phenotypic plasticity as a mechanism for understanding how various cell populations contribute to the formation of the heart—a topic that will be addressed further below.

RECRUITMENT OF CIRCULATING CELLS INTO THE MYOCARDIUM

A subject that has been mired in controversy concerns whether adult stem cells have the capacity to generate new cardiomyocytes in the diseased or damaged adult heart.[37–42] Experiments have been reported showing that various tissue-derived stem cells will integrate into the myocardium and exhibit a myocyte phenotype when inserted directly into the adult heart.[43–46] For example, bone marrow-derived hematopoietic stem cells injected into the hearts of adult mice suffering from induced myocardial infarctions helped repair the damaged myocardium and restore cardiac function.[43,47] Whereas claims have been made that these results are not repeatable,[39,48] other data strongly suggest that extracardiac stem cells may be a regenerative cell source for the adult myocardium.[42] In a study examining cardiac biopsies obtained from human female patients who had undergone sex-mismatched bone marrow transplantation, high rates of myocardial cell recruitment were observed.[49] From a total of 80,000 cardiomyocytes examined from four patients, greater than 1 in 500 cells were Y-chromosome-positive, indicating the presence of myocardial cells of bone marrow origin. It is also relevant to this discussion that blood-lineage cells from the bone marrow exhibit the capacity to undergo myocardial differentiation *in vitro*, both in direct response to growth factor stimulation and by incorporation into co-cultured myocardium.[37,50]

If the studies described above prove correct, then circulating cells may have a normal physiologic role in replenishing muscle cells in the adult heart. This raises the intriguing possibility that circulating progenitor cells may contribute to the development of the embryonic myocardium (FIG. 2B). Since changes in tissue composition within the developing embryo are more dynamic than within the adult, it may be predicted that circulating cells play a significant role in the development of embryonic tissues. In an effort to investigate this possibility, we introduced labeled stem cells into the embryonic circulation by a single injection and examined whether they incorporated into the heart and showed evidence of myocardial differentiation.[51] These experiments employed the multipotential stem cell line QCE6,[52,53] which was labeled and injected into the yolk sac vessel leading to the sinus venosus of embryonic-day-9 whole-mouse embryos.[51] After allowing the embryos to develop for 24 h *ex utero*, individual labeled cells were detected that integrated into the myocardium and stained positive for sarcomeric proteins. However, we were aware that the protocol used in these experiments was flawed, as single injections of labeled cells poorly imitated the continuous supply of endogenous circulating cells that are exposed to the heart throughout embryogenesis. Nonetheless, these initial results were encouraging and gave credence to the idea that circulating cells contribute to the development of the myocardium.[1,51]

The presence of circulating myocardial progenitors has been investigated recently, with much more impressive results, by Ricardo Moreno. In these experiments, Dr. Moreno used parabiosis to combine the circulation of quail and chick embryos, and allowed the organisms to develop together in an eggshell culture (FIG. 3). Since quail cells can be distinguished from their chick counterparts by the species-specific QCPN antibody,[54] quail-derived myocytes can be readily identified within the hearts of the conjoined chick embryos. What was shown from these studies is that the hearts in the parabiotically combined chick embryos contained significant numbers of quail cells as constituents of the myocardium (FIG. 4). Because chimeric tissues could arise only via the circulation, these results demonstrate for the first time that circulating cells are a normal contributor to the developing myocardium.

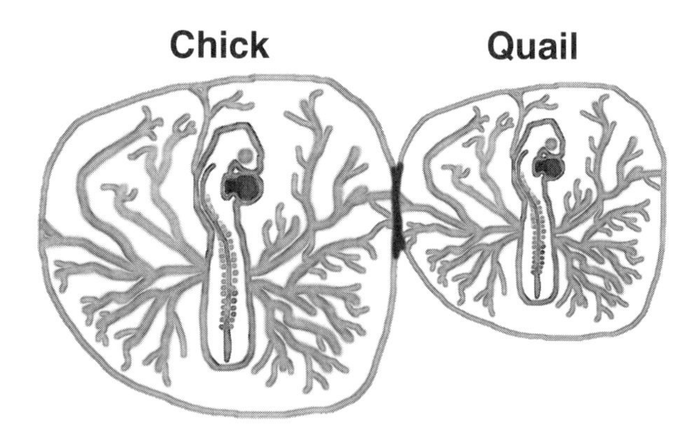

FIGURE 3. Schematic diagram showing parabiosis between chick and quail embryos. To produce these embryos, individual chick and quail eggs are incubated for 2 days. An opening is then made in the chick egg and half the volume of albumen removed. Thereafter, the contents of the 2-day quail egg is added to the host chick egg, which is subsequently sealed tightly with tape and the embryos allowed to develop further in a 37°C incubator. As has been shown in previous studies,[59] integration of extra-embryonic blood vessels of the chick and quail embryos will occur over the next 2 days of incubation, thus producing chimeric blood circulation. These embryos will remain healthy and continue to develop for an additional 10 days.

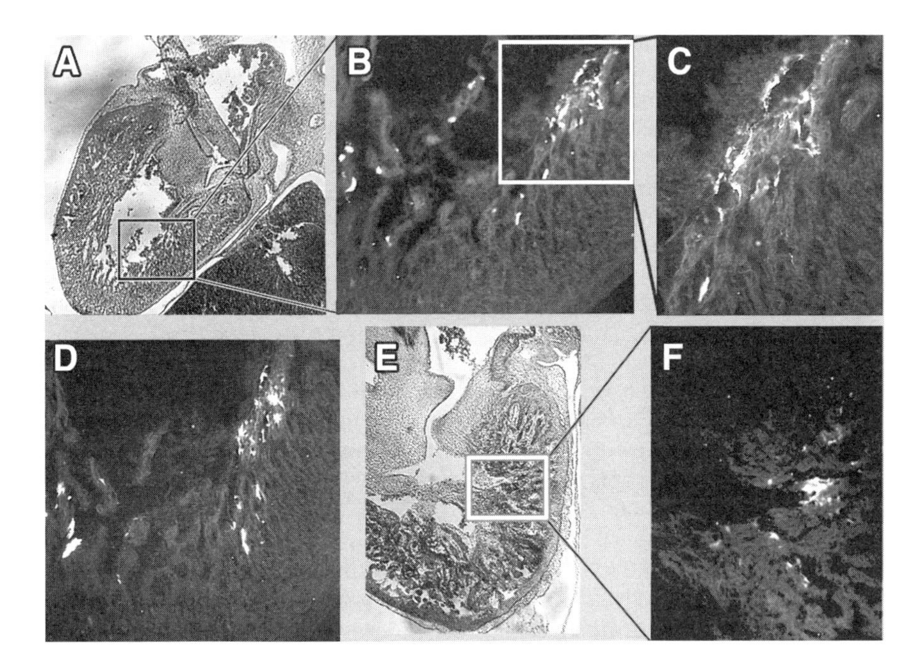

FIGURE 4. Chimeric myocardium of parabiosed chick and quail embryos. Parabiotic embryos produced by co-incubating 2-day-old chick and quail embryos in a chick host egg were allowed to develop for an additional 6 days. Afterward, the hearts from the conjoined chick embryos were excised, sectioned, and fluorescently stained for the quail cell-specific antibody QCPN. (**A–C**) A brightfield image of a sectioned chick heart, shown in sequence, imaged at successively higher magnifications (indicated by the boxed area) for QCPN-specific staining. (**D**) An additional example of a QCPN-stained conjoined chick heart. (**E**) Another sectioned chick heart imaged by brightfield microscopy, with the box indicating the area of the higher magnified view in (**F**) showing QCPN immunofluorescence. Note that this data demonstrates the presence of quail embryo-derived cells that have fully integrated into the myocardium of chick embryos that are linked by a chimeric blood circulation.

PHENOTYPIC PLASTICITY AND THE MYOCARDIAL STEM CELL

From a cellular perspective, the development of the myocardium is a story whose major theme concerns the continual recruitment of new cells as components of this tissue. Once a primitive contractile heart emerges by the merging of the bilaterally separated primary heart fields, the myocardium relies on new cellular recruits to help support its subsequent development. Among the reasons for the existence of multiple myocardial cell sources, one part of the answer would elicit little debate. The myocardium is an intricately arranged tissue embracing diverse regions of an organ whose composition is second only to the brain in its complexity. Considering the substantial growth and remodeling that occurs during the maturation of the heart, the construction of the myocardium in its entirety from the proliferation of an individual tissue primordium would appear to be a very difficult enterprise developmentally.

Another reason for the existence of multiple myocardial progenitor populations may be related to the issue of phenotypic plasticity, and whether this property is a general trait of embryonic cells. Over the past several years, various tissue-derived stem cell populations of the adult have been shown to exhibit a high degree of phenotypic plasticity.[40,55–57] Nonetheless, phenotypic plasticity is still rarely invoked in discussions of the embryo. Instead, embryonic development is usually presented as a series of progressive and unilateral steps as cells become specified, committed, and then fully differentiated. Typically, cell lineage acquisition is portrayed as a fixed pathway in which phenotypic plasticity is not considered part of the equation. Yet, let us briefly consider the cellular components of the primary heart fields. In previous papers, we have shown that the stem cells of the precardiac mesoderm are neither uniquely specified to become heart nor fully committed to a myocardial cell fate.[51,58] Although posterior mesodermal tissue of the gastrula stage embryo does not contribute cells to the developing heart *in situ*, it will demonstrate the potential to differentiate into cardiomyocytes if removed from its normal environment.[58] In addition, cells of the primary heart field will produce blood cells if placed under hematopoietic-promoting culture conditions,[51] even though this precardiac tissue does not provide a cellular contribution to the embryonic blood. These data indicate that cells within the mesodermal layer of the early embryo can be shifted readily to or from a cardiac phenotype by changing their extracellular environment. Even though fate mapping studies have located precisely the cellular sources that supply the myocardium, it appears that many embryonic cell subpopulations may differ very little in their phenotypic potential.

If phenotypic plasticity is a basic property of embryonic cells, what is the mechanism that directs their fate to form specific tissue? Again, studies on the adult heart may provide a clue, as it is the exposure to the adult myocardium, either by direct injection or via the circulation, that prompts the differentiation of extracardiac stem cells into cardiomyocytes.[43,47,49] Moreover, the embryonic myocardium appears to possess the ability to steer bone marrow-derived cells toward the myocardial lineage.[50] These observations indicate that the differentiation of putative noncardiac progenitors to cardiac muscle is provoked by their interactions with the myocardial environment. Accordingly, the recruitment of new cells to the developing myocardium does not require progenitors that are uniquely specified to form cardiac muscle. Instead, the major requirement met by the pharyngeal mesoderm, septal mesenchyme, venous pole mesenchyme, zebrafish neural crest, and circulating blood cells for assuming a myocardial phenotype is their proximity to a developing myocardium. What this may mean is that the driving force behind cardiogenesis is not necessarily specification of cell fate, but specification of the myocardial tissue environment.

These observations have significant implications for both cardiac embryology and stem cell biology. If the growth and remodeling of the myocardium is powered by the emergence of a molecular environment that recruits multipotential cells of various origins to contribute to the development of the heart, then the inference is that cardiac competency is not restricted to small subsets of stem cells. Moreover, the observation that circulating cells may be significant contributors to the myocardium in the embryo lends great credence to the claims that hematopoietic stem cells have clinical utility for repairing diseased or damaged myocardium in the adult.

ACKNOWLEDGMENTS

This work was supported by American Heart Association grant 0355766U; NASA EPSCoR grant HEDS6; and NIH Grants PO1 HD-52813 and P01 HD-39946.

REFERENCES

1. EISENBERG, L.M. & R.R. MARKWALD. 2004. Cellular recruitment and the development of the myocardium. Dev. Biol. **274:** 225–232.
2. STALSBERG, H. & R.L. DEHAAN. 1969. The precardiac areas and formation of the tubular heart in the chick embryo. Dev. Biol. **19:** 128–159.
3. REDKAR, A., M. MONTGOMERY & J. LITVIN. 2001. Fate map of early avian cardiac progenitor cells. Development **128:** 2269–2279.
4. EISENBERG, L.M. 2002. Belief versus scientific observation: the curious story of the precardiac mesoderm. Anat. Rec. **266:** 194–197.
5. DEHAAN, R.L. 1963. Organization of the cardiogenic plate in the early chick embryo. Acta Emb. Morphol. Exp. **6:** 26–38.
6. DEHAAN, R.L. 1965. Morphogenesis of the vertebrate heart. *In* Organogenesis. R.L. DeHaan & H. Ursprung, Eds.: 377–419. Holt, Rinehart and Winston. New York.
7. LINASK, K.K. 1992. N-cadherin localization in early heart development and polar expression of Na^{+}, K^{+}-ATPase, and integrin during pericardial coelom formation and epithelization of the differentiating myocardium. Dev. Biol. **151:** 213–224.
8. EISENBERG, L.M. & C.A. EISENBERG. 2002. Onset of a cardiac phenotype in the early embryo. *In* Cardiovascular Molecular Morphogenesis: Myofibrillogenesis. D.K. Dube, Ed.: 181–205. Springer Verlag. New York.
9. TOKUYASU, K.T. & P.A. MAHER. 1987. Immunocytochemical studies of cardiac myofibrillogenesis in early chick embryos. I. Presence of immunofluorescent titin spots in premyofibril stages. J. Cell Biol. **105:** 2781–2793.
10. TOKUYASU, K.T. & P.A. MAHER. 1987. Immunocytochemical studies of cardiac myofibrillogenesis in early chick embryos. II. Generation of alpha-actinin dots within titin spots at the time of the first myofibril formation. J. Cell Biol. **105:** 2795–2801.
11. RUZICKA, D.L. & R.J. SCHWARTZ. 1988. Sequential activation of alpha-actin genes during avian cardiogenesis: vascular smooth muscle alpha-actin gene transcripts mark the onset of cardiomyocyte differentiation. J. Cell Biol. **107:** 2575–2586.
12. BISAHA, J.G. & D. BADER. 1991. Molecular analysis of myogenic differentiation: isolation of a cardiac-specific myosin heavy chain. Dev. Biol. **148:** 335–364.
13. COLAS, J.-F., A. LAWSON & G.C. SCHOENWOLF. 2000. Evidence that translation of smooth muscle alpha-actin mRNA is delayed in the chick promyocardium until fusion of the bilateral heart-forming regions. Dev. Dyn. **218:** 316–330.
14. HAN, Y., J.E. DENNIS, L. COHEN-GOULD, *et al.* 1992. Expression of sarcomeric myosin in the presumptive myocardium of chicken embryos occurs within six hours of myocyte commitment. Dev. Dyn. **193:** 257–265.
15. DE JONG, F., W.J. GEERTS, W.H. LAMERS, *et al.* 1990. Isomyosin expression during pattern formation of the tubular heart: A three-dimensional immunohistochemical analysis. Anat. Rec. **226:** 213–227.
16. DE LA CRUZ, M.V. & R.R. MARKWALD. 1998. Living morphogenesis of the heart. *In* Cardiovascular Molecular Morphogenesis, Vol. 1. R.R. Markwald, Ed.: 382 pages. Birkhauser/Springer-Verlag. Boston/New York.
17. MJAATVEDT, C.H., T. NAKAOKA, R. MORENO-RODRIGUEZ, *et al.* 2001. The outflow tract of the heart is recruited from a novel heart-forming field. Dev. Biol. **238:** 97–109.
18. WALDO, K.L., D.L. KUMISKI, K.T. WALLIS, *et al.* 2001. Conotruncal myocardium arises from a secondary heart field. Development **128:** 3179–3188.
19. KELLY, R.G. & M.E. BUCKINGHAM. 2002. The anterior heart-forming field: voyage to the arterial pole of the heart. Trends Genet. **18:** 210–216.
20. PEXIEDER, T. 1995. Conotruncus and its septation at the advent of the molecular biol-

ogy era. *In* Developmental Mechanisms of Heart Disease. E.B Clark, R.R. Markwald & A. Takao, Eds.: 227–247. Futura. Armonk, NY.
21. Kelly, R.G., N.A. Brown & M.E. Buckingham. 2001. The arterial pole of the mouse heart forms from Fgf10-expressing cells in pharyngeal mesoderm. Dev. Cell. **1:** 435–440.
22. Cai, C.-L., X. Liang, Y. Shi, *et al.* 2003. Isl1 identifies a cardiac progenitor population that proliferates prior to differentiation and contributes a majority of cells to the heart. Dev. Cell **5:** 877–889.
23. Moorman, A., S. Webb, N.A. Brown, *et al.* 2003. Development of the heart: (1) formation of the cardiac chambers and arterial trunks. Heart **89:** 806–814.
24. Anderson, R.H., S Webb, N.A. Brown, *et al.* 2003. Development of the heart: (2) septation of the atrium and ventricles. Heart **89:** 949–958.
25. Kruithof, B.P.T, M.J.B. Van Den Hoff, S. Tesink-Taekema & A.F.M. Moorman. 2003. Recruitment of intra- and extracardiac cells into the myocardial lineage during mouse development. Anat. Rec. **271A:** 303–314.
26. Kruithof, B.P.T., M.J.B. Van Den Hoff, A. Wessels & A.F.M. Moorman. 2003. Cardiac muscle cell formation after development of the linear heart tube. Dev. Dyn. **227:** 1–13.
27. Van den Hoff, M.J.B., A.F.M. Moorman, J.M. Ruijter, *et al.* 1999. Myocardialization of the cardiac outflow tract. Dev. Biol. **212:** 477–490.
28. Van den Hoff, M.J.B., B.P.T. Kruithof, A.F.M. Moorman, *et al.* 2001. Formation of myocardium after the initial development of the linear heart tube. Dev. Biol. **240:** 61–76.
29. Gammill, L.S. & M. Bronner-Fraser. 2003. Neural crest specification: migrating into genomics. Nat. Rev. Neurosci. **4:** 795–805.
30. Graham, A. 2003. The neural crest. Curr. Biol. **13:** R381–R384.
31. Le Douarin, N.M. & E. Dupin. 2003. Multipotentiality of the neural crest. Curr. Opin. Genet. Dev. **13:** 529–536.
32. Hutson, M.R. & M.L. Kirby. 2003. Neural crest and cardiovascular development: a 20-year perspective. Birth Defects Res. C Embryo Today **69:** 2–13.
33. Sieber-Blum, M. 2004. Cardiac neural crest stem cells. Anat. Rec. **276A:** 34–42.
34. Sato, M. & H.J. Yost. 2003. Cardiac neural crest contributes to cardiomyogenesis in zebrafish. Dev. Biol. **257:** 127–139.
35. Li, Y.-X., M. Zdanowicz, L. Young, *et al.* 2003. Cardiac neural crest in zebrafish embryos contributes to myocardial cell lineage and early heart function. Dev. Dyn. **226:** 540–550.
36. Anderson, D.J. 1997. Cellular and molecular biology of neural crest cell lineage determination. Trends Genet. **13:** 276–280.
37. Eisenberg, L.M. & C.A. Eisenberg. 2004. Adult stem cells and their cardiac potential. Anat. Rec. **276A:** 103–112.
38. Anversa, P., J. Kajstura, B. Nadal-ginard & A. Leri. 2003. Primitive cells and tissue regeneration. Circ. Res. **92:** 579–582.
39. Murry, C.E., M.H. Soonpaa, H. Reinecke., *et al.* 2004. Haematopoietic stem cells do not transdifferentiate into cardiac myocytes in myocardial infarcts. Nature **428:** 664–668.
40. Quesenberry, P.J., M. Abedi, J. Aliotta, *et al.* 2004. Stem cell plasticity: an overview. Blood Cells Mol. Dis. **32:** 1–4.
41. Rodic, N., M.S. Rutenberg & N. Terada. 2004. Cell fusion and reprogramming: resolving our transdifferences. Trends Mol. Med. **10:** 93–96.
42. Sussman, M.A. & P. Anversa. 2004. Myocardial aging and senescence: where have the stem cells gone? Annu. Rev. Physiol. **66:** 29–48.
43. Orlic, D., J. Kajstura, S. Chimenti, *et al.* 2001. Bone marrow cells regenerate infarcted myocardium. Nature **410:** 701–705.
44. Toma, C., M.F. Pittenger, K.A. Cahill, *et al.* 2002. Human mesenchymal stem cells differentiate to a cardiomyocyte phenotype in the adult murine heart. Circulation **105:** 93–98.
45. Tomita, A., R.-K. Li, R.D. Weisel, *et al.* 1999. Autologous transplantation of bone marrow cells improves damaged heart function. Circulation **100:** II247–II256.

46. WANG, J.S., D. SHUM-TIM, J. GALIPEAU, *et* al. 2000. Marrow stromal cells for cellular cardiomyoplasty: feasibility and potential clinical advantages. J. Thorac. Cardiovasc. Surg. **120:** 999–1006.
47. KAJSTURA, J., M. ROTA, B. WHANG, *et al.* 2005. Bone marrow cells differentiate in cardiac cell lineages after infarction independently of cell fusion. Circ. Res. **96:** 127–137.
48. BALSAM, L.B., A.J. WAGERS, J.L. CHRISTENSEN, *et al.* 2004. Haematopoietic stem cells adopt mature haematopoietic fates in ischaemic myocardium. Nature **428:** 668–673.
49. DEB, A., S. WANG, K.A. SKELDING, *et al.* 2003. Bone marrow-derived cardiomyocytes are present in the adult human heart. A study of gender-mismatched bone marrow transplantation recipients. Circulation **107:** 1247–1249.
50. EISENBERG, L.M., L. BURNS & C.A. EISENBERG. 2003. Hematopoietic cells from the bone marrow have the potential to differentiate into cardiomyocytes *in vitro*. Anat. Rec. **274A:** 870–882.
51. EISENBERG, L.M., S.W. KUBALAK & C.A. EISENBERG. 2004. Stem cells and the formation of the myocardium in the vertebrate embryo. Anat. Rec. **276A:** 2–12.
52. BRANDON, C., L.M. EISENBERG & C.A. EISENBERG. 2000. WNT signaling modulates the diversification of hematopoietic cells. Blood **96:** 4132–4141.
53. EISENBERG, C.A. & R.R. MARKWALD. 1997. Mixed cultures of avian blastoderm cells and the quail mesoderm cell line QCE6 provide evidence for the pluripotentiality of early mesoderm. Dev. Biol. **191:** 167–181.
54. SELLECK, A.J. & M. BRONNER-FRASER. 1995. Origins of the avian neural crest: the role of neural plate-epidermal interactions. Development **121:** 525–538.
55. EISENBERG, L.M. & C.A. EISENBERG. 2003. Stem cell plasticity, cell fusion, and transdifferentiation. Birth Defects Res. C Embryo Today **69:** 209–218.
56. GRAF, T. 2002. Differentiation plasticity of hematopoietic cells. Blood **99:** 3089–3099.
57. Poulsom, R., M.R. Alison, S.J. Forbes & N.A. Wright, *et al.* 2002. Adult stem cell plasticity. J. Pathol. **197:** 441–456.
58. EISENBERG, L.M. & C.A. EISENBERG. 2004. An in vitro analysis of myocardial potential indicates that phenotypic plasticity is an innate property of early embryonic tissue. Stem Cells Dev. **13:** 614–624.
59. WONG, T.S. & C.P. ORDAHL. 1996. Troponin T gene switching is developmentally regulated by plasma-borne factors in parabiotic chicks. Dev. Biol. **180:** 732–744.

Differentiation Pathways in Human Embryonic Stem Cell–Derived Cardiomyocytes

SOPHIE LEV, IZHAK KEHAT, AND LIOR GEPSTEIN

The B. Rappaport Institute in the Medical Sciences, Faculty of Medicine, Technion, Israel Institute of Technology, Haifa, Israel

ABSTRACT: Human embryonic stem (hES) cells are pluripotent cell lines derived from the inner cell mass of the blastocyst-stage embryo. These unique cell lines can be propagated in the undifferentiated state in culture, while retaining the capacity to differentiate into derivatives of all three germ layers, including cardiomyocytes. The derivation of the hES cell lines presents a powerful tool to explore the early events of cardiac progenitor cell specification and differentiation, and it also provides a novel cell source for the emerging field of cardiovascular regenerative medicine. A spontaneous differentiation system of these stem cells to cardiomyocytes was established and the generated myocytes displayed molecular, structural, and functional properties of early-stage heart cells. In order to follow the *in vitro* differentiation process, the temporal expression of signaling molecules and transcription factors governing early cardiac differentiation was examined throughout the process. A characteristic pattern was noted recapitulating the normal *in vivo* cardiac differentiation scheme observed in other model systems. This review discusses the known pathways involved in cardiac specification and the possible factors that may be used to enhance cardiac differentiation of hES cells, as well as the steps required to fully harness the enormous potential of these unique cells.

KEYWORDS: cardiac differentiation; cell therapy; human embryonic stem cells

INTRODUCTION

Human embryonic stem (hES) cells may provide a new powerful tool to explore the early events involved in cardiac progenitors' specification and differentiation. In addition, these cells may potentially be used in the future for grafting in patients suffering from heart failure. Understanding of the differentiation processes in embryonic stem cells may enhance these efforts by providing strategies to augment differentiation of cells toward a cardiomyocyte lineage. Here we review the early events leading to differentiation of stem cells to cardiomyocytes and explore potential strategies to induce cardiac differentiation.

Address for correspondence: Lior Gepstein, M.D, Ph.D. The Shonis Family Research Laboratory for the Regeneration of Functional Myocardium. The Bruce Rappaport Faculty of Medicine, Technion, Israel Institute of Technology, 2 Efron St., POB 9649, Haifa, 31096, Israel. Voice: 972-4-8295303; fax: 972-4-8524758.
mdlior@tx.technion.ac.il

Ann. N.Y. Acad. Sci. 1047: 50–65 (2005).
doi: 10.1196/annals.1341.005

HEART DEVELOPMENT

Three germ layers—endoderm, ectoderm, and mesoderm—are formed during gastrulation. Cardiac myocytes originate from the mesodermal germ layer. In the course of gastrulation, cardiac progenitors move through the node region and primitive streak to form the cardiac crescent, which begins to express cardiac transcription factors. Results from experiments with mesodermal tissue explants from quail and chick embryos indicate that the specification of cardiomyocytes occurs just before or during the formation of cardiac crescent, and differentiation markers become expressed shortly after that. In a second cell migration process, established heart progenitor cells move ventrally and fuse at the midline to create the linear heart tube. The heart tube consists of an inner endothelial tube enclosed in a layer of myocardium that remains attached to the ventral foregut. The elongating heart tube then loses its connection to the foregut and folds rightward in a complex process termed cardiac looping. Looping of the heart transforms the single heart tube to a more complex structure with two atria and two ventricles, which closely resembles the adult organ (reviewed in Ref. 1).

Signaling Molecules and Transcription Factors Involved in Early Cardiac Development

Bone morphogenic proteins (BMPs) belong to the transforming growth factor β (TGF-β) superfamily. BMPs are expressed in the ectoderm and endoderm adjacent to the heart-forming region and play an inductive role in cardiogenesis. In mouse embryonic carcinoma cells (P19CL6), BMP induces expression of the early cardiac transcription factor genes Nkx2-5 and GATA4 through activation of MAP kinase kinase kinase Tak1 and the transcription factors of Smad family.[2,3] Signaling glycoproteins of the Wnt family are secreted by the neural plate during heart field formation. Wnts act primarily through the canonical signal transduction pathway involving inhibition of glycogen synthase kinase-3 (GSK3), which phosphorylates β-catenin and targets it to degradation. Activation of this pathway results in stabilization of β-catenin, which interacts with transcription factors of TCF/LEF family and activates gene expression. The role of canonical Wnt signaling pathway in vertebrate cardiogenesis is somewhat controversial: deletion of β-catenin in the endoderm of mouse embryo led to the formation of multiple hearts all along the anterior-posterior axis, implicating the repressive role of the canonical Wnt signaling pathway.[4] On the other hand, in pluripotent mouse P19CL6 cells, the canonical Wnt/β-catenin signaling pathway was essential for cardiogenic differentiation.[5] Previous studies in *Xenopus* and chick embryos have indicated that canonical Wnt signaling represses heart formation, whereas Wnt antagonists have been shown to stimulate cardiogenesis. In contrast, the activation of the canonical Wnt signaling pathway in *Drosophila* was essential for heart formation (reviewed in Ref. 6).

Some of the Wnt family members (e.g., Wnt11) act through the noncanonical signaling pathway, which involves protein kinase C and Jun N-terminal kinase. It was shown that Wnt11 stimulates cardiogenesis in mouse P19 cells, avian explants, quail mesodermal cell line, and *Xenopus*.[7,8]

Complex interaction of inductive and repressive signals from surrounding tissues shapes the heart-forming region of early embryo. In response to these signals, the cardiac precursor field activates transcriptional regulators of the cardiac program—including Gata4/Gata5/Gata6, Nkx2-5 (NK2 transcription-factor related, locus 5); myocyte enhancer factor (Mef2b/Mef2c); eHand/dHand (heart and neural crest derivatives expressed transcripts e/d); and T-box 5/20 (Tbx5/Tbx20)—forming a positive cardiac cross-regulatory network.

HUMAN EMBRYONIC STEM CELLS

Human embryonic stem cells are pluripotent cell lines derived from the inner cell mass of blastocyst-stage embryos.[9,10] These cell lines are characterized by their ability to proliferate in the undifferentiated state in culture for a prolonged period, and by their capacity to differentiate into derivatives of all three germ layers. Cell populations grow as compact colonies of undifferentiated cells on mouse or human feeder cell layers or in feeder-free conditions using matrix and conditioned medium.[11,12] Human embryonic stem cell lines have been shown to keep a stable diploid karyotype and continuously express a high level of telomerase activity.[9,10,13]

The pluripotency of the hES cells was established by demonstration of their ability to differentiate into all three germ layers. hES cells injected into immunodeficient mice formed benign teratomas containing advanced differentiated tissue types, representing all three germ layers.[9,10,13] Another approach establishes ES pluripotency by *in vitro* differentiation. A variety of studies have described *in vitro* spontaneous and directed differentiation of hES cells to different lineages: cardiomyocytes,[14,15] neurons and glia,[16–18] endothelial cells,[19] hematopoietic precursors,[20] trophoblast,[15] and hepatocyte-like cells.[21] The most common method used for *in vitro* differentiation is to remove the hES cells from the feeder layer and culture in suspension in absence of mouse embryonic fibroblasts (MEFs). Following culturing in suspension, hES cells aggregate into three-dimensional structures termed embryoid bodies (EBs). The aggregation process itself triggers initial cell differentiation. It is thought that the EBs consist of derivatives of all three germ layers,[10,22] which interact and cross-induce each other, resulting in complex differentiation into the various lineages. This process is considered to recapitulate early embryonic development from the blastocyst stage to the egg-cylinder stage.[23]

Spontaneous Development of ES-Derived Cardiomyocytes

In order to generate a cardiomyocyte-differentiating system from the hES cells (single-cell clone H9.2[13]), small clumps of 3–20 cells were grown in suspension for 8 days. The EBs were then plated on gelatin-coated culture dishes and observed microscopically for the appearance of spontaneous contraction. Rhythmically contracting areas appeared at 6 to 12 days after plating. Cells isolated from the beating areas expressed cardiac-specific structural genes, such as cardiac troponin I and T, atrial natriuretic peptide (ANP), and atrial and ventricular myosin light chains (MLCs). Immunostaining studies demonstrated the presence of the cardiac-specific sarcomeric proteins myosin heavy chain, α-actinin, desmin, and cardiac troponin I, as well as ANP. Electron microscopy revealed varying degrees of myofibrillar organization, consistent with the ultrastructural properties reported for early-stage cardiomyocyte.[14] Further studies in our laboratory demonstrated a reproducible temporal pattern of hES cell-derived cardiomyocyte proliferation, cell-cycle withdrawal, and ultrastructural maturation.

Controlled Expression of Transcription Factors during Embryoid Body Development

In order to follow the development of the embryoid bodies (EBs) during culturing in suspension and following plating, we examined the expression of signaling proteins and transcription factors governing cardiogenic differentiation throughout the process, using semi-quantitative reverse trascriptase PCR (FIG. 1). A number of interesting findings could be noted in these initial studies.

The pluripotent state-specific transcription factors Oct4 was highly expressed in undifferentiated hES cells, and its expression gradually decreased during culturing of cells in suspension and after plating.

Significant expression of Wnt3A was detected in undifferentiated ES cells prior to culturing in suspension. Recently, it has been shown that activation of canonical signaling pathway by Wnt3A in mouse and human ES cells maintains undifferentiated phenotype and sustains expression of the pluripotent state-specific transcription

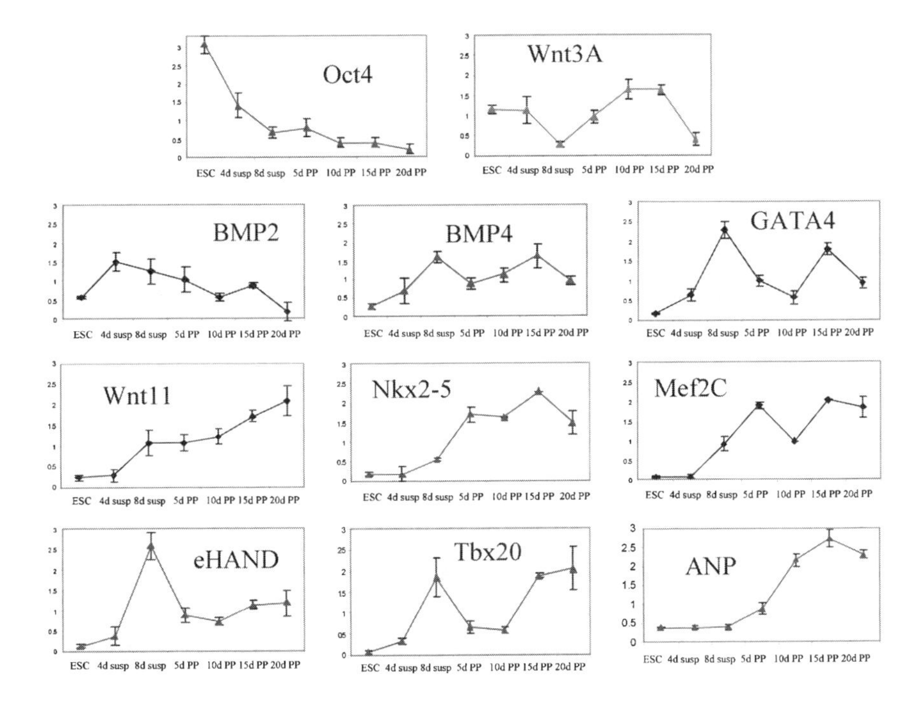

FIGURE 1. Expression of genes involved in cardiac differentiation during development of human embryonic stem cell–derived embryoid bodies (EBs). Samples were collected from embryonic stem cells (ESC); after 4 and 8 days of culturing of EBs in suspension (4d and 8d susp.); and 5, 10, 15, and 20 days after plating of the EBs on gelatin-coated plates (5d, 10d, 15d, and 20d post-plating [PP]). RT-PCR products were quantified and normalized by β-actin expression values. Data are expressed as average of 2–4 RT-PCR gels ±SD.

factors.[24] In another study it was shown that early activation of canonical signaling pathway by Wnt3A is indispensable for cardiac differentiation of mouse PC19CL6 cells.[5] The initial expression of Wnt3A in ES cells and in the newly formed EBs (4-day-old suspension) in our system was followed by what seems to be down-regulation on day 8 of culturing in suspension—the same day when the expression of Wnt11 acting through the non-canonical signaling pathway is activated (FIG. 1). In P19 mouse embryonic carcinoma cells, Wnt11 was shown to repress canonical signaling pathway activated by Wnt3A and promote expression of cardiac transcription factors and formation of cardiomyocytes.[8,25] After plating, the expression of Wnt3A gradually increases until day 15. In addition to cardiogenesis, Wnt3A is involved in neurogenic differentiation of mouse ES cells and osteoblast differentiation of mesenchymal stem cells.[26,27] Therefore, the rise in Wnt3A expression may be attributed to its involvement in other lineages differentiation.

BMP2, BMP4, and GATA4 gene expression increased during the first 4 days of culturing in suspension. The expression of BMP2 reached its highest level on day 4, whereas BMP4 and GATA4 were expressed maximally on day 8 of culturing (FIG. 1). It was shown that BMP4 and GATA4 positively regulate transcription of one another, creating a positive feedback loop which may serve to amplify the BMP4/GATA4 pathway.[28] BMP downstream transducers Smad1/4 and GATA4 cooperatively activate Nkx2-5 gene transcription,[29] which was upregulated on day 8, together with Mef2C, eHand, and Tbx20. Nkx2-5 and Mef2C also cross-react to upregulate the expression of both transcription factors, further contributing to the process of cardiogenic differentiation.[30] Tbx20 was shown to physically interact with the transcription factors Nkx2-5, GATA4, and GATA5, and synergistically activate cardiac gene expression.[31] Interestingly, the expression of genes GATA4, BMP4, eHand, and Tbx20 was significantly downregulated at 5 days post-plating, and increased again during the subsequent 15 days of incubation. GATA4 and Nkx2-5 physically interact and synergistically activate transcription from the ANP promoter.[32] In our system, expression of ANP was upregulated starting from 5 days post-plating, and the mRNA levels continued to increase during subsequent incubation.

In summary, our preliminary RT-PCR results corroborate data obtained from the mouse ES and other model systems that the expression of signaling proteins, cardiogenesis-related transcription factors, and structural genes in EBs occurs in a developmentally controlled manner.[33] To some extent, this process recapitulates the early development of human embryo and, therefore, might be used as a model for further studies.

Development of Excitability during Early Cardiogenesis

Several functional assays—including extracellular and intracellular electrophysiological recordings, calcium imaging, and pharmacological studies—clearly demonstrated that the contracting areas within the EBs also displayed physiological properties consistent with an early-stage cardiac phenotype. Hence, all the components of normal cardiac excitation–contraction (E-C) coupling were demonstrated to be present within this tissue; namely the typical electrical activation, increase in $[Ca^{2+}]_i$, and the resulting contraction.[14] To allow long-term, high-resolution electrophysiological recordings from the EBs, we microdissected spontaneously contracting areas and plated them on the micro-electrode array plates (MEA). These measurements demonstrated the presence of a functional syncytium with stable spontaneous pacemaking activity and synchronous action-potential propagation. Both the site of earliest focal activation and the conduction properties within each EB were relatively reproducible during short-term (3 h) and long-term (10 ± 5 days) recordings.[34] The spontaneous activity and the relatively rapid conduction in these studies suggested the presence of a high-density inward current via voltage-activated channels. Therefore, we next tested the effects of Na^+ channel and Ca^{2+} channel blockers on spontaneous activity and conduction.[35] The Na^+ channel blocker, tetradotoxin (TTX), caused a significant slowing of the conduction. In contrast to TTX, the Ca^{2+} channel blocker diltiazem (or nifedipine) neither slowed nor blocked action-potential conduction and had no effect on automaticity. These results suggested that Na^+ channels, but not Ca^{2+} channels, are essential for initiation and efficient conduction of action potentials through the hES cardiomyocyte syncytium. The ionic mechanism responsible for the automaticity of the hES derived cardiomyocytes was further explored in our lab. A prominent Na^+ current and a hyperpolarization-activated current (HCN)—but no inward rectifier K^+ current—were found in isolated cardiomyocytes. The absence of background K^+ current creates conditions for spontaneous activity that is sensitive to TTX in the same range of partial block of the Na^+ channel; thus, this channel is important for initiating spontaneous excitability in hES-derived heart cells.

Collectively, these results show that the hES-derived cardiomyocyte tissue may serve as the first long-term *in vitro* model for human cardiac tissue. Such a model can be used to further explore conduction in human cardiomyocytes. Moreover, this model can be used as a platform to study the effect of drugs on the electrophysiological properties of human cardiac tissue, as well as to assess the response of human cardiac tissue to several physiological, pathological and pharmacological, manipulations.

Myocardial Cell Therapy Using Human Embryonic Stem Cells

Recent reports suggest that myocyte transplantation may improve cardiac function in animal models of myocardial infarction.[36–39] However, it is not clear whether this functional improvement is due to direct contribution to contractility by the transplanted myocytes, attenuation of the remodeling process, amplification of an endogenous repair process, or induction of angiogenesis. Even the presence of such gap junctions between host and donor cardiomyocyte tissues, as observed in some studies,[40,41] does not guarantee functional integration. For such integration to occur, currents generated in one cell passing through gap junctions must be sufficient to depolarize neighboring cells.

To explore the ability of hES-derived cardiomyocytes to functionally integrate with host tissue, we used the microelectrode array (MEA) platform to assess, *ex vivo*, whether the hES-derived cardiomyocytes were able to structurally and functionally integrate with pre-existing cardiac cultures.[42] The contracting areas within the EBs were dissected and added to neonatal rat cardiomyocyte cultures. Within 24 h after grafting we could already detect microscopically, in all twenty-two cocultures studied, synchronous mechanical activity. We mapped their electrical activity with a high-resolution MEA mapping technique and documented synchronous activity in the cocultures. We also showed electromechanical connections and structural integration as evident by the presence of gap junctions between the human and rat cardiomyocytes. Next, we also provide evidence for the *in vivo* functional integration of donor cells by demonstrating the ability of hES-cell cardiomyocytes to pace the heart in swine with complete atrioventricular block. Electroanatomical mapping and subsequent pathological examination confirmed that the source of the new ventricular ectopic rhythm was the site of cell transplantation.

DIRECTING EMBRYONIC STEM CELL DIFFERENTIATION

Although cardiomyocyte tissue can be reproducibly generated from ES cells using the EB differentiation system in both the murine and human models, the differentiating cardiomyocytes typically account for only a minority of the cells within the EBs. Similarly, spontaneously contracting areas are not observed in all EBs, and even less so in the human model. Since the number of cardiomyocytes generated may have an important effect on the ultimate success of future cell-grafting procedures, cardiomyocyte enrichment of the EB differentiating system may be of crucial importance. Possible strategies for increasing the cardiomyocyte yield during ES cell differentiation may include the use of different ES lines or clones, modification of the protocols used during ES propagation and EB formation, use of different growth factors, transcription factors, feeder layers, and physical conditions.

Genetic Manipulations

The ability to genetically manipulate the ES lines, either transiently or by stable transfection, may play an important role in studying the mechanisms' underlying lineage commitment and differentiation, and in establishing directed differentiation systems. For example, introduction of reporter genes to hES cells enables us to study the expression profile of genes during differentiation and to select specific cell populations within differentiating EBs.[43,44] Knockout of genes by homologous recombination or by RNA interference provides a powerful tool to study the function of specific genes in the differentiation process.[45] Overexpression of transcriptional regulators might affect the differentiation process and direct it to the preferred route.[46] For example, overexpression of either transcription factor, GATA4, Nkx2-5, or MEF2C, in mouse P19 embryonal carcinoma cells, resulted in differentiation into cardiomyocytes in absence of any inductive agent, in contrast to untransfected cells which differentiate into cardiomyocytes only following dimethyl sulfoxide (DMSO) treatment.[30] On the other hand, inhibition of GATA-4 expression in P19 cells by antisense transcripts blocked transcription of terminal cardiac differentiation markers and formation of beating cardiomyocytes.[47]

Although genetic modification of the mouse ES cells can be easily performed and is one of the essential tools in several research areas, achieving similar results in the hES model have proven to be more difficult. Nevertheless, stable hES cell transformants expressing enhanced green fluorescent protein (eGFP) under control of murine Rex1 promoter have been generated using ExGen, a cationic polymer transfection reagent. The transfected cells expressed GFP in an undifferentiated state, and the expression was dramatically reduced during differentiation.[44] ExGene was also used to downregulate Oct4 and β-microglobulin transcripts in hES cells by RNA interference. The knockdown of Oct4 resulted in ES cells differentiation.[43] Efficient transduction of undifferentiated hES cells was also obtained using viral vectors, including adenovirus, adeno-associated virus, and lentivirus.[48–50]

Efficient transient transfection of hES cells was achieved by nucleofection, a novel electroporation-based method.[51] Modified mouse ES electroporation protocol was used to yield a homologous recombination in human ES cells.[45] HPRT1 gene located on X-chromosome was partially removed by homologous recombination. In another experiment, promoterless GFP reporter gene, fused to viral internal ribosomal entry site (IRES), was introduced into the untranslated region of Oct4. The presence of reporter allowed the monitoring of Oct4 gene expression during differentiation and the isolation of Oct4 expressing cells from mixed cell populations by flow cytometry.[45]

Coculture with Visceral Endoderm–Like Cells

Data from amphibians and chicks, and more recently from mice, have suggested that signals emanating from embryonic endoderm may be involved in induction of cardiogenesis. Visceral endoderm–like cell lines induced mouse P19 cells to aggregate spontaneously in coculture and differentiate to cultures containing beating muscle.[52,53] This induction was also observed when aggregates of P19 cells were grown in conditioned medium from a visceral endoderm–like cell line, END-2. Coculture of hES cells—which do not undergo cardiogenesis spontaneously—with endothelial cell line END-2 cells induced differentiation to cardiac and epithelial lineages.[54]

Growth Factors

It is well known that members of the transforming growth factor β1 (TGFβ1) superfamily play a central role in stimulating cardiogenesis. Application of TGFβ and BMP2 to mouse ES in the presence of LIF, a suppressor of differentiation, resulted in upregulation of mRNA of mesodermal transcription factor Brachyury and cardiac transcription factors Nkx2.5 and MEF2C. Embryoid bodies generated from stem cells treated with these growth factors demonstrated significant increase in beating areas and enhanced myofibrillogenesis.[55] Treatment of mouse ES cells with BMP2 and/or FGF2 efficiently enhanced cardiomyocyte differentiation.[56]

Positive and negative influence of Wnt family signaling proteins on cardiogenesis has been reported. Wnt3A and Wnt8 act through the canonical Wnt/β-catenin signaling pathway. In mouse P19CL6 cells, 1% DMSO induced expression of Wnt3A and Wnt8 together with FGF8 two days after the treatment, whereas BMP2, BMP4, and GATA4 were induced only after four days, and Nkx2-5 was induced after 6 days. The expression of Wnt3A and Wnt8 was quickly downregulated on day 4, but it was essential for the subsequent cardiogenesis because inhibition of the canonical pathway by soluble Wnt antagonist inhibited the induction of GATA4, TBX5, BMP2/4, FGF8, and α-myosin heavy chain (α-MHC) genes. Likewise, differentiation was inhibited by constitutively active glycogen synthase kinase 3β, an intracellular inhibitor of the Wnt/β-catenin pathway. Conversely, lithium chloride, which inhibits glycogen synthase kinase 3β and Wnt3A-conditioned medium, upregulated early cardiac markers and the proportion of differentiated cells.[5] In the EBs formed from the mouse ES cells, transient expression of the Wnt3A at days 3 to 5 was followed by expression of Wnt11 and Nkx2-5 at day 6.[57] Wnt11 activates a non-canonical signaling pathway involving protein kinase C and Jun N-terminal kinase. Mouse EBs were treated by Wnt11 conditioned medium, resulting in an increased number of developing cardiomyocytes.[57] Treatment of mouse P19 cells with Wnt11 conditioned medium in absence of DMSO resulted in downregulation of Oct4 expression and induction of GATA4, Nkx2.5, α-MHC, and ANP.[8] It is possible that two Wnt signaling systems participate in the vertebrate cardiogenesis: first, Wnt3A and Wnt8 activate a canonical signaling pathway; and second, this pathway is repressed by Wnt11 acting through non-canonical cascade.[25]

Cardiotrophin and Reactive Oxygen Species

Cardiotrophin-1 is a cytokine that promotes growth and survival of cardiac cells. It was shown that treatment of mouse EBs with cardiotrophin stimulates generation of reactive oxygen species (ROS) by NADPH oxidase. In EBs, the expression of cardiotrophin, as well as generation or ROS, is upregulated during differentiation. Treatment of developing EBs with cardiotrophin or by prooxidants hydrogen peroxide and menadione significantly stimulated cardiogenesis, whereas the radical scavengers exerted inhibitory effect.[58,59] Likewise, ROS generation induced by externally applied electrical field was demonstrated to stimulate cardiogenesis.[60]

Nitric oxide (NO) is a short-lived, free radical gas generated by NO-synthase that acts as a diffusible messenger mediating cell-to-cell communication. Expression of both inducible NO synthase and endothelial NO synthase proteins has been observed in the murine heart during embryonic development starting on day 9. This prominent expression decreases before birth, suggesting that a period of NO exposure is essential for normal development.[61] EBs from mouse ES cells were supplemented with NO donor at the plating stage, resulting in an increased expression of cardiac-specific genes and an increase in the number and the size of beating foci. Overexpression of inducible NO synthase also increased the incidence of beating EBs and the cluster size of beating cells in EB outgrowths. Interestingly, NO not only induced cardiogenic differentiation, but also enhanced apoptosis in undifferentiated cells.[62] Signal transduction cascades activated by ROS and NO overlap in some types of cells, suggesting the involvement of the same signaling system activating cardiogenesis (reviewed by Refs. 63 and 64).

Dimethyl Sulfoxide

Dimethyl sulfoxide (DMSO), which is widely used as cryoprotectant during cell freezing, triggers cardiogenesis in aggregated mouse P19 cells. In the absence of DMSO, P19 cells aggregate and even express a mesoderm-specific gene (Brachyury T), but do not continue to differentiate.[65] Addition of DMSO almost instantaneously triggers the release of Ca^{2+} from intracellular stores. This large, DMSO-induced Ca^{2+} spike may play a role in the induction of differentiation in mouse embryonal carcinoma (EC) cells.[66] Interestingly, DMSO treatment did not enhance cardiogenesis in the hES cells in our experience or in the experience of others.[14,15]

Retinoic Acid and Oxytocin

Retinoic acid is an active form of vitamin A (retinal) that acts as a hormone with important regulatory functions during embryonic development. *In vitro* retinoic acid induces differentiation of mouse ES and EC cells into specific cell types in a time-

and concentration-dependent manner (reviewed by Ref. 67). Treatment of mouse ES-derived EBs during the first two days of development with 10^{-7} to 10^{-8} retinoic acid induced neurogenesis, whereas cardiogenesis was strongly inhibited. From day 2 to 5, treatment of EBs with 10^{-8} M retinoic acid still resulted in induction of neurogenesis and inhibition of cardiogenesis. However, in addition, skeletal muscle cells and adipocytes were induced. When EBs were treated with 10^{-8} M retinoic acid after day 5, myogenic and adipogenic differentiation was inhibited, but differentiation of cardiac and vascular smooth muscle cells was induced.[67] Interestingly, in another study conducted on the teratocarcinoma cell line (P19S1801A1), the differentiation route was found to be dependent on the concentration of retinoic acid used in the treatment.[68] In contrast to mouse embryonic carcinoma and embryonic stem cells, treatment of hES cells with retinoic acid did not stimulate cardiogenic differentiation, emphasizing the differences between two species.[15]

Recently it was suggested that the retinoic acid effect leading to differentiation of P19 cells into cardiomyocytes is mediated by activation of oxytocin/oxytocin receptor signaling system.[69] Oxytocin is a growth factor largely expressed in the hypothalamus, which causes uterine contractions and stimulates lactation. Activation of cardiac oxytocin receptor stimulates the release of ANP, which is involved in the regulation of blood pressure and cell growth. Mouse P19 cells treated with oxytocin differentiate into cardiomyocytes in the absence of DMSO even faster then DMSO-treated cells. One of the mechanisms by which oxytocin and DMSO induce cardiogenesis involves the oxytocin receptor, since both agents upregulated the expression of this receptor, and the oxytocin antagonist blocked their cardiogenic action.[70] In addition to the induction of differentiation of P19 cells, oxytocin was postulated to induce the cardiogenic differentiation of adult stem/progenitor cells. Sca-1-positive cells (stem cell antigen) isolated from the adult mouse heart expressed cardiac transcription factors and contractile proteins and showed sarcomeric structure and spontaneous beating when treated with oxytocin.[71]

The oxytocin gene contains response elements for retinoic acid. It was shown that the rat oxytocin promoter is stimulated by retinoic acid through two distinct regions in the 5′-flanking region.[72] Retinoic acid–mediated cardiomyocyte differentiation of P19 cells was inhibited by the presence of 10^{-7} M oxytocin antagonist.[69] Both retinoic acid and oxytocin induced an increase in oxytocin receptor expression, which was neutralized in the presence of the oxytocin antagonist. These data suggest that the retinoic acid effect, leading to differentiation of P19 cells into a cardiomyocyte phenotype, is mediated by the oxytocin system.[69]

Thyroid hormone, like retinoic acid, binds to the nuclear receptors of the steroid-thyroid-retinoic superfamily and is capable of regulation of oxytocin gene expression.[72] It was demonstrated that thyroid hormone (3,3′,5-triiodo-L-thyronine) induced cardiomyogenesis in P19 cells in a dose-dependent manner. Furthermore, there is evidence that this induction was inhibited by oxytocin antagonist.[69,73]

Dynorphin B

Mouse GTR1 (a derivative of R1 ES cells) and P19 cells express the prodynorphin gene encoding the opioid peptide dynorphin B.[74,75] Treatment of P19 cells with DMSO induced expression of prodynorphin gene and dynorphin B synthesis and secretion, followed by GATA-4 and Nkx-2.5 gene expression. In the absence of

DMSO, dynorphin B triggered GATA-4 and Nkx-2.5 gene expression and led to the appearance of both α-MHC and MLC-2V transcripts. DMSO-induced cardiogenesis was inhibited by κ-opioid receptor antagonist Mr1452 and by a prodynorphin antisense phosphorothioate oligonucleotide. DMSO-triggered activation of prodynorphin gene expression, followed by induction of cardiogenic genes, suggests that initiation of cardiogenesis may be controlled by dynorphin B–mediated autocrine signaling.[75] There is evidence that dynorphin B binding to κ–opioid receptors induces activation and changes in expression and subcellular distribution of protein kinase C isozymes.[74,76] Interestingly, the prodynorphin gene is a likely target of the cardiogenic effect of magnetic field on differentiating ES cells. Extremely low frequency magnetic field applied to GTR1 cells following LIF removal enhanced prodynorphin, GATA4 and Nkx2-5 gene expression, and resulted in a consistently increased yield of cardiomyocytes.[77]

A novel ester of hyaluronan linked to butyric and retinoic acids (HBR) was developed to facilitate cardiogenic differentiation. In mouse ES cells, HBR enhanced prodynorphin gene expression and the synthesis and secretion of dynorphin B. HBR also induced the expression of cardiac-specific transcripts and highly enhanced the yield of spontaneously beating cardiomyocytes.[78]

5-Azacytidine

5-Azacytidine and 5-aza-2′-deoxycytidine are cytosine analogues that are incorporated into the DNA and cause a decrease in DNA methylation, as a result of the formation of stable covalent complexes between DNA methyltransferases and 5-azacytidine-containing DNA. Treatment with 5-azacytidine facilitated cardiogenic differentiation of bone marrow and adipose tissue-derived mesenchymal stem cells,[79–81] as well as cardiac progenitor cells isolated from the mouse adult heart.[82] Moreover, 5-aza-2′-deoxycytidine enhanced cardiomyocyte differentiation of hES cells in some of the studies.[15]

Cardiogenol

Screening of a combinatorial library of small heterocyclic molecules resulted in isolation of several differentiation-inducing compounds: TWS119, a 4,6-disubstituted pyrrolopyrimidine that induces neurogenesis in murine ES cells; and purmorphamine, a 2,6,9-trisubstituted purine that differentiates multipotent mesenchymal progenitor cells into an osteoblast lineage.[83–85] Promoter of the ANP gene fused to luciferase reporter was used to screen the library for inductive cardiogenic activity in mouse embryonic carcinoma cells seeded in monolayer. Immunostaining with the anti-MHC antibody was used as a second assay to find the most potent cardiogenic compounds, resulting in isolation of four structurally similar diaminopyrimidines (termed cardiogenol A-D). The most potent of these molecules, cardiogenol C, induced effective cardiogenic differentiation in mouse ES cells in the absence of aggregation and EB formation.[86]

Ascorbic Acid

Screening of library of bioactive molecules approved for human use with a fluorescent reporter gene fused to α-MHC promoter in monolayer mouse ES cells resulted in isolation of ascorbic acid (vitamin C) as a cardiogenesis-promoting compound. Treatment with ascorbic acid markedly increased the expression of cardiac markers including GATA4, Nkx2.5, α-MHC, β-MHC, and ANP, and proportion of beating cardiomyocytes.[87] It has been reported that ascorbic acid also enhances differentiation of mouse embryonic stem cells into neurons.[88]

CONCLUSION

We have discussed the known pathways and factors involved in early cardiac differentiation in different *in vitro* and *in vivo* models. The establishment of the hES *in vitro* cardiogenic model may provide a unique tool to study the mechanisms underlying early cardiac lineage commitment, differentiation, and maturation. A better understanding of these pathways may eventually also facilitate the development of directed and more efficient cardiac differentiation systems from the hES cells. The latter effect may have important implications for several research and applied cardiovascular fields, including early human embryonic development, functional genomics, drug development and screening, and the developing areas of myocardial cell therapy and tissue engineering.

REFERENCES

1. HARVEY, R.P. 2002. Patterning the vertebrate heart. Nat. Rev. Genet. **3:** 544–556.
2. MONZEN, K., Y. HIROI, S. KUDOH, *et al.* 2001. Smads, TAK1, and their common target ATF-2 play a critical role in cardiomyocyte differentiation. J. Cell Biol. **153:** 687–698.
3. MONZEN, K., I. SHIOJIMA, Y. HIROI, *et al.* 1999. Bone morphogenetic proteins induce cardiomyocyte differentiation through the mitogen-activated protein kinase TAK1 and cardiac transcription factors Csx/Nkx-2.5 and GATA-4. Mol. Cell. Biol. **19:** 7096–7105.
4. LICKERT, H., S. KUTSCH, B. KANZLER, *et al.* 2002. Formation of multiple hearts in mice following deletion of beta-catenin in the embryonic endoderm. Dev. Cell **3:** 171–181.
5. NAKAMURA, T., M. SANO, Z. SONGYANG, *et al.* 2003. A Wnt- and beta-catenin-dependent pathway for mammalian cardiac myogenesis. Proc. Natl. Acad. Sci. U.S.A. **100:** 5834–5839.
6. SOLLOWAY, M.J. & R.P. HARVEY. 2003. Molecular pathways in myocardial development: a stem cell perspective. Cardiovasc. Res. **58:** 264–277.
7. EISENBERG, C.A. & L.M. EISENBERG. 1999. WNT11 promotes cardiac tissue formation of early mesoderm. Dev. Dyn. **216:** 45–58.
8. PANDUR, P., M. LASCHE, L.M. EISENBERG, *et al.* 2002. Wnt-11 activation of a non-canonical Wnt signalling pathway is required for cardiogenesis. Nature **418:** 636–641.
9. THOMSON, J.A., J. ITSKOVITZ-ELDOR, S.S. SHAPIRO, *et al.* 1998. Embryonic stem cell lines derived from human blastocysts. Science **282:** 1145–1147.
10. REUBINOFF, B.E., M.F. PERA, C.Y. FONG, *et al.* 2000. Embryonic stem cell lines from human blastocysts: somatic differentiation *in vitro*. Nat. Biotechnol. **18:** 399–404.
11. XU, C., J. JIANG, V. SOTTILE, *et al.* 2004. Immortalized fibroblast-like cells derived from human embryonic stem cells support undifferentiated cell growth. Stem Cells **22:** 972–980.
12. XU, C., M.S. INOKUMA, J. DENHAM, *et al.* 2001. Feeder-free growth of undifferentiated human embryonic stem cells. Nat. Biotechnol. **19:** 971–974.
13. AMIT, M., M.K. CARPENTER, M.S. INOKUMA, *et al.* 2000. Clonally derived human embryonic stem cell lines maintain pluripotency and proliferative potential for prolonged periods of culture. Dev. Biol. **227:** 271–278.

14. KEHAT, I., D. KENYAGIN-KARSENTI, M. SNIR, *et al.* 2001. Human embryonic stem cells can differentiate into myocytes with structural and functional properties of cardiomyocytes. J. Clin. Invest. **108:** 407–414.
15. XU, C., S. POLICE, N. RAO, *et al.* 2002. Characterization and enrichment of cardiomyocytes derived from human embryonic stem cells. Circ. Res. **91:** 501–508.
16. REUBINOFF, B.E., P. ITSYKSON, T. TURETSKY, *et al.* 2001. Neural progenitors from human embryonic stem cells. Nat. Biotechnol. **19:** 1134–1140.
17. ZHANG, S.C., M. WERNIG, I.D. DUNCAN, *et al.* 2001. *In vitro* differentiation of transplantable neural precursors from human embryonic stem cells. Nat. Biotechnol. **19:** 1129–1133.
18. CARPENTER, M.K., M.S. INOKUMA, J. DENHAM, *et al.* 2001. Enrichment of neurons and neural precursors from human embryonic stem cells. Exp. Neurol. **172:** 383–397.
19. LEVENBERG, S., J.S. GOLUB, M. AMIT, *et al.* 2002. Endothelial cells derived from human embryonic stem cells. Proc. Natl. Acad. Sci. U.S.A. **99:** 4391–4396.
20. KAUFMAN, D.S., E.T. HANSON, R.L. LEWIS, *et al.* 2001. Hematopoietic colony-forming cells derived from human embryonic stem cells. Proc. Natl. Acad. Sci. U.S.A. **98:** 10716–10721.
21. RAMBHATLA, L., C.P. CHIU, P. KUNDU, *et al.* 2003. Generation of hepatocyte-like cells from human embryonic stem cells. Cell Transplant. **12:** 1–11.
22. ITSKOVITZ-ELDOR, J., M. SCHULDINER, D. KARSENTI, *et al.* 2000. Differentiation of human embryonic stem cells into embryoid bodies compromising the three embryonic germ layers. Mol. Med. **6:** 88–95.
23. HAMAZAKI, T., M. OKA, S. YAMANAKA, *et al.* 2004. Aggregation of embryonic stem cells induces Nanog repression and primitive endoderm differentiation. J. Cell Sci. **117:** 5681–5686.
24. SATO, N., L. MEIJER, L. SKALTSOUNIS, *et al.* 2004. Maintenance of pluripotency in human and mouse embryonic stem cells through activation of Wnt signaling by a pharmacological GSK-3-specific inhibitor. Nat. Med. **10:** 55–63.
25. MAYE, P., J. ZHENG, L. LI, *et al.* 2004. Multiple mechanisms for Wnt11-mediated repression of the canonical Wnt signaling pathway. J. Biol Chem. **279:** 24659–24665.
26. LUO, Q., Q. KANG, W. SI, *et al.* 2004. Connective tissue growth factor (CTGF) is regulated by Wnt and BMP signaling in osteoblast differentiation of mesenchymal stem cells. J. Biol. Chem. **279:** 55958–55968.
27. MURASHOV, A.K., E.S. PAK, W.A. HENDRICKS, *et al.* 2005. Directed differentiation of embryonic stem cells into dorsal interneurons. FASEB J. **19:** 252–254.
28. NEMER, G. & M. NEMER. 2003. Transcriptional activation of BMP-4 and regulation of mammalian organogenesis by GATA-4 and -6. Dev. Biol. **254:** 131–148.
29. BROWN, C.O., III, X. CHI, E. GARCIA-GRAS, *et al.* 2004. The cardiac determination factor, Nkx2-5, is activated by mutual cofactors GATA-4 and Smad1/4 via a novel upstream enhancer. J. Biol. Chem. **279:** 10659–10669.
30. SKERJANC, I.S., H. PETROPOULOS, A.G. RIDGEWAY, *et al.* 1998. Myocyte enhancer factor 2C and Nkx2-5 upregulate each other's expression and initiate cardiomyogenesis in P19 cells. J. Biol. Chem. **273:** 34904–34910.
31. STENNARD, F.A., M.W. COSTA, D.A. ELLIOTT, *et al.* 2003. Cardiac T-box factor Tbx20 directly interacts with Nkx2-5, GATA4, and GATA5 in regulation of gene expression in the developing heart. Dev. Biol. **262:** 206–224.
32. DUROCHER, D. & M. NEMER. 1998. Combinatorial interactions regulating cardiac transcription. Dev. Genet. **22:** 250–262.
33. BOHELER, K.R., J. CZYZ, D. TWEEDIE, *et al.* 2002. Differentiation of pluripotent embryonic stem cells into cardiomyocytes. Circ. Res. **91:** 189–201.
34. KEHAT, I., A. GEPSTEIN, A. SPIRA, *et al.* 2002. High-resolution electrophysiological assessment of human embryonic stem cell–derived cardiomyocytes: a novel *in vitro* model for the study of conduction. Circ. Res. **91:** 659–661.
35. SATIN, J., I. KEHAT, O. CASPI, *et al.* 2004. Mechanism of spontaneous excitability in human embryonic stem cell derived cardiomyocytes. J. Physiol. **559:** 479–496.
36. TAYLOR, D.A., B.Z. ATKINS, P. HUNGSPREUGS, *et al.* 1998. Regenerating functional myocardium: improved performance after skeletal myoblast transplantation. Nat. Med. **4:** 929–933.

37. ORLIC, D., J. KAJSTURA, S. CHIMENTI, *et al.* 2001. Bone marrow cells regenerate infarcted myocardium. Nature **410:** 701–705.
38. LI, R.K., D.A. MICKLE, R.D. WEISEL, *et al.* 1997. Natural history of fetal rat cardiomyocytes transplanted into adult rat myocardial scar tissue. Circulation **96:** 179–186.
39. LI, R.K., G. LI, D.A. MICKLE, *et al.* 1997. Overexpression of transforming growth factor-beta1 and insulin-like growth factor-I in patients with idiopathic hypertrophic cardiomyopathy. Circulation **96:** 874–881.
40. SOONPAA, M.H., G.Y. KOH, M.G. KLUG, *et al.* 1994. Formation of nascent intercalated disks between grafted fetal cardiomyocytes and host myocardium. Science **264:** 98–101.
41. REINECKE, H., M. ZHANG, T. BARTOSEK, *et al.* 1999. Survival, integration, and differentiation of cardiomyocyte grafts: a study in normal and injured rat hearts. Circulation **100:** 193–202.
42. KEHAT, I., L. KHIMOVICH, O. CASPI, *et al.* 2004. Electromechanical integration of cardiomyocytes derived from human embryonic stem cells. Nat. Biotechnol. **22:** 1282–1289.
43. MATIN, M.M., J.R. WALSH, P.J. GOKHALE, *et al.* 2004. Specific knockdown of Oct4 and beta2-microglobulin expression by RNA interference in human embryonic stem cells and embryonic carcinoma cells. Stem Cells **22:** 659–668.
44. EIGES, R., M. SCHULDINER, M. DRUKKER, *et al.* 2001. Establishment of human embryonic stem cell-transfected clones carrying a marker for undifferentiated cells. Curr Biol. **11:** 514–518.
45. ZWAKA, T.P. & J.A. THOMSON. 2003. Homologous recombination in human embryonic stem cells. Nat. Biotechnol. **21:** 319–321.
46. GREPIN, C., G. NEMER & M. NEMER. 1997. Enhanced cardiogenesis in embryonic stem cells overexpressing the GATA-4 transcription factor. Development **124:** 2387–2395.
47. GREPIN, C., L. ROBITAILLE, T. ANTAKLY, *et al.* 1995. Inhibition of transcription factor GATA-4 expression blocks in vitro cardiac muscle differentiation. Mol. Cell. Biol. **15:** 4095–4102.
48. SMITH-ARICA, J.R., A.J. THOMSON, R. ANSELL, *et al.* 2003. Infection efficiency of human and mouse embryonic stem cells using adenoviral and adeno-associated viral vectors. Cloning Stem Cells **5:** 51–62.
49. MA, Y., A. RAMEZANI, R. LEWIS, *et al.* 2003. High-level sustained transgene expression in human embryonic stem cells using lentiviral vectors. Stem Cells **21:** 111–117.
50. GROPP, M., P. ITSYKSON, O. SINGER, *et al.* 2003. Stable genctic modification of human embryonic stem cells by lentiviral vectors. Mol. Ther. **7:** 281–287.
51. LAKSHMIPATHY, U., B. PELACHO, K. SUDO, *et al.* 2004. Efficient transfection of embryonic and adult stem cells. Stem Cells **22:** 531–543.
52. MUMMERY, C.L., T.A. VAN ACHTERBERG, A.J. VAN DEN EIJNDEN-VAN RAAIJ, *et al.* 1991. Visceral-endoderm–like cell lines induce differentiation of murine P19 embryonal carcinoma cells. Differentiation **46:** 51–60.
53. MUMMERY, C., D. WARD, C.E. VAN DEN BRINK, *et al.* 2002. Cardiomyocyte differentiation of mouse and human embryonic stem cells. J. Anat. **200:** 233–242.
54. MUMMERY, C., D. WARD-VAN OOSTWAARD, P. DOEVENDANS, *et al.* 2003. Differentiation of human embryonic stem cells to cardiomyocytes: role of coculture with visceral endoderm–like cells. Circulation **107:** 2733–2740.
55. BEHFAR, A., L.V. ZINGMAN, D.M. HODGSON, *et al.* 2002. Stem cell differentiation requires a paracrine pathway in the heart. FASEB J. **16:** 1558–1566.
56. KAWAI, T., T. TAKAHASHI, M. ESAKI, *et al.* 2004. Efficient cardiomyogenic differentiation of embryonic stem cell by fibroblast growth factor 2 and bone morphogenetic protein 2. Circulation **68:** 691–702.
57. TERAMI, H., K. HIDAKA, T. KATSUMATA, *et al.* 2004. Wnt11 facilitates embryonic stem cell differentiation to Nkx2.5-positive cardiomyocytes. Biochem. Biophys. Res. Commun. **325:** 968–975.
58. SAUER, H., W. NEUKIRCHEN, G. RAHIMI, *et al.* 2004. Involvement of reactive oxygen species in cardiotrophin-1-induced proliferation of cardiomyocytes differentiated from murine embryonic stem cells. Exp. Cell Res. **294:** 313–324.
59. SAUER, H., G. RAHIMI, J. HESCHELER, *et al.* 2000. Role of reactive oxygen species and

phosphatidylinositol 3-kinase in cardiomyocyte differentiation of embryonic stem cells. FEBS Lett. **476:** 218–223.
60. Sauer, H., G. Rahimi, J. Hescheler, *et al.* 1999. Effects of electrical fields on cardiomyocyte differentiation of embryonic stem cells. J. Cell. Biochem. **75:** 710–723.
61. Bloch, W., B.K. Fleischmann, D.E. Lorke, *et al.* 1999. Nitric oxide synthase expression and role during cardiomyogenesis. Cardiovasc. Res. **43:** 675–684.
62. Kanno, S., P.K. Kim, K. Sallam, *et al.* 2004. Nitric oxide facilitates cardiomyogenesis in mouse embryonic stem cells. Proc. Natl. Acad. Sci. U.S.A. **101:** 12277–12281.
63. Marshall, H.E., K. Merchant & J.S. Stamler. 2000. Nitrosation and oxidation in the regulation of gene expression. FASEB J. **14:** 1889–1900.
64. Pfeilschifter, J., W. Eberhardt & K.F. Beck. 2001. Regulation of gene expression by nitric oxide. Pfleugers Arch. **442:** 479–486.
65. Skerjanc, I.S. 1999. Cardiac and skeletal muscle development in P19 embryonal carcinoma cells. Trends Cardiovasc. Med. **9:** 139–143.
66. Morley, P. & J.F. Whitfield. 1993. The differentiation inducer, dimethyl sulfoxide, transiently increases the intracellular calcium ion concentration in various cell types. J. Cell Physiol. **156:** 219–225.
67. Rohwedel, J., K. Guan & A.M. Wobus. 1999. Induction of cellular differentiation by retinoic acid in vitro. Cells Tissues Organs **165:** 190–202.
68. Edwards, M.K. & M.W. McBurney. 1983. The concentration of retinoic acid determines the differentiated cell types formed by a teratocarcinoma cell line. Dev. Biol. **98:** 187–191.
69. Jankowski, M., B. Danalache, D. Wang, *et al.* 2004. Oxytocin in cardiac ontogeny. Proc. Natl. Acad. Sci. U.S.A. **101:** 13074–13079.
70. Paquin, J., B.A. Danalache, M. Jankowski, *et al.* 2002. Oxytocin induces differentiation of P19 embryonic stem cells to cardiomyocytes. Proc. Natl. Acad. Sci. U.S.A. **99:** 9550–9555.
71. Matsuura, K., T. Nagai, N. Nishigaki, *et al.* 2004. Adult cardiac Sca-1-positive cells differentiate into beating cardiomyocytes. J. Biol. Chem. **279:** 11384–11391.
72. Adan, R.A., J.J. Cox, T.V. Beischlag, *et al.* 1993. A composite hormone response element mediates the transactivation of the rat oxytocin gene by different classes of nuclear hormone receptors. Mol. Endocrinol. **7:** 47–57.
73. Rodriguez, E.R., C.D. Tan, U.S. Onwuta, *et al.* 1994. 3,5,3′-Triiodo-L-thyronine induces cardiac myocyte differentiation but not neuronal differentiation in P19 teratocarcinoma cells in a dose dependent manner. Biochem. Biophys. Res. Commun. **205:** 652–658.
74. Ventura, C., E. Zinellu, E. Maninchedda, *et al.* 2003. Protein kinase C signaling transduces endorphin-primed cardiogenesis in GTR1 embryonic stem cells. Circ. Res. **92:** 617–622.
75. Ventura, C. & M. Maioli. 2000. Opioid peptide gene expression primes cardiogenesis in embryonal pluripotent stem cells. Circ. Res. **87:** 189–194.
76. Ventura, C., E. Zinellu, E. Maninchedda, *et al.* 2003. Dynorphin B is an agonist of nuclear opioid receptors coupling nuclear protein kinase C activation to the transcription of cardiogenic genes in GTR1 embryonic stem cells. Circ. Res. **92:** 623–629.
77. Ventura, C., M. Maioli, Y. Asara, *et al.* 2005. Turning on stem cell cardiogenesis with extremely low frequency magnetic fields. FASEB J. **19:** 155–157.
78. Ventura, C., M. Maioli, Y. Asara, *et al.* 2004. Butyric and retinoic mixed ester of hyaluronan. A novel differentiating glycoconjugate affording a high throughput of cardiogenesis in embryonic stem cells. J. Biol. Chem. **279:** 23574–23579.
79. Makino, S., K. Fukuda, S. Miyoshi, *et al.* 1999. Cardiomyocytes can be generated from marrow stromal cells *in vitro*. J. Clin. Invest. **103:** 697–705.
80. Fukuda, K. 2002. Reprogramming of bone marrow mesenchymal stem cells into cardiomyocytes. C. R. Biol. **325:** 1027–1038.
81. Rangappa, S., C. Fen, E.H. Lee, *et al.* 2003. Transformation of adult mesenchymal stem cells isolated from the fatty tissue into cardiomyocytes. Ann. Thorac. Surg. **75:** 775–779.
82. Oh, H., S.B. Bradfute, T.D. Gallardo, *et al.* 2003. Cardiac progenitor cells from

adult myocardium: homing, differentiation, and fusion after infarction. Proc. Natl. Acad. Sci. U.S.A. **100:** 12313–12318.
83. Wu, X., S. Ding, Q. Ding, *et al.* 2002. A small molecule with osteogenesis-inducing activity in multipotent mesenchymal progenitor cells. J. Am. Chem. Soc. **124:** 14520–14521.
84. Ding, S., T.Y. Wu, A. Brinker, *et al.* 2003. Synthetic small molecules that control stem cell fate. Proc. Natl. Acad. Sci. U.S.A. **100:** 7632–7637.
85. Ding, S., N.S. Gray, X. Wu, *et al.* 2002. A combinatorial scaffold approach toward kinase-directed heterocycle libraries. J. Am. Chem. Soc. **124:** 1594–1596.
86. Wu, X., S. Ding, Q. Ding, *et al.* 2004. Small molecules that induce cardiomyogenesis in embryonic stem cells. J. Am. Chem. Soc. **126:** 1590–1591.
87. Takahashi, T., B. Lord, P.C. Schulze, *et al.* 2003. Ascorbic acid enhances differentiation of embryonic stem cells into cardiac myocytes. Circulation **107:** 1912–1916.
88. Bibel, M., J. Richter, K. Schrenk, *et al.* 2004. Differentiation of mouse embryonic stem cells into a defined neuronal lineage. Nat. Neurosci. **7:** 1003–1009.

Functional Properties of Human Embryonic Stem Cell–Derived Cardiomyocytes

KATYA DOLNIKOV, MARK SHILKRUT, NAAMA ZEEVI-LEVIN, ASAF DANON, SHARON GERECHT-NIR, JOSEPH ITSKOVITZ-ELDOR, AND OFER BINAH

Rappaport Institute, Ruth and Bruce Rappaport Faculty of Medicine, Technion, Israel Institute of Technology, Haifa, Israel

ABSTRACT: Regeneration of the diseased myocardium by cardiac cell transplantation is an attractive therapeutic modality. Yet, because the transplanted cardiomyocytes should functionally integrate within the diseased myocardium, it is preferable that their properties resemble those of the host. To determine the functional adaptability of human embryonic stem cell–derived cardiomyocytes (hESC-CM) to the host myocardium, the authors investigated the excitation–contraction (E-C) coupling and the responsiveness to common physiological stimuli. The main findings are: (1) hESC-CM readily respond to electrical pacing and generate corresponding $[Ca^{2+}]_i$ transients (measured by fura-2 fluorescence) and contractions (measured by video edge detector). (2) In contrast to the mature myocardium, hESC-CM display negative force-frequency relations. (3) The hESC-CM contraction is dependent on $[Ca^{2+}]_o$ and blocked by verapamil. (4) Surprisingly, ryanodine, the sarcoplasmic-endoplasmic reticulum Ca^{2+}-ATPase inhibitor thapsigargin, and caffeine do not affect the $[Ca^{2+}]_i$ transient or contraction. Collectively, these results indicate that at the developmental stage of 45 to 60 days, the contraction is largely dependent on $[Ca^{2+}]_o$ rather than on sarcoplasmic reticulum (SR) Ca^{2+} stores. The results show for the first time that the E-C coupling properties of hESC-CM differ from the adult myocardium, probably due to immature SR function. Based on these findings, genetic manipulation of hESC-CM toward the adult myocardium should be considered.

KEYWORDS: $[Ca^{2+}]_i$ transients; contraction; excitation–contraction coupling; human embryonic stem cell–derived cardiomyocytes

INTRODUCTION

Cardiovascular diseases are the most frequent cause of death in the industrialized world. In the United States alone, congestive heart failure affects approximately five million patients, with 400,000 new cases each year.[1] The main contributor to the development of this condition is myocardial infarction, affecting about 1.1 million Americans yearly. End-stage heart failure still has a poor prognosis, underlined by the fact that only 50% of these patients survive to the following year. The current pharmacotherapy for congestive heart failure, including neurohormonal inhibition

Address for correspondence: A. Prof. Ofer Binah, Ph.D., Rappaport Institute, Technion, Israel Institute of Technology, P.O. Box 9697, Haifa 31096, Israel. Voice: 972-4829-5262; fax: 972-4851-3919. binah@tx.technion.ac.il

Ann. N.Y. Acad. Sci. 1047: 66–75 (2005).
doi: 10.1196/annals.1341.006

with angiotensin converting enzyme inhibitors and β-blockers, improves clinical outcomes. Nevertheless, other treatment options including various interventional and surgical therapeutic methods are limited in preventing ventricular remodeling because of their inability to repair or replace damaged myocardium. Given the high morbidity and mortality rates associated with congestive heart failure, shortage of donor hearts for transplantation, complications resulting from immunosuppression, and long-term failure of transplanted organs,[2] novel therapeutic modalities for improving cardiac function and preventing heart failure are in critical demand.[3] As discussed below, regeneration or repair of infarcted or ischemic myocardium may be accomplished by the use of cell therapy, that is, the transplantation of healthy, functional, and propagating cells to restore the viability or function of deficient tissues.[4–6]

Embryonic Stem Cells and Cardiomyocytes

Stem cells are fundamentally characterized by prolonged self-renewal and long-term potential to form differentiated cell types. In most adult tissues, stem or progenitor cells are mobilized in response to environmental stimuli. However, stem cells present in adult tissues form only a limited number of cell types. In early mammalian embryos at the blastocyst stage, the inner cell mass is pluripotent, and therefore, the identification, derivation, and characterization of human embryonic stem cells (hESC) may open the door to the rapidly progressing field of therapeutic cell transplantation. Human ESC, which are derived from human preimplantation embryos at the blastocyst stage,[7–9] were demonstrated to fulfill all the criteria defining embryonic stem cells: immortality, capability to proliferate indefinitely in culture while maintaining the undifferentiated phenotype, and the capacity to form derivatives of all three germ layers (see Ref. 8 for example). Therefore, hESC can serve as a source of numerous types of differentiated cells. Human ESC remain in the undifferentiated state when cultured on inactivated mouse embryonic fibroblasts (MEF)[5] human fetal fibroblasts and adult epithelial cells,[9] or human feeder layer.[10] Human ESC can also be grown on matrigel with media conditioned by MEF[11] or on fibronectin with media containing transforming growth factor β1 (TGF-β1) and basic fibroblast growth factor (bFGF).[12] Undifferentiated hESC are characterized by the expression of surface markers typical of undifferentiated primate embryonic stem cells, such as stage-specific embryonic antigen 3 (SSEA3), SSEA4, Tra-1-60 and Tra-1-81, expression of the embryonic transcription factor Oct-4, and high levels of telomerase activity.

Cell Therapy with Human Embryonic Stem Cell–Derived Cardiomyocytes

The most attractive application of hESC is cell replacement therapy aimed at substituting healthy cells for diseased, missing, or degenerated tissue. The use of cell replacement therapy is of particular significance in the field of regenerative cardiovascular medicine, because massive loss of terminally differentiated adult cardiomyocytes, as occurs, for example, during myocardial infarction, *is irreversible*. Recent experimental observations suggest that cell transplantation has a potential therapeutic value for treating heart diseases.[6,13–17] This notion is based on the fundamental concept that the transplanted donor cardiomyocytes will integrate with the host myocardium and contribute directly to the mechanical cardiac performance. Preferably, cell transplantation allows the replacement of non-functional myocytes and scar tissue with new, fully functional contracting cells, thereby improving cardiac function and relieving the symptoms of heart failure. A detailed consideration of the advantages and disadvantages of the various sources of cardiomyocytes is beyond the scope of this chapter and is provided in several reviews.[4,18,19] In recent years, several studies have shown that improvement of myocardial function can be achieved in experimental animal models of heart failure and infarction by transplanting exogenous adult and embryonic cell types into the compromised myocardium. The transplanted cells include embryonic, fetal, and neonatal cardiomyocytes from rodents and pigs, as well as human adult and fetal cardiomyocytes, autologous adult atrial cells, and dermal fibroblasts.[4,5,16–19,20–24] Independent of the cell type used, the improved myocardial function accompanying the transplantation may result from either the ability of the muscle cells to contract or from the passive contribution of mechanical support to the myocardial architecture. In order to improve the prospects of cardiac cell transplantation, it is widely realized that the functional properties and hormonal responsiveness of hESC-cardiomyocytes (hESC-CM) should be thoroughly investigated. In view of that, our overall aim is to establish a comprehensive scientific infrastructure that will significantly improve our abilities to use hESC-CM for cell therapy in heart diseases. Because it is desired that the transplanted cells fully integrate within the diseased myocardium, contribute to its contractile performance, and respond appropriately to various stimuli (e.g., β-adrenergic stimulation), it is important to decipher their adaptability with the host myocardium. In the first stage of this long-term project we investigated the fundamental properties of the excitation–contraction (E-C) coupling machinery of 45- to 60-day-old hESC-CM.

RESULTS AND DISCUSSION

E-C Coupling in hESC-CM

Electrophysiological Properties

The functional properties of hESC-CM were studied by characterizing the E-C coupling in spontaneously beating or in electrically paced embryoid bodies (EB) (FIG. 1A). The first component of the E-C coupling measured from hESC-CM was the extracellular electrical activity, which was monitored using our Micro-Electrode-Array (MEA) data acquisition system (Multi Channel Systems, Reutlingen, Germany).[25] The MEA consist of a 50 × 50 mm glass substrate, in the center of which is an embedded 8 × 8 matrix of 60 titanium-nitride, 30-μm-diameter electrodes, with an interelectrode distance of 200 μm (FIG. 1B). During the recording sessions, the MEA was kept at 37 ± 0.1°C. As seen in FIGURE 1C and D, electrograms recorded from hESC-CM resemble the electrograms recorded from cultured neonatal mouse and rat ventricular myocytes (FIG. 1E). This technology will enable us to investigate important aspects of hESC-CM physiology, such as the time course of electrophysiological development and maturation, the responsiveness to hormones, neurotransmitters, and drugs, as well as the effects of pathological states such as hypoxia. An important utilization of the MEA setup is that it enables us to study the developmental changes in hESC-CM electrophysiological properties, focusing on

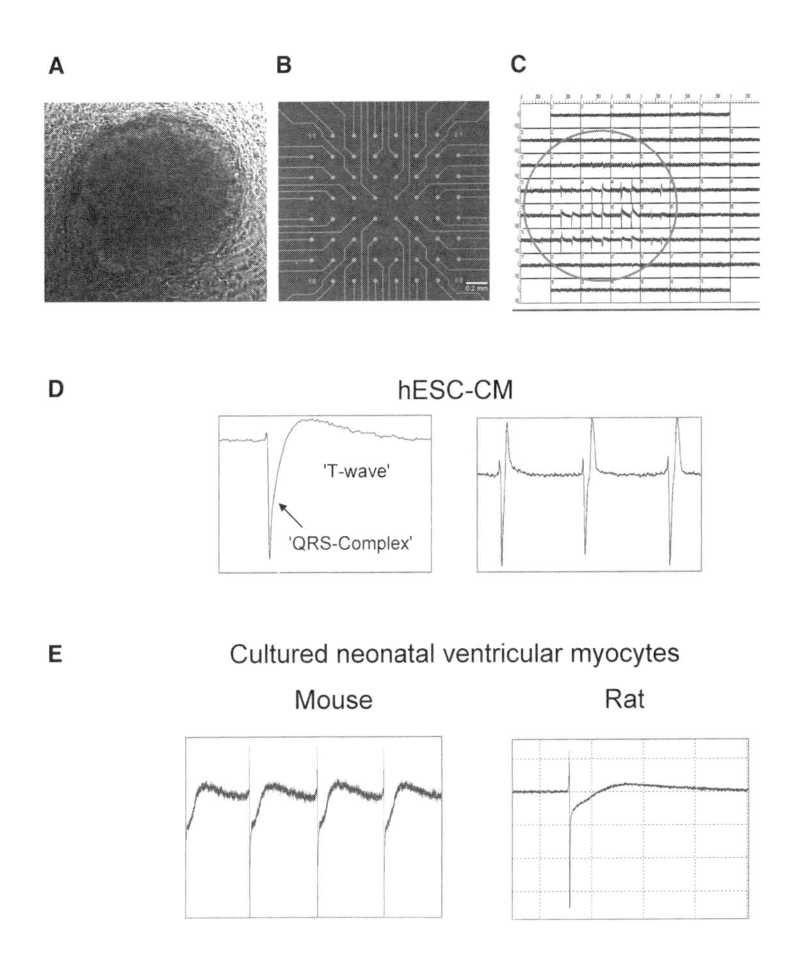

FIGURE 1. Electrophysiological recordings from hESC-CM. (**A**) A typical EB containing a spontaneously contracting area. (**B**) The Micro-Electrode-Array (MEA) electrode layout. (**C**) Electrograms recorded from hESC-CM, which occupied the central area of the electrode array. (**D**) Two representative electrograms recorded from 45- to 50-day-old hESC-CM. (**E**) Electrograms recoded from cultured neonatal mouse and rat ventricular myocytes [in color in Annals Online].

rhythm, activation, and repolarization. In the experiment depicted in FIGURE 2, EB were created from H9.2 embryonic stem cell line, and kept in suspension in low attachment petri dishes containing EB differentiation medium. Thereafter, seven-day-old EB were plated onto 24-well plates coated with collagen type I. Contracting areas (which usually appeared on day 9) were mechanically dissected with a microscalpel and transferred to MEA for electrophysiological recordings. As seen in FIGURE 2, the electrogram QRS amplitude, QRS dV/dt, and conduction velocity increased between day one and day three after plating, indicating that the electrophysiological maturation proceeds on the MEA.

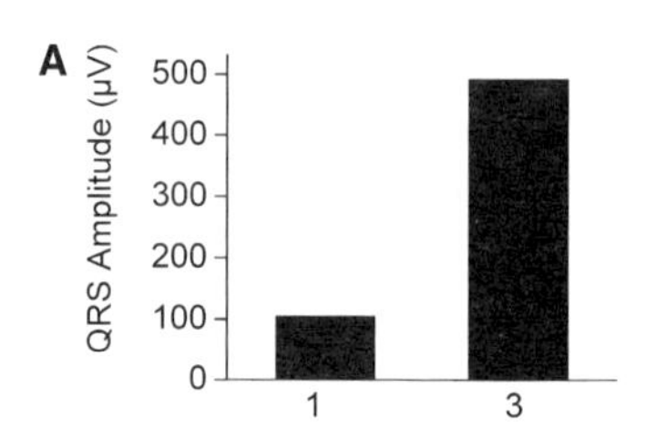

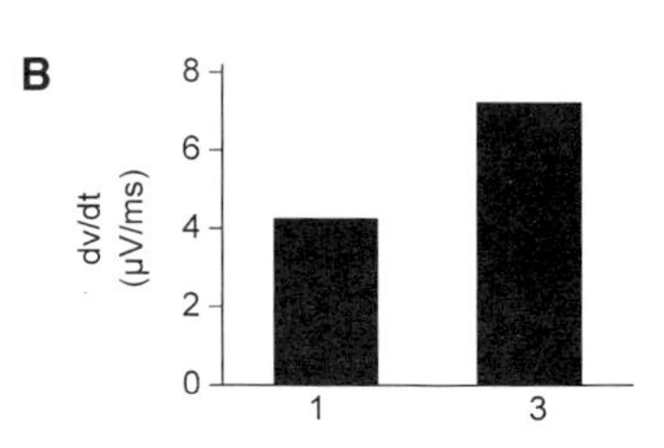

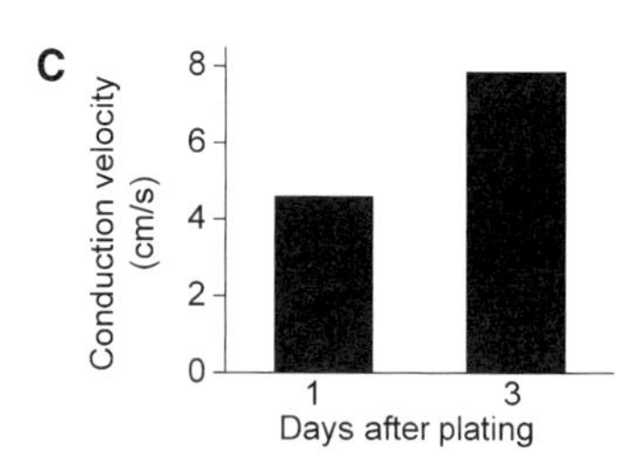

FIGURE 2. The electrogram's QRS amplitude (**A**), QRS dV/dt (**B**), and the conduction velocity (**C**) increase between day 1 and 3 after plating. See text for more details.

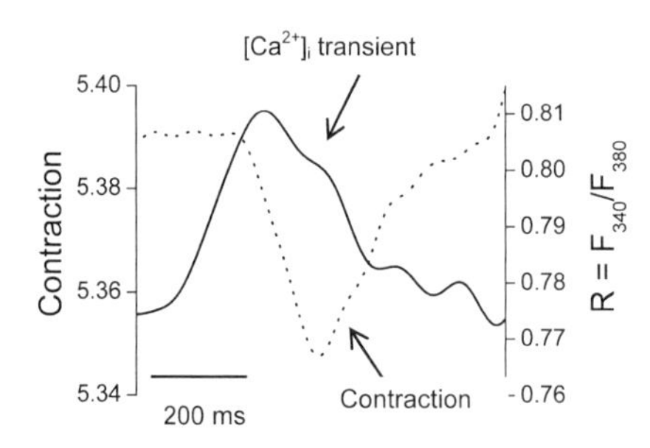

FIGURE 3. $[Ca^{2+}]_i$ transients and length changes recorded simultaneously from a spontaneously contracting EB.

The Ca^{2+} Handling Machinery: Does Sarcoplasmic Reticulum Ca^{2+} Release Contribute to the Contraction in hESC-CM?

To investigate the Ca^{2+} handling machinery in hESC-CM, we measured key events of the E-C coupling in spontaneously beating and in electrically paced EB: the $[Ca^{2+}]_i$ transient (measured by fura-2 fluorescence) and contraction (measured by motion edge detector). As seen in FIGURE 3, the 45-day-old EB displays normal functionality of the E-C coupling machinery. In the mature myocardium, the major source for Ca^{2+} ions required for contraction is Ca^{2+} induced Ca^{2+} release (CICR),

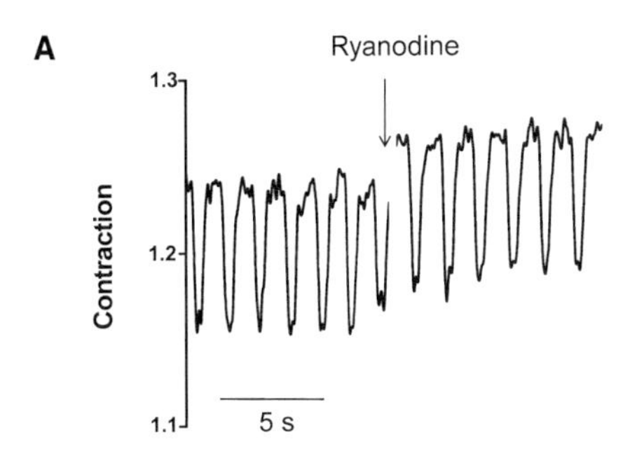

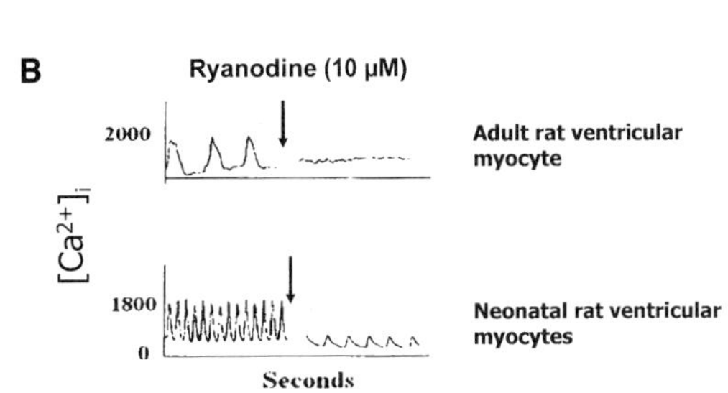

FIGURE 4. The effect of ryanodine on the contraction in hESC-CM. (**A**) Representative length changes of hESC-CM stimulated at 0.5 Hz before and 25 min after superfusion with ryanodine (10 μmol/l) dissolved in Tyrode's solution. (**B**) Ryanodine markedly reduces $[Ca^{2+}]_i$ transient amplitude in adult rat ventricular myocytes and in cultured neonatal rat ventricular myocytes. (Adapted from Takekura *et al.*[27])

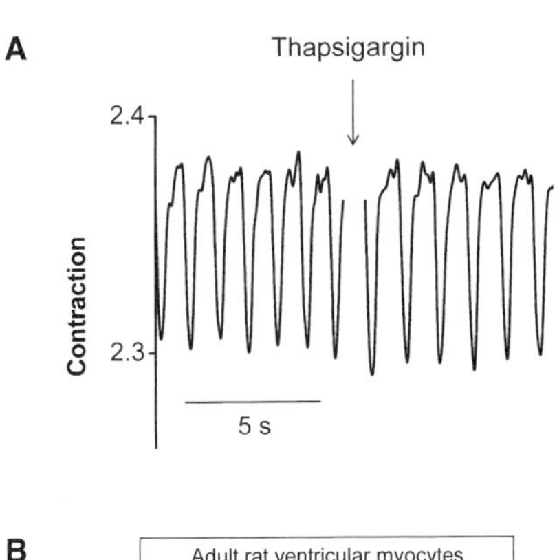

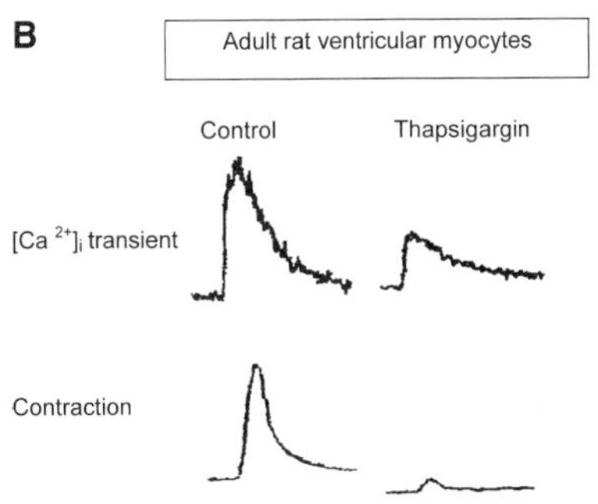

FIGURE 5. The effect of thapsigargin on the contraction in hESC-CM. (**A**) Representative length changes of hESC-CM stimulated at 0.5 Hz before and 25 min after superfusion with thapsigargin (10 nmol/l) dissolved in Tyrode's solution. (**B**) Thapsigargin markedly reduces $[Ca^{2+}]_i$ transient amplitude in adult rat ventricular myocytes. (Adapted from Meissner.[28])

which in the adult mammalian myocardium accounts for ~70% of the entire Ca^{2+} ions utilized by the contractile machinery (reviewed in Ref. 26). To test the contribution of Ca^{2+} release from the SR to the contraction in hESC-CM, we tested the effects of (1) ryanodine (the RyR blocker), causing a prominent negative inotropic effect in mature myocardium;[27] and (2) thapsigargin, the SERCA inhibitor causing marked negative inotropy in the adult myocardium by blocking Ca^{2+} uptake.[28] As shown in FIGURES 4 and 5, neither ryanodine nor thapsigargin affected the contraction of hESC-CM, suggesting that CICR does not contribute to the contraction in hESC-CM. Furthermore, in contrast to a variety of cardiac preparations, in hESC-CM caffeine did not cause the commonly seen increase in $[Ca^{2+}]_i$ (data not shown).

Force-Frequency Relationship

A fundamental property of the adult myocardium (for example, guinea pig, rabbit, and human[29–32]) are positive force-frequency relations, which constitute a mechanism utilized by the heart to increase cardiac output under exercise or stress conditions. To generate the force-frequency relations of hESC-CM, spontaneously beating EB were stimulated at 0.5, 1.0, 1.5, 2.0, and 2.5 Hz, and contraction measured. As depicted in FIGURE 6B, in contrast to the mature myocardium (FIG. 6A), hESC-CM displayed negative force-frequency relations. The mechanisms underlying the positive/negative force-frequency relations have been thoroughly investigated in a variety of animal species,[26,33,34] and it is commonly thought that the positive relations result from augmented SR $[Ca^{2+}]_i$ release at higher rates, which

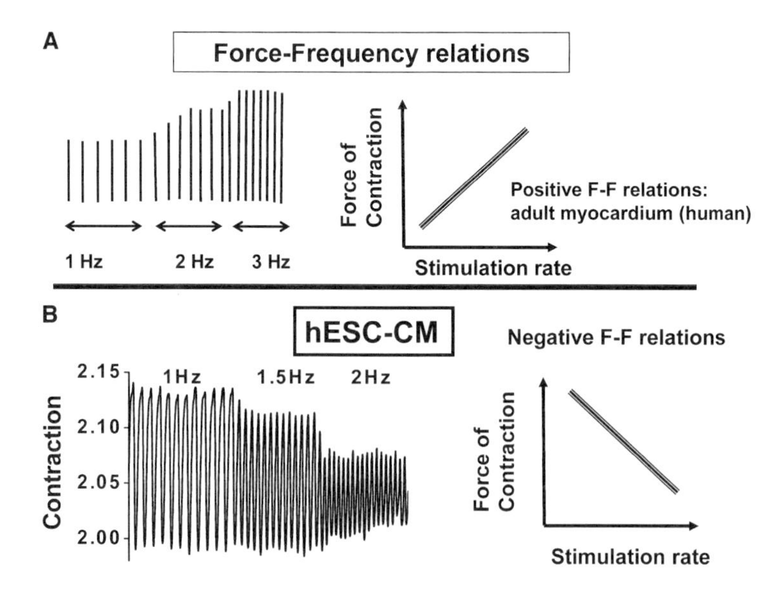

FIGURE 6. Force-frequency relations in adult and embryonic cardiomyocytes. (**A**) Schematic illustration of the positive force-frequency relations in the adult myocardium. (**B**) Negative force-frequency relations in a 40-day-old hESC-CM. The left side depicts actual recording of the hESC-CM contractions at different stimulation frequencies.

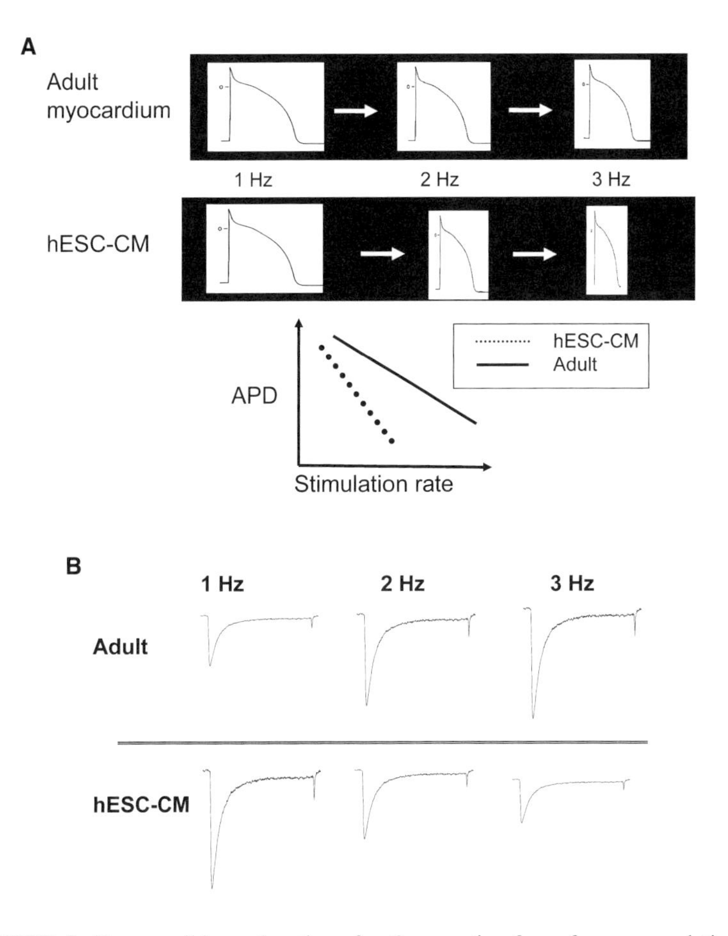

FIGURE 7. Two possible explanations for the negative force-frequency relations in hESC-CM. (**A**) In hESC-CM, the action potential duration (APD) versus stimulation rate relations are much steeper than in the adult myocardium. (**B**) In hESC-CM, in contrast to the adult myocardium, increasing the rate of stimulation decreases the L-type Ca^{2+} current ($I_{Ca,L}$).

generates increased force. Hence, the negative force-frequency relations in hESC-CM is in agreement of the ineffectiveness of ryanodine, thapsigargin, and caffeine, and supports the notion that at this embryonic stage the contraction depends only on extracellular Ca^{2+} rather than on SR Ca^{2+} release.

Evidently, because the SR appears to be dysfunctional in hESC-CM, it is not involved in the negative force-frequency relations in hESC-CM. Therefore, two possibilities can be considered (FIG. 7). (1) *Increased stimulation rate steeply decreases action potential duration in hESC-CM*. Numerous studies have shown that action potential duration (APD) is inversely related to the rate of stimulation, which in the absence of a concomitant increase in L-type Ca^{2+} ($I_{Ca,L}$), may decrease the force of contraction (FIG. 7A). If indeed in hESC-CM APD versus stimulation rate relations are much steeper than in the adult myocardium, such relations may result in negative force-frequency relations. (2) *Increased stimulation rate decreases the L-type Ca^{2+} current in hESC-CM*. Studies in experimental animals and human cardiac specimens have shown that the $I_{Ca,L}$ current is augmented by increasing the rate of stimulation—a mechanism contributing to the positive force-frequency relations. If in hESC-CM $I_{Ca,L}$, density is inversely dependent on the stimulation rate, then increasing the rate may cause negative force-frequency relations. Thus, determining the mechanisms underlying the negative force-frequency relations in hESC-CM and the process of switching into the mature behavior (positive force-frequency relations) are important aspects of our research on hESC-CM.

SUMMARY

In the present article, we discuss several key issues related to the utilization of hESC-CM for cardiac regeneration following irreversible myocardial damage, and show that the fundamental functional properties of hESC-CM differ from the mature myocardium. In view of the prominent limitations of current therapies (e.g., heart transplantation), novel therapeutic modalities for improving cardiac function and preventing heart failure are in critical demand and, therefore, hESC-CM should be considered as a viable option.

ACKNOWLEDGMENTS

This work was supported by a grant to Ofer Binah from the Rappaport Family Institute for Research in the Medical Sciences, and by a grant to Joseph Itskovitz-Eldor from the Ministry of Science and Technology, Israel.

REFERENCES

1. COHN, J.N., M.R. BRISTOW, K.R. CHIEN, *et al*. 1997. Report of the National Heart, Lung, and Blood Institute Special Emphasis Panel on Heart Failure Research. Circulation **95:** 766–770.
2. HUNT, S.A. 1998. Current status of cardiac transplantation. J. Am. Med. Assoc. **280:** 1692–1698.
3. LEE, M.S. & R.R. MAKKAR. 2004. Stem-cell transplantation in myocardial infarction: a status report. Ann. Inter. Med. **140:** 729–737.
4. GOLDENTHAL, M.J. & J. MARIN-GARCIA. 2003. Stem cells and cardiac disorders: an appraisal. Cardiovasc. Res. **58:** 369–377.
5. GERECHT-NIR, S., R. DAVID, M. ZARUBA, *et al*. 2003. Human embryonic stem cells for cardiovascular repair. Cardiovasc. Res. **58:** 313–323.
6. HASSINK, R.J., A.B. DE LA RIVIERA, C.L. MUMMERY, *et al*. 2003. Transplantation of cells for cardiac repair. J. Am. Coll. Cardiol. **41:** 711–717.
7. AMIT, M. & J. ITSKOVITZ-ELDOR. 2002. Derivation and spontaneous differentiation of human embryonic stem cells. J. Anat. **200:** 225–232.
8. AMIT, M., M.K. CARPENTER, M.S. INOKUMA, *et al*. 2000. Clonally derived human embryonic stem cell lines maintain pluripotency and proliferative potential for prolonged periods of culture. Dev. Biol. **227:** 271–278.
9. RICHARDS, M., C.Y. FONG, W.K. CHAN, *et al*. 2002. Human feeders support prolonged undifferentiated growth of human inner cell masses and embryonic stem cells. Nat. Biotechnol. **20:** 933–936.

10. AMIT, M., V. MARGULETS, H. SEGEV, *et al.* 2003. Human feeder layers for human embryonic stem cells. Biol. Reprod. **68:** 2150–2156.
11. XU, C., S. POLICE, N. RAO, *et al.* 2002. Characterization and enrichment of cardiomyocytes derived from human embryonic stem cells. Circ. Res. **91:** 501–508.
12. AMIT, M., C. SHARIKI, V. MARGULETS, *et al.* 2004. Feeder layer- and serum-free culture of human embryonic stem cells. Biol. Reprod. **70:** 837–845.
13. BARBASH, I.M., P. CHOURAQUI, J. BARON, *et al.* 2003. Systemic delivery of bone marrow-derived mesenchymal stem cells to the infarcted myocardium. Circulation **108:** 863–868.
14. DOWELL, J.D., M. RUBART, B.S. KISHORE, *et al.* 2003. Myocyte and myogenic stem cell transplantation in the heart. Cardiovasc. Res. **58:** 336–350.
15. HE, J.Q., Y. MA, Y. LEE, *et al.* 2003. Human embryonic stem cells develop into multiple types of cardiac myocytes. Circ. Res. **93:** 32–39.
16. PASSIER, R. & C. MUMMERY. 2003. Origin and use of embryonic and adult stem cells in differentiation and tissue repair. Cardiovasc. Res. **58:** 324–335.
17. KEHAT, I., L. KHIMOVICH, O. CASPI, *et al.* 2004. Electromechanical integration of cardiomyocytes derived from human embryonic stem cells. Nat. Biotechnol. **22:** 1282–1289.
18. LOVELL, M.J. & A. MATHUR. 2004. The role of stem cells for treatment of cardiovascular disease. Cell Prolif. **37:** 67–87.
19. STRAUER, B.E. & R. KORONOWSKI. 2003. Stem cell therapy in perspective. Circulation **107:** 929–934.
20. VAN DER HYDEN, M.A.G., J. HESCHELER, & C. MUMMERY. 2003. Spotlight on stem cells—makes old hearts fresh. Cardiovasc. Res. **58:** 241–245.
21. LI, R.K., Z.Q. JIA, R.D. WEISEL, *et al.* 1996. Cardiomyocyte transplantation improves heart function. Ann. Thorac. Surg. **62:** 654–660.
22. LI, R.K., T.M. YAU, R.D. WEISEL, *et al.* 2000. Construction of a bioengineered cardiac graft. J. Thorac. Cardiovasc. Surg. **119:** 368–375.
23. SAKAI, T., R.K. LI, R.D. WEISEL, *et al.* 1999. Autologous heart cell transplantation improves cardiac function after myocardial injury. Ann. Thorac. Surg. **68:** 2074–2080.
24. SCORSIN, M., F. MAROTTE, A. SABRI, *et al.* 1996. Can grafted cardiomyocytes colonize peri-infarct myocardial areas? Circulation **94:** II337–340.
25. MEIRY, G., Y. REISNER, Y. FELD, *et al.* 2001. Evolution of action potential propagation and repolarization in cultured neonatal rat ventricular myocytes. J. Cardiovasc. Electrophysiol. **12:** 1269–1277.
26. BERS, D.M. 2002. Cardiac excitation-contraction coupling. Nature **415:** 198–205.
27. POINDEXTER, B.J., J.R. SMITH, L.M. BUJA, *et al.* 2001. Calcium signaling mechanisms in dedifferentiated cardiac myocytes: comparison with neonatal and adult cardiomyocytes. Cell Calcium **30:** 373–382.
28. KIRBY, M.S., Y. SAGARA, S. GAA, *et al.* 1992. Thapsigargin inhibits contraction and Ca^{2+} transient in cardiac cells by specific inhibition of the sarcoplasmic reticulum Ca^{2+} pump. J. Biol. Chem. **267:** 12545–12551.
29. FAGGIANO, P. & W.S. COLUCCI. 1996. The force-frequency relation in normal and failing heart. Cardiology **41:** 1155–1164.
30. GWATHMEY, J.K., M.T. SLAWSKY, R.J. HAJJAR, *et al.* 1990. Role of intracellular calcium handling in force-interval relationships of human ventricular myocardium. J. Clin. Invest. **85:** 1599–1613.
31. MULIERI, L.A., G. HASENFUSS, B. LEAVITT, *et al.* 1992. Altered myocardial force-frequency relation in human heart failure. Circulation **85:** 1743–1750.
32. PIESKE, B., B. KRETSCHMANN, M. MEYER, *et al.* 1995. Alterations in intracellular calcium handling associated with the inverse force-frequency relation in human dilated cardiomyopathy. Circulation **92:** 1169–1178.
33. ALPERT, N.R., B.J. LEAVITT, F.P. ITTLEMAN, *et al.* 1998. A mechanistic analysis of the force-frequency relation in non-failing and progressively failing human myocardium. Basic Res. Cardiol. **93**(Suppl. 1): 23–32.
34. BORZAK, S., S. MURPHY & J.D. MARSH. 1991. Mechanisms of rate staircase in rat ventricular cells. Am. J. Physiol. **260:** H884–H892.

The Assembly of Calcium Release Units in Cardiac Muscle

CLARA FRANZINI-ARMSTRONG,[a] FELICIANO PROTASI,[a,b]
AND PIERRE TIJSKENS[a]

[a] *Department of Cell and Developmental Biology, University of Pennsylvania, Philadelphia, Pennsylvania 19104, USA*

[b] *CeSI, Laboratory of Cellular Physiology, Center for Research on Aging, University G. d'Annunzio School of Medicine, 66023 Chieti, Italy*

ABSTRACT: Calcium release units (CRUs) are constituted of specialized junctional domains of the sarcoplasmic reticulum (jSR) that bear calcium release channels, also called ryanodine receptors (RyRs). In cardiac muscle, CRUs come in three subtypes that differ in geometry, but have common molecular components. Peripheral couplings are formed by a junction of the jSR with the plasmalemma; dyads occur where the jSR is associated with transverse (T)-tubules; corbular SR is a jSR domain that is located within the cells and bears RyRs but does not associate with either plasmalemma or T-tubules. Using transmission electron microscopy, this study followed the formation of CRUs and their accrual of four components: the L-type channel dihydropyridine receptors (DHPRs) of plasmalemma/T-tubules; the RyRs of jSR; triadin (Tr) and junctin (JnC), two homologous components of the jSR membrane; and calsequestrin (CSQ), the internal calcium binding proteins. During differentiation, peripheral couplings are formed first and the others follow. RyRs and DHPRs are targeted to subdomains of the CRUs that face each other and are acquired in a concerted manner. Overexpressions of either junction (JnC or Tr) and of CSQ, singly or in conjunction, shed light on the specific role of JnC in the structural development, organization, and maintenance of jSR cisternae and on the independent synthetic pathways and targeting of JnC and CSQ. In addition, the structural cues provided by the overexpression models allow us to define sequential steps in the synthetic pathway for JnC and CSQ and their targeting to the CRUs of differentiating myocardium.

KEYWORDS: calsequestrin; dihydropyridine; excitation–contraction coupling; junctin; ryanodine; sarcoplasmic reticulum; SERCA; transverse tubules; triadin

INTRODUCTION

All cells have internal calcium storage compartments that accumulate calcium by means of calcium pumps of the sarco/endoplasmic reticulum Ca^{2+}-ATPase (SERCA) family.[1,2] Muscle fibers have an extensive Ca^{2+} storage compartment, the sarcoplasmic reticulum (SR). The SR in both skeletal and cardiac muscle is rich in ryanodine

Address for correspondence: Prof. Clara Franzini-Armstrong, University of Pennsylvania, Cell and Developmental Biology, Philadelphia, PA 19104, USA. Voice: 215-898-3345; fax: 215-573-2170.
armstroc@mail.med.upenn.edu

Ann. N.Y. Acad. Sci. 1047: 76–85 (2005). © 2005 New York Academy of Sciences.
doi: 10.1196/annals.1341.007

receptors (RyRs),[3] mostly localized in specialized junctional domains (jSR).[4] In addition, cardiac SR has detectable levels of 1,4,5-trisphosphate (IP3) receptors.[5–7] Thus, release of calcium from the cardiac SR occurs through the two related calcium release channels, the IP3 receptors[8] and the ryanodine receptors, or RyRs,[9,10] but the latter predominates in most myocardial cells.

Calcium release via RyRs from the SR of cardiac muscle is part of a process termed excitation–contraction (E-C) coupling. This is initiated by depolarization of the plasma membrane and its invaginations the T-tubules. Depolarization is sensed by calcium channels, the dihydropyridine receptors (DHPRs), located in the plasma membrane and T-tubules. In cardiac, as in skeletal, muscle DHPRs and RyRs are in close proximity to each other because they are clustered at sites where jSR, bearing arrays of RyRs, are closely apposed either to junctional domains of the plasma membrane (jPM) or to junctional domains of the T-tubules (jT), containing DHPRs.[11–14] jSR–jPM junctions are called peripheral couplings; jSR–jT junctions are usually in the form of dyads. In addition, myocardial cells contain corbular SR, an SR domain that is located within the cells and bears RyRs but is not junctional, that is, it does not associate with either plasmalemma or T-tubules.[15–17]

The junctions between jSR and plasmalemma/T-tubules function as discrete calcium release units (CRUs) and are the sites at which calcium sparks are detected.[18,19] Corbular SR domains also function as CRUs, although release of calcium in these jSR domains is not directly initiated by the action of DHPRs.

In addition to the two calcium channels (DHPRs and RyRs), CRUs contain a set of interrelated and interacting proteins calsequestrin (CSQ), triadin (Tr), and junctin (JnC) that are located in the jSR membrane where RyRs are present and in the jSR lumen immediately adjacent to it. CSQ is responsible for binding calcium in the SR lumen thus increasing its capacity as a calcium sink. Tr and JnC are SR membrane proteins that link to RyRs and to CSQ. RyRs, CSQ, Tr, and JnC form a supramolecular quaternary complex in the junctional sarcoplasmic reticulum.[20–25]

In the work summarized here, we took advantage of the fact that RyRs, DHPRs, and CSQ are directly visible in the electron microscope[26–28] and that the presence of Tr/JnC has a specific affect on the architecture of the jSR cisternae.[29] The cytoplasmic domains of RyRs form the "feet," linking jSR to jT or jPM.[26] DHPRs have been identified with large intramembrane particles seen in freeze-fracture replicas of jPM and jT in skeletal muscle.[26,27] A similar identification has been proposed for cardiac muscle.[12]

We address the following questions: (1) It is known that RyRs are detectable at very early stages in cardiac muscle.[29] Is the formation of calcium release units also an early developmental event? (2) Is there a developmental evolution in the type of CRU that is deployed during development? (3) What are the sequential steps involved in the association of RyRs and DHPRs with the junctions? (4) Is there an obligatory relationship between DHPRs and RyRs during the formation of calcium release units? (5) What steps occur during the maturation of the jSR cisternae from their initial formation to the adult arrangement?

METHODS

Standard techniques of thin section and freeze-fracture for electron microscopy were used. See Reference 12 for details.

RESULTS

The three forms of CRUs found in cardiac muscle are illustrated in FIGURE 1a. Peripheral couplings, dyads, and corbular SR, as defined in the introduction, are shown. All three junctions are formed by specialized cisternae of the SR, the jSR, that bear RyRs on one surface[30] and are filled by CSQ. The cytoplasmic domains of RyRs are visible as periodically disposed "feet"; CSQ forms a meshwork in the lumen of the jSR. The feet of corbular SR are constituted of the same RyR isoform (RyR2) as the feet of the other two types of CRUs.[31] The relative proportion of the three types of CRUs is quite variable. In frog myocardium, both atrium and ventricle, peripheral couplings are the only CRUs present.[32–34] The small diameter of the cells in the heart of this species permits activation by diffusion of calcium from the periphery. In the atrial myocardium of mammals and in all avian myocardium, there are no T-tubules, but corbular SR is quite abundant.[31,33] In these cells, internal release of calcium occurs both at the cell periphery, via the peripheral couplings, and internally, via corbular SR.[35] However, internal release requires fuller SR loading[36] and is secondary to the initial release from peripheral couplings. Finally, in the ventricle of adult mammalian heart, both peripheral couplings and T-tubule–based dyads are present, in addition to some corbular SR.

During myocardial differentiation, the various types of CRUs appear in sequential steps. The initial step in the differentiation of the membrane system is the formation of an internal network of membranes, the SR, which quite early docks to the plasma membrane to form peripheral couplings. With further development, the frequency of peripheral couplings increases, but internal CRUs do not appear until after birth (or hatching in birds) when the E-C coupling system adapts itself to the higher demands of larger and more active cells. As in skeletal muscle, we can define a maturation phase. This involves an increase in size and frequency of junctions, the acquisition of a regular disposition of junctions relative to the sarcomere, and, finally, the development of the internal extended jSR and/or of the dyads. The latter, of course, needs development of T-tubules. With the development of T-tubules and dyads, ventricular myocardium in larger mammals reduces its frequency of peripheral couplings. However, some peripheral couplings are always present, and they are more frequent in smaller mammals. Avian hearts and mammalian atria maintain a high level of peripheral couplings in the adult. An increase in size and frequency of all junctions and the acquisition of a regular disposition relative to the bands of the sarcomere are further maturation steps of the system.

Our results indicate that the initial docking of the SR membrane to the plasmalemma, and the formation of arrays of feet accompanied by the clustering of large particles (presumably DHPRs) in the junctional plasma membrane domains are not simultaneous events.[29,37] The initial peripheral couplings in both avian and mammalian myocardium contain few feet (FIG. 1b). Arrays of feet and clusters of large particles grow in parallel within these preformed junctions, until the junctional gap is completely filled by feet (FIG. 1c). An obligatory relationship between the locations and quantities of DHPRs and RyRs is apparently established early during development and maintained throughout.

The molecular machinery of CRUs is also differentially expressed in various myocardia and is assembled gradually in a complex process that involves the initial targeting of the necessary proteins, followed by their assembly and the maturation

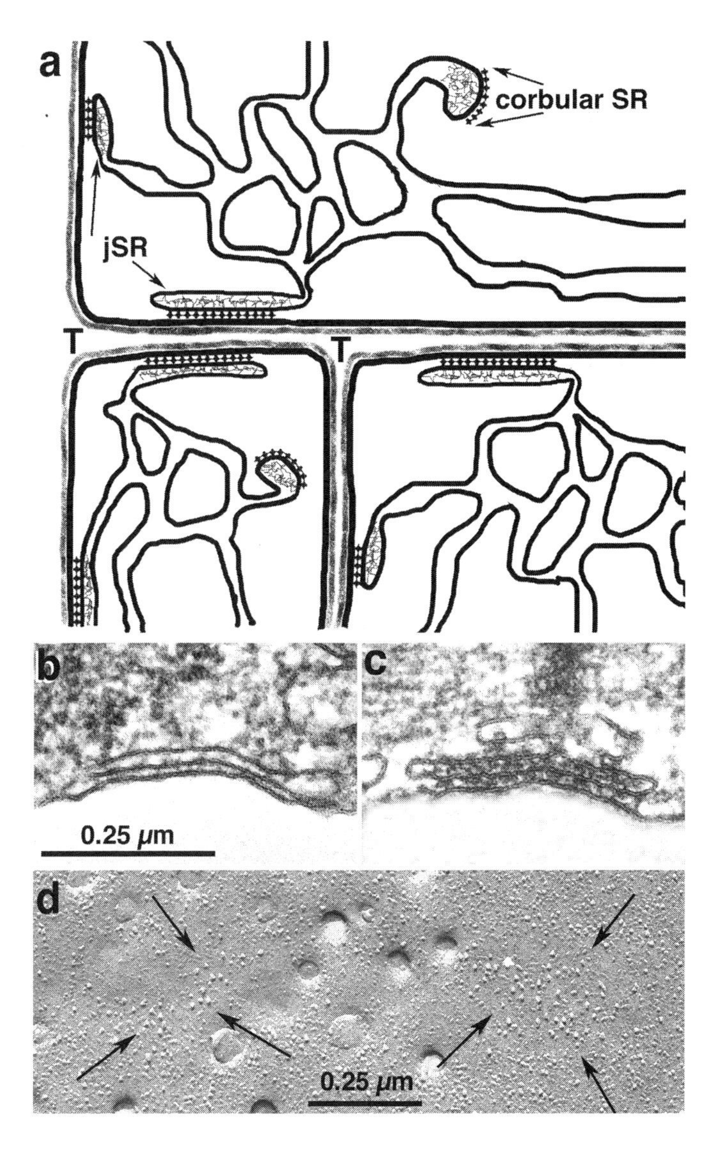

FIGURE 1. (**a**) Diagram illustrating the three types of calcium release units in cardiac muscle. Specialized junctional domains of the sarcoplasmic reticulum (jSR) containing calsequestrin (CSQ) and bearing feet on one surface are associated either with the plasmalemma at left, to form peripheral couplings or with T-tubules (T) to form dyads. Corbular sarcoplasmic reticulum (SR) contains feet and CSQ but is not associated with external membranes. (**b, c**) Peripheral couplings in chick myocardium. (**b**) A jSR vesicle that has docked but is poorly differentiated, from a chick embryo at 2.5 days after fertilization. (**c**) A fully differentiated jSR at 6 days after hatching. A nearby myocardial cell has been erased form the image for clarity. (**d**) Freeze-fracture of the plasmalemma from a myocardial cell in the left ventricle of a young chicken. *Arrows* surround junctional domains containing DHR particles.

of the CRUs. In most myocardia, peripheral couplings and dyads contain co-localized RyRs and DHPRs, but the ratio between the two proteins can be quite different. In avian ventricle from chicken and finch, well visible groups of DHPRs (detected as intramembranous particles in freeze-fracture replicas) are visible at sites of peripheral couplings (FIG. 1d; J.R. Sommer, F. Protasi, and C. Franzini-Armstrong, unpublished data).[12] A rough estimate indicates that the DHPR/RyR ratio in chicken myocardium is ~1:1, or somewhat lower than in skeletal muscle. In mammalian myocardium, however, even though a co-localization of DHPR and RyR has been detected,[11] very few particles bearing the signature of DHPRs are found at presumed sites of peripheral couplings (C. Franzini-Armstrong, unpublished data) suggesting a very low DHPR/RyR ratio. In frog myocardium, on the other hand, clusters of DHPRs are frequent at sites of peripheral couplings, while the feet content is very low in the atrium and basically nil in the ventricle, so that the DHPR/RyR ratio is quite high.[34] Conversely, at sites of corbular SR, RyRs exist without any relationship to DHPRs, so the DHPR/RyR ratio is zero.

Co-localization of DHPR and RyR was demonstrated by immunolabeling at 6 through 10 days *in ovo* in chick myocardium.[37] Feet (RyR) and clustered DHPR "particles" were detected by electron microscopy at even earlier ages (as little as 2.5 days *in ovo*).[37] Increase in the frequency and size of peripheral couplings and accrual of RyRs, DHPRs, and CSQ proceed gradually throughout the *in ovo* incubation, and it reaches a plateau just before hatching. SR/surface docking and acquisition of feet/DHPRs, however, are not simultaneous events: in the early stages, the junctions contain few feet/DHPR particles (FIG. 1b). Later, the junctional gap gets fully occupied by feet, and the arrays of particles reach the same size as the full jSR surface (FIG. 1c). At hatching, only 80% of the junctions are fully occupied by feet and the full complement of feet is present in all junctions 14 days later. This clearly shows that docking is the first step in the formation of CRUs and that accrual of the two E-C coupling proteins comes later and gradually. Interestingly, the area of jSR membrane that is occupied by feet and the apposed area of plasmalemma-occupied DHPRs grow in concert, indicating that the two proteins are not only added in approximately equal ratios, but that they occupy the same subdomain of the developing CRU.[37]

The jSR proteins CSQ, JnC, and Tr are also added gradually, but their detection is less direct. Whereas the presence of CSQ in the lumen of the SR is visible in electron micrographs due to its electron dense appearance after appropriate fixation and contrast enhancement,[28,38] Tr and JnC are not directly visible because they are small and partially embedded in the membrane. However, the two proteins link to CSQ and the ratio of Tr/JnC to CSQ strongly affects the disposition of CSQ itself and the overall shape of the jSR cisternae. This was shown in a series of studies originated in Larry R. Jones' laboratory, in which transgenic mice with a targeted overexpression of one of the three proteins were engineered. Comparison of the jSR vesicles in the three overexpressing mice, and in mice with double and triple overexpressions generated by further mating, is highly informative regarding both the structural roles of Tr and JnC and the delivery of CSQ and the two intrinsic proteins to the jSR cisternae.

The basal reference structure is that of the jSR cisternae in the left ventricle of adult wild-type mice. Characteristically, the cisternae are small and flat (resembling flat pancakes); they have an array of feet on one surface and they contain a small amount of CSQ that is condensed into a line parallel to the junctional membrane.[39] The line shows condensations at periodic intervals into nodules spaced with a peri-

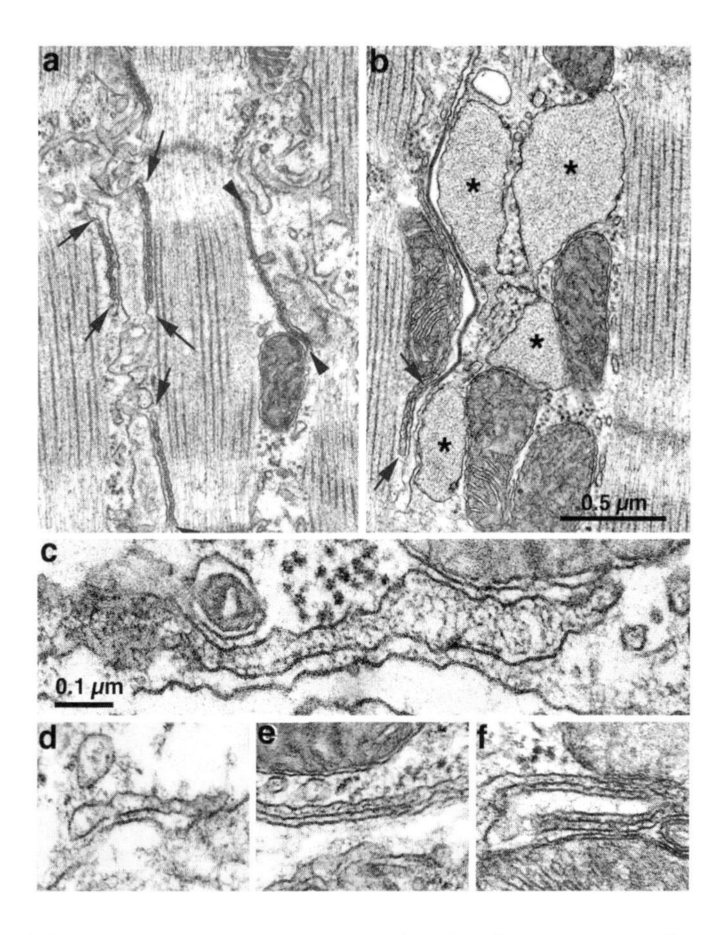

FIGURE 2. (**a**) Specialized junctional domains (jSR) in mouse myocardium overexpressing junctin (JnC). jSR (*between arrows*) is narrow and its calsequestrin (CSQ) content is tightly arranged. (**b**) jSR in mouse myocardium overexpressing both JnC and CSQ. *Asterisks* indicate sarcoplasmic reticulum vesicles that are filled by a fine meshwork of CSQ. One jSR vesicle (*between arrows*) shows a vesicle in which the effect of JnC prevails. (**c**) In the jSR of myocardium from mice that overexpress JnC, triadin (Tr), and CSQ, the latter protein is condensed at the edges, presumably by being linked to JnC and/or Tr. (**d–f**) Images of peripheral couplings from 3 days, 1-week- and 2-week-old mice, illustrating structural changes during the jSR maturation. (**d**) The jSR has a diffuse content, presumably CSQ, that is not well anchored. (**e, f**) CSQ becomes more tightly arranged, presumably due to the presence of JnC and/or Tr.

odicity equal to that of the feet. Overexpression of JnC induces a subtle but consistent tightening of the junctional cisternae of the SR, which acquire a narrower lumen and a very tightly clustered content of CSQ, resulting in a dense appearance (FIG. 2a).[23] Tr has the same structural effect as JnC when overexpressed (L.R. Jones, P. Tijskens, and C. Franzini-Armstrong, unpublished data).

Overexpression of CSQ increases the ratio of this protein to the other two and has essentially the opposite effect as overexpression of JnC/Tr. Excess of CSQ results in a considerable widening of the jSR cisternae and the change in appearance of their content, which becomes more finely granular and evenly dispersed (FIG. 2b).[40] Essentially, the excess CSQ is not bound to the membrane by Tr/JnC and it acquires a more diffuse disposition. Double overexpression of JnC and CSQ combines the effects of the two proteins: while the effect of the CSQ dominates, and the jSR vesicles are swollen with a finely granular content, there also is a clustering of the CSQ at the edges of the vesicles, presumably brought about by JnC.[29] Triple overexpression of JnC, Tr, and CSQ produces a slightly more marked affect of CSQ clustering and results in a decrease in average vesicle size (FIG. 2c) (L.R. Jones, P. Tijskens, and C. Franzini-Armstrong, unpublished data).

The visual clues given above allow a dissection of events in the targeting and delivery of CSQ and JnC within the SR of the differentiating cardiac myocyte and show different pathways for distribution and delivery to the jSR for CSQ and JnC. In CSQ^+ myocardium, CSQ filled vesicles first appear with high frequency in the trans-Golgi region of the cell, near the nuclei. In JnC^+ myocardium, on the other hand, flat, dense vesicles with the "JnC^+ phenotype" are present only in proximity of the surface membrane and T-tubules, even at early time points. In the double mutant, the structural effects of JnC overexpression described above are not seen in the large CSQ filled vesicles that are near the Golgi region. Thus, CSQ goes through the Golgi, while JnC may flow directly from the rough endoplasmic reticulum (RER) to the SR membrane and is directly targeted to the jSR.

A second observation is that at early time points, overexpressed CSQ does not uniformly fill the SR network but can be seen to fill vesicles that bud off from the SR and become junctional. Thus, CSQ is delivered to the SR, but within that compartment it is targeted to specific regions, the ones that develop into junctional SR.

The JnC and CSQ overexpressing myocardium further allow exploration of the relative timing for the delivery of the two junctional proteins to the jSR. Using the double clue of jSR vesicle width and of the condensed status of the luminal content, we find that 26% of jSR vesicles at 3 days (n, total vesicles = 170) show evidence of extra JnC, and that the percentage of "tight" vesicles increases to 86% (n = 268) at 1 week. For CSQ, on the other hand, we find that 69% of the jSR vesicles (n = 269) are enlarged at 3 days and 84% (n = 173) at 1 week. Thus, CSQ reaches the jSR in precedence to JnC. Further evidence for this statement comes from observations on normal development, described below.

Finally, the clues obtained from overexpression are useful in understanding the normal maturation of CRUs during myocardial differentiation. jSR profiles from sections of myocardium from postnatal day 3, 1-week- and 2-week-old wild-type mice show that at the earlier ages the majority of jSR vesicles have a variable diameter and are filled with a diffuse content and have few feet (FIG. 2d). A minority of vesicles at this age has a smaller width and is associated with an array of evenly spaced feet (not shown). At 1 and 2 weeks there is a clear maturation of the jSR cisternae involving both a reduction in the average width and an increasing condensation of the content, in addition to the increased visibility and extent of junctional feet arrays (FIG. 2e and f). Taken in the context of the structural effects of CSQ and JnC overexpression, these data indicate that CSQ is present in the jSR at earlier times than JnC or Tr.

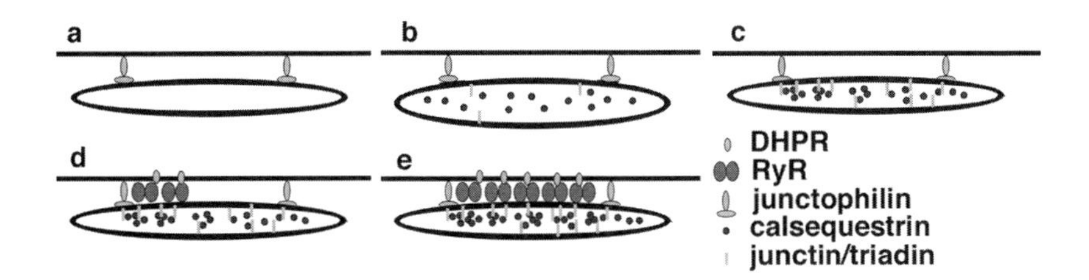

FIGURE 3. Diagrammatic representation of sequential steps in the assembly of the junctional complex proteins of myocardium (see discussion).

DISCUSSION AND SUMMARY

Using the morphological clues described above, it is clear that two sets of events are involved in the differentiation of the machinery dedicated to internal Ca^{2+} handling. One is the overall formation and positioning of CRUs. During the embryonic stages, CRUs are mostly restricted to the fiber edge and are in the form of peripheral couplings. Invaginations of T-tubules and the formation of internally located dyads, as well as the development of corbular SR, are later events.

The second set of events is the maturation of the CRUs. The molecular machinery of CRUs is assembled in a series of sequential steps, as illustrated in FIGURE 3. It is not surprising that an early step is the association of the SR with the plasmalemma. In the very early embryonic myocardium, this SR docking occurs in the virtual absence of RyRs and at a time when very little CSQ and even less Tr/JnC are present, suggesting that a different docking protein is involved (FIG. 3a). The junctophilins are very strong candidates for this role.[41,42] Junctophilin 2 is the isoform specifically expressed in cardiac muscle and it is necessary for its development. In the absence of RyR2, embryonic cardiac muscle shows abnormal peripheral couplings and calcium transients indicating that these two proteins are essential both to the formation and to the function of CRUs. Additionally, this also indicates that CRUs play a major role in activation of immature myocardium.[43] Ectopic expression of junctophilin 1 (the skeletal muscle isoforms) in myocardium increases the incidence of SR docking.[44] Expression and targeting of CSQ to CRUs is also an early embryonic event, (FIG. 3b) and this is followed more slowly by the arrival of JnC/Tr and the more tight organization of CSQ (FIG. 3c). RyR and DHPR are delivered to the CRUs in concert, and they tend to aggregate initially on one side of the junction (FIG. 3d). Since the two proteins are not in apparent direct association, this indicates that both are possibly interacting with the same, perhaps yet unidentified, component of the junction. The final step is the complete assembly of the DHPR/RyR and the RyR/Tr/JnC/CSQ macromolecular assemblies (FIG. 3e), allowing a more efficient coupling between surface membrane depolarization and calcium delivery to the myofibrils.

ACKNOWLEDGMENT

This work was supported by National Institutes of Health Grant RO1 HL-48093.

REFERENCES

1. MacLennan, D.H. & T. Toyofuku. 1992. Structure-function relationships in sarcoplasmic or endoplasmic reticulum type Ca^{++} pumps. Ann. N.Y. Acad. Sci. **671:** 1–10.
2. Pozzan, T., R. Rizzuto, P. Volpe, *et al.* 1994. Molecular and cellular physiology of intracellular calcium stores. Physiol. Rev. **74:** 595–636.
3. Meissner, G. 1994. Ryanodine receptor/Ca2+ release channels and their regulation by endogenous effectors. Annu. Rev. Physiol. **56:** 485–508.
4. Franzini-Armstrong, C. & A.O. Jorgensen. 1994. Structure and development of E-C coupling units in skeletal muscle. Annu. Rev. Physiol. **56:** 509–534.
5. Gozza, I., S. Schiaffino & P. Volpe. 1993. Inositol 1,4,5-trisphosphate receptor in heart: evidence for its concentration in Purkinje myocytes of the conduction system. J. Cell Biol. **121:** 345–353.
6. Perez P.J., J. Ramos-Franco, M. Fill, *et al.* 1997. Identification and functional reconstitution of the type 2 inositol 1,4,5-trisphosphate receptor from ventricular cardiac myocytes. J. Biol. Chem. **272:** 23961–23969.
7. Garcia, K.D., T. Shah & J. Garcia. 2004. Immunolocalization of type 2 inositol 1,4,5-trisphosphate receptors in cardiac myocytes from newborn mice. Am. J. Physiol. **287:** C1048–C1057.
8. Berridge, M.J. & R.F. Irvine. 1989. Inositol phosphates and cell signaling. Nature **341:** 197–205.
9. Inui, M., A. Saito & S. Fleischer. 1987. Isolation of the ryanodine receptor from cardiac sarcoplasmic reticulum and identity with the feet structure. J. Biol. Chem. **262:** 15637–15642.
10. Lai, F.A., H.P. Erickson, E. Rosseau, *et al.* 1988. Purification and reconstitution of the calcium release channel from skeletal muscle. Nature **331:** 315–319.
11. Carl, S.L., K. Felix, A.H. Caswell, *et al.* 1995. Immunolocalization of sarcolemmal dihydropyridine receptor and sarcoplasmic reticular triadin and ryanodine receptor in rabbit ventricle and atrium. J. Cell Biol. **129:** 673–682.
12. Sun, X-H., F. Protasi, M. Takahashi, *et al.* 1995. Molecular architecture of membranes involved in excitation–contraction coupling of cardiac muscle. J. Cell Biol. **129:** 659–671.
13. Scriven, D.R., P. Dan & E.D. Moore. 2000. Distribution of proteins implicated in excitation–contraction coupling in rat ventricular myocytes. Biophys. J. **79:** 2682–2691.
14. Scriven, D.R., A. Klimek, K.L. Lee, *et al.* 2002. The molecular architecture of calcium microdomains in rat cardiomyocytes. Ann. N.Y. Acad. Sci. **976:** 488–499.
15. Bossen, E.H. & J.R. Sommer. 1984. Comparative stereology of the lizard and frog myocardium. Tissue Cell **16:** 173–178.
16. Jorgensen, A.O., A.C. Shen, W. Arnold, *et al.* 1993. The Ca2+-release channel/ ryanodine receptor is localized in junctional and corbular sarcoplasmic reticulum in cardiac muscle. J. Cell Biol. **120:** 969–980.
17. Yamasaki, Y., Y. Furuya, K. Araki, *et al.* 1997. Ultra-high-resolution scanning electron microscopy of the sarcoplasmic reticulum of the rat atrial myocardial cells. Anat. Rec. **248:** 70–75.
18. Cheng, H., M.R. Lederer, W.J. Lederer, *et al.* 1996. Calcium sparks and [Ca2+]i waves in cardiac myocytes. Am. J. Physiol. **270:** C148–C159.
19. Kirk, M.M., L.T. Izu, Y. Chen-Izu, *et al.* 2003. Role of the transverse-axial tubule system in generating calcium sparks and calcium transients in rat atrial myocytes. J. Physiol. **547:** 441–451.
20. Knudson, C.M., K.K. Stang, A.O. Jorgensen, *et al.* 1993. Biochemical characterization and ultrastructural localization of a major junctional sarcoplasmic reticulum glycoprotein (triadin). J. Biol. Chem. **268:** 12637–12645.
21. Jones, L.R., L. Zhang, K. Sanborn, *et al.* 1995. Purification, primary structure, and immunological characterization of the 26 kDa calsequestrin binding protein (Junctin) from cardiac junctional sarcoplasmic reticulum. J. Biol. Chem. **270:** 30787–30796.
22. Guo, W., A.O. Jorgensen, L.R. Jones & K.P. Campbell. 1996. Biochemical characterization and molecular cloning of cardiac triadin. J. Biol. Chem. **271:** 458–465.

23. ZHANG, L., C. FRANZINI-ARMSTRONG, V. RAMESH, *et al.* 2000. Structural alterations in cardiac calcium release units resulting from overexpression of junctin. J. Mol. Cell. Cardiol. **33:** 233–247.
24. KOBAYASHI, Y.M., B.A. ALSEIKHAN & L.R. JONES. 2000. Localization and characterization of the calsequestrin-binding domain of triadin 1: evidence for a charged β-strand in mediating the protein–protein interaction. J. Biol. Chem. **275:** 17639–17646.
25. GLOVER L., K. CULLIGAN, S. CALA, *et al.* 2001. Calsequestrin binds to monomeric and complexed forms of key calcium-handling proteins in native sarcoplasmic reticulum membranes from rabbit skeletal muscle. Biochim. Biophys. Acta **1515:** 120–132.
26. BLOCK, B.A., T. IMAGAWA, K.P. CAMPBELL, *et al.* 1988. Structural evidence for direct interaction between the molecular components of the transverse tubule/sarcoplasmic reticulum junction in skeletal muscle. J. Cell Biol. **107:** 2587–2600.
27. TAKEKURA, H., L. BENNETT, T. TANABE, *et al.* 1994. Restoration of junctional tetrads in dysgenic myotubes by dihydropyridine receptor cDNA. Biophys. J. **67:** 793–804.
28. MEISSNER, G. 1975. Isolation and characterization of two types of sarcoplasmic reticulum vesicles. Biochim. Biophys. Acta **389:** 51–68.
29. TIJSKENS, P., L.R. JONES & C. FRANZINI-ARMSTRONG. 2003. Junctin and calsequestrin overexpression in cardiac muscle: the role of junctin and the synthetic and delivery pathways for the two proteins. J. Mol. Cell. Cardiol. **35:** 961–974.
30. DUTRO, S.M., J.A. AIREY, C.F. BECK, *et al.* 1993. Ryanodine receptor expression in embryonic avian cardiac muscle. Dev. Biol. **155:** 431–441.
31. JUNKER, J., J.R. SOMMER, M. SAR & G. MEISSNER. 1994. Extended junctional sarcoplasmic reticulum of avian cardiac muscle contains functional ryanodine receptors. J. Biol. Chem. **269:** 1627–1634.
32. PAGE, S.G. & R. NIEDERGERKE. 1972. Structures of physiological interest in the frog heart ventricle. J. Cell Sci. **11:** 179–203.
33. BOSSEN, E.H., J.R. SOMMER & R.A. WAUGH. 1978. Comparative stereology of the mouse and finch left ventricle. Tissue Cell **10:** 773–779.
34. TIJSKENS, P., G. MEISSNER & C. FRANZINI-ARMSTRONG. 2003. Location of ryanodine and dihydropyridine receptors in frog myocardium. Biophys. J. **84:** 1079–1092.
35. WOO, S.H., L. CLEEMANN & M. MORAD. 2003. Spatiotemporal characteristics of junctional and nonjunctional focal Ca2+ release in rat atrial myocytes. Circ. Res. **92:** 1–11.
36. HUSER, J., S.L. LIPSIUS & L.A. BLATTER. 1996. Calcium gradients during excitation–contraction coupling in cat atrial myocytes. J. Physiol. **494:** 641–651.
37. PROTASI, F., X-H SUN & C. FRANZINI-ARMSTRONG. 1996. Formation and maturation of calcium release units in developing and adult avian myocardium. Dev. Biol. **173:** 265–278.
38. JORGENSEN, A.O., A.C. SHEN & K.P. CAMPBELL. 1985. Ultrastructural localization of calsequestrin in adult rat atrial and ventricular muscle cells. J. Cell Biol. **101:** 257–268.
39. SOMMER, J.R. & E.A. JOHNSON. 1969. Cardiac muscle. A comparative ultrastructural study with special reference to frog and chicken hearts. Z. Zellforsch. Mikrosk. Anat. **98:** 437–468.
40. JONES, L.R., Y.J. SUZUKI, W. WANG, *et al.* 1998. Regulation of Ca2+ signaling in transgenic mouse cardiac myocytes overexpressing calsequestrin. J. Clin. Invest. **101:** 1385–1393.
41. TAKESHIMA, H., S. KOMAZAKI, M. NISHI, *et al.* 2000. Junctophilins: a novel family of junctional membrane complex proteins. Mol. Cell **6:** 11–22.
42. ITO, K., S. KOMAZAKI, K. SASAMOTO, *et al.* 2001. Deficiency of triad junction and contraction in mutant skeletal muscle lacking junctophilin type 1. J. Cell Biol. **154:** 1059–1067.
43. TAKESHIMA, H. 2002. Intracellular Ca2+ store in embryonic cardiac myocytes. Front. Biosci. **7:** d1642–d1652.
44. KOMAZAKI, S., M. NISHI & H. TAKESHIMA. 2003. Abnormal junctional membrane structures in cardiac myocytes expressing ectopic junctophilin type 1. FEBS Lett. **542:** 69–73.

Calcium Signaling in Cardiac Ventricular Myocytes

DONALD M. BERS AND TAO GUO

Loyola University Chicago, Department of Physiology, Maywood, Illinois 60153, USA

ABSTRACT: Calcium (Ca) is a multifunctional regulator of diverse cellular functions. In cardiac muscle Ca is a direct central mediator of electrical activation, ion channel gating, and excitation–contraction (E-C) coupling that all occur on the millisecond time scale. The key amplification step in E-C coupling is under tight control of very local [Ca]. Ca also directly activates signaling via kinases and phosphatases (e.g., Ca-calmodulin-dependent protein kinase [CaMKII] and calcineurin) that occur over a longer time scale (seconds to minutes), and the co-localization of these Ca-dependent modulators to their targets and to Ca is also critical in distinct signaling pathways. Finally, Ca-dependent signaling is also involved in long-term (minutes to hours/days) alterations in gene expression (or excitation-transcription coupling). These pathways are involved in hypertrophy and heart failure, and they can alter the expression of some of the key Ca regulatory proteins involved in E-C coupling and their regulation by kinases and phosphatases. There may again be physical microenvironments involved in this nuclear transcription, such that they sense a discrete Ca signal that is distinct from that involved in E-C coupling. In this way cells can use Ca signaling in multiple ways that function in spatially and temporally distinct manners.

KEYWORDS: calcium; calmodulin; coupling; dependent protein kinase; excitation–contraction coupling

INTRODUCTION

Calcium (Ca) functions as a ubiquitous intracellular messenger that regulates many different cellular processes in cardiac myocytes. In cardiac muscle, Ca directly contributes to electrical and contractile activity and is central in excitation–contraction (E-C) coupling.[1,2] Ca itself exerts important direct regulatory effects on events during E-C coupling. Ca can also modulate E-C coupling by activation of kinases and phosphatases, which can modulate gene expression responsible for hypertrophic signaling and heart failure.[3] This latter effect is sometimes called excitation–transcription (E-T) coupling. Changes in $[Ca]_i$ produce both acute and chronic effects on cardiac function. This raises questions about how cardiac myocytes distinguish different Ca signaling pathways during the large repetitive Ca transients during E-C coupling. The answer might lie in different localized Ca signaling complexes that

Address for correspondence: Prof. Donald Bers, Ph.D., Department of Physiology, Loyola University Chicago, 2160 S. First Ave, Maywood, IL 60153, USA. Voice: 708-216-1018; fax: 708-216-6308.
dbers@lumc.edu

Ann. N.Y. Acad. Sci. 1047: 86–98 (2005). © 2005 New York Academy of Sciences.
doi: 10.1196/annals.1341.008

have strategic locations and time frames of action. This subcellular localization and temporal coding of Ca signaling can be effectively coupled to specific physiological or pathological function. Among Ca-dependent proteins, calmodulin (CaM) serves as a multipurpose intracellular Ca sensor, decoding Ca signals and mediating many Ca regulated processes.[4]

Here we review the spatiotemporal control of Ca signaling occurring in three different time frames: (1) E-C coupling within milliseconds to seconds, (2) phosphorylation and dephosphorylation over seconds to minutes, and (3) E-T coupling on an even longer time scale. We will focus on Ca-CaM related modulation related to E-C and E-T coupling.

EXCITATION–CONTRACTION COUPLING

It is well known that Ca plays a pivotal role in cardiac E-C coupling.[1,2] Upon membrane depolarization, Ca entry through L-type Ca channels (or dihydropyrodine receptors [DHPRs]) triggers the Ca-induced Ca release from the sarcoplasmic reticulum (SR) via ryanodine receptors (RyRs) to amplify Ca current (I_{Ca}). The resultant rise in global $[Ca]_i$ activates the myofilaments to produce cardiac contraction. To allow cardiac muscle to relax, cytosolic [Ca] must decline quickly. This is accomplished mainly by transport mediated by the SR Ca ATPase (SERCA), which pumps Ca back into SR, and the Na/Ca exchanger (NCX), which extrudes Ca out of myocytes.[1]

The apparent global E-C coupling, however, is composed of thousands of synchronized local events. Both DHPRs and RyRs are co-localized in the junctional microdomain between the SR and sarcolemmal membrane (SL). At each junctional cleft, a cluster of ~100 individual RyRs and ~25 DHPRs constitute a local SR Ca-release unit called a junction (or couplon). This local functional unit concept is substantiated by identification of Ca sparks which originate from a local cluster of RyRs in the SR, occurring spontaneously in resting myocytes, which can also be spatially resolved during the E-C coupling upon block of most of the I_{Ca} (so the individual events are less overlapping spatially).[5–7] Ca influx through I_{Ca} (even a single channel) raises local $[Ca]_i$ from 0.1 to >10 μM, and local Ca release from a cluster of RyRs further increases local cleft [Ca] to >100 μM, whereas global $[Ca]_i$ only reaches ~1 μM (at a later time). A critical aspect of this discrete local signaling is that as local $[Ca]_i$ declines between junctions, it is not sufficient to trigger SR Ca release at neighboring junctions 1–2 μm away.[1] However, physiologically these local SR Ca release events are synchronized by action potentials and simultaneous activation of I_{Ca} at all junctions to produce a relatively homogenous increase in $[Ca]_i$ throughout the cytosol.[1] Therefore, Ca sparks are the elementary units of SR release both at rest and during the normal Ca transient during E-C coupling.[1,5,8]

ACUTE MODULATION OF E-C COUPLING

Ca is not only the central player of ECC, but is also a critical modulator of its own signaling pathway via Ca-dependent proteins. In this section we focus on acute regulatory mechanisms in the millisecond-second time frame during E-C coupling. One of the regulated proteins is the myofilament protein troponin C, which is the main

Ca target responsible for switching on the contractile machinery in cardiac muscles.[2] Another ubiquitous Ca sensor is CaM, which deciphers Ca signals and modulates E-C coupling directly and indirectly through downstream pathways.[4]

The L-type Ca channel is one of the targets modulated by Ca-CaM. I_{Ca} carries most of the Ca entry into the cardiac myocyte during the action potential, and it inactivates rapidly during each pulse. I_{Ca} inactivation is highly Ca-dependent and affected by both Ca influx and SR Ca release, such that it provides a negative feedback to limit the integrated I_{Ca}.[4,9] Since this Ca-dependent inactivation also occurs even with the presence of cytosolic Ca buffering, it is clear that Ca influx through the DHPR itself can exert a highly local inactivation on the channels. It is now clear that Ca-dependent I_{Ca} inactivation is mediated by CaM, which is pre-bound during diastole to the carboxy-terminal region (a region between an EF-hand and IQ domain) of the $\alpha 1$ subunit of L-type Ca channel.[10–16] Upon depolarization, Ca that enters and is released by RyRs binds to the C-terminal of pre-bound CaM, which then interacts with IQ domain of the Ca channel to produce inactivation.[10–16] Ca-CaM also produces Ca-dependent I_{Ca} facilitation via Ca-calmodulin-dependent protein kinase II (see following text). Both of these are very local microdomain regulatory phenomena.

RyRs are homotetramers, with each of the 4 subunits having a molecular weight of ~560 kDa.[2,8] RyR2 is the predominant form in heart, and it is both the SR Ca release channel and a scaffolding protein that strategically anchors numerous regulatory proteins to the junctional complex (the critical site where ECC occurs).[17] These regulatory proteins include CaM, Ca-calmodulin dependent protein kinase II (CaMKII), cAMP-dependent protein kinase (PKA, anchored via mAKAP), phosphatase 1 (PP1 anchored via spinophilin), phosphatase 2A (PP2A anchored via PR130), FK-506 binding protein, sorcin, calsequestrin, junctin, triadin, and homer.[17] All of these proteins are anchored to RyR to form a macromolecular complex, and some of them may modulate the function of RyR itself highly locally.[2,17] CaM is such an example. CaM binding to RyR2 is Ca-dependent, because higher [Ca] increases CaM:RyR2 stoichiometry and slows down the dissociation of CaM from the RyR (from ~40 s at low [Ca] to ~9 min at 100 μM [Ca]).[18,19] CaM is reported to decrease RyR2 opening at all [Ca] (even though not all reports support that) and shift the Ca-dependence of RyR2 activation to higher [Ca].[19,20] This is also the case when Mg-ATP is included and when redox-state is altered, but the difference is overcome at very high [Ca] and [ATP].[19] In our hands, CaM inhibits Ca spark frequency in saponin-permeabilized mouse ventricular myocytes in a dose-dependent manner and K_i is ~100 nM (unpublished data). CaM may also affect RyR activity via CaMKII, which we will address later.

Ca also activates the RyR2 directly from the cytosolic side, but at very high [Ca] it causes inactivation. Although part of this could be related to CaM, the mM levels of Ca (or Mg) required suggest that it is not CaM-dependent. Intra-SR [Ca] also modulates RyR gating (high SR [Ca] activates) and this may be important in the triggering of "spontaneous" SR Ca release associated with Ca overload and arrhythmias.[21–25] This effect may be mediated by intra-SR Ca-binding protein calsequestrin.[26] Regulation of RyR gating by luminal SR [Ca] may also be critical in the termination of SR Ca release, which is otherwise a positive feedback of Ca-induced Ca release.[27] That is, as intra-SR [Ca] declines, the RyR shuts off without completely emptying the SR of Ca.

$[Ca]_i$ also regulates NCX function by binding to an allosteric regulatory site on the large intracellular loop of the NCX.[28,29] Ca binding at this site (possibly to the NCX itself) with an apparent K_m of ~150 nM $[Ca]_i$ in myocytes is required for activation of the protein to participate in Ca transport. As for the Ca channel, this regulation may be part of a Ca feedback system, where the NCX is activated more strongly to extrude Ca when $[Ca]_i$ is relatively high. It may also be regulated by local subsarcolemmal $[Ca]_i$ rather than global $[Ca]_i$.[1]

PHOSPHORYLATION OF PROTEINS IN E-C COUPLING

Three Ca-CaM-dependent enzymes that mediate phosphorylation and dephosphorylation reactions have significant effects on cardiac function: CaMKII, calcineurin (protein phosphatase 2B), and myosin light chain kinase (MLCK). They have different Ca sensitivity and localization, and therefore mediate different functions related to E-C coupling. The kinase or phosphatase reactions also incur a temporal delay, such that these mechanisms tend to work on the time frame of seconds to minutes.

MLCK is associated with actomyosin and phosphorylates regulatory myosin light chain (MLC2). This Ca-CaM-dependent phosphorylation of MLC2 in cardiac myocytes has two apparent consequences: (1) increase in Ca sensitivity of the myofilament and therefore an improvement in cardiac contractility,[30] and (2) facilitation of the actin and myosin filaments' incorporation into the scaffold to form organized sarcomeres during the cardiac hypertrophy.[31] MLCK is a crucial central mediator of E-C coupling in smooth muscle, but in cardiac muscle this pathway is thought to be a minor modulator.

CaMKII and ECC

CaMKII is a multifunctional CaMK that may modulate cardiac E-C coupling. The CaMKII holoenzyme consists of homomultimers (or heteromultimers) of 6–12 kinase subunits (and there are four closely related but distinct genes: α, β, γ, and δ). In cardiac myocytes the δ isoform is the main isoform (with some CaMKIIγ).[32–34] Each CaMKII subunit has an amino-terminal catalytic domain, a central regulatory domain (containing partially overlapping autoinhibitory and CaM-binding regions), and a C-terminal association domain responsible for oligomerization. Resting CaMKII is inhibited by the autoinhibitory domain that acts as a pseudosubstrate, preventing substrates from binding.[4,32] Upon activation by increased $[Ca]_i$, CaMKII locks itself into the activated state by auto-phosphorylation on the conserved Thr^{286} in the auto-inhibitory segment of adjacent CaMKII subunits.[32–35] Autophosphorylation at Thr^{286} results in an autonomous form of CaMKII (even after the dissociation of CaM) and increases CaM affinity of the kinase-CaM complex.[35] CaMKII remains fully active with CaM bound regardless of $[Ca]_i$ level[35] and 20% to 80% active after CaM dissociation from this autonomous state.[36,37] In heart, CaMKIIδ is the predominant isoform including δ_B (contains a nuclear localization signal [NLS]) localized to nucleus and δ_C localized to the cytosol. Different subcellular localizations may suggest that they have different Ca-CaM sensitivity, specific mode of activation, and distinct functions. In support of this idea, transient overexpression of wild-type CaMKIIδ_B (but not CaMKIIδ_C) in neonatal rat ventricular myocytes increased expression of atrial natriuretic peptide (a marker of cardiac hypertrophic signaling) and led to an enhanced transcriptional response to phenylephrine.[38] This suggests that the δ_B isoform is more related to gene regulation, which will be addressed later. Also, Bayer *et al.*[39] demonstrated that different CaMKII splice variants have different sensitivity to Ca oscillation. However, the relative roles of CaMKII isoforms and their splice variants in regulating cardiac function still remain to be extensively studied.[34]

CaMKII regulates almost all of components related to cardiac ECC. First, CaMKII phosphorylates L-type Ca channel and results in Ca-dependent I_{Ca} facilitation, which is typically observed as an increased I_{Ca} amplitude and slower inactivation over 2 to 5 pulses.[40–42] This positive I_{Ca} staircase is Ca-dependent and CaMKII-dependent, because this I_{Ca} dependent facilitation could be abolished by CaMKII inhibitory peptides and when Ba was the charge carrier. I_{Ca} facilitation is also a local event because it is still seen with 10 mM EGTA in the patch pipette. The molecular mechanism of facilitation is not clear as to whether the tethered CaM involved in I_{Ca} inactivation is the same one activating CaMKII involved in I_{Ca} facilitation. Indeed, site-specific mutation of isoleucine of the IQ domain that abolishes Ca/CaM dependent inactivation can either increase (Ile to Ala) or abolish (Ile to Glu) Ca-dependent I_{Ca} facilitation.[11] I_{Ca} inactivation is a rapid negative feedback limiting the Ca entry at a single heart beat, while I_{Ca} facilitation may limit the decrease of Ca channel availability from beat to beat at higher heart rate (on a longer time scale). Hence, I_{Ca} is tightly controlled by local Ca-CaM-CaMKII signaling.

Phosphorylation of Thr17 on PLB by CaMKII (or Ser16 by PKA) releases SERCA from the inhibitory influence of PLB and enhances SR Ca uptake.[43] It has also been suggested that CaMKII phosphorylation of PLB may be involved in frequency dependent acceleration of relaxation (FDAR).[44,45] Some studies even showed that direct CaMKII phosphorylation of SERCA stimulates its activity, although others failed to show the significant stimulatory function of CaMKII on SERCA.[4]

Finally, CaMKII also co-immunoprecipitates with RyR[46] and may alter its activity.[17] The specific sites on RyR2 phosphorylated by CaMKII are controversial. Witcher *et al.*[47] reported that the unique phosphorylation site was Ser-2809 on RyR2 which regulated channel activity, whereas Rodriguez *et al.*[48] suggested that CaMKII may phosphorylate at least four sites in addition to Ser-2809. Recently, Wehrens *et al.*[49] identified a CaMKII site on RyR2 at Ser2815 and a stoichiometry of RyR2 phosphorylation (~1.5) that is lower than that identified in Rodriguez *et al.*[48] Similarly, the real effect of phosphorylation of CaMKII is still unclear. Most studies are in isolated systems, including single channel recording and SR vesicles, and these showed CaMKII to either increase[47,49,50] or decrease[51] RyR opening. It is important to study RyR behavior in its native cellular environment. In intact voltage clamped ventricular myocytes endogenous CaMKII increased the amount of SR Ca release for a given SR Ca content and I_{Ca} trigger.[52] This effect was Ca- and CaMKII-specific because smaller conditioning Ca transients (which may not activate CaMKII) prevented KN-93 from having any effect. In addition, in transgenic mice overexpressing CaMKIIδ_C, fractional SR Ca release was enhanced and resting spontaneous SR Ca spark frequency was dramatically increased despite lower SR Ca load and diastolic $[Ca]_i$.[53] Moreover, both acute adenoviral overexpression of CaMKIIδ_C in adult rabbit myocytes and direct application of pre-activated CaMKII to permeabilized mouse myocytes showed the same activating effect on Ca sparks and fractional Ca release.[54,55] However, Wu *et al.*[56] showed that constitutively active CaMKII inhibited Ca transients, while a CaMKII inhibitor increased Ca transients. If CaMKII increases diastolic Ca leak (by activating RyR), Ca transients might be low secondary to decreased Ca content. Taken together, CaMKII-dependent RyR phosphorylation seems to have strong stimulatory effect on RyR activity.

Phosphatases and ECC

Ca-CaM-dependent phosphatase calcineurin may act on several proteins related to E-C coupling. Calcineurin was found to co-localize with RyR2 and inositol (1,4,5)-trisphosphate receptor ($InsP_3R$).[57,58] This interaction seemed to be Ca-dependent and mediated by FKBP, since it could be abolished by EGTA and FK506. Overexpression of a constitutively active calcineurin reduced caffeine-induced Ca release in myocytes, while overexpression of an endogenous calcineurin inhibitor or inactive form enhanced caffeine-induced Ca release.[59] Inhibition of calcineurin by CsA or deltamethrin led to repetitive Ca oscillation.[58] Similar results were observed with FK506 or rapamycin, which dissociate FKBP12 from RyR. Therefore, calcineurin may act as an endogenous phosphatase of RyR and fulfill a regulatory role on RyR activity. However, Marx *et al.*[60] did not find calcineurin to be part of the RyR macromolecular complex containing other phosphatases. Hence, the role of calcineurin on RyR activity still remains controversial.

Other phosphatases, PP1 & PP2A (which could participate in RyR dephosphorylation) are also associated with RyR2 via leucine zipper motifs.[61] PP1 and PP2A reduced SR Ca release channel activity in bilayers[62] and cellular SR Ca release for a given SR load and I_{Ca},[63] while phosphatase inhibitors increased the relationship.[64–66] This suggested that phosphatases decreased E-C coupling gain (when normalized for SR load & I_{Ca}). However, Terentyev *et al.*[67] found that exposure of permeabilized cardiac myocytes to PP1 and PP2A caused an acute increase in Ca spark frequency followed by depletion of the SR content. Therefore, the effect of phosphatases on RyR activity is still controversial.

Several lines of evidence have showed that PP1/PP2A may associate with L-type Ca channels and modulate their function. Intracellular dialysis of rat ventricular myocytes with exogenous PP2A decreased steady state L-type I_{Ca}.[63] Application of PP1/PP2A inhibitors such as okadaic acid and calyculin A activated L-type Ca channels[64,65] and net I_{Ca} reflected the balance of activity of endogenous PP1 and kinase activity (possibly PKC).[66] Moreover, PP1/PP2A dephosphrylation of PLB may also contribute to the reduced SR Ca content by decreasing SERCA Ca transport.[68–70] It is very important to consider the function of kinases and phosphatases in the local environment. For example, despite an increase in global phosphatase activity in heart failure, Marx *et al.*[60] found that the local phosphatase activity associated with RyR was reduced. In summary, almost all components of cardiac ECC are directly or indirectly subjected to local phosphorylation-dephosphorylation regulation, even though the mechanisms remain to be determined. This intermediate level of Ca-dependent regulation of E-C coupling thus also depends on local signaling complexes.

EXCITATION–TRANSCRIPTION COUPLING

Ca is believed to be a critical regulator of gene transcription and expression.[71] In heart, Ca signaling mediates hypertrophic responses of cardiomyocytes,[38,72] and it must be transduced to the nucleus. Recent studies indicated that Ca-CaM dependent pathways act as central mediators in this process. Overexpression of CaM in the heart of transgenic mice leads to severe cardiac hypertrophy.[72] Furthermore, W-7, the CaM inhibitor, blocked hypertrophy of primary cultured cardiac myocytes in response to electrical pacing and adrenergic stimulation.[73] Moreover, the expression of mutant CaM which could bind Ca but not activate downstream targets failed to produce cardiac hypertrophy. This suggests that Ca-CaM dependent pathways are critical in hypertrophic signaling. Among key Ca-CaM target pathways are Ca-CaM-CaMK-HDAC (histone deacetylase) and Ca-CaM-Calcineurin-NFAT (nuclear factor of T-cell activation), and these have attracted a lot of attention.[74–77]

Ca/CaM-CaMK-HDAC Pathway

CaMK has been suggested to modulate gene expression via different transcription factors, including activation protein 1 (AP-1), CAAT-enhancer binding protein, activating transcription factor (ATF-1), serum response factor (SRF), cAMP-response element binding protein (CREB), and myocytes enhancer factor 2 (MEF2).[34] CREB is a ubiquitous transcription factor which can be phosphorylated at Ser 133 by both CaMKII & CaMKIV, as well as at Ser 142 by CaMKIV. However, CREB phosphorylation was unaltered in transgenic mice overexpressing CaMKIIδ_B or CaMKIV, suggesting that phosphorylated CREB was not responsible for cardiac hypertrophy in these mice.[74,78] However, studies in crossbred transgenic mice expressing cardiac CaMKIV and a MEF2-reporter supported the idea that MEF2 can be a downstream target for CaMK signaling in hypertrophic heart.[74] Despite the fact that CaMK can phosphorylate MEF2D, it did not phosphorylate other MEF2 isoforms.[79] Therefore, direct MEF2 phosphorylation cannot be the main mechanism by which CaMK regulates gene transcription. But recent studies showed that HDACs, a family of transcriptional repressors, served as functional links between CaMK and MEF2.[80,81]

Class II HDACs, including HDAC4, 5, 7, and 9, are enriched in the heart and have a unique MEF2 binding domain. They contain a C-terminal catalytic domain and N-terminal extension that mediates interaction with MEF2. The N-terminal extensions also contain two conserved serines that when phosphorylated bind to a chaperone protein (14-3-3).[58] The actions of HDACs are opposed by histone acetyltransferases (HATs), which acetylate the N-terminal tails of core histones and result in chromatin relaxation and transcriptional activation. Therefore, the association of HDACs with MEF2 causes repression of MEF2 targeted genes and may be responsible for the low transcriptional activity of MEF2 in the adult myocardium.[74]

Ca/CaM can compete with the MEF2-binding domain in class II HDACs and result in the dissociation of MEF2-HDAC complex.[82] The MEF2-HDAC interaction can also be disrupted by phosphorylation of the two conserved serine residues on HDAC, resulting in nuclear export of the HDAC-14-3-3 complex, freeing MEF2 to drive downstream gene transcription. CaMK is such a class II HDAC kinase,[83] and in neurons the CaMK inhibitor KN-93 blocked nuclear export of HDAC5 in response to depolarization.[84] However, little is known about how CaMKIIδ (especially CaMKIIδ_B), the predominant cardiac isoform, regulates HDAC in heart. Also, a CaMK-independent mechanism to control class II HDAC function was suggested to exist in heart, in that the cardiac HDAC kinase(s) was resistant to pharmacological inhibitors of CaMK.[85]

We also have new data, which indicate that CaMKII may be part of a local signaling complex at the nuclear envelope that is involved in activating HDAC nuclear export. CaMKII associates with $InsP_3R$ type 2 in the nuclear envelope.[86] This creates a situation where hypertrophic neurohumoral agents (e.g., endothelin-1) can cause production of $InsP_3$ at the sarcolemma, which diffuses to the nuclear envelope to cause local Ca release from $InsP_3R$, and the activation of CaMKII which phosphorylates HDAC triggering its nuclear export. We have preliminary data that support most aspects of this novel local signaling pathway in E-T coupling,[87] although further study is warranted.

Ca-CaM-Calcineurin-NFAT Pathway

In contrast to CaMKII, which responds to high and transient Ca oscillation and its activity is dependent on Ca spike frequency, calcineurin has relative high CaM affinity (K_m~ 0.1 nM for calcineurin vs. 20–100 nM for CaMKII) and can be activated by low but sustained Ca plateaus.[88] In cardiomyocytes, a constant rise in intracellular Ca activates calcineurin, which dephosphorylates NFAT transcription factor. Dephosphorylated NFAT is imported into the nucleus where it works cooperatively with cardiac-restricted zinc finger transcription factor GATA4 to activate hypertrophic genes (e.g. B-type natriuretic factor).[83]

It has been shown that calcineurin was required for hypertrophy in cardiac myocytes induced by phenylephrine, angiotensin II, and endothelin-1, since cyclosporin A can block the hypertrophic effects of these hypertrophic agonists.[83] In addition, transgenic overexpression of constitutively activated calcineurin or NFATc4/ NFAT3 was also sufficient to induce massive cardiac hypertrophy that progressed to heart failure and sudden death.[89] Furthermore, many studies showed that calcineurin was activated in pressure overload- and exercise-induced hypertrophy.[75] However, others found either no change or decrease in cardiac calcineurin activity in those models.[75] Using systemic administration of calcineurin inhibitors to determine the role of calcineurin in cardiac hypertrophy induced by pressure overload has also produced contradictory results.[71] Using more specific calcineurin inhibitors yielded more clear results. Cain/Cabin-1, or A-kinase anchoring protein 79 (AKAP79), inhibited calcineurin activity and have been shown to reduce phenylephrine and angiotensin II-induced cardiomyocyte hypertrophy *in vitro*.[76] In addition, the response to hypertrophic stimuli is impaired in calcineurin-deficient mice and calcineurin dominant-negative transgenic mice.[76] Moreover, cardiac-specific overexpression of modulatory calcineurin-interacting protein 1 (MCIP1) prevents hypertrophy in response to calcineurin overexpression, isoproterenol infusion, exercise, and pressure overload.[76,90] It was also suggested that MEF2 is involved in the calcineurin pathway via a post-translational mechanism, and transcription factors NF-κB and Elk-1 may also be regulated by calcineurin.[34] In addition, the calcineurin/NFAT pathway synergized with CaMK/HDAC pathway to evoke an especially profound hypertrophy.[74]

This sort of E-T coupling discussed above may also involve altered transcription of key Ca transport and regulatory proteins such as SERCA, PLB, NCX, RyR2, CaM, and CaMKII. Thus, this can complete a longer time-frame feedback loop, where altered Ca signaling triggers the altered expression of genes which may either feed back to restore normal cardiac myocyte function or could contribute to exacerbation of hypertrophic or heart failure phenotypes.[1]

SUMMARY

Calcium works as a critical signaling molecule on very different time scales (milliseconds to hours/days) and in different processes including E-C and E-T coupling. There are key issues related to signaling partners and spatially discrete signaling which play major roles in how a myocyte can distinguish between Ca signals that activate SR Ca release in E-C coupling, activate muscle contraction, and activate gene transcription.

ACKNOWLEDGMENTS

This work was supported by National Institutes of Health Grants (HL-30077 and HL-64724 to D.M.B.) and an American Heart Association Student Fellowship (T.G.).

REFERENCES

1. BERS, D.M. 2002. Cardiac excitation-contraction coupling. Nature **415:** 198–205.
2. BERS, D.M. 2001. Excitation-Contraction Coupling and Cardiac Contractile Force. 2nd ed. Kluwer. Dordrecht, Netherlands.
3. BERRIDGE, M.J. 2003. Cardiac calcium signaling. Biochem. Soc. Trans. **31:** 930–933.
4. MAIER, L.S. & D.M. BERS. 2002. Calcium, calmodulin and calcium-calmodulin kinase II: heartbeat to heartbeat and beyond. J. Mol. Cell. Cardiol. **34:** 919–939.
5. CHENG, H., W.J. LEDERER & M.B. CANNELL. 1993. Calcium sparks: elementary events underlying excitation-contraction coupling in heart muscle. Science **262:** 740–744.
6. LOPEZ-LOPEZ, J.R., P.S. SHACKLOCK, C.W. BALKE, *et al.* 1995. Local calcium transients triggered by single L-type calcium channel currents in cardiac cells. Science **268:** 1042–1045.
7. LOPEZ-LOPEZ, J.R., P.S. SHACKLOCK, C.W. BALKE, *et al.* 1994. Local, stochastic release of Ca^{2+} in voltage clamped rat heart cells: visualization with confocal microscopy. J. Physiol. **480:** 21–29.
8. FILL, M. & J.A. COPELLO. 2002. Ryanodine receptor calcium release channels. Physiol. Rev. **82:** 893–922.
9. PUGLISI, J.L., W. YUAN, J.W. BASSANI, *et al.* 1999. Ca^{2+} influx through Ca^{2+} channels in rabbit ventricular myocytes during action potential clamp. Circ. Res. **85:** e7–e16.
10. ZUHLKE, R.D. & H. REUTER. 1998. Ca^{2+}-sensitive inactivation of L-type Ca^{2+}-channels depends on multiple cytoplasmic amino acid sequences of the α1C subunit. Proc. Natl. Acad. Sci. U.S.A. **95:** 3287–3294.
11. ZUHLKE, R.D., G.S. PITT, K. DEISSEROTH, *et al.* 1999. Calmodulin supports both inactivation and facilitation of L-type calcium channels. Nature **399:** 159–162.
12. DE LEON, M., Y. WANG, L. JONES, *et al.* 1995. Essential Ca^{2+} binding motif for Ca^{2+}-sensitive inactivation of L-type Ca^{2+} channels. Science **270:** 1502–1506.
13. PETERSON, B.Z., J.S. LEE, J.G. MULLE, *et al.* 2000. Critical determinants of Ca^{2+}-dependent inactivation within an EF-hand motif of L-type Ca^{2+} channels. Biophys. J. **78:** 1906–1920.
14. PATE, P., J. MOCHCA-MORALES, Y. WU, *et al.* 2000. Determinants for calmodulin binding on voltage-dependent Ca^{2+} channels. J. Biol. Chem. **275:** 39786–39792.
15. ERICKSON, M.G., B.A. ALSEIKHAN, B.Z. PETERSON, *et al.* 2001. Preassociation of calmodulin with voltage-gated Ca^{2+} channels revealed by FRET in single living cells. Neuron **31:** 973–985.
16. PITT, G.S., R.D. ZUHLKE, A. HUDMON, *et al.* 2001. Molecular basis of calmodulin tethering and Ca^{2+}-dependent inactivation of L-type Ca^{2+} channels. J. Biol. Chem. **276:** 30794–30802.

17. BERS, D.M. 2004. Macromolecular complexes regulating cardiac ryanodine receptor function. J. Mol. Cell Cardiol. **37:** 417–429.
18. FRUEN, B.R., J.M. BARDY, T.M. BYREM, *et al.* 2000. Differential Ca^{2+} sensitivity of skeletal and cardiac muscle ryanodine receptors in the presence of calmodulin. Am. J. Physiol. **279:** C724–C733.
19. BALSHAW, D.M., L. XU, N. YAMAGUCHI, *et al.* 2001. Calmodulin binding and inhibition of cardiac muscle calcium release channel (ryanodine receptor). J. Biol. Chem. **276:** 20144–20153.
20. YAMAGUCHI, N., L. XU, D.A. PASEK, *et al.* 2003. Molecular basis of calmodulin binding to cardiac muscle Ca^{2+} release channel (ryanodine receptor). J. Biol. Chem. **278:** 23480–23486.
21. BASSANI, J.W., W. YUAN & D.M. BERS. 1995. Fractional SR Ca release is altered by trigger Ca and SR Ca content in cardiac myocytes. Am. J. Physiol. **268:** C1313–C1319.
22. LUKYANENKO, V., I. GYÖRKE & S. GYÖRKE. 1996. Regulation of calcium release by calcium inside the sarcoplasmic reticulum in ventricular myocytes. Pfluegers Arch. Eur. J. Physiol. **432:** 1047–1054.
23. GYÖRKE, I. & S. GYÖRKE. 1998. Regulation of the cardiac ryanodine receptor channel by luminal Ca^{2+} involves luminal Ca^{2+} sensing sites. Biophys. J. **75:** 2801–2810.
24. SHANNON, T.R., K.S. GINSBURG, & D.M. BERS. 2000. Potentiation of fractional SR Ca release by total and free intra-SR Ca concentration. Biophys. J. **78:** 334–343.
25. POGWIZD, S.M., K. SCHLOTTHAUER, L. LI, *et al.* 2001. Sodium-Calcium exchange contributes to mechanical dysfunction and triggered arrhythmias in heart failure. Circ. Res. **88:** 1159–1167.
26. GYÖRKE, I., N. HESTER, L.R. JONES, *et al.* 2004. The role of calsequestrin, triadin, and junctin in conferring cardiac ryanodine receptor responsiveness to luminal calcium. Biophys. J. **86:** 2121-2128.
27. SHANNON, T.R., T. GUO & D.M. BERS. 2003. Ca scraps: local depletions of free [Ca] in cardiac sarcoplasmic reticulum during contractions leave substantial Ca reserve. Circ. Res. **93:** 40–45.
28. MATSUOKA, S., D.A. NICOLL, L.V. HRYSHKO, *et al.* 1995. Regulation of the cardiac Na^{+}-Ca^{2+} exchanger by Ca^{2+}. Mutational analysis of the Ca^{2+}-binding domain. J. Gen. Physiol. **105:** 403–420.
29. WEBER, C.R., K.S. GINSBURG, K.D. PHILIPSON, *et al.* 2001. Allosteric regulation of Na/Ca exchange current by cytosolic Ca in intact cardiac myocytes. J. Gen. Physiol. **117:** 119–131.
30. MORANO, I., C. BACHLE-STOLZ, H.A. KATUS, *et al.* 1985. Increased calcium sensitivity of chemically skinned human atria by myosin light chain kinase. FEBS Lett. **189:** 221–224.
31. AOKI, H., J. SADOSHIMA & K.S. IZUMO. 2000. Myosin light chain kinase mediates sarcomere organization during cardiac hypertrophy in vitro. Nature Med. **6:** 183–188.
32. HUDMON, A. & H. SCHULMAN. 2002. Structure-function of the multifunctional Ca/calmodulin-dependent protein kinase II. Biochem. J. **364:** 593–611.
33. HANSON, P.I., T. MEYER, L. STRYER, *et al.* 1994. Dual role of calmodulin in autophosphorylation of multifunctional CaM kinase may underlie decoding of calcium signals. Neuron **12:** 934–956.
34. ZHANG, T. & J.H. BROWN. 2004. Role of Ca^{2+}/calmodulin-dependent protein kinase II in cardiac hypertrophy and heart failure. Cardiovasc. Res. **15:** 476–486.
35. SCHULMAN H, P.I. HANSON & T. MEYER. 1992. Decoding calcium signals by multifunctional CaM kinase. Cell Calcium **13:** 401–411.
36. MEYER, T., P.I. HANSON, L. STRYER, *et al.* 1992. Calmodulin trapping by calcium-calmodulin-dependent protein kinase. Science **256:** 1199–1201.
37. MILLER, S.B. & M.B. KENNEDY. 1986. Regulation of brain type II Ca/calmodulin-dependent protein kinase by autophosphorylation: a Ca-trigger molecular switch. Cell **44:** 861–870.
38. RAMIREZ, M.T., X. ZHAO, H. SCHULMAN, *et al.* 1997. The nuclear δ_B isoform of Ca^{2+}/calmodulin-dependent protein kinase II regulates atrial natriuretic factor gene expression in ventricular myocytes. J. Biol. Chem. **272:** 31203–31208.

39. BAYER, K.U., P. DE KONINCK & H. SCHULMAN. 2002. Alternative splicing modulates the frequency-dependent response of CaMKII to Ca^{2+} oscillations. EMBO J. **21:** 3590–3597.
40. YUAN, W. & D.M. BERS. 1994. Ca-dependent facilitation of cardiac Ca current is due to Ca-calmodulin dependent protein kinase. Am. J. Physiol. **267:** H982–H993.
41. XIAO, R.P., H. CHENG, W.J. LEDERER, *et al.* 1994. Dual regulation of Ca/calmodulin kinase II activity by membrane voltage and by calcium influx. Proc. Natl. Acad. Sci. U.S.A. **91:** 9659–9663.
42. WU, Y., I. DZHURA, R.J. COLBRAN, *et al.* 2001. Calmodulin kinase and a calmodulin-binding 'IQ' domain facilitate L-type Ca^{2+} current in rabbit ventricular myocytes by a common mechanism. J. Physiol. **535:** 679–687.
43. MATTIAZZI, A. HOVE-MADSEN L & D.M. BERS. 1994. Protein kinase inhibitors reduce SR Ca transport in permeabilized cardiac myocytes. Am. J. Physiol. **267:** H812-H820.
44. BASSANI, R.A., A. MATTIAZZI & D.M. BERS. 1995. CaMKII is responsible for activity-dependent acceleration of relaxation in rat ventricular myocytes. Am. J. Physiol. **268:** H703–H712.
45. DESANTIAGO, J., L.S. MAIER & D.M. BERS. 2002. Frequency-dependent acceleration of relaxation (FDAR) in heart depends on CaMKII, but not phospholamban. J. Molec. Cell Cardiol. **34:** 975–984.
46. ZHANG, T, L.S. MAIER, N.D. DALTON, *et al.* 2003. The δC isoform of CaMKII is activated in cardiac hypertrophy and induces dilated cardiomyopathy and heart failure. Circ. Res. **92:** 912–919.
47. WITCHER, D.R., R.J. KOVACS, H. SCHULMAN, *et al.* 1991. Unique phosphorylation site on the cardiac ryanodine receptor regulates calcium channel activity. J. Biol. Chem. **266:** 11144–11152.
48. RODRIGUEZ, P., M.S. BHOGAL & J. COYLER. 2003. Stoichiometric phosphorylation of cardiac ryanodine receptor on serine-2809 by calmodulin-dependent kinase II and protein kinase A. J. Biol. Chem. **278:** 38593–38600.
49. WEHRENS, X.H.T., S.E. LENHART, S.R. REIKEN, *et al.* 2004. CaMKII phosphorylation regulates the cardiac ryanodine receptor. Circ. Res. **94:** e61–e70.
50. HAIN, J., H. ONOUE, M. MAYRLEITNER, *et al.* 1995. Phosphorylation modulates the function of the calcium release channel of sarcoplasmic reticulum from cardiac muscle. J. Biol. Chem. **270:** 2074–2081.
51. LOKUTA, A.J., T.B. ROGERS, W.J. LEDERER, *et al.* 1995. Modulation of cardiac ryanodine receptors of swine and rabbit by a phosphorylation–deposphorylation mechanism. J. Physiol. (London) **487:** 609–622.
52. LI, L., H. SATOH, K.S. GINSBURG, *et al.* 1997. The effect of Ca-calmodulin-dependent protein kinase II on cardiac excitation-contraction coupling in ferret ventricular myocytes. J. Physiol. **501:** 17–32.
53. MAIER, L.S., T. ZHANG, L. CHEN, *et al.* 2003. Transgenic CaMKIIδC overexpression uniquely alters cardiac myocyte Ca^{2+} handling, reduced SR Ca^{2+} load and activated SR Ca^{2+} release. Circ. Res. **92:** 904–911.
54. MAIER, L.S., T. ZHANG, T. SEIDLER, *et al.* 2003. CaMKδ_c alters calcium handling in isolated rabbit cardiac myocytes. Circulation **108:** IV–54.
55. GUO, T., T. ZHANG, J.H. BROWN, *et al.* 2004. Effects of CaMKII on cardiac Ca release channels in myocytes. Biophys. J. **86:** 241a.
56. WU, Y.J., R.J. COLBRAN & M.E. ANDERSON. 2001. Calmodulin kinase is a molecular switch for cardiac excitation-contraction coupling. Proc. Natl. Acad. Sci. U.S.A. **98:** 2877–2881.
57. CAMERON, A.M., J.P. STEINER, A.J. ROSKAMS, *et al.* 1995. Calcineurin associated with the inositol 1,4,5-trisphosphate receptor-FKBP12 complex modulates Ca^{2+} flux. Cell **83:** 463–472.
58. BANDYOPADHYAY, A., D.W. SHIN, J.O. AHN, *et al.* 2000. Calcineurin regulates Ryanodine receptor/Ca^{2+}-release channels in rat heart. Biochem. J. **352:** 61–70.
59. BULTYNCK, G., E. VERMASSEN, K. SZLUFCIK, *et al.* 2003. Calcineurin and intracellular Ca^{2+}-release channels: regulation or association? Biochem. Biophys. Res. Commun. **311:** 1181–1193.

60. MARX, S.O., S. REIKEN, Y. HISAMATSU, *et al.* 2000. PKA phosphorylation dissociates FKBP12.6 from the calcium release channel (ryanodine receptor): defective regulation in failing hearts. Cell **101:** 365–376.
61. MARX, S.O., S. REIKEN, Y. HISAMATSU, *et al.* 2001. Phosphorylation-dependent regulation of ryanodine receptors: a novel role for leucine/isoleucine zippers. J. Cell Biol. **153:** 699–708.
62. HESCHELER, J., G. MIESKES, J.C. RUEGG, *et al.* 1988. Effects of a protein phosphatase inhibitor, okadaic acid, on membrane currents of isolated guinea-pig cardiac myocytes. Pfluegers Arch. Eur. J. Physiol. **412:** 248–252.
63. DUBELL, W.H., W.J. LEDERER & T.B. ROGERS. 1996. Dynamic modulation of excitation-contraction coupling by protein phosphatases in rat ventricular myocytes. J. Physiol. **493:** 793–800.
64. NEUMANN, J., P. BOKNIK, S. HERZIG, *et al.* 1993. Evidence for physiological functions of protein phosphatases in the heart: evaluation with okadaic acid. Am. J. Physiol. Heart Circ. Physiol. **265:** H257–H266.
65. DUBELL, W.H., M.S. GIGENA, S. GUATIMOSIM, *et al.* 2002. Effects of PP1/PP2A inhibitor calyculin A on the E-C coupling cascade in murine ventricular myocytes. Am. J. Physiol. Heart Circ. Physiol. **282:** H38–H48.
66. DUBELL, W.H. & T.B. ROGERS. 2004. Protein phosphatase 1 and an opposing protein kinase regulate steady-state L-type Ca^{2+} current in mouse cardiac myocytes. J. Physiol. **556:** 79–93.
67. TERENTYEV, D., S. VIATCHENKO-KARPINSKI, I. GYORKE, *et al.* 2003. Protein phosphatases decrease sarcoplasmic reticulum calcium content by stimulating calcium release in cardiac myocytes. J. Physiol. **552:** 109–118.
68. MACDOUGALL, L.K., L.R. JONES & P. COHEN. 1991. Identification of the major protein phosphatases in mammalian cardiac muscle which dephosphorylate phospholamban. Eur. J. Biochem. **196:** 725–734.
69. STEENAART, N.A., J.R. GANIM, J. DI SALVO, *et al.* 1992. The phospholamban phosphatase associated with cardiac sarcoplasmic reticulum is a type 1 enzyme. Arch. Biochem. Biophys. **293:** 17–24.
70. BOKNIK, P., M. FOCKENBROCK, S. HERZIG, *et al.* 2000. Protein phosphatase activity is increased in a rat model of long-term β-adrenergic stimulation. Naunyn Schmiedebergs Arch. Pharmacol. **362:** 222–231.
71. HARDINGHAM, G.E. & H. BADING. 1998. Nuclear calcium: a key regulator of gene expression. Biometals **11:** 345–358.
72. GRUVER, C.L., F. DEMAYO, M.A. GOLDSTEIN, *et al.* 1993. Targeted developmental overexpression of calmodulin induces proliferative and hypertrophic growth of cardiomyocytes in transgenic mice. Endocrinology **133:** 376–388.
73. MCDONOUGH, P.M. & C.C. GLEMBOTSKI. 1992. Induction of natriuretic factor and myosin light chain-2 gene expression in cultured ventricular myocytes by electrical stimulation. J. Biol. Chem. **267:** 11665–11668.
74. PASSIER, R., H. ZENG, N. FREY, *et al.* 2000. CaM kinase signaling induces cardiac hypertrophy and activates the MEF2 transcription factor in vivo. J. Clin. Invest. **105:** 1395–1406.
75. MOLKENTIN, J.D. 2000. Calcineurin and beyond: cardiac hypertrophic signaling. Circ. Res. **87:** 731–738.
76. WILKINS, B.J. & J.D. MOLKENTIN. 2002. Calcineurin and cardiac hypertrophy: where have we been? Where are we going? J. Physiol. **541:** 1–8.
77. OLSON, E.N. & M.D. SCHNEIDER. 2003. Sizing up the heart: development redux in disease. Genes & Dev. **17:** 1937–1956.
78. ZHANG, T., E.N. JOHNSON, Y. GU, *et al.* 2002. The cardiac-specific nuclear delta(B) isoform of Ca^{2+}/calmodulin-dependent protein kinase II induces hypertrophy and dilated cardiomyopathy associated with increased protein phosphatase 2A activity. J. Biol. Chem. **277:** 1261–1267.
79. BLAESER, F., N. HO, R. PRYWES, *et al.* 2000. Ca^{2+}-dependent gene expression mediated by MEF2 transcription factors. J. Biol. Chem. **275:** 197–209.
80. LU, J., T.A. MCKINSEY, R.L. NICOL, *et al.* 2000. Signal-dependent activation of the MEF2 transcription factor by dissociation from histone deacetylases. Proc. Natl. Acad. Sci. U.S.A. **97:** 4070–4075.

81. Sparrow, D.B., E.A. Miska, E. Langley, *et al.* 1999. MEF-2 function is modified by a novel co-repressor, MITR. EMBO J. **18:** 5085–5098.
82. Berger, I., C. Bieniossek, C. Schaffitzel, *et al.* 2003. Direct interaction of Ca^{2+}/calmodulin inhibits histone deacetylase 5 repressor core binding to myocyte enhancer factor 2. J. Biol. Chem. **278:** 17625–17635.
83. Frey, N., T.A. McKinsey, E.N. Olson, *et al.* 2000. Decoding calcium signals involved in cardiac growth and function. Nature Med. **6:** 1221–1227.
84. Linseman, D.A., C.M. Bartley, S.S. Le, *et al.* 2003. Inactivation of the myocyte enhancer factor-2 repressor histone deacetylase-5 by endogenous Ca^{2+}/calmodulin-dependent kinase II promotes depolarization-mediated cerebellar granule neuron survival. J. Biol. Chem. **278:** 41472–41481.
85. Zhang, C.L., T.A. McKinsey & S. Chang. 2002. Class II histone deacetylases act as signal responsive repressors of cardiac hypertrophy. Cell **110:** 479–488.
86. Bare, D., C.S. Kettlun, M. Liang, D.M. Bers & G.A. Mignery. 2005. Cardiac type-2 Inositol tris phosphate receptor : interaction and modulation by CaMKII. J. Biol. Chem. **280:** 15912-15920.
87. Wu, X., J. Bossuyt, T. Zhang, *et al.* 2004. Excitation-Transcription coupling in adult myocytes: local InsP3-dependent perinuclear signaling activates HDAC nuclear export. Circ. **110:** III-285.
88. Dolmetsch, R.E., R.S. Lewis, C.C. Goodnow, *et al.* 1997. Differential activation of transcription factors induced by Ca^{2+} response amplitude and duration. Nature **386:** 855–858.
89. Molkentin, J.D., J.R. Lu, C.L. Antos, *et al.* 1998. A calcineurin-dependent transcriptional pathway for cardiac hypertrophy. Cell **93:** 215–228.
90. Taigen, T., L.J. De Wind, H.W. Lim, et al. 2000. Targeted inhibition of calcineurin prevents agonist-induced cardiomyocyte hypertrophy. Proc. Natl. Acad. Sci. U.S.A. **97:** 1196–1201.

Calcium Biology of the Transverse Tubules in Heart

LONG-SHENG SONG,[a] SILVIA GUATIMOSIM,[a,b] LETICIA GÓMEZ-VIQUEZ,[a] ERIC A. SOBIE,[a,c] ANDREW ZIMAN,[a,d] HALI HARTMANN,[a] AND W.J. LEDERER[a,d]

[a] *Medical Biotechnology Center and the Institute of Molecular Cardiology, University of Maryland Biotechnology Institute, Baltimore, Maryland 21201, USA*

[b] *Department of Physiology and Biophysics, Federal University of Minas Gerais, Belo Horizonte, Brazil*

[c] *Department of Pediatrics, New York University School of Medicine, New York, New York 10016, USA*

[d] *Department of Physiology, University of Maryland, Baltimore, Maryland 21201, USA*

ABSTRACT: Ca^{2+} sparks in heart muscle are activated on depolarization by the influx of Ca^{2+} through dihydropyridine receptors in the sarcolemmal (SL) and transverse tubule (TT) membranes. The cardiac action potential is thus able to synchronize the $[Ca^{2+}]_i$ transient as Ca^{2+} release is activated throughout the cell. Increases in the amount of Ca^{2+} within the sarcoplasmic reticulum (SR) underlie augmented Ca^{2+} release globally and an increase in the sensitivity of the ryanodine receptors (RyRs) to be triggered by the local $[Ca^{2+}]_i$. In a similar manner, phosphorylation of the RyRs by protein kinase A (PKA) increases the sensitivity of the RyRs to be activated by local $[Ca^{2+}]_i$. Heart failure and other cardiac diseases are associated with changes in SR Ca^{2+} content, phosphorylation state of the RyRs, $[Ca^{2+}]_i$ signaling defects and arrhythmias. Additional changes in transverse tubules and nearby junctional SR may contribute to alterations in local Ca^{2+} signaling. Here we briefly discuss how TT organization can influence Ca^{2+} signaling and how changes in SR Ca^{2+} release triggering can influence excitation–contraction (EC) coupling. High speed imaging methods are used in combination with single cell patch clamp experiments to investigate how abnormal Ca^{2+} signaling may be regulated in health and disease. Three issues are examined in this presentation: (1) normal Ca^{2+}-induced Ca^{2+} release and Ca^{2+} sparks, (2) abnormal SR Ca^{2+} release in disease, and (3) the triggering and propagation of waves of elevated $[Ca^{2+}]_i$.

KEYWORDS: Ca^{2+} sparks; Ca^{2+} waves; ryanodine receptors; transverse tubules

INTRODUCTION

Mammalian heart muscle cells distinguish themselves from reptilian, bird, and fish cardiac ventricular myocytes by the presence of an extensive transverse tubule

Address for correspondence: Prof. W.J. Lederer, M.D., Ph.D., Medical Biotechnology Center, UMBI, 725 W. Lombard Street, Baltimore, MD 21201, USA. Voice: 410-706-8181; fax: 410-510-1545.

lederer@umbi.umd.edu

Ann. N.Y. Acad. Sci. 1047: 99–111 (2005). © 2005 New York Academy of Sciences.
doi: 10.1196/annals.1341.009

(TT) system as shown in FIGURE 1. This system of tubules permits each of the larger heart cells of the mammalian ventricular myocardium to nearly synchronously trigger the release of Ca^{2+} from the sarcoplasmic reticulum (SR) in response to its depolarization.

The propagated action potential (AP) originates in the sino-atrial (SA) node, is conducted through the atria, the atrio-ventricular (AV) node, and the His-Purkinje fiber system before it reaches the septum and the endocardial surfaces of the ventricles. This conducted AP then propagates from cell to cell within the myocardium moving from endocardium to epicardium. In this manner, cardiac myocytes are sequentially activated as the wave of depolarization spreads. The TT system is largely responsible for the synchrony of local Ca^{2+} release within each cell. It would appear to be advantageous for each cell to generate force in an organized manner so that one region of a cell does not serve to stretch another region of the cell as would tend to occur if the Ca^{2+} release and contractions were asynchronous. The cellular contractions within the entire heart are coordinated so that the contraction of the myocardium in mammals squeezes blood out of the heart as the intraventricular pressure increases and volume is reduced. Here we briefly examine the TT system in heart with respect to its role in Ca^{2+} signaling in mammalian heart muscle. In addition, we discuss issues related to TT development, de-differentiation, and disease. Recent work has made the local signaling within the ventricular myocyte of extraordinary interest because it underlies normal excitation–contraction (EC) coupling.[1,2] When

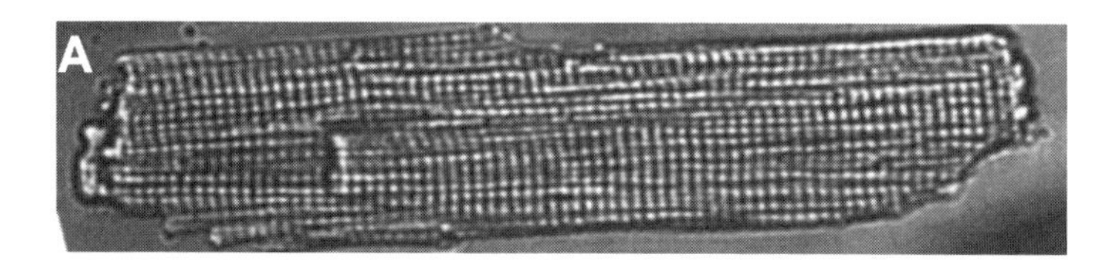

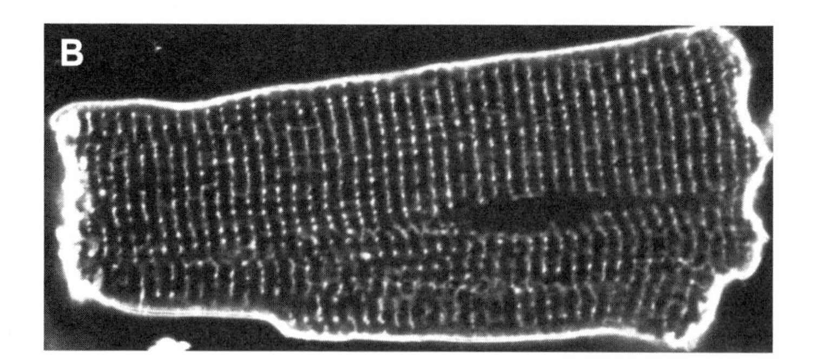

FIGURE 1. Transverse tubules (TTs) in rat heart cells. (**A**) A wide field transmitted light image of a rat ventricular myocyte. (**B**) A confocal fluorescence image of the TTs in a living rat ventricular myocyte stained with the membrane marker di-8 ANEPPS imaged with 488 nm excitation. The myocyte was exposed to 10 μM dye for 10 min. The striation or TT separation along the long axis of the cell is 1.8 μm.

abnormalities develop in this process, local Ca^{2+} overload may occur and this may lead to abnormal Ca^{2+} release, which can underlie contractile dysfunction and arrhythmogenesis.

TRANSVERSE TUBULES

TTs are physical invaginations (100 to 300 nm in diameter[3]) that occur at regular intervals (every 1.2 microns along the Z-line[4]) and run deep into the ventricular cell as shown in FIGURE 1. One TT is separated from the next TT along the long axis of the cell by junctional SR (jSR) and a nearly 2 micron long mitochondrion, as diagrammed in FIGURE 2. With a roughly 2 micron Z-line spacing on the long axis of the cell, there is virtually no region inside a ventricular myocyte more than about 1.2 microns from a TT or SL membrane. This is illustrated in FIGURE 3. At every cross-section of the ventricular myocyte along its long axis at a Z-line there is a grid of TTs that conducts the cardiac AP into the depths of the cell. In this manner, the TTs form a network of membranes within the heart cell, and this extension of the surface sarcolemmal (SL) membrane contains many channels and transporters[5] (see FIGS. 3 and 4). This includes all major channel types in heart: L-type Ca^{2+} channels (also known as dihydropyridine receptors, or DHPRs), Na^+ channels, and diverse K^+ channels. Critical transporters include the Na^+/Ca^{2+} exchanger (NCX), multiple isoforms of the Na^+, K^+ ATPase, transporters that regulate pH, and those that move chloride. Of particular importance is the spatial organization of these proteins along the TT, because it affects local Ca^{2+} signaling which underlies EC coupling.[6–9]

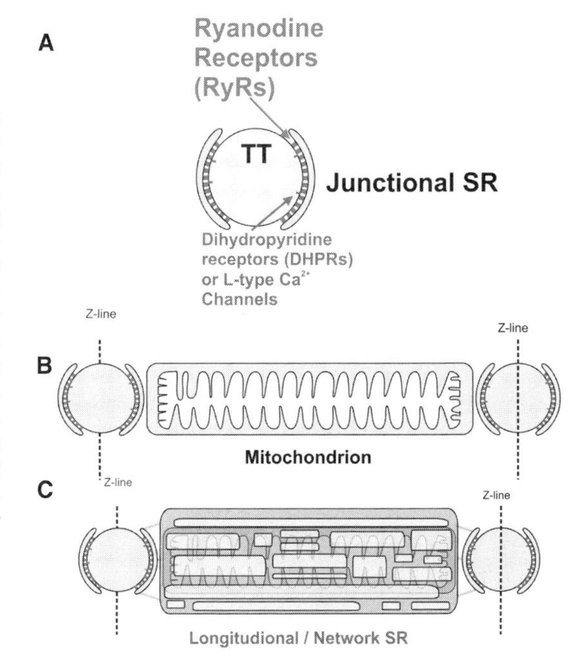

FIGURE 2. Transverse tubules (TTs) and other related structures in the sarcomere. (**A**) The TT, schematic diagram of the TT, ryanodine receptor (RyR) cluster, and junctional sarcoplasmic reticulum (jSR). (**B**) The intermyofibrillar mitochondrion. Relationship between TT structures and the intermyofibrillar mitochondrion. (**C**) The reticular longitudinal sarcoplasmic reticulum (SR). The longitudinal SR wraps around the mitochondrion and is linked to the jSR through diffusion-restricted connections [in color in Annals Online].

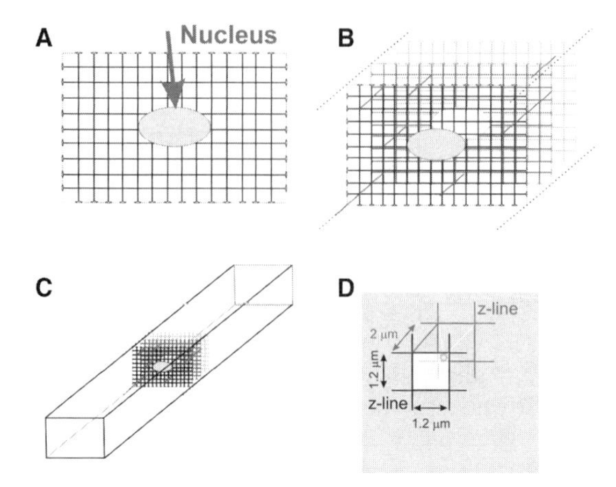

FIGURE 3. The transverse tubule (TT) array in a mammalian ventricular myocyte. (**A**) Diagram of a fully populated TT system (array) at the level of the nucleus (*arrow*). The TTs flare open at the sarcolemmal surface and do not penetrate the nucleus. The TTs are approximately 1.2 microns apart. (**B**) An array of TTs are found at every Z-line and are sparsely connected to the neighboring Z-line array by longitudinal tubules (lines at 45 degrees, shown in red in Annals Online). There are three Z-line arrays shown in this diagram. (**C**) Diagram of the placement of the array of TTs in an intact cell. (**D**) Diagram of a TT elementary unit. An elementary unit of TTs is composed of a square (1.2 microns on an edge) in successive Z-line arrays. The small sphere (shown in green online) represents the most distant location in the elementary unit from a TT, and it is approximately 1.17 microns from the nearest TT, assuming the distance between Z-lines is 2 microns. Yellow lines (seen online) connect opposite corners and go through the center of the small sphere [in color in Annals Online].

Because of the critical roles played by the TTs and their array of local proteins, the TTs have become an area of considerable research interest.

Soeller and Cannell[10] have examined the organization of TTs in rat ventricular myocytes and have revealed the grid-like structure noted above. These investigators have also shown clearly that there are occasional missing elements of the grid as well as longitudinal or axial tubules that connect the TT grid at one Z-line with that at another Z-line (see, for example, solid lines at 45 degrees in FIG. 3). A similar pattern was also seen in the TT distribution of the Na^+/Ca^{2+} exchanger in rat ventricular myocytes[4] and can be observed in the TT image shown in FIGURE 1. In contrast, atrial myocytes have few TTs,[11–13] but they can have some which show a sparse distribution and, in many cases, a more prominent longitudinal component.

Sarcoplasmic Reticulum

The SR is an organelle that consists of multiple interconnected Ca^{2+}-containing components. Each component is a membrane-enclosed volume that is Ca^{2+}-rich and is connected to the others.[14] The jSR is joined to the network SR (or longitudinal SR) through diffusion-limiting connections. The jSR is found close to the Z-line and the

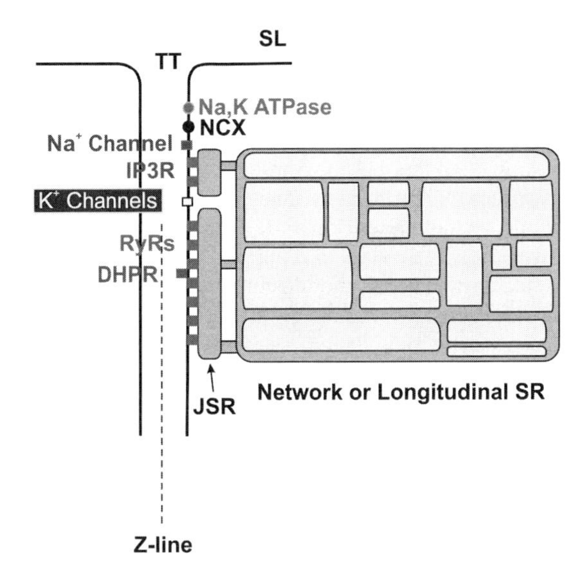

FIGURE 4. Organization of key channel and transporter proteins and sarcoplasmic reticulum (SR) structures near the transverse tubule (TT). The IP3 receptors (IP3Rs) are separated from the ryanodine receptors (RyRs) in this diagram because of recent immunofluorescence data. However, both IP3Rs and RyRs are thought to be SR structures. The key proteins are labeled in the figure. The extensive array of cytoskeletal proteins, signaling receptors, adaptor proteins, and contractile filaments are not shown. SL, sacrolemmal; NCX, Na^+/Ca^{2+} exchanger; DHPR, dihydropyridine receptors; JSR, junctional sarcoplasmic reticulum [in color in Annals Online].

TTs and has a very small volume, while the longitudinal SR spans the distance across the sarcomere. Some evidence suggests that each SR component within a cell may be directly or indirectly connected to all of the other SR components within the cell. SR Ca^{2+}-release channels (ryanodine receptors, or RyRs) span the approximately 15 nm gap between jSR and the TT membrane and are found primarily at the SR-TT junction. At these junctions there are clusters of RyRs (between 10 and 300).[15] The RyRs are sensitive to local $[Ca^{2+}]_i$ and are activated to release Ca^{2+} from the SR when there is an elevation of $[Ca^{2+}]_i$ in the gap or "subspace" between the TT membrane and the SR. When activated by $[Ca^{2+}]_i$ a RyR opens as an SR Ca^{2+}-release channel and floods the subspace with Ca^{2+} from the jSR. Elevation of subspace $[Ca^{2+}]$ from Ca^{2+} entry across the TT membrane (mainly by DHPRs facing the subspace) or across the jSR membrane by RyRs will tend to activate all RyRs facing the subspace. This regenerative RyR activation and consequent Ca^{2+} efflux from the SR produces a Ca^{2+} spark. The large efflux of Ca^{2+} from the small jSR volume is possible because of the SR "lumenal" buffer, calsequestrin, and the extensive network of the longitudinal SR. Termination of the Ca^{2+} spark is thought to occur largely due to depletion of Ca^{2+} from the SR, thereby decreasing RyR sensitivity to $[Ca^{2+}]_i$. The termination may also depend on a cooperative interaction between RyRs within the cluster.[16]

Ryanodine Receptors

RyRs are large homotetrameric proteins of 2 mega-Daltons. The bulk of the protein is in the cytoplasmic space and spans the gap or subspace between the jSR and the TT membrane.[17,18] The RyRs form a cluster of interacting channels[15,19,20] that are triggered by the Ca^{2+} influx into the subspace through the L-type Ca^{2+} channels and other channels and transporters.[16] The RyRs are phosphorylated by protein kinase A (PKA) and by calmodulin-dependent protein kinase II (CaMKII),[20–23] and this phosphorylation modifies the sensitivity of the RyR to $[Ca^{2+}]$.[23] Phosphatases de-phosphorylate the RyR and provide balanced modulation of this important protein.[24] There is significant evidence that indicates RyRs are regulated by the SR lumenal Ca^{2+}. Higher $[Ca^{2+}]_{lumen}$ leads to the increased sensitivity of RyRs[25] so that RyRs are more readily activated by $[Ca^{2+}]_{subspace}$. The SR Ca^{2+} ATPase (SERCA) transports Ca^{2+} from the cytoplasm into the SR so that the RyRs can release it. While SERCA is clearly localized primarily to the Z-line region based on immunofluorescence data, SERCA is also found on the network SR. In a similar manner, RyRs, which are primarily found at the Z-line, are also seen between the Z-lines. This raises a question not yet fully addressed: Where exactly are RyRs that are not at the SR-TT junctions and not part of the junctional clusters?

The mitochondria span the region from one jSR to the next and are surrounded by the longitudinal SR (see FIG. 2). This intimate connection does not make it easy to separate—based on function or by experimental intervention—RyRs that may be found along the longitudinal SR from those that are on the mitochondria. Thus the question raised above remains unanswered. A second and equally important question is this: If RyRs are found elsewhere, then what do they do there?

IP3 Receptors

IP3 receptors (IP3Rs) are found in mammalian ventricular myocytes at the Z-lines.[5] They appear to co-localize with NCX and Na^+, K^+ ATPase proteins but not with RyRs and DHPRs. The RyRs and DHPRs are co-localized and also found along the Z-line.[5,26] This observation made by several groups suggests that there may be close but distinct groups of channels and transporters that form separate but interacting elements. Since both IP3Rs and RyRs are thought to reside in the SR, it is not clear in which functional parts of the SR each receptor or set of receptors is found. Atrial myocytes have more IP3Rs than ventricular myocytes, and it has been suggested that activation of IP3Rs may contribute to the Ca^{2+} signal by influencing RyRs.[27,28] Mohler *et al.*[5] also show that the IP3Rs are found at sites that contain the Na^+/Ca^{2+} exchanger. It is still not clear, however, how the IP3Rs are activated and what they may do in ventricular myocytes. Since IP3Rs are "Ca^{2+} release channels" that are activated by IP3 and Ca^{2+}, along with other possible agonists, the question that remains is this: What is their function in ventricular myocytes? It is tempting to speculate that they may influence local $[Ca^{2+}]_i$ and hence influence EC coupling at nearby RyRs, as Zima & Blatter[27] suggested for atrial myocytes; however, as yet there is no compelling evidence. FIGURE 4 shows how IP3Rs may be located close to RyRs as suggested by the current information available.

Proteins Located On or Near The Transverse Tubules

The TTs and Z-line region have many proteins including many cytoskeletal proteins, the adaptor protein ankyrin B[5], transporters (e.g., NCX and the various isoforms of the Na^+, K^+ ATPase [Na^+ pump]), SL channels including DHPRs, Na^+ channels, diverse K^+ channels, the Ca^{2+} release channels RyR and IP3R, and the SR Ca^{2+} ATPase (SERCA), among others. There are several specific questions that are relevant for this examination of Ca^{2+} signaling along the TT. Each of these questions centers on the spatial organization of the channels, transporters, and other proteins and the special role(s) they may play.

The presence of the cardiac isoform of the Na^+ channel (Nav 1.5) on the SL and TT membranes underlies the rapid upstroke of the conducted cardiac action potential. Analysis of membrane currents suggest that about 300 channels per square micron may exist[29] in ventricular myocytes. Tetrodotoxin (TTX) sensitivity suggests that the cardiac isoform is the dominant isoform of the Na^+ channel found in ventricular myocytes,[30] but some immunofluorescence findings raise the possibility that there may be neuronal isoforms of the TTX-sensitive I_{Na} in ventricular myocytes.[30] It is not clear what role, if any, would be played by these other Na^+ channel isoforms in heart cells, and further identification and functional investigations must still be carried out.

A question that has been raised by diverse groups is whether the Na^+ channels may contribute to local Ca^{2+} signaling. There is some evidence that Na^+ flux through Na^+ channels affects Na^+/Ca^{2+} exchanger and Ca^{2+} signaling.[31] Immunofluorescence imaging at high resolution suggests that while the Na^+ channels and Na^+/Ca^{2+} exchanger may be close to each other, they are not co-localized with the RyRs at the jSR.[26,32] This observation is consistent with investigations of other Z-line proteins.[5] That leaves open the question of how Na^+ influx via Na^+ channels can influence local Ca^{2+} signaling as suggested by recent physiological studies.[31]

DEVELOPMENT, DE-DIFFERENTIATION, AND DISEASE AND THE TT SYSTEM: SR CA^{2+} OVERLOAD

All investigations to date show that "control" or "wild type" ventricular myocytes have $[Ca^{2+}]_i$ transients that are virtually synchronously triggered upon depolarization everywhere within the cell.[33] FIGURE 5 shows the measurement of SR Ca^{2+} release under normal conditions using confocal imaging in linescan mode. However, when TTs are lost (or become dysfunctional in some disease states), Ca^{2+} release dyssynchrony may develop. Similarly, during cardiac development or while cardiac myocytes are in culture, Ca^{2+}-induced Ca^{2+}-release (CICR) may spread throughout the cell as a propagating wave of elevated $[Ca^{2+}]_i$,[34] depending on diffusion distances, cell morphology, and electrical behavior. Dyssynchrony has been shown to occur in myocytes in primary culture,[35] in disease,[36–38] and under conditions where the triggering Ca^{2+} is reduced.[39] Dyssynchrony is often characterized by a propagation delay of the triggered release of Ca^{2+} and may contribute to dysfunction of cellular contraction[35,38] and/or $[Ca^{2+}]_i$ transient "alternans."[39] FIGURE 5E shows that the half-time of the increase in $[Ca^{2+}]_i$ goes up as the square of distance that Ca^{2+} must diffuse. Electrical activation of EC coupling is rapid, but Ca^{2+} diffusion-dependent activation of Ca^{2+} release is slow. This raises a clear question: How do Ca^{2+} sparks get triggered when the local $[Ca^{2+}]_i$ is not elevated the way it is when DHPR activate the apposed RyR cluster in the jSR? Under these conditions, a propagating "wave" can develop from a site of normal activation to the untriggered SR. For this

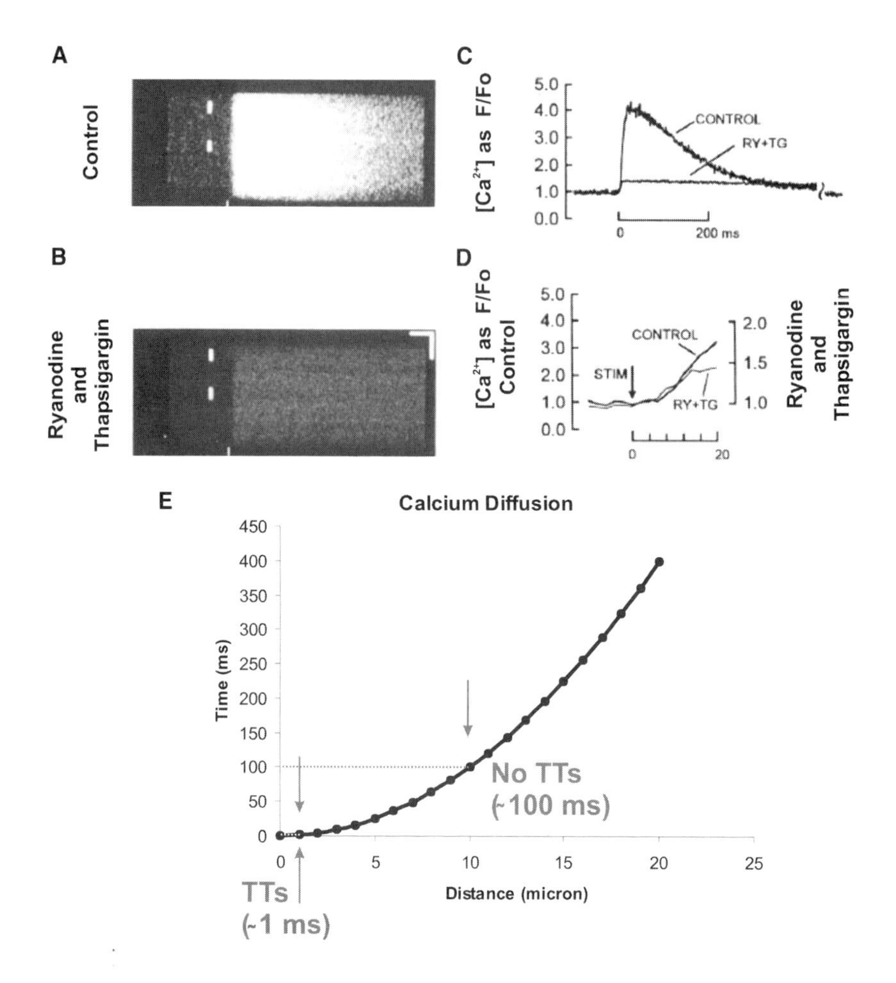

FIGURE 5. Calcium profile across the diameter of the cell during excitation–contraction (EC) coupling. (**A**) A $[Ca^{2+}]_i$ transient shown using a linescan image of a heart cell taken across the diameter of the cell. Note that there is virtually synchronous activation of Ca^{2+} release at all points. (**B**) Same as **A** but after the application of ryanodine (RY) and thapsigargin (TG) to completely block sarcoplasmic reticulum (SR) Ca^{2+} release and reuptake respectively. (**C**) The $[Ca^{2+}]_i$ transients from **A** and **B**. (**D**) Time-course of the $[Ca^{2+}]_i$ transient due to I_{Ca} alone and I_{Ca} plus triggered release are nearly identical during the early phase. (**E**) Effect of diffusion distance and diffusion half-time. (**A–C** modified from Cheng *et al.*[33] Used with permission.) TT, transverse tubule; STIM, stimulus [in color in Annals Online.]

"orphaned SR" to be triggered, it must be in a state of local Ca^{2+} overload with high EC coupling gain. This is clear since isolated Ca^{2+} sparks do not normally trigger other Ca^{2+} sparks,[40–43] even when there is modest elevation of the SR Ca^{2+} load.[44] Thus, as TTs de-differentiate (as may occur in primary culture), the dyssynchrony that develops is associated with local SR Ca^{2+} overload. Some have speculated that similar conditions develop in diseases such as heart failure.[36,37] This alteration may not be an aberration but instead may reflect a cellular adaptation to a loss of TTs. Consistent with this, EC coupling resulting from propagation of Ca^{2+} release is a common finding in atrial myocytes,[11–13] cells which often lack TTs. Because of the smaller cross-section of these cells, negative functional consequences of this release are minimal.

MODELING LOCAL AND GLOBAL CA^{2+} SIGNALING

The great stability of Ca^{2+}-induced Ca^{2+}-release in heart arises largely because of two factors: (1) the relative insensitivity of RyRs to be triggered by $[Ca^{2+}]_i$ changes, and (2) local Ca^{2+} triggering.[45] This investigation proposed that local Ca^{2+} signaling solved the problem posed a few years earlier:[46] How can myocytes have both apparent high gain and EC coupling stability? Additional modeling investigations have supported the earlier work and explored diverse local control models.[47] Importantly, the discovery of Ca^{2+} sparks[43] provided experimental support for the local signaling and accounted for the transition to the increased sensitivity of the SR Ca^{2+} release process that underlies propagating waves of elevated Ca^{2+} and arrhythmogenesis. FIGURE 6 shows a sequence of images that reveals how such a propagating wave of elevated $[Ca^{2+}]_i$ is initiated by Ca^{2+} sparks. This imaging is obtained at high resolution with a high-speed Zeiss confocal microscope, the LSM 5 Live. While normal modeling of Ca^{2+} sparks and SR Ca^{2+} release provides stability,[16] when SR Ca^{2+} content becomes elevated, the "gain" of the SR Ca^{2+} release mechanism increases and the process becomes unstable.

MAJOR QUESTIONS UNDER INVESTIGATION RELATED TO TTS AND SR CA^{2+} SIGNALING

One of the primary questions that remains open is this: Why does SERCA expression decrease in many cardiac pathologies? The cellular and molecular signals that underlie this change and the decrease in average SR Ca^{2+} content have not yet been identified. Further, if only SERCA were to decrease, then the SR Ca^{2+} content would also decline. But it is also clear that the SR Ca^{2+} content is not always uniformly decreased within the cell. Despite the decrease in global averaged $[Ca^{2+}]_i$ during EC coupling, cells retain the ability to develop Ca^{2+}-dependent (or Ca^{2+}-triggered) arrhythmias. This raises the second major question: What underlies the development of the local SR Ca^{2+} overload (even when global SR Ca^{2+} may be depleted)? A related question is whether the non-uniform decreases in SR Ca^{2+} are related to the decrease in SERCA. One possible answer suggested by circumstantial evidence is that there is local Ca^{2+} overload in some regions. Could the local increase of Ca^{2+} within the SR provide a signal that leads to the decrease in SERCA globally? This local Ca^{2+} overload is partly a consequence of the "orphaned" SR[48] that can develop as TTs become disorganized with respect to the Z-lines and SR. This hypothesis continues to be examined by several groups. A third major question focuses on the RyRs in heart. How does modulation of the RyRs by phosphorylation and by other interacting proteins influence CICR, EC coupling, and SR Ca^{2+} overload? Direct

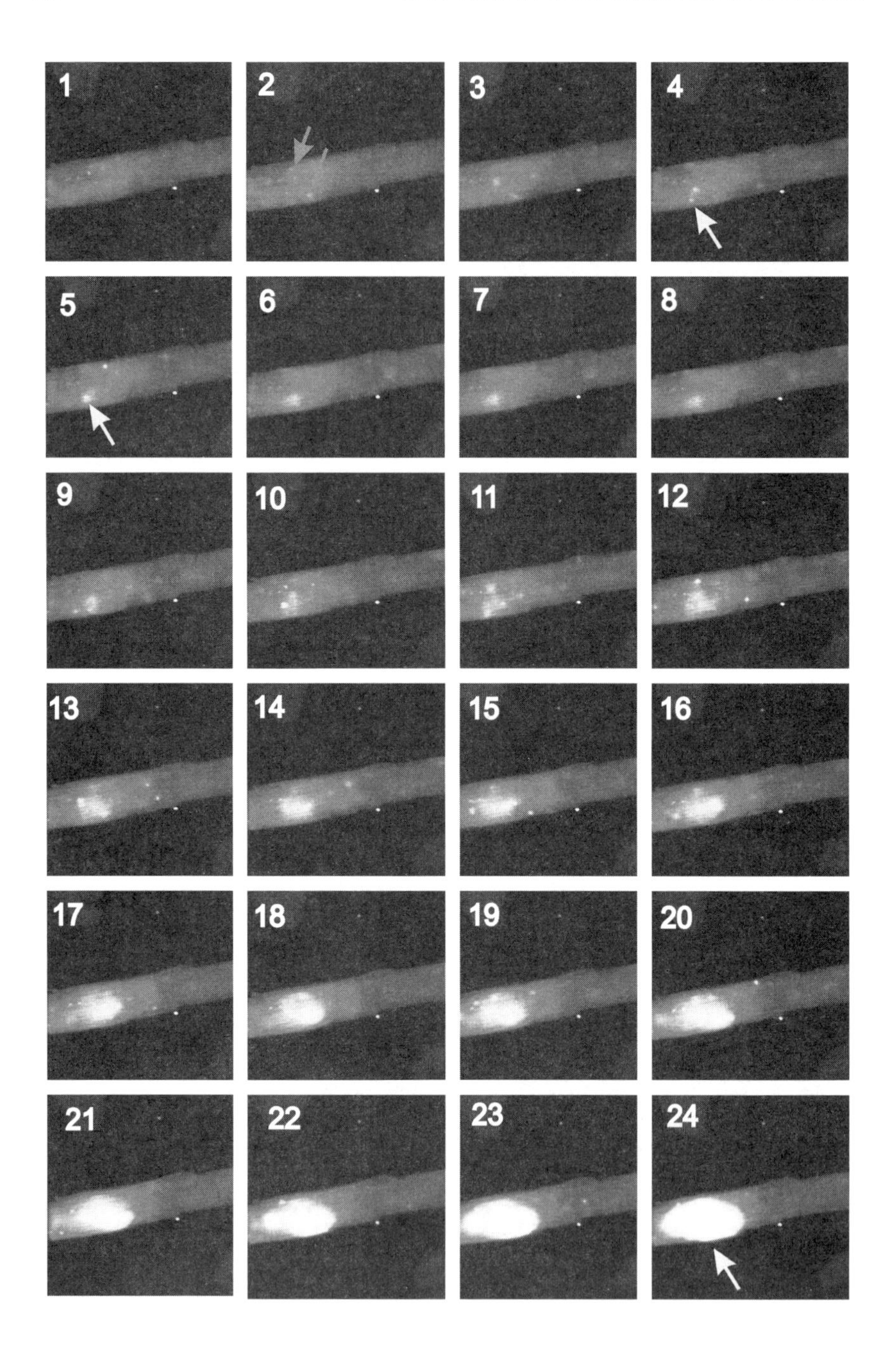
1
2
3
4
5
6
7
8
9
10
11
12
13
14
15
16
17
18
19
20
21
22
23
24

evidence obtained in simple systems (e.g., isolated channels) is very provocative, and circumstantial evidence suggests that these modulations of the RyRs are important in intact cells.

SUMMARY

Ca^{2+} signaling in mammalian ventricular myocytes depends on local signaling between DHPRs and RyR clusters to produce a high-gain Ca^{2+} signaling system with great stability. Disruption of the TT organization or weak Ca^{2+} triggering underlies an increase in SR Ca^{2+} content (and thus in EC coupling gain) and produces instability. Continuing investigations of cellular control of SERCA and the organization of channels, transporters, cytoskeletal proteins, adaptor proteins, intracellular organelles, and signaling cascades should provide improved understanding of Ca^{2+} signaling in mammalian ventricular myocytes.

ACKNOWLEDGMENTS

This work has been supported by the National Heart, Lung, and Blood Institute; the National Science Foundation; the American Heart Association; and the Whitaker Foundation.

REFERENCES

1. WANG, S.Q., L.S. SONG, E.G. LAKATTA, *et al.* 2001. Ca^{2+} signaling between single L-type Ca^{2+} channels and ryanodine receptors in heart cells. Nature **410:** 592–596.
2. SONG, L.S., S.Q. WANG, R.P. XIAO, *et al.* 2001. β-adrenergic stimulation synchronizes intracellular Ca^{2+} release during excitation-contraction coupling in cardiac myocytes. Circ. Res. **88:** 794–801.
3. BERS, D.M. 2001. Excitation-Contraction Coupling and Cardiac Contractile Force. 2nd ed. Kluwer. Boston.
4. KIEVAL, R.S., R.J. BLOCH, G.E. LINDENMAYER, *et al.* 1992. Immunofluorescence localization of the Na-Ca exchanger in heart cells. Am. J. Physiol. **263:** C545–C550.
5. MOHLER, P.J., J.J. SCHOTT, A.O. GRAMOLINI, *et al.* 2003. Ankyrin-B mutation causes type 4 long-QT cardiac arrhythmia and sudden cardiac death. Nature **421:** 634–639.

FIGURE 6. Images of the initiation of a propagated wave of elevated $[Ca^{2+}]_i$ in a rat ventricular myocyte. Frames are numbered sequentially 1–24. Ca^{2+} sparks can be seen in virtually every frame. In frame 2, two of several Ca^{2+} sparks are identified with *arrows*. In Frames 4 and 5, the *arrows* (shown in yellow in Annals Online) mark a Ca^{2+} spark that initiates the wave of propagating elevated $[Ca^{2+}]_i$. It is clear in the frames (starting in frame 4) that there is a local instability over a region of about 8 microns, where sparks activate neighboring regions and then form an increasing cloud of activating and spreading elevated $[Ca^{2+}]_i$ that by frame 24 has started to exit the region. This wave propagates throughout the cell in frames that follow these first 24. The cell was loaded with the Ca^{2+} sensitive indicator fluo-4 and was exposed to a normal extracellular solution with $[Ca^{2+}]_o = 3$ mM. Sequential high-resolution images acquired at 30 ms intervals using the very fast confocal microscope, the Zeiss LSM 5 Live [in color in Annals Online].

6. WANG, S.Q., C. WEI, G. ZHAO, *et al.* 2004. Imaging microdomain Ca^{2+} in muscle cells. Circ. Res. **94:** 1011–1022.
7. WEHRENS, X.H., S.E. LEHNART, F. HUANG, *et al.* 2003. FKBP12.6 deficiency and defective calcium release channel (ryanodine receptor) function linked to exercise-induced sudden cardiac death. Cell **113:** 829–840.
8. MARKS, A.R., S.O. MARX & S. REIKEN. 2002. Regulation of ryanodine receptors via macromolecular complexes. A novel role for leucine/isoleucine zippers. Trends Cardiovasc. Med. **12:** 166–170.
9. MARX, S.O., J. KUROKAWA, S. REIKEN, *et al.* 2002. Requirement of a macromolecular signaling complex for β adrenergic receptor modulation of the KCNQ1-KCNE1 potassium channel. Science **295:** 496–499.
10. SOELLER, C. & M.B. CANNELL. 1999. Examination of the transverse tubular system in living cardiac rat myocytes by 2-photon microscopy and digital image-processing techniques. Circ. Res. **84:** 266–275.
11. BLATTER, L.A., J. KOCKSKAMPER, K.A. SHEEHAN, *et al.* 2003. Local calcium gradients during excitation-contraction coupling and alternans in atrial myocytes. J. Physiol. **546:** 19–31.
12. KOCKSKAMPER, J, K.A. SHEEHAN, D.J. BARE, *et al.* 2001. Activation and propagation of Ca^{2+} release during excitation-contraction coupling in atrial myocytes. Biophys. J. **81:** 2590–2605.
13. KIRK, M.M., L.T. IZU, Y. CHEN-IZU, *et al.* 2003. Role of the transverse-axial tubule system in generating calcium sparks and calcium transients in rat atrial myocytes. J. Physiol. **547:** 441–451.
14. BROCHET D.X., D YANG, A. DI MAIO, *et al.* 2005. Ca^{2+} blinks: rapid nanoscopic store calcium signaling. Proc. Natl. Acad. Sci. U.S.A. **102:** 3099–3104.
15. FRANZINI-ARMSTRONG, C., F. PROTASI & V. RAMESH. 1999. Shape, size, and distribution of Ca^{2+} release units and couplons in skeletal and cardiac muscles. Biophys. J. **77:** 1528–1539.
16. SOBIE, E.A., K.W. DILLY, C.J. DOS SANTOS, *et al.* 2002. Termination of cardiac Ca^{2+} sparks: an investigative mathematical model of calcium-induced calcium release. Biophys. J. **83:** 59–78.
17. WAGENKNECHT, T. & M. SAMSO. 2002. Three-dimensional reconstruction of ryanodine receptors. Front. Biosci. **7:** d1464–d1474.
18. SHARMA, M.R., P. PENCZEK, R. GRASSUCCI, *et al.* 1998. Cryoelectron microscopy and image analysis of the cardiac ryanodine receptor. J. Biol. Chem. **273:** 18429–18434.
19. FRANZINI-ARMSTRONG, C., F. PROTASI & V. RAMESH. 1998. Comparative ultrastructure of Ca^{2+} release units in skeletal and cardiac muscle. Ann. N.Y. Acad. Sci. **853:** 20–30.
20. MARX, S.O., J. GABURJAKOVA, M. GABURJAKOVA, *et al.* 2001. Coupled gating between cardiac calcium release channels (ryanodine receptors). Circ. Res. **88:** 1151–1158.
21. MARX, S.O., S. REIKEN, Y. HISAMATSU, *et al.* 2000. PKA phosphorylation dissociates FKBP12.6 from the calcium release channel (ryanodine receptor): defective regulation in failing hearts. Cell **101:** 365–376.
22. WEHRENS, X.H., S.E. LEHNART, S.R. REIKEN, *et al.* 2004. Ca^{2+}/calmodulin-dependent protein kinase II phosphorylation regulates the cardiac ryanodine receptor. Circ. Res. **94:** e61–e70.
23. VALDIVIA, H.H., J.H. KAPLAN, G.C.R. ELLIS-DAVIES, *et al.* 1995. Rapid adaptation of cardiac ryanodine receptors: modulation by Mg^{2+} and phosphorylation. Science **267:** 1997–2000.
24. DU BELL, W.H., W.J. LEDERER & T.B. ROGERS. 1996. Dynamic modulation of excitation-contraction coupling by protein phosphatases in rat ventricular myocytes. J. Physiol. (London). **493:** 793–800.
25. GYORKE, I. & S. GYORKE. 1998. Regulation of the cardiac ryanodine receptor channel by luminal Ca^{2+} involves luminal Ca^{2+} sensing sites. Biophys. J. **75:** 2801–2810.
26. SCRIVEN, D.R., P. DAN & E.D. MOORE. 2000. Distribution of proteins implicated in excitation-contraction coupling in rat ventricular myocytes. Biophys. J. **79:** 2682–2691.
27. ZIMA, A.V. & L.A. BLATTER. 2004. Inositol-1,4,5-trisphosphate-dependent Ca^{2+} signalling in cat atrial excitation-contraction coupling and arrhythmias. J Physiol. **555:** 607–615.

28. MACKENZIE, L, M.D. BOOTMAN, & M. LAINE, *et al.* 2002. The role of inositol 1,4,5-trisphosphate receptors in Ca^{2+} signalling and the generation of arrhythmias in rat atrial myocytes. J. Physiol. **541:** 395–409.
29. MAKIELSKI, J.C., M.F. SHEETS, D.A. HANCK, *et al.* 1987. Sodium current in voltage clamped internally perfused canine cardiac Purkinje cells. Biophys. J. **52:** 1–11.
30. MAIER, S.K., R.E. WESTENBROEK, T.T. YAMANUSHI, *et al.* 2003. An unexpected requirement for brain-type sodium channels for control of heart rate in the mouse sinoatrial node. Proc. Natl. Acad. Sci. U.S.A. **100:** 3507–3512.
31. SU, Z., K. SUGISHITA, M. RITTER, *et al.* 2001. The sodium pump modulates the influence of I_{Na} on $[Ca^{2+}]_i$ transients in mouse ventricular myocytes. Biophys. J. **80:** 1230–1237.
32. SCRIVEN, D.R., A. KLIMEK, K.L. LEE, *et al.* 2002. The molecular architecture of calcium microdomains in rat cardiomyocytes. Ann. N.Y. Acad. Sci. **976:** 488–499.
33. CHENG, H, M.B. CANNELL & W.J. LEDERER. 1994. Propagation of excitation-contraction coupling into ventricular myocytes. Pfluegers. Arch. Eur. J. Physiol. **428:** 415–417.
34. LIPP, P., & E. NIGGLI. 1994. Modulation of Ca^{2+} release in cultured neonatal rat cardiac myocytes: insight from subcellular release patterns revealed by confocal microscopy. Circ. Res. **74:** 979–990.
35. LOUCH, W.E., V. BITO, F.R. HEINZEL, *et al.* 2004. Reduced synchrony of Ca^{2+} release with loss of T-tubules-a comparison to Ca^{2+} release in human failing cardiomyocytes. Cardiovasc. Res. **62:** 63–73.
36. HATEM, S. 2004. Does the loss of transverse tubules contribute to dyssynchronous Ca^{2+} release during heart failure? Cardiovasc. Res. **62:** 1–3.
37. GÓMEZ, A.M., H.H. VALDIVIA, H. CHENG, *et al.* 1997. Defective excitation-contraction coupling in experimental cardiac hypertrophy and heart failure. Science **276:** 800–806.
38. LITWIN, S.E., D. ZHANG & J.H. BRIDGE. 2000. Dyssynchronous Ca^{2+} sparks in myocytes from infarcted hearts. Circ. Res. **87:** 1040–1047.
39. DIAZ, M.E., S.C. O'NEILL & D.A. EISNER. 2004. Sarcoplasmic reticulum calcium content fluctuation is the key to cardiac alternans. Circ. Res. **94:** 650–656.
40. SANTANA, L.F., H. CHENG, A.M. GÓMEZ, *et al.* 1996. Relation between the sarcolemmal Ca^{2+} current and Ca^{2+} sparks and local control theories for cardiac excitation-contraction coupling. Circ. Res. **78:** 166–171.
41. CANNELL, M.B., H. CHENG & W.J. LEDERER. 1995. The control of calcium release in heart muscle. Science **268:** 1045–1050.
42. CHENG, H., M.B. CANNELL & W.J. LEDERER. 1995. Partial inhibition of Ca^{2+} current by methoxyverapamil (D600) reveals spatial nonuniformities in $[Ca^{2+}]_i$ during excitation-contraction coupling in cardiac myocytes. Circ. Res. **76:** 236–241.
43. CHENG, H., W.J. LEDERER & M.B. CANNELL. 1993. Calcium sparks: Elementary events underlying excitation-contraction coupling in heart muscle. Science **262:** 740–744.
44. SANTANA, L.F., E.G. KRANIAS & W.J. LEDERER. 1997. Calcium sparks and excitation-contraction coupling in phospholamban-deficient mouse ventricular myocytes. J. Physiol. (Lond). **503:** 21–29.
45. NIGGLI, E. & W.J. LEDERER. 1990. Voltage-independent calcium release in heart muscle. Science **250:** 565–568.
46. CANNELL, M.B., J.R. BERLIN & W.J. LEDERER. 1987. Effect of membrane potential changes on the calcium transient in single rat cardiac muscle cells. Science **238:** 1419–1423.
47. STERN, M.D. 1992. Theory of excitation-contraction coupling in cardiac muscle. Biophys. J. **63:** 497–517.
48. GOMEZ, A.M., S. GUATIMOSIM, K.W. DILLY, *et al.* 2001. Heart failure after myocardial infarction: altered excitation-contraction coupling. Circulation **104:** 688–693.

Multimodality of Ca^{2+} Signaling in Rat Atrial Myocytes

MARTIN MORAD, ASHKAN JAVAHERI, TIM RISIUS, AND STEVE BELMONTE

Department of Pharmacology, Georgetown University, Washington, DC 20057, USA

ABSTRACT: It has been suggested that the multiplicity of Ca^{2+} signaling pathways in atrial myocytes may contribute to the variability of its function. This article reports on a novel Ca^{2+} signaling cascade initiated by mechanical forces induced by "puffing" of solution onto the myocytes. Ca_i transients were measured in fura-2 acetoxymethyl (AM) loaded cells using alternating 340- and 410-nm excitation waves at 1.2 kHz. Pressurized puffs of bathing solutions, applied by an electronically controlled micro-barrel system, activated slowly (~300 ms) developing Ca_i transients that lasted 1,693 ± 68 ms at room temperature. Subsequent second and third puffs, applied at ~20 s intervals activated significantly smaller or no Ca_i transients. Puff-triggered Ca_i transients could be reactivated once again following caffeine (10 mM)-induced release of Ca^{2+} from sarcoplasmic reticulum (SR). Puff-triggered Ca_i transients were independent of $[Ca^{2+}]_o$, and activation of voltage-gated Ca^{2+} or cationic stretch channels or influx of Ca^{2+} on Na^+/Ca^{2+}exchanger, because puffing solution containing no Ca^{2+}, 10 μM diltiazem, 1 mM Cd^{2+}, 5 mM Ni^{2+}, or 100 μM Gd^{3+} failed to suppress them. Puff-triggered Ca_i transients were enhanced in paced compared to quiescent myocytes. Electrically activated Ca_i transients triggered during the time course of puff-induced transients were unaltered, suggesting functionally separate Ca^{2+} pools. Contribution of inositol 1,4,5-triphosphate (IP_3)-gated or mitochondrial Ca^{2+} pools or modulation of SR stores by nitric oxide/nitric oxide synthase (NO/NOS) signaling were evaluated using 0.5 to 500 μM 2-aminoethoxydiphenyl borate (2-APB) and 0.1 to 1 μM carbonylcyanide-p-trifluoromethoxyphenylhydrazone (FCCP), and 1 mM Nω-Nitro-L-arginine methyl ester (L-NAME) and 7-nitroindizole, respectively. Only FCCP appeared to significantly suppress the puff-triggered Ca_i transients. It was concluded that neither Ca^{2+} influx nor depolarization was required for activation of this signaling pathway. These studies suggest that pressurized puffs of solutions activate a mechanically sensitive receptor, which signals in turn the release of Ca^{2+} from a limited Ca^{2+} store of mitochondria. How mechanical forces are sensed and transmitted to mitochondria to induce Ca^{2+} release and what role such a Ca^{2+} signaling pathway plays in the physiology or pathophysiology of the heart remain to be worked out.

KEYWORDS: atrial Ca^{2+} signaling; CICR (Ca^{2+}-induced Ca^{2+} release); puff-triggered Ca^{2+} release

Address for correspondence: Prof. Martin Morad, Ph.D., Director, Cell Physiology Program, Pharmacology, Georgetown University, Washington, DC 20057, USA. Voice: 202-687-8464; fax: 202-687-8458.

moradm@georgetown.edu

Ann. N.Y. Acad. Sci. 1047: 112–121 (2005). © 2005 New York Academy of Sciences.
doi: 10.1196/annals.1341.010

INTRODUCTION

The functional diversity of the atrium has been long recognized. Recent data also points to significant diversity in Ca^{2+} signaling mechanisms between different atrial myocytes and within the same myocyte.[1] On the ultrastructural level there is also great diversity; for example, t-tubules appear to be heterogeneously expressed, mostly missing, or very rudimentary in atria of most species, or within different cells of the same atrium.[2] This ultrastructural diversity is reflected in the immunofluorescence measurements of Ca^{2+} channel proteins, showing essentially little or no evidence for the expression of Ca^{2+} channels in the central zones of atrial myocytes.[3] Quantification of local Ca^{2+} release in the periphery and center of atrial myocytes also suggests diversity among cells, such that in about 60% of cells, peripheral Ca^{2+} release is significantly larger and faster than the slower and smaller central release, whereas in approximately 40% of the cells, central Ca^{2+} release is equivalent to peripheral release. Rapid, 2-D confocal imaging of rat atrial myocytes also suggests that in almost 50% of the cells, the central Ca^{2+} sparks are activated within 4 to 8 ms of the activation of I_{Ca} on the surface membrane. Such an activation time is significantly faster than the diffusion of Ca^{2+} wave traveling at ~230 μm/s across the atrial myocyte.[1] Whether such rapid Ca^{2+} signaling is related to the extent of development of t-tubules in individual myocytes or is mediated by other mechanisms (Ca^{2+} release from IP_3-gated endoplasmic reticulum [ER], or mitochondria) is not clear. Because it is well recognized that atrial contraction occurs faster than ventricular rise of tension,[4] it is somewhat puzzling that activation of the central zone of the atrial myocyte can occur as fast (~4 ms) if the diffusion of Ca^{2+} to its center (30 to 50 ms)[1] were the primary mechanism of atrial activation.

In this report, we provide evidence for yet another Ca^{2+} signaling pathway that is activated mechanically, is independent of extracellular Ca^{2+} or activation of stretch or Ca^{2+} channels, but may serve to amplify the Ca^{2+} induced Ca^{2+} release (CICR) signaling pathway. A small, surface-membrane–associated compartment, independent of IP_3-gated ER stores or NO/NOS signaling pathway that trigger the endothelial shear-stress responses in vascular tree,[5] may provide the Ca^{2+} for the puff-triggered Ca_i-transients.

RESULTS

Puff-Triggered Release of Ca^{2+}

FIGURE 1 shows the paradigm used to examine the puff-triggered Ca_i transients in a representative fura-2 AM loaded, intact atrial myocyte. In well attached and electrically paced atrial myocytes, rapid (~20 ms) application of pulses of pressurized (~30 cm H_2O) bathing solution, delivered via an electronically controlled multi-barreled microperfusion system—with its tip 200 μm from the experimental cell, and producing flow forces of ~25 dynes/cm^2 at its tip (250 μm diameter)—activate slowly developing (~300 ms rise time) Ca_i transients lasting ~1.5 s. The effect of application of such pulses of solutions was monitored optically and the Ca_i transients were measured at 1.2 kHz frequency (with alternating excitation waves of 340 and 410 nm), providing a critical check on the fidelity of the Ca^{2+} signal and its alteration by possible movement artifacts.

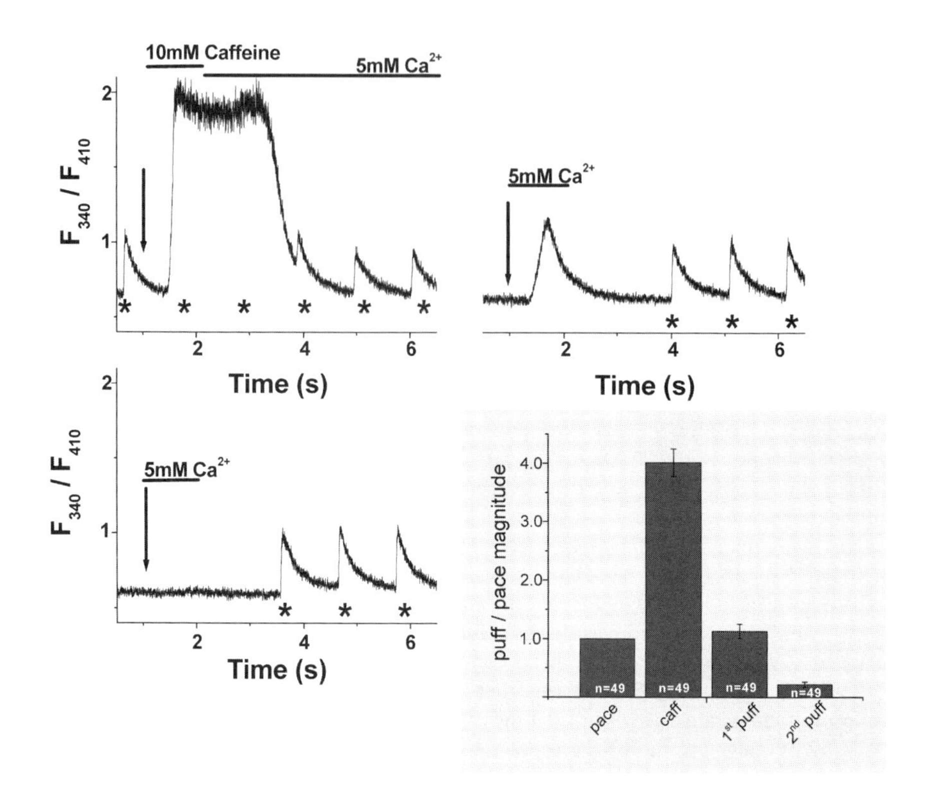

FIGURE 1. Experimental paradigm for puff-triggered Ca_i transients. Atrial myocytes were first loaded with 1 to 5 μM fura-2 AM. Then cells were electrically stimulated (*asterisk*) and treated with caffeine (1 s) that was subsequently washed out with 5 mM Ca^{2+} bathing solution (*upper left panel*). After 15 s, electrical stimulation was turned off and cells were subjected to 1 s pressurized puff of control 5 mM Ca^{2+} bathing solution, activating a slowly developing Ca_i transient (*upper right panel*). Pacing of the cell resumed shortly after the puffing episode. A second puffing episode generated a significantly smaller Ca_i transient (*lower left panel*). Among different atrial myocytes ($n = 49$), the puff-triggered Ca_i transients had an average magnitude similar to that of paced Ca_i transients, but smaller than caffeine Ca_i transients (*lower right panel*) [in color in Annals Online].

FIGURE 1 compares the time course of electrically and mechanically activated Ca_i transients in the same myocyte. Note that unlike electrically activated Ca_i transients with a rapidly rising (~20 ms) and slow decaying (200 to 300 ms) phase, the mechanically activated Ca_i transient developed in ~300 ms and decayed within ~1 s. Note also that the second puff-triggered Ca_i transient, applied 25 to 30 s after the first, generated significantly smaller Ca_i transient. In the vast majority of trials, a third puff failed to activate any transients.

Puff-triggered Ca_i transients could be repeatedly reactivated if the paradigm of electrical stimulation and/or caffeine application were repeated, suggesting that a transient rise of cytosolic Ca^{2+} induced by higher stimulation frequency or 2 s caffeine exposures effectively load the puff-triggered stores. Nevertheless, this store appears to sustain only one or two episodes of puff-triggered Ca^{2+} release.

One likely mechanism for activation of Ca_i transients is that the pressurized puffs of solution cause local depolarization that activate Ca^{2+} influx either through Ca^{2+} channels, the Na^+/Ca^{2+} exchanger, or stretch activated channels. We tested for this possibility by omitting Ca^{2+} from the puffing solution or exposing the myocyte to 5 mM Ni^{2+} or 100 μM Gd^{3+} or 10 μM diltiazem (Ca^{2+} channel blocker). Surprisingly, not only were the puff-triggered Ca_i transients not suppressed, but they were somewhat enhanced in the presence of Ni^{2+} or Gd^{3+}, most likely by suppression of Ca^{2+} efflux on the Na^+/Ca^{2+} exchanger following release of the intracellular Ca^{2+} pool. Thus, it was concluded that the puff-induced Ca^{2+} release requires neither extracellular Ca^{2+} nor the activation of standard Ca^{2+} transporting proteins. Interestingly, puff-triggered Ca_i transients were significantly larger in paced compared to resting myocytes, consistent with the idea that higher resting cytosolic Ca^{2+} may be required to load the intracellular pool responsible for the puff-triggered Ca^{2+} release.

Nature of Puff-Triggered Ca^{2+} Pool

Three possible pools of Ca^{2+} have been either identified in atrial myocytes or are thought to be potentially functional. The SR pool of Ca^{2+} gated by ryanodine receptor (RyR) is universally considered as the major source of Ca^{2+} release for beat-to-beat activation and regulation of contraction. This pool is the dominant Ca^{2+} pool of myocardium and is trigger-released by influx of Ca^{2+} primarily via the Ca^{2+} channel and activation of RyRs (CICR), or it can be released directly by rapid application of 5 to 10 mM caffeine.

Somewhat more controversial and significantly smaller pool of Ca^{2+} in atrium, is that associated with ER and gated by IP_3Rs. The evidence for such a pool is primarily structural (immunofluorescent studies[6]) though in some studies endothelin exposures are reported to activate focal Ca^{2+} releases (sparks) associated with such Ca^{2+} pools.[7] IP_3 receptors have been reported to be enhanced in a number of cardiac pathologies,[7,8] suggesting a possible role for IP_3R in Ca^{2+} signaling of diseased myocytes.

The third pool of Ca^{2+} is thought to be associated with the mitochondria. Though this pool contributes little to the beat-to-beat regulation of contraction, there is significant evidence for activation of "Ca^{2+} marks" associated with spontaneous release of Ca^{2+} from mitochondria[9] or for the existence of ryanodine receptors on mitochondrial membrane that could play a critical role in Ca^{2+} signaling of metabolically compromised cells. Thus, mitochondria may store significant Ca^{2+}, but whether such Ca^{2+} is released following depolarization of membrane or contributes to the amplification of CICR signal is not yet clear.

Considering the RyRs-gated SR as the immediate source of Ca^{2+} for puff-triggered Ca^{2+} release is not consistent with the finding that no more than two episodes of Ca^{2+} release could be activated following caffeine-induced release of Ca^{2+}, even though electrically paced Ca_i transients continue to be activated. More direct support for the idea that electrically paced and puff-triggered Ca_i transients originate from different cellular compartments comes from the finding that electrically paced Ca_i transients activated during the time course of puff-triggered Ca_i transients were equivalent in size to those activated under control conditions (FIG. 2). Interestingly,

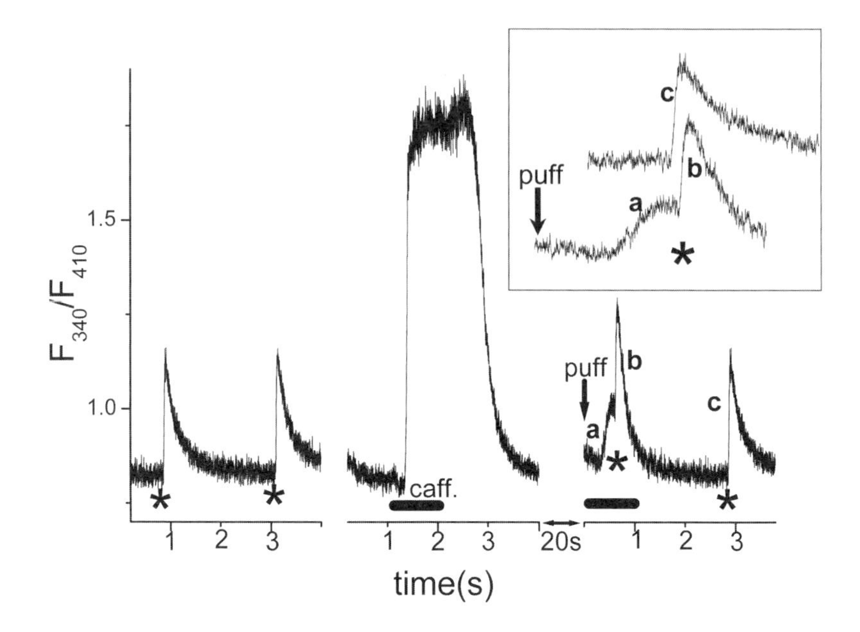

FIGURE 2. Interaction of puff-triggered and electrically activated Ca^{2+} stores. The traces show the kinetics and magnitude of Ca_i transients triggered electrically by caffeine and by puffing in a continuously paced atrial myocyte. (**a**) represents puff-triggered response while (**b**) represents activation of electrically stimulated Ca_i transient at the peak of the puff response; (**c**) is the subsequent pace (control). The inset demonstrates not only the clear differences in kinetics of the two Ca_i transients (**a** versus **b**), but also that the two responses can be independently activated at the same time, without significant attenuation of either response. Our data consistently shows that full release of Ca^{2+} from the SR or rapid pacing rates are capable of refilling the puff-triggered store. Thus we concluded that although the SR is not the immediate source of Ca^{2+} for puff-triggered Ca_i transients, the Ca^{2+} content of SR is critical in determining the magnitude of puff-triggered response. Consistent with this idea, agents that depleted the SR (thapsigargin or ryanodine) abolished the puff-triggered responses, somewhat similar to the well-known effect of these agents on electrically activated Ca_i transients [in color in Annals Online].

in some myocytes puff-triggered Ca_i transients generated a paced-like response at the peak of puff-triggered Ca_i transients without being electrically stimulated, which we assumed was related to activation of an inward Na^+/Ca^{2+} exchanger current triggering an action potential (extra systolic beat).

We also examined the possibility that following caffeine-induced release of Ca^{2+} or a period of rapid electrical pacing, the Ca^{2+} concentrations of mitochondria or IP_3-gated Ca^{2+} stores increases. To probe this idea, we used pharmacological agents that are known to inhibit IP_3R, or mitochondrial oxidative metabolism. Though 2-APB is known to be a non-specific blocker of IP_3R, we found little or no effect of this agent on puff-triggered Ca_i transients at concentrations between 0.1 to 500 μM (TABLE 1). At higher concentrations, the cells deteriorated rapidly and the effect of the drug could not be examined critically.

TABLE 1. Differential modulation of electrically stimulated and puff-triggered Ca^{2+} releases by compounds that target different Ca^{2+} pools

Ca_i transients	Cd^{2+} (1 mM)	Diltiazem (10 μM)	Gd^{3+} (100 μM)	Ni^{2+} (5 mM)	
Paced response	↓↓	↓↓	↓↓	↓↓	
Puffed response	↔, ↑	↔	↔	↔, ↑	
	2-APB (0.1–500 μM)	FCCP (0.1–1 μM)	Genistein (5–25 μM)	L-NAME (1 mM)	7-Nitroindazole (10 μM)
Paced response	↔	↓	↔	↔	↔
Puffed response	↔	↓↓	↔	↔	↔
	CPP (10–50 μM)	Isoproterenol (1 μM)	Zero Ca^{2+}	Spermine (10 μM)	Spermine (1 mM)
Paced response	↔	↑↑	↓↓	↔	↔
Puffed response	↓	↔, ↑	↔, ↑	↑	↓

NOTE: ↑ = enhanced, ↓ = reduced, ↔ = minimal change.

Similarly, we examined the effects of mitochondrial uncoupler FCCP at 0.1 and 1 μM concentrations. Short exposures (10 to 15 s) of 1 μM FCCP reversibly suppressed the puff-triggered and electrically paced Ca_i transients. In other cells, a significant suppressive effect of FCCP was also noted at 0.1 μM following the 10 to 15 s exposure periods, with only a small suppression of paced transients (FIG. 3). Thus, it appears that mitochondrial Ca^{2+} may serve as the immediate source of Ca^{2+} for the puff-triggered Ca_i transients. In this respect, puffs of bathing solutions that wash off caffeine effects generated a secondary transient rise in Ca^{2+} prior to full relaxation of the signal. This finding is consistent with the idea that a fraction of Ca^{2+} released by caffeine may be taken up by mitochondria. Alternatively, following release of SR Ca^{2+} stores by caffeine, activation of a store-operated channel could induce Ca^{2+} influx across the surface membrane to replenish the mitochondrial Ca^{2+}.

Likely Mechanisms that Sense the Flow-Induced Shear/Pressure Forces

In a series of comprehensive reports, Solott and Hare groups[10,11] recently reported that moderate, maintained longitudinal stretches (~18%) enhanced the electrically paced Ca_i transients and increased the frequency of occurrence of spontaneous Ca^{2+} sparks in rat and murine ventricular myocytes. These authors reported that the effect of such stretches was strongly suppressed by NOS inhibitors L-NAME and in mice deleted of NOS-1. Since the shear response in endothelial cells is thought to be mediated by activation of NO/NOS signaling pathway,[5] we attempted to pharmacologically suppress the NO/NOS signaling pathway by using L-NAME (1 mM) or more specific blockers of NOS-1 (7-nitroindizole). FIGURE 4 shows that L-NAME does not alter the puff-triggered Ca_i transients. TABLE 1, in addition, provides a list of other pharmacological agents that failed to alter the puff-triggered Ca_i transients, suggesting that the standard proteins associated with excitation–contraction (E-C)

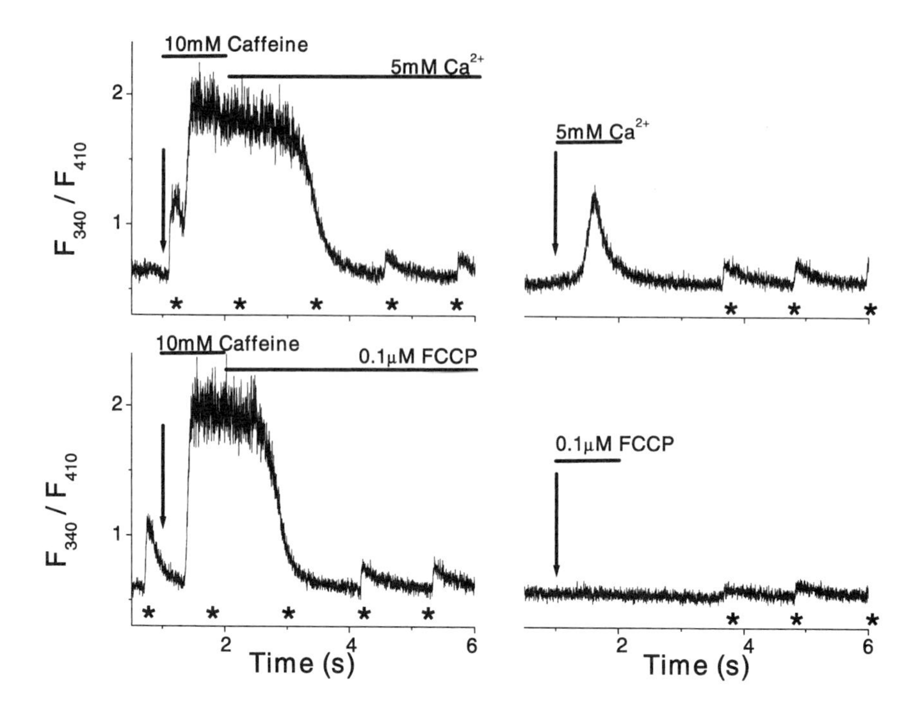

FIGURE 3. FCCP suppresses puff-triggered Ca_i transients. The *upper panel* show control caffeine and puff-triggered Ca_i transients. Subsequent treatment with 0.1 μM of the mitochondrial uncoupler FCCP (*lower panel*) significantly suppressed both the electrically activated and puff-triggered Ca_i transients.

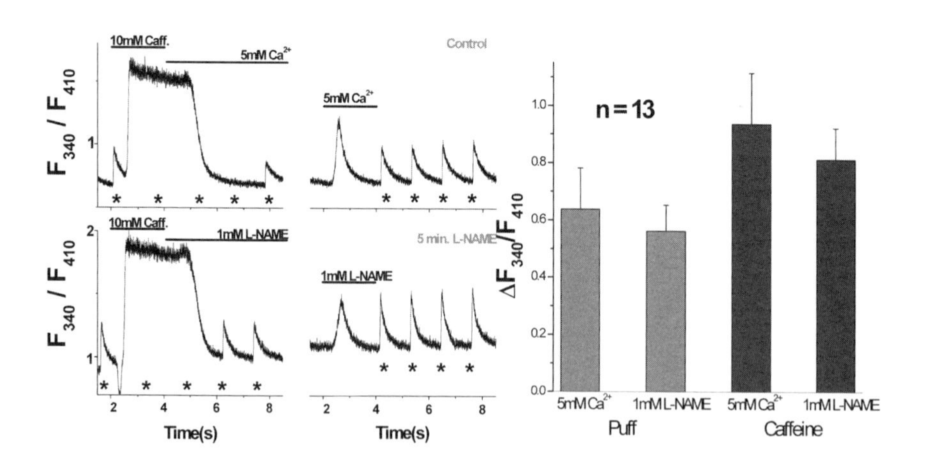

FIGURE 4. L-NAME (nitric oxide synthase inhibitor) does not suppress puff-triggered Ca_i transients. The *upper left panel* shows control caffeine and puff-triggered Ca_i transients. The bath in which the cell sits was then perfused with 1 mM L-NAME for 5 min (*lower left panel*). The magnitude of caffeine and puff-triggered Ca_i transients in the presence of L-NAME was not significantly reduced compared to control (*right panel*) [in color in Annals Online].

coupling are not directly involved with the pathway that senses the shear/pressure force.

Interestingly, we found that acidic (pH 6.0) puffing solutions suppressed significantly the puff-triggered Ca_i transients. Because protons interact at many stages of cell function, we considered the likely surface membrane–associated proteins that are highly sensitive to moderate acidification besides the L-type Ca^{2+} channels. Surprisingly, we found that two agents well known to inhibit the N-methyl-D-aspartate (NMDA) receptor, also significantly suppressed the puff-triggered response. For example, at concentrations of 10 to 50 μM 3-(2-carboxypiperazin-4-yl)propyl-1-phosphonic acid (CPP), a specific inhibitor of NR2B subunit of NMDA receptor reversibly suppressed the puff-triggered Ca_i transients. Similarly, spermine at 0.5 to 1 mM also reversibly suppressed the Ca_i transients. Because spermine has been reported to have an enhancing effect on NMDA receptor at lower (100 nM) concentrations, we also tested concentrations of spermine between 10 and 100 nM and similarly found significant enhancement of puff-triggered transients, but not electrically paced Ca_i transients. As neither NMDA nor glutamate was able to generate any current in either atrial or ventricular myocytes, we concluded that, as it has been reported previously, there is no significant functional NMDA receptor in cardiac myocyte. Nevertheless, we tested for possible presence of this protein using NR2B subunit specific antibody generated in our department. In myocytes isolated on three different days, we found evidence (Western blots) for a 100 kDa protein that interacted specifically with the polyclonal rabbit antibody raised against the neuronal NMDA receptor. There have been previous reports[12] that NMDA receptor protein is expressed (Western blots) in neonatal hearts, but that this signal disappears in adult myocytes. These authors immunoprecipitated the NMDA protein with the RyRs in embryonic myocytes and suggested a possible cytoskeletal structural role for the NMDA receptor in embryonic heart,[12] as they found no evidence for glutamate-activated current in the embryonic myocytes. We speculate, therefore, that a smaller component of NMDA receptor may have survived in atrial myocytes, which may be involved in sensing the flow-induced shear/pressure response, thereby transmitting the shear signal via its cytoskeletal domain directly or via an intermediary protein such as Homer—also reported to be expressed in cardiac myocytes[13]—to mitochondrial Ca^{2+} signaling apparatus or the mitochondrial RyRs,[9] in a manner somewhat similar to direct interaction of RyRs with 1,4-dihydropyridine receptors (DHPRs) in skeletal muscle.

CONCLUSIONS AND PHYSIOLOGICAL IMPLICATIONS OF PUFF-TRIGGERED Ca^{2+} RELEASE

The signaling events described above, terminating in activation of slowly developing and relaxing Ca_i transient, somewhat reminiscent of slow propagation of Ca^{2+} waves reported by numerous investigators in cardiac myocytes, still needs to be worked out. Many details and signaling steps and their regulation remain unknown. Most puzzling is why such a system has developed in cardiac myocytes when the dominant Ca^{2+} signaling mechanism is mediated via the CICR pathway. Perhaps even more puzzling is why does the activation of this signaling system not fully activate CICR-based release of Ca^{2+} from the SR? What this set of experiments do show, however, is that there are two parallel Ca^{2+} signaling systems in atrial myo-

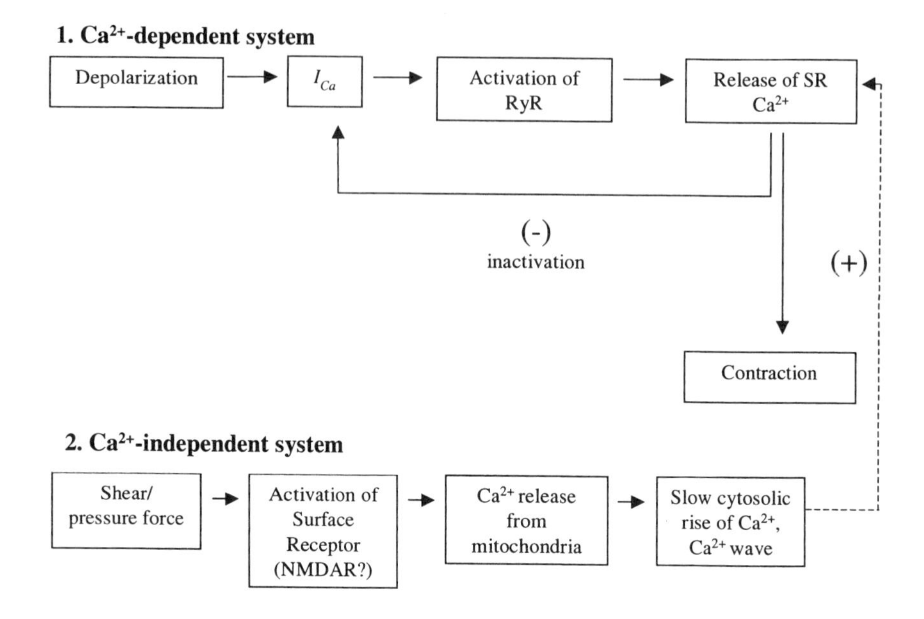

FIGURE 5. Ca^{2+} signaling pathways in atrium.

cytes (see FIG. 5), one of which regulates E-C coupling, while the other may be involved in functions that range from regulation of cardiac compliance in response to quick volume changes to secretion of atrial natriuretic peptide (ANP). It is tempting to propose that puff-triggered Ca_i transients may be specifically developed in atrial myocytes to amplify the I_{Ca}-gated Ca^{2+} release mechanism across the cell diameter, thus allowing rapid and homogeneous release of Ca^{2+} across the atrial myocyte lacking t-tubules. Whether the puff-triggered stores are also alternatively gated by other mechanisms, in addition to the mechanical forces imposed by shear stress, remain to be explored.

ACKNOWLEDGMENT

This study was supported by NIH Grant RO1-16152. A.J. and T.R. contributed equally to this article.

REFERENCES

1. WOO, S.H., L. CLEEMANN & M. MORAD. 2002. Ca2+ current-gated focal and local Ca2+ release in rat atrial myocytes: evidence from rapid 2-D confocal imaging. J. Physiol. **543:** 439–453.
2. KIRK, M.M., L.T. IZU, Y. CHEN-IZU, *et al.* 2003. Role of the transverse-axial tubule system in generating calcium sparks and calcium transients in rat atrial myocytes. J. Physiol. **547:** 441–451.
3. CARL, S. L., K. FELIX, A.H. CASWELL, *et al.* 1995. Immunolocalization of sarcolemmal

dihydropyridine receptor and sarcoplasmic reticular triadin and ryanodine receptor in rabbit ventricle and atrium. J. Cell Biol. **129:** 672–682.
4. LUSS, I., P. BOKNIK, L.R. JONES, *et al.* 1999. Expression of cardiac calcium regulatory proteins in atrium v ventricle in different species. J. Mol. Cell. Cardiol. **31:** 1299–1314.
5. DIMMELER, S., I. FLEMING, B. FISSLTHALER, *et al.* 1999. Activation of nitric oxide synthase in endothelial cells by Akt-dependent phosphorylation. Nature **399:** 601–605.
6. MOSCHELLA, M.C. & A.R. MARKS. 1993. Inositol 1,4,5-trisphosphate receptor expression in cardiac myocytes. J. Cell Biol. **120:** 1137–1146.
7. MACKENZIE, L., M.D. BOOTMAN, M. LAINE, *et al.* 2002. The role of inositol 1,4,5-trisphosphate receptors in Ca(2+) signalling and the generation of arrhythmias in rat atrial myocytes. J. Physiol. **541:** 395–409.
8. WOODCOCK, E.A., S.J. MATKOVICH & O. BINAH. 1998. Ins(1,4,5)P3 and cardiac dysfunction. Cardiovasc. Res. **40:** 251–256.
9. PACHER, P., A.P. THOMAS & G. HAJNOCZKY. 2002. Ca^{2+} marks: miniature calcium signals in single mitochondria driven by ryanodine receptors. Proc. Natl. Acad. Sci. U.S.A. **99:** 2380–2385.
10. PETROFF, M.G., S.H. KIM, S. PEPE, *et al.* 2001. Endogenous nitric oxide mechanisms mediate the stretch dependence of Ca^{2+} release in cardiomyocytes. Nat. Cell Biol. **3:** 867–873.
11. BAROUCH, L.A., R.W. HARRISON, M.W. SKAF, *et al.* 2002. Nitric oxide regulates the heart by spatial confinement of nitric oxide synthase isoforms. Nature **416:** 337–339.
12. SEEBER, S., A. HUMENY, M. HERKERT, *et al.* 2004. Formation of molecular complexes by N-methyl-D-aspartate receptor subunit NR2B and ryanodine receptor 2 in neonatal rat myocard. J. Biol. Chem. **279:** 21062–21068.
13. WESTHOFF, J.H., S.Y. HWANG, R.S. DUNCAN, *et al.* 2003. Vesl/Homer proteins regulate ryanodine receptor type 2 function and intracellular calcium signaling. Cell Calcium **34:** 261–269.

Effects of Na^+-Ca^{2+} Exchange Expression on Excitation–Contraction Coupling in Genetically Modified Mice

JOSHUA I. GOLDHABER,[a] SCOTT A. HENDERSON,[a] HANNES REUTER,[b] CHRISTIAN POTT,[a] AND KENNETH D. PHILIPSON[a]

[a] *Cardiovascular Research Laboratories, David Geffen School of Medicine at UCLA, Los Angeles, California 90095, USA*

[b] *Laboratory for Muscle Research and Molecular Cardiology, Clinic III Internal Medicin, University of Köln, Joseph-Stelzmann-Str. 9, 50924 Köln, Germany*

ABSTRACT: We have created genetically altered mice to investigate how expression of the Na^+-Ca^{2+} exchange protein alters excitation–contraction (E-C) coupling. Whereas low levels of exchanger overexpression have minimal effects on E-C coupling properties, high levels of overexpression in homozygous animals results in susceptibility to hypertrophy and heart failure, along with a significant reduction in E-C coupling gain. While global knockout of the exchanger in mice is embryonic-lethal, conditional knockout mice live to adulthood. Cardiac function is surprisingly normal in seven-week-old mice, but E-C coupling gain is apparently increased. Thus, genetic modification of exchanger expression has a major influence on E-C coupling.

KEYWORDS: cardiac excitation–contraction coupling; Cre-lox; L-type calcium current; sodium-calcium exchange; transgenic mice

INTRODUCTION

In order to maintain intracellular calcium (Ca^{2+}) homeostasis, Ca^{2+} entering cardiac cells during the process of excitation–contraction (E-C) coupling must be balanced by Ca^{2+} removal. The most dominant Ca^{2+} efflux mechanism in cardiac cells is the sarcolemmal sodium-calcium (Na^+-Ca^{2+}) exchanger.[1] This electrogenic transport protein, composed of nine transmembrane segments and a large cytoplasmic loop, removes one intracellular Ca^{2+} ion in exchange for three extracellular Na^+ ions.[2] Na^+-Ca^{2+} exchange can also function in reverse, and in this fashion contributes to Ca^{2+} influx during the early phase of the action potential. Exactly how this component of Ca^{2+} influx participates in the physiology of E-C coupling is uncertain and controversial. Modulation of intracellular Ca^{2+} levels by Na^+-Ca^{2+} exchange is thought to mediate the inotropic response to cardiac glycosides such as digitalis. Pathological increases in exchanger activity and expression have been implicated in the pathogenesis of sarcoplasmic reticulum (SR) Ca^{2+} depletion characterizing some

Address for correspondence: Dr. Joshua I. Goldhaber, Geffen School of Medicine at UCLA, Division of Cardiology, 47-123 CHS, 10833 LeConte Avenue, Los Angeles, CA 90095, USA. Voice: 310-206-1432; fax: 310-206-5777.
jgoldhaber@mednet.ucla.edu

Ann. N.Y. Acad. Sci. 1047: 122–126 (2005). © 2005 New York Academy of Sciences.
doi: 10.1196/annals.1341.011

models of heart failure (HF).[3] Na^+-Ca^{2+} exchange currents are also thought to be a critical component underlying the pathophysiology of triggered arrhythmias in the heart.[4] Understanding the influence of the exchanger on E-C coupling has been challenging because of the lack of specific pharmacological agonists and antagonists. To better understand how the exchanger contributes to E-C coupling, we have created several lines of mice in which the exchanger is either overexpressed or knocked out. In this review, we will discuss how these mice have both contributed to our understanding of E-C coupling and introduced new and exciting areas for exploration.

OVEREXPRESSION

Our first transgenic mouse was a heterozygous overexpressor of the Na^+-Ca^{2+} exchanger (2.3-fold increase).[5] Heterozygous overexpression has little effect on myocardial contractility, Ca^{2+} current, or Ca^{2+} release.[6,7] More detailed analysis has suggested that the exchanger is in a functionally separate microdomain from the L-type Ca^{2+} channel (LTCC).[5] We subsequently bred Na^+-Ca^{2+} overexpressing mice to homozygosity, achieving a 3.1-fold increase in exchanger activity.[8] These animals exhibit modest hypertrophy at baseline and are more susceptible to cardiac failure during the stress of pregnancy or aortic banding than their wild-type littermates. In contrast to heterozygous overexpressors, the homozygous animals display significant increases in LTCC amplitude (by 42% compared with wild type) consistent with some but not all reports of experimental and human hypertrophy[9–11] (FIG. 1). There is also a marked slowing of Ca^{2+} channel inactivation kinetics. Interestingly, these characteristics can be altered by reversible block of the exchanger, suggesting that homozygous overexpression forces some exchangers into the same microdomain as the LTCC (unlike the situation in the heterozygous overexpressors). Despite the increase in LTCC amplitude, there is little increase in the Ca^{2+} transient measured with fura-2 during voltage clamps, consistent with a decrease in E-C coupling gain. Decreases in E-C coupling gain have been observed in other models of hypertrophy and failure, including the Dahl salt-sensitive rat[12] and aortic banded rat hearts.[13] Many of these models demonstrate increased Na^+-Ca^{2+} exchange as well. Though increased Na^+-Ca^{2+} exchange could conceivably reduce gain by rapidly removing Ca^{2+} as it enters the dyadic cleft via LTCC during systole, we were unable to completely restore gain to normal in overexpressors by acutely blocking the exchanger. Thus, reduced E-C coupling gain appears to be an inherent property of the model, which may be due to changes in the physical relationship between LTCCs and the ryanodine receptors (RyRs), or other compensatory changes in the response of RyRs to trigger Ca^{2+}.

KNOCKOUT

Several groups, including ours, have developed a global knockout of the Na^+-Ca^{2+} exchanger in mice.[14–16] This transgenic model is embryonic-lethal at day 10.5 post coitum. Nevertheless, Ca^{2+} transients can be measured in heart tubes from day 9.5 embryos loaded with fura-2. Surprisingly, global knockouts exhibit relatively normal spontaneous and electrically stimulated Ca^{2+} transients. As expected, the absence of Na^+-Ca^{2+} exchange prevents Ca^{2+} overload in response to removal of

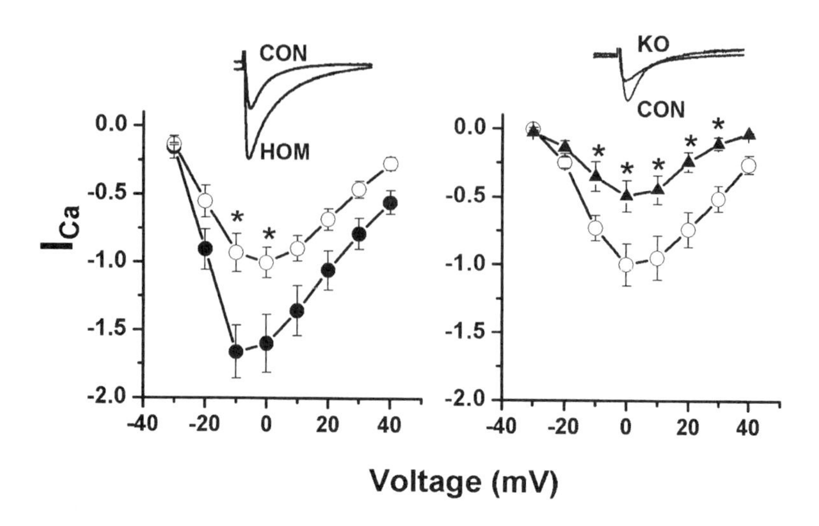

FIGURE 1. Current traces and I-V plots comparing Ca^{2+} current amplitude in control (CON) and either homozygous overexpressing transgenic (HOM) myocytes (*left panel*) or NCX knockout (KO) myocytes (*right panel*). Current traces were obtained during 200 ms voltage clamps from –40 to 0 mV and were normalized to the amplitude in control myocytes at 0 mV. I_{Ca} is normalized to cell capacitance and peak amplitude at 0 mV in control myocytes. Note the marked increase in I_{Ca} in homozygous transgenic overexpressors and the marked decrease in conditional knockout cells. (From Reuter *et al.*[8] and Henderson *et al.*[19] Redrawn by permission.)

extracellular Na^+. Notably, the cardiac glycoside digitalis has no effect on Ca^{2+} transients in knockout myocytes, confirming the importance of the exchanger in the pharmacology of this still commonly used inotrope.[17] Knockout myotubes are also a convenient tool for determining specificity of drugs designed to block the exchanger. For example, we have found that the putative Na^+-Ca^{2+} exchange blockers, SEA0400 and KBR7943, both depress Ca^{2+} transients in knockout heart tubes, consistent with non-specific effects on E-C coupling that could complicate interpretation of effects in wild-type preparations.[18]

Isolating and measuring ionic currents in embryonic heart tubes is technically challenging. Thus, we have developed a tissue specific knockout of the Na^+-Ca^{2+} exchanger in mice using Cre-lox technology.[19] These mice live into adulthood and at 6–8 weeks of age have nearly normal cardiac function based on echocardiography. Adult myocytes can be isolated using standard enzymatic digestion so that patch clamp techniques can be employed to assess ionic currents in conjunction with measurements of Ca^{2+} transients using Ca^{2+} sensitive indicators such as fluo-3 and fura-2. Of the cells isolated from these animals, 80% to 90% have no functional Na^+-Ca^{2+} exchanger, a typical percentage for this technique. Ca^{2+} transients in isolated electrically stimulated fluo-3 acetoxymethyl (AM) loaded knockout cells are indistinguishable from control, as is their response to increases in pacing frequency and stimulation with isoproterenol. SR Ca^{2+} stores appear unaffected when assessed by rapid caffeine application, but there is very limited Ca^{2+} efflux during prolonged caffeine application. This suggests that no alternative efflux mechanisms, for example the sarcolemmal Ca^{2+} pump, compensate fully for the absence of the exchanger. Thus, we were initially surprised that there was no evidence of diastolic Ca^{2+} overload in these cells. One explanation may be related to our finding that patch clamped myocytes from knockout animals show a significant reduction in the amplitude of the LTCC (FIG. 1) and yet a preserved Ca^{2+} transient. This effectively increases E-C coupling gain (in contrast to homozygous overexpressing myocytes where we observed a decrease in gain), while reducing the transsarcolemmal flux or recirculating fraction of Ca^{2+}. Furthermore, it appears that action potentials in knockout myocytes tend to have shorter duration, which may also serve to limit Ca^{2+} influx during each cardiac cycle. These knockout animals die prematurely compared to control. Thus, operating at a higher than normal E-C coupling gain may be deleterious in the long run.

SUMMARY

Genetic manipulation of Na^+-Ca^{2+} exchanger expression has revealed important aspects of this protein's role in E-C coupling and the development of pathology. Whereas modest overexpression of the exchanger in heterozygous transgenics has minimal effects on E-C coupling or cardiac function, higher overexpression in homozygotes leads to pathological hypertrophy and failure, and major changes in E-C coupling, including increased interaction between LTCC and Na^+-Ca^{2+} exchange, increased LTCC current and transsarcolemmal flux and reduced E-C coupling gain. Cardiac myocytes adapt surprisingly well to knockout of the exchanger, apparently by reducing transsarcolemmal fluxes of Ca^{2+} and increasing E-C coupling gain. Nevertheless, as these knockout animals age, they develop more cardiac pathology and die prematurely.

Thus, the degree of Na^+-Ca^{2+} exchanger expression has a major effect on the gain of E-C coupling. Deviations away from normal expression levels result in cardiac pathologies. It is not known whether these abnormalities are a direct consequence of altered expression levels or are secondary to compensatory changes in Ca^{2+} regulation. Future studies will need to address this important question.

ACKNOWLEDGMENTS

This work was supported by NIH Grant HL-70828 (JIG), by NIH Grant HL-48509 (KDP), and by the Laubisch Foundation.

REFERENCES

1. BERS, D.M. 2002. Cardiac excitation-contraction coupling. Nature **415:** 198–205.
2. PHILIPSON, K.D., D.A. NICOLL, M. OTTOLIA, *et al.* 2002. The Na+/Ca2+ exchange molecule: an overview. Ann. N. Y. Acad. Sci. **976:** 1–10.
3. DIPLA, K., J.A. MATTIELLO, K.B. MARGULIES, *et al.* 1999. The sarcoplasmic reticulum and the Na+/Ca2+ exchanger both contribute to the Ca2+ transient of failing human ventricular myocytes. Circ. Res. **84:** 435–444.
4. POGWIZD, S.M., K. SCHLOTTHAUER, L. LI, *et al.* 2001. Arrhythmogenesis and contrac-

tile dysfunction in heart failure: roles of sodium-calcium exchange, inward rectifier potassium current, and residual beta-adrenergic responsiveness. Circ. Res. **88:** 1159–1167.
5. ADACHI-AKAHANE, S., L. LU, Z. LI, *et al.* 1997. Calcium signaling in transgenic mice overexpressing cardiac Na+-Ca2+ exchanger. J. Gen. Physiol. **109:** 717–729.
6. YAO, A., Z. SU, A. NONAKA, *et al.* 1998. Effects of overexpression of the Na+-Ca2+ exchanger on [Ca2+]i transients in murine ventricular myocytes. Circ. Res. **82:** 657–665.
7. TERRACCIANO, C.M., A.I. SOUZA, K. D. PHILIPSON & K. T. MACLEOD. 1998. Na+-Ca2+ exchange and sarcoplasmic reticular Ca2+ regulation in ventricular myocytes from transgenic mice overexpressing the Na+-Ca2+ exchanger. J. Physiol. **512:** 651–667.
8. REUTER, H., T. HAN, C. MOTTER, *et al.* 2004. Mice overexpressing the cardiac sodium-calcium exchanger: defects in excitation-contraction coupling. J. Physiol. **554:** 779–789.
9. KEUNG, E.C. 1989. Calcium current is increased in isolated adult myocytes from hypertrophied rat myocardium. Circ. Res. **64:** 753–763.
10. XIAO, Y.F. & J.J. MCARDLE. 1994. Elevated density and altered pharmacologic properties of myocardial calcium current of the spontaneously hypertensive rat. J. Hypertens. **12:** 783–790.
11. MUKHERJEE, R. & F.G. SPINALE. 1998. L-type calcium channel abundance and function with cardiac hypertrophy and failure: a review. J. Mol. Cell Cardiol. **30:** 1899–1916.
12. GOMEZ, A.M., H.H. VALDIVIA, H. CHENG, *et al.* 1997. Defective excitation-contraction coupling in experimental cardiac hypertrophy and heart failure. Science **276:** 800–806.
13. AHMMED, G.U., P.H. DONG, G. SONG, *et al.* 2000. Changes in Ca2+ cycling proteins underlie cardiac action potential prolongation in a pressure-overloaded guinea pig model with cardiac hypertrophy and failure. Circ. Res. **86:** 558–570.
14. KOUSHIK, S.V., J. WANG, R. ROGERS, *et al.* 2001. Targeted inactivation of the sodium-calcium exchanger (Ncx1) results in the lack of a heartbeat and abnormal myofibrillar organization. FASEB J. **15:** 1209–1211.
15. CHO, C. H., S.S. KIM, M.J. JEONG, *et al.* 2000. The Na+ -Ca2+ exchanger is essential for embryonic heart development in mice. Mol. Cells **10:** 712–722.
16. REUTER, H., S.A. HENDERSON, T. HAN, *et al.* 2003. Cardiac excitation-contraction coupling in the absence of Na(+) - Ca2+ exchange. Cell Calcium **34:** 19–26.
17. REUTER, H., S.A. HENDERSON, T. HAN, *et al.* 2002. The Na+-Ca2+ exchanger is essential for the action of cardiac glycosides. Circ. Res. **90:** 305–308.
18. REUTER, H., S.A. HENDERSON, T. HAN, *et al.* 2002. Knockout mice for pharmacological screening: testing the specificity of Na+-Ca2+ exchange inhibitors. Circ. Res. **91:** 90–92.
19. HENDERSON, S.A., J. I. GOLDHABER, J. M. SO, *et al.* 2004. Functional adult myocardium in the absence of Na+-Ca2+ exchange: cardiac-specific knockout of NCX1. Circ. Res. **95:** 604–611.

Mitochondrial Calcium Signaling and Energy Metabolism

MY-HANH T. NGUYEN[a] AND M. SALEET JAFRI[a,b]

[a] *George Mason University, Manassas, Virginia 20110, USA*

[b] *University of Maryland Biotechnology Institute, Baltimore, Maryland 21205, USA*

ABSTRACT: A computational model of energy metabolism in the mammalian ventricular myocyte is developed to study the effect of cytosolic calcium (Ca^{2+}) transients on adenosine triphosphate (ATP) production. The model couples the Jafri-Dudycha model for tricarboxylic acid cycle regulation to a modified version of the Magnus-Keizer model for the mitochondria. The fluxes associated with Ca^{2+} uptake and efflux (i.e., the Ca^{2+} uniporter and Na^+-Ca^{2+} exchanger) and the F_1F_0-ATPase were modified to better model heart mitochondria. Simulations were performed at steady state and with Ca^{2+} transients at various pacing frequencies generated by the Rice-Jafri-Winslow model for the guinea pig ventricular myocyte. The effects of the Ca^{2+} transients for mitochondria both adjacent to the dyadic space and in the bulk myoplasm were studied. The model shows that Ca^{2+} activation of both the tricarboxylic acid cycle and the F_1F_0-ATPase are necessary to produce increases in ATP production. The model also shows that in mitochondria located near the subspace, the large Ca^{2+} transients can depolarize the mitochondrial membrane potential sufficiently to cause a transient decline in ATP production. However, this transient is of short duration, minimizing its impact on overall ATP production.

KEYWORDS: **calcium; computational model; energy metabolism; mitochondria**

INTRODUCTION

The primary function of the heart is to pump blood. As such, the energy demands of the heart are great and can vary from a basal level at rest to very high levels during exercise due mainly to increased adenosine triphosphate (ATP) usage by the myofilaments and increased ion cycling by the ion-motive ATPases to maintain ionic homeostasis. To accommodate this, there is effective regulation of energy metabolism in cardiac ventricular myocytes. Several factors such as adenosine diphosphate (ADP), ATP, nicotinamide adenine dinucleotide (NAD^+ and NADH), inorganic phosphate (P_i), Ca^{2+}, pH, and creatine phosphate have been suggested to play a role in regulation of energy metabolism.

A computational model of mitochondrial energy metabolism has been developed to study how the various regulatory factors work together to control ATP production. Computational models have been a valuable tool to help understand the complex

Address for correspondence: Prof. Saleet Jafri, Ph.D., School of Computational Sciences, George Mason University, 10900 University Blvd., Occoquan Building, Room 328G, MSN 5B3, Manassas, VA 20110, USA. Voice: 703-993-8420; fax: 703-993-8401.

sjafri@gmu.edu

Ann. N.Y. Acad. Sci. 1047: 127–137 (2005).
doi: 10.1196/annals.1341.012

dynamics of mitochondrial function.[1,2] The Magnus-Keizer model includes mitochondrial calcium (Ca^{2+}) dynamics and its effects on energy metabolism in pancreatic β-cells.[3–5] It includes the following fluxes: Ca^{2+} uptake by the uniporter, Ca^{2+} extrusion by the Na^+-Ca^{2+} exchanger, the ADP/ATP translocase, a proton leak, ATP formation by the F_1F_0-ATPase, the respiration-driven proton pumps, and H^+ flux through the F_1F_0-ATPase. It contains differential equations describing the mitochondrial Ca^{2+} concentration ($[Ca^{2+}]_m$), the mitochondrial ADP concentration ($[ADP]_m$), and the mitochondrial membrane potential (Ψ_m).

The Ca^{2+} uniporter in the Magnus-Keizer model is based on data from Wingrove and co-workers from liver mitochondria, which shows steady-state mitochondrial Ca^{2+} dynamics.[6] Furthermore, the model shows high steady-state mitochondrial $[Ca^{2+}]$ (15 μM) in response to myoplasmic $[Ca^{2+}]$ (1 μM), which is uncharacteristic for cardiac myocytes. Recent work by Wan and co-workers and Chalmers and Nicholls have shown more modest levels in steady state mitochondrial $[Ca^{2+}]$ in response to elevated myplasmic $[Ca^{2+}]$.[7,8]

Other recent work has suggested that mitochondrial Ca^{2+} dynamics can be quite fast and refer to a second mode of Ca^{2+} transport named "rapid uptake mode."[9] In fact, Trollinger and co-workers have observed a beat to beat mitochondrial $[Ca^{2+}]$ transient in conjunction with myoplasmic Ca^{2+} transient.[10] They attribute the slow kinetics seen in earlier work to contamination of the fluorescence signal with measurement of lysosomal Ca^{2+} uptake and release. Also, accompanying the local myoplasmic Ca^{2+} transient, local depolarization of mitochondrial membrane potential has been observed in agreement with the Magnus-Keizer prediction that elevated Ca^{2+} can depolarize the mitochondria.[11] Others have attributed the "rapid uptake mode" to local elevated domains of Ca^{2+} and have measured mitochondrial $[Ca^{2+}]$ transients, named marks, in response to Ca^{2+} sparks.[12]

To address the new data on mitochondrial Ca^{2+} dynamics, this modeling effort has endeavored to incorporate the most recent available data on mitochondrial Ca^{2+} dynamics. To this end, the recent experimental characterization of the Ca^{2+} uniporter performed by Kirichok and co-workers has been used to develop an accurate model of the uniporter.[13] Accompanying this, a revised formulation of the Na^+-Ca^{2+} exchanger is also included, based on the experimental observations of Wingrove and Gunter, and modified to include the 3:1 stoichiometry of the exchanger found by Baysal and co-workers and Jung and co-workers.[14–16] Both these formulations are used to replace the former descriptions to better simulate Ca^{2+} dynamics in the cardiac ventricular cell.

METHODOLOGY

The model integrates a model for the tricarboxylic acid cycle (TCA) developed by Dudycha and Jafri with a modified version of the model for mitochondrial Ca^{2+} handling and respiration developed by Magnus and Keizer.[3,17,18] A schematic of the model is shown in FIGURE 1.

The TCA cycle model consists of a system of differential equations for the intermediates in the cycle. It is assumed that substrate enters the cycle as acetyl-CoA. The key regulatory reactions in the TCA cycle—namely isocitrate dehydrogenase,

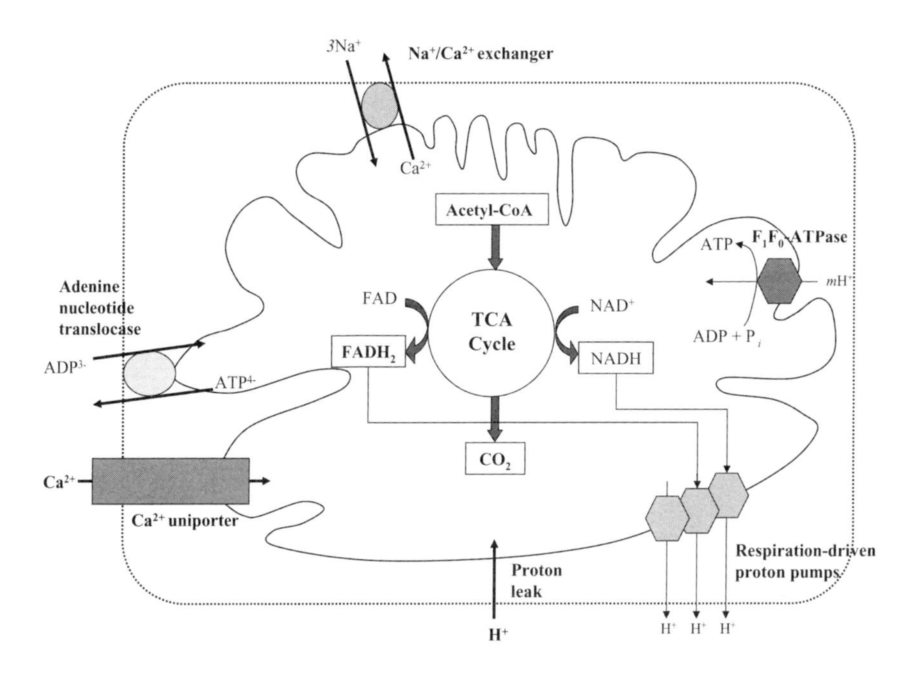

FIGURE 1. Schematic diagram of the model for mitochondrial energy metabolism.

α-ketoglutarate dehydrogenase, citrate synthase, and malate dehydrogenase—are modeled in detail based on experimental measurement of enzyme kinetics in the presence of different concentrations of regulatory factors. The remaining reactions are modeled using the law of mass action. The TCA cycle produces reducing equivalents in the form of NADH and flavin adenine dinucleotide ($FADH_2$) as well as CO_2. NADH and $FADH_2$ then enter the electron transport chain described by the Magnus-Keizer model.

Several modifications were made to the Magnus-Keizer model to better represent the physiology of mitochondria found in cardiac ventricular myocytes: (1) an improved version of Ca^{2+} dynamics was incorporated including new formulations of the uniporter and the Na^+-Ca^{2+} exchanger; (2) a new formulation of the F_1F_0-ATPase was also included; and (3) to integrate the model, the units of the Magnus-Keizer model was converted to mM/ms.

The model was solved using the Runge-Kutta-Merson method, which includes an adaptive time step. The code was written in FORTRAN and solved on a HP Visualize Unix Workstation using HP FORTRAN or GNU g77 compilers. Results were graphed using Xmgr software by Paul J. Turner or PV-WAVE by Visual Numerics, Inc. (San Ramon, CA) Final figures were generated using PV-WAVE.

Myoplasmic [Ca^{2+}] transients were generated by the Jafri-Rice-Winslow model of the guinea pig ventricular myocyte at various frequencies.[19] The time courses for myoplasmic [Ca^{2+}] were used in the mitochondrial model for simulating mitochondrial Ca^{2+} dynamics in a beating heart. The response of the mitochondria near the dyad was

TABLE 1. Model parameters, definitions, and values

Parameter	Definition	Value	Reference
P_{Ca}	Uniporter Ca^{2+} permeability	4.2×10^{-18} cm s^{-1}	Fit to Refs. 10 and 13
z	Valence of Ca^{2+}	2	Known
F	Faraday's constant	96500 coul (mol $e^{-})^{-1}$	Known
V	Mitochondrial matrix volume	10.336×10^{-18} l	Rcf. 3
R	Ideal gas constant	8.314 J mol^{-1} K^{-1}	Known
T	Absolute temperature	310	Body temp
α_m	Mitochondrial Ca^{2+} activity coefficient	0.2	Fit to Ref. 13
α_i	Cytosolic Ca^{2+} activity coefficient	0.341	Ref. 19
V_{nc}	Na^+-Ca^{2+} exchanger maximal velocity	140.0 mM ms^{-1}	Fit to Ref. 10
K_{Na}	Na^+-Ca^{2+} exchanger Na^+ affinity	9.4 mM	Ref. 14
K_{Ca}	Na^+-Ca^{2+} exchanger Ca^{2+} affinity	4.54 mM	Ref. 14
V_p	F_1F_0-ATPase maximal velocity	1.125×10^{-3} mM ms^{-1}	Fit to Ref. 3
$K_{v,ATP}$	F_1F_0-ATPase potential constant	70.0 mV	Fit to Ref. 20
$K_{Ca,ATP}$	F_1F_0-ATPase Ca^{2+} binding constant	165.0 nM	Ref. 22

simulated by exposing the mitochondria to the myoplasmic [Ca^{2+}] time course seen as subspace [Ca^{2+}] in the Jafri-Rice-Winslow model. In both cases, the mitochondria were exposed to [Ca^{2+}] transients as a fixed frequency until a steady state pattern was reached. The parameter values, their definitions, and their sources are shown in TABLE 1.

RESULTS

The Model

One of the goals of the model was to simulate the recent experimental observations on mitochondrial Ca^{2+} dynamics in cardiac ventricular myocytes. A new formulation of the uniporter was developed to describe the data observed by Kirichok and co-workers.[13] The uniporter was assumed to be an ion channel permeable only to Ca^{2+}. The model was based on a Goldman-Hodgkin-Katz formulation and closely matches membrane potential ramp data at different levels of myoplasmic [Ca^{2+}] (not shown). The uniporter can be described by

$$J_{uni} = -P_{\mathrm{Ca}} \frac{z\Psi_m F}{VRT} \frac{\alpha_m[\mathrm{Ca}^{2+}]_m \exp\left(\frac{-z\Psi_m F}{RT}\right) - \alpha_i[\mathrm{Ca}^{2+}]_i}{\exp\left(\frac{-z\Psi_m F}{RT}\right) - 1} . \tag{1}$$

The Na^+-Ca^{2+} exchanger in the Magnus-Keizer model shows a voltage dependence; however, the Na^+-Ca^{2+} exchanger in cardiac ventricular myocyte mitochondria shows no such dependence. Any effect of membrane potential is reflected by changes in the mitochondrial Na^+ concentration due to Na^+-H^+ exchange. Hence, the model incorporated the formulation of the exchanger developed by Wingrove and Gunter with the modification that the stoichiometry is 3:1 for Na^+:Ca^{2+} as observed by Baysal and co-workers and Jung and co-workers.[14–16] The Na^+-Ca^{2+} exchanger can be described by

$$J_{nc} = V_{nc} \frac{[\mathrm{Na}^+]_i^3}{K_{\mathrm{Na}}^3 + [\mathrm{Na}^+]_i^3} \frac{[\mathrm{Ca}^{2+}]_m}{K_{\mathrm{Ca}} + [\mathrm{Ca}^{2+}]_m} . \tag{2}$$

After incorporating these changes, it was observed that during the small depolarizations of the mitochondrial membrane potential (~6 mV) that occur during a $[Ca^{2+}]$ transient (see FIG. 3), the activity of the F_1F_0-ATPase showed an unrealistically large transient decrease. Analysis of the model indicated that this was due to a decrease in proton flux through the F_1F_0-ATPase. Experimental observations indicate that ATP production by F_1F_0-ATPase is mainly dependent on the mitochondrial membrane potential and not on the proton gradient.[20] This is due to the property that with large membrane potential, positive ions, such as protons, is driven into the mitochondria and can do so even up a concentration gradient. The formulation in the Magnus-Keizer model, which demonstrates a strong dependence on the proton gradient, was replaced by simpler formulation based on the work of Kaim and Dimroth.[20] The model agrees with the experimental finding that cardiac ventricular myocyte mitochondria have been shown to have a linear Ca^{2+} dependence in the physiological range.[21]

Along with a more accurate dependence on membrane potential, recent experimental work has indicated that the F_1F_0-ATPase activity is stimulated by Ca^{2+}.[22] This relationship has been explored quantitatively by Balaban and co-workers and has been incorporated into the model.[22] The resultant model equations for the F_1F_0-ATPase are

$$J_p = V_p \frac{[\Psi_m]^8}{K_{V,\mathrm{ATP}}^8 + [\Psi_m]^8} \left[1 - e^{\frac{-[\mathrm{Ca}^{2+}]_m}{K_{\mathrm{Ca,ATP}}}}\right], \tag{3}$$

$$J_{hf} = 3V_p \frac{[\Psi_m]^8}{K_{V,\mathrm{ATP}}^8 + [\Psi_m]^8} \left[1 - e^{\frac{-[\mathrm{Ca}^{2+}]_m}{K_{\mathrm{Ca,ATP}}}}\right]. \tag{4}$$

Simulation Results

The model was used to simulate both steady-state behavior and transient behavior as described above. FIGURE 2 shows the model responses to simulated myoplasmic [Ca^{2+}] transients. In FIGURE 2A, myoplasmic and mitochondrial [Ca^{2+}] transients are shown for a frequency of 1 Hz. The results match the experimental work of Trollinger and co-workers as the mitochondrial [Ca^{2+}] transients are ~½ the amplitude of the myoplasmic [Ca^{2+}] transients, have similar rise time, and have delayed decay.[10] Simulated pacing at 2 Hz is shown in FIGURE 2B. Once again, the mitochondrial [Ca^{2+}] transients have reduced amplitude, similar rise time, and delayed decay when compared to myoplasmic [Ca^{2+}] transients.

Steady-state data for the relationship between mitochondrial and myoplasmic [Ca^{2+}] concentrations indicate an almost linear behavior as shown in FIGURE 2C. The behavior is similar to the experimental observations by Wan and co-workers in mitoplasts derived from cardiac myocytes and by Chalmers and Nicholls in isolated cardiac myocyte mitochondria.[7,8] The main difference between the simulated and the experimental data is that both groups observe a flattening of the curve at the lower end of their measured data. However, each group noted this as different myoplasmic Ca^{2+} levels. This is most probably due to the limitations of the methods used for characterizing mitochondrial [Ca^{2+}] and was ignored when formulating the model.

The next set of simulations explores how energy production in the mitochondria

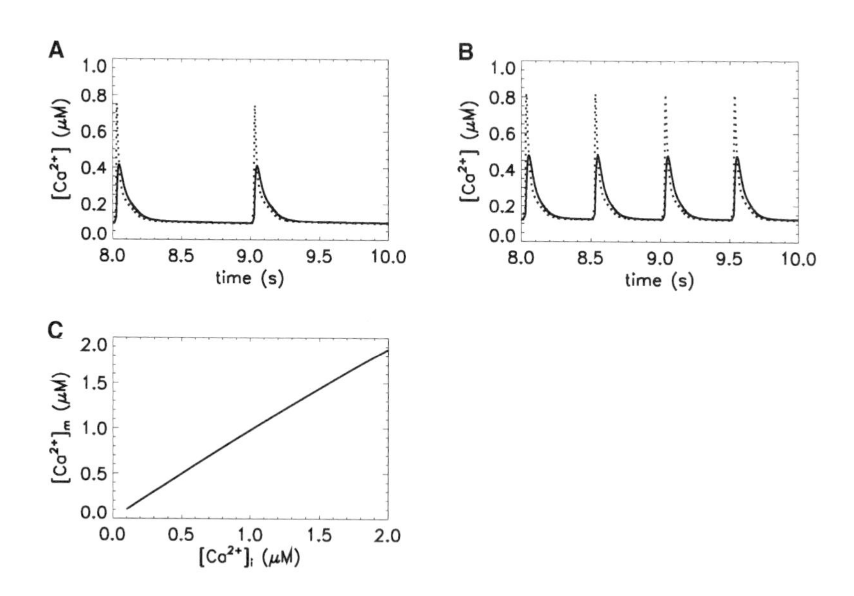

FIGURE 2. (**A**) Myoplasmic (*dotted line*) and mitochondrial (*solid line*) Ca^{2+} concentration transients at a pacing rate of 1.0 Hz. (**B**) Myoplasmic (*dotted line*) and mitochondrial (*solid line*) Ca^{2+} transients at a pacing rate of 2.0 Hz. (**C**) The steady-state relationship between myoplasmic Ca^{2+} concentration ($[Ca^{2+}]_m$) and cytosolic Ca^{2+} concentration ($[Ca^{2+}]_i$).

might be influenced by transient changes in myoplasmic $[Ca^{2+}]$, and hence mitochondrial $[Ca^{2+}]$ that would be seen during normal cardiac function. The model explores the importance of parallel activation of the TCA cycle and the F_1F_0-ATPase for achieving increased ATP production in the beating heart. The $[Ca^{2+}]$ transients from the simulation in FIGURE 3 are the same as those seen in FIGURE 2A. FIGURE 3A shows the changes in the net NADH production flux (i.e., the difference between production and consumption). In the case of parallel activation, the net NADH flux is negative, indicating a net consumption of NADH. Net NADH flux is diminished when the stimulatory effects of Ca^{2+} are removed from the F_1F_0-ATPase by setting $[Ca^{2+}]_m$ in EQUATIONS 3 and 4 to its resting value of 0.0945 μM. FIGURE 3B shows that during parallel activation, the ATP production flux increases during the $[Ca^{2+}]$ transients. If the Ca^{2+} dependence of the F_1F_0-ATPase is removed, stimulation is not significant. The next question to explore is why there seems to be little stimulation. FIGURE 3C shows the changes to mitochondrial membrane potential (Ψ_m). In both cases, influx of Ca^{2+} through the Ca^{2+} uniporter provides an inward current that depolarizes the

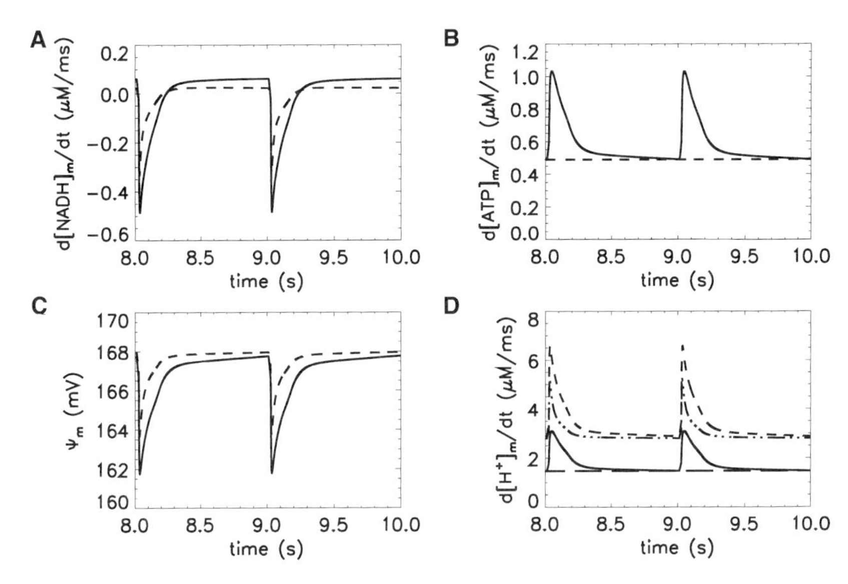

FIGURE 3. (**A**) The net nicotinamide adenine dinucleotide (NADH) production flux (d[NADH]/dt) at 1.0 Hz pacing rate with a $[Ca^{2+}]_m$ dependent F_1F_0-ATPase (*solid line*) and with a $[Ca^{2+}]_m$ independent F_1F_0-ATPase (*dashed line*). (**B**) The adenosine triphosphate (ATP) production flux from the F_1F_0-ATPase (d[ADP]/dt) given by EQUATION 3 (J_p) with a $[Ca^{2+}]_m$ dependent F_1F_0-ATPase (*solid line*) and with a $[Ca^{2+}]_m$ independent F_1F_0-ATPase (*dashed line*). (**C**) Mitochondrial membrane potential (Ψ_m) with a $[Ca^{2+}]_m$ dependent F_1F_0-ATPase (*solid line*) and with a $[Ca^{2+}]_m$ independent F_1F_0-ATPase (*dashed line*). (**D**) Proton fluxes ($d[H^+]$/dt) for the respiration driven H^+ pumps (*dashed line*). A $[Ca^{2+}]_m$ dependent F_1F_0-ATPase with a $[Ca^{2+}]_m$ independent F_1F_0-ATPase (*dot dashed line*) and for the F_1F_0-ATPase (*solid line*) with a $[Ca^{2+}]_m$ dependent F_1F_0-ATPase and with a $[Ca^{2+}]_m$ independent F_1F_0-ATPase (*long dashed line*).

mitochondria. Also during the depolarization, the activity of the respiration driven proton pumps increases due to less resistance because of the diminished membrane potential as suggested by Murphy and Brand.[23] This is shown in FIGURE 3D. During transient depolarization, the proton flux through the F_1F_0-ATPase increases in the case with parallel activation. In the absence of parallel activation, the proton flux through the F_1F_0-ATPase remains relatively constant. FIGURE 3D also suggests that the NADH net production flux shown in FIGURE 3A indicates a transient decrease due to increased proton pump activity even if the F_1F_0-ATPase activity does not change.

Experimental studies have suggested that mitochondria are also located near the Ca^{2+} release sites (i.e., near the cardiac dyadic subspace). In the case of Ca^{2+} sparks, localized mitochondrial [Ca^{2+}] transients known as marks have been observed.[12] Local depolarization of the mitochondrial membrane potential have also been observed.[11] The model was used to assess how these local Ca^{2+} signaling events affect energy metabolism. To do so, the mitochondrial model was exposed to the elevated Ca^{2+} likely to be seen near the dyadic subspace generated from the Jafri-Rice-Winslow model for the guinea pig ventricular myocyte. FIGURE 4A shows the elevated local myoplasmic [Ca^{2+}] near the subspace and the resulting mitochondrial [Ca^{2+}]. The

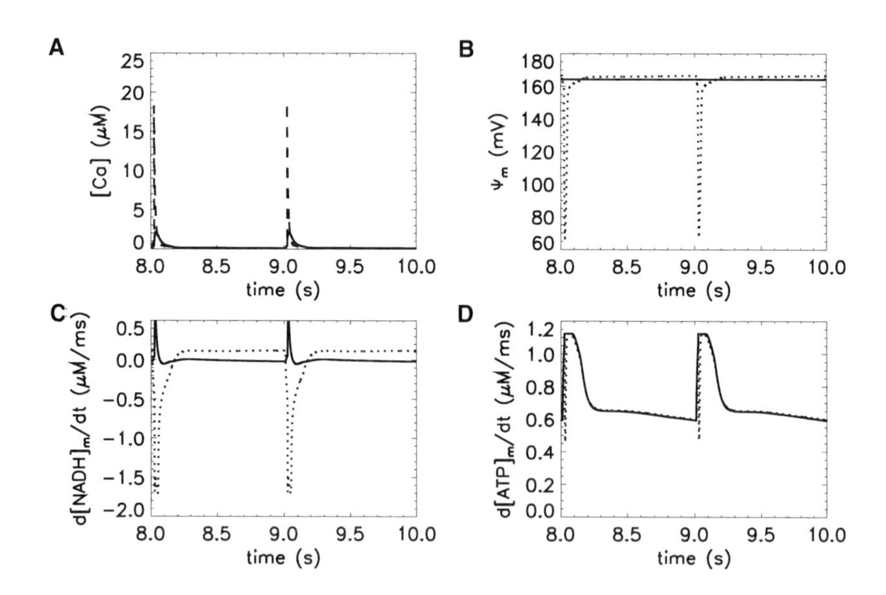

FIGURE 4. (**A**) Dyadic subspace calcium (*dashed line*) and local mitochondrial calcium (*solid line*) concentrations at 1.0 Hz pacing rate. (**B**) Membrane potential changes during the calcium transient (*dotted line*) and under membrane potential clamp (*solid line*). (**C**) The net nicotinamide adenine dinucleotide (NADH) production flux (d[NADH]/dt) in mitochondria local to the dyadic space at 1.0 Hz pacing rate during membrane depolarization (*dotted line*) and membrane potential clamp (*solid line*). (**D**) The adenosine triphosphate (ATP) production flux from the F_1F_0-ATPase (d[ADP]/dt) during membrane depolarization (*dotted line*) and under membrane potential clamp (*solid line*).

resulting transient decrease in mitochondrial membrane potential (Ψ_m) in FIGURE 4B shows large depolarization similar to the local depolarization observed by Duchen and co-workers.[11] Also shown is the case when the mitochondrial membrane potential is held constant.

FIGURE 4C shows that a large transient decrease in net NADH production (production by TCA cycle minus consumption by the respiration driven proton pumps) occurs during depolarization. If there is no depolarization, there is actually a transient increase in the net NADH production due to activation of the Ca^{2+}-dependent dehydrogenases in the TCA cycle. This is followed by a transient decrease below the baseline as the proton pumps catch up and exceed the production of NADH. FIGURE 4D shows that during the $[Ca^{2+}]$ transients there is a transient increase in the ADP production flux. Note that there is a marked reduction in ADP production in the early transient due to the depolarization of the membrane potential. When the membrane potential is clamped, the slight inhibition is relieved.

DISCUSSION

In this work, the Magnus-Keizer model for the mitochondria is modified and coupled to the Dudycha-Jafri model for the TCA cycle. While this enhances the model so that it better represents energy metabolism in the cardiac ventricular myocyte mitochondria, there are some issues that should be discussed. These have to do with the new formulation of the fluxes and assumptions behind the TCA cycle model. First, the fluxes associated with the F_1-F_0-ATPase, namely J_p and J_{hf}, are replaced with the new formulations. The new formulations use voltage dependence derived from bacterial ATPases and relate them to cardiac data.[20,21] However, the fact that the mitochondrial membrane potential dominates function will still be relevant unless the mitochondria almost fully depolarizes. In doing so, the presence of "slip," or a non-exact stoichiometry between the protons needed to produce ATP, has been replaced by an exact 3:1 stoichiometry.

The new formulation of the uniporter is based on the recently published characterization by Kirichok and co-workers.[13] The new formulation of the Na^+-Ca^{2+} exchanger used the empirical formula derived from data on liver by Wingrove and Gunter, with the stoichiometry altered to reflect a 3:1 stoichiometry observed by more recent experiments.[14–16] Once the voltage and Ca^{2+}-dependent behavior were described by a mathematical representation, the total flux was adjusted to match the steady-state and dynamic data observed in heart mitochondria. The new formulation reflects the correct stoichiometry and lack of voltage-dependence. The effects of changes in membrane potential would affect the exchanger through Na^+-H^+ exchange. In the current model Na^+ is assumed to be constant. The model could be improved by inclusion of the exchanger and equations for Na^+ and H^+ balance.

The TCA cycle model by Dudycha and Jafri uses enzyme regulator kinetic data obtained from multiple sources.[17,18] These experimental studies were all performed with isolated enzymes. In the mitochondria the enzymes are found in complexes and are associated with membranes that would most probably affect their behavior. The rates for the enzyme reactions were fit using MRI data from Sherry and co-workers.[24–26] While this data is from intact perfused heart, it has a time resolution of 10 to 20 s and hence gives basically steady-state kinetics.

CONCLUSIONS

A computational model of mitochondrial Ca^{2+} handling and energy metabolism in the cardiac ventricular myocyte is presented that integrates an improved version of the Magnus-Keizer mitochondrial model with the Dudycha-Jafri model for regulation of the tricarboxylic acid cycle. The model takes as input myoplasmic Ca^{2+}concentration data generated by the Jafri-Rice-Winslow guinea pig ventricular myocyte model. The model simulated mitochondrial Ca^{2+} handling both for steady-state conditions as well as change during normal beating of the heart. The dynamics are studied both for changes to bulk myoplasmic [Ca^{2+}] and local changes seen near the Ca^{2+} release sites in the dyadic space.

The model suggests that depolarization of the mitochondrial membrane potential occur during myoplasmic [Ca^{2+}] transients and that these changes are increased in mitochondria that lie adjacent to the dyadic space. Furthermore, it predicts that Ca^{2+}-dependent activation of the F_1F_0-ATPase is necessary in parallel with activation of the TCA cycle to achieve significant increases in ATP production. It also suggests that in mitochondria located at the dyadic junction, the membrane depolarization that occurs from Ca^{2+} uptake is enough to cause a very transient decline in ATP production that recovers, quickly minimizing its impact on total ATP production. In ways such as this, the model can give insight into the complex dynamics of mitochondrial function.

ACKNOWLEDGMENT

This work was supported in part by the Whitaker Foundation.

REFERENCES

1. JAFRI, M.S., S.J. DUDYCHA & B. O'ROURKE. 2001. Cardiac energy metabolism: models of cellular respiration. Annu. Rev. Biomed. Eng. **3:** 57–81.
2. CORTASSA, S., M.A. AON, E. MARBAN, *et al.* 2003. An intergrated model of cardiac mitochondrial energy metabolism and calcium dynamics. Biophys. J. **84:** 2734–2755.
3. MAGNUS, G. & J. KEIZER. 1997. Minimal model of beta-cell mitochondrial Ca^{2+} handling. Am. J. Physiol. **273:** C717–C733.
4. MAGNUS, G. & J. KEIZER. 1998. Model of beta-cell mitochondrial calcium handling and electrical activity. II. Mitochondrial variables. Am. J. Physiol. **274:** C1174–C1184.
5. MAGNUS, G. & J. KEIZER. 1998. Model of beta-cell mitochondrial calcium handling and electrical activity. I. Cytoplasmic variables. Am. J. Physiol. **274:** C1158–C1173.
6. WINGROVE, D.E., J.M. AMATRUDA & T.E. GUNTER. 1984. Glucagon effects on the membrane potential and calcium uptake rate of rat liver mitochondria. J. Biol. Chem. **259:** 9390–9394.
7. WAN, B., K.F. LANOUE, J.Y. CHEUNG & R.C. SCADUTO, JR. 1989. Regulation of citric acid cycle by calcium. J. Biol. Chem. **264:** 13430–13439.
8. CHALMERS, S. & D.G. NICHOLLS. 2003. The relationship between free and total calcium concentrations in the matrix of liver and brain mitochondria. J. Biol. Chem. **278:** 19062–19070.
9. BUNTINAS, L., K.K. GUNTER, G.C. SPARAGNA & T.E. GUNTER. 2001. The rapid mode of calcium uptake into heart mitochondria (RaM): comparison to RaM in liver mitochondria. Biochim. Biophys. Acta. **1504:** 248–261.

10. TROLLINGER, D.R., W.E. CASCIO & J.J. LEMASTERS. 2000. Mitochondrial calcium transients in adult rabbit cardiac myocytes: inhibition by ruthenium red and artifacts caused by lysosomal loading of Ca^{2+}-indicating fluorophores. Biophys. J. **79:** 39–50.
11. DUCHEN, M.R., A. LEYSSENS & M. CROPMPTON. 1998. Transient mitochondrial depolarizations reflect focal sarcoplasmic reticular calcium release in single rat cardiomyocytes. J. Cell Biol. **142:** 975–988.
12. PACHER, P., A.P. THOMAS & G. HAJNOCZKY. 2002. Ca^{2+} marks: miniature calcium signals in single mitochondria driven by ryanodine receptors. Proc. Natl. Acad. Sci. U.S.A. **99:** 2380–2385.
13. KIRICHOK, Y., G. KRAPIVINSKY & D.E. CLAPMAN. 2004. The mitochondrial calcium uniporter is a highly selective ion channel. Nature **427:** 360–364.
14. WINGROVE, D.E. & T.E. GUNTER. 1986. Kinetics of mitochondrial calcium transport. II. A kinetic description of the sodium-dependent calcium efflux mechanism of liver mitochondria and inhibition by ruthenium red and by tetraphenylphosphonium. J. Biol. Chem. **261:** 15166–15171.
15. BAYSAL, K., D.W. JUNG, K. GUNTER, *et al.* 1994. Na^+-dependent Ca^{2+} efflux mechanism of heart mitochondria is not a passive $Ca^{2+}/2Na^+$ exchanger. Am. J. Physiol. **266:** C800–C808.
16. JUNG, D.W., K. BAYSAL & G.P. BRIERLEY. 1995. The sodium-calcium antiport of heart mitochondria is not electroneutral. J. Biol. Chem. **270:** 672–678.
17. DUDYCHA, S. 2000. A detailed model of the tricarboxylic acid cycle in heart cells. Master's thesis, Johns Hopkins University.
18. DUDYCHA, S. & M.S. JAFRI. 2001. Increases in mitochondrial Ca^{2+} can result in increases in NADH production. Biophys. J. **80:** 589.
19. JAFRI, M.S., J.J. RICE & R.L. WINSLOW. 1998. Cardiac Ca^{2+} dynamics: the roles of ryanodine receptor adaptation and sarcoplasmic reticulum load. Biophys. J. **74:** 1149–1168.
20. KAIM, G. & P. DIMROTH. 1999. ATP synthesis by F-type ATP synthase is obligatorily dependent on the transmembrane voltage. EMBO J. **18:** 4118–4127.
21. KNOX, B. & T. TWONG. 1984. Voltage-driven ATP synthesis by beef heart mitochondrial F_1F_0-ATPase. J. Biol. Chem. **259:** 4757–4763.
22. BALABAN, R.S., S. BOSE, S.A. FRENCH & P.R. TERRITO, *et al.* 2003. Role of calcium in metabolic signaling between cardiac sarcoplasmic reticulum. Am. J. Physiol. Cell Physiol. **278:** C285–C293.
23. MURPHY, M. & M. BRAND. 1988. The stoichiometry of charge translocation by cytochrome oxidase and the cytochrome bc1 complex of mitochondria at high membrane potential. Eur. J. Biochem. **173:** 645–651.
24. MALLOY, C.R., A.D. SHERRY & F.M. JEFFREY. 1990. Analysis of tricarboxylic acid cycle of the heart using ^{13}C isotope isomers. Am. J. Physiol. **259:** H987–H995.
25. SHERRY, A.D., P. ZHAO, A.J. WIETHOFF, *et al.* 1998. Effects of aminooxyacetate on glutamate compartmentation and TCA cycle kinetics in rat hearts. Am. J. Physiol. **274:** H591–H599.
26. JEFFREY, F.M, A. RESHETOV, C.J. STOREY, *et al.* 1999. Use of a single ^{13}C NMR resonance of glutamate for measuring oxygen consumption in tissue. Am. J. Physiol. **277:** E1111–E1121.

Rhythmic Ca^{2+} Oscillations Drive Sinoatrial Nodal Cell Pacemaker Function to Make the Heart Tick

TATIANA M. VINOGRADOVA, VICTOR A. MALTSEV, KONSTANTIN Y. BOGDANOV, ALEXEY E. LYASHKOV, AND EDWARD G. LAKATTA

Laboratory of Cardiovascular Science, Intramural Research Program, National Institute on Aging, National Institutes of Health, Baltimore, Maryland 21224, USA

ABSTRACT: Excitation-induced Ca^{2+} cycling into and out of the cytosol via the sarcoplasmic reticulum (SR) Ca^{2+} pump, ryanodine receptor (RyR) and Na^{+}-Ca^{2+} exchanger (NCX) proteins, and modulation of this Ca^{2+}cycling by β-adrenergic receptor (β-AR) stimulation, governs the strength of ventricular myocyte contraction and the cardiac contractile reserve. Recent evidence indicates that heart rate modulation and chronotropic reserve via β-ARs also involve intracellular Ca^{2+} cycling by these very same molecules. Specifically, sinoatrial nodal pacemaker cells (SANC), even in the absence of surface membrane depolarization, generate localized rhythmic, submembrane Ca^{2+} oscillations via SR Ca^{2+} pumping-RyR Ca^{2+} release. During spontaneous SANC beating, these rhythmic, spontaneous Ca^{2+} oscillations are interrupted by the occurrence of an action potential (AP), which activates L-type Ca^{2+} channels to trigger SR Ca^{2+} release, unloading the SR Ca^{2+} content and inactivating RyRs. During the later part of the subsequent diastolic depolarization (DD), when Ca^{2+} pumped back into the SR sufficiently replenishes the SR Ca^{2+} content, and Ca^{2+}-dependent RyR inactivation wanes, the spontaneous release of Ca^{2+} via RyRs again begins to occur. The local increase in submembrane [Ca^{2+}] generates an inward current via NCX, enhancing the DD slope, modulating the occurrence of the next AP, and thus the beating rate. β-AR stimulation increases the submembrane Ca^{2+} oscillation amplitude and reduces the period (the time from the prior AP triggered SR Ca^{2+} release to the onset of the local Ca^{2+} release during the subsequent DD). This increased amplitude and phase shift causes the NCX current to occur at earlier times following a prior beat, promoting the earlier arrival of the next beat and thus an increase in the spontaneous firing rate. Ca^{2+} cycling via the SR Ca^{2+} pump, RyR and NCX, and its modulation by β-AR stimulation is, therefore, a general mechanism of cardiac chronotropy and inotropy.

KEYWORDS: beta-adrenergic receptor stimulation; local Ca^{2+} release; ryanodine receptor; sinoatrial nodal cells

Address for correspondence: Edward G. Lakatta, M.D., Laboratory of Cardiovascular Science, Intramural Research Program, National Institute on Aging, NIH, 5600 Nathan Shock Drive, Baltimore, MD 21224, USA. Voice: 410-558-8202; fax: 410-558-8150.
lakattae@grc.nia.nih.gov

**Ann. N.Y. Acad. Sci. 1047: 138–156 (2005). © 2005 New York Academy of Sciences.
doi: 10.1196/annals.1341.013**

INTRODUCTION

Rhythmic intracellular Ca^{2+} cycling—not directly dependent on surface membrane potential—is an integral component of biological clocks that regulate diverse vital functions throughout nature.[1–5] A capacity to generate rhythmic changes in intracellular Ca^{2+} is imparted by organelles that can sequester and release Ca^{2+} at specific intervals. One type of intracellular Ca^{2+} cycling involves the endoplasmic reticulum or sarcoplasmic reticulum (SR), which is endowed with a Ca^{2+} pump and release channels that permit rhythmic cycling of Ca^{2+} into and out of this compartment. The periodicity of such Ca^{2+} cycling is regulated by the Ca^{2+} pump and Ca^{2+} release channel characteristics, their acute modulation by various signaling pathways, and the quantity of Ca^{2+} available for pumping. The gain of multiple cell signaling pathways can be modulated by changes in the periodicity of these Ca^{2+} rhythms (i.e., frequency modulation of an amplitude function).

The heart's pacemaker cells generate spontaneous rhythmic changes of their membrane potential, thereby producing roughly periodic electrical depolarizations or action potentials (APs). Pacemaker cells within the heart having "clocks" with the briefest periods "capture" or drive other excitable cells to depolarize the myocardium. Sinoatrial nodal pacemaker cells (SANCs) are the dominant cardiac pacemaker cells because they exhibit shorter periods between spontaneous APs than do atrioventricular nodal, or His-Purkinje cells. A role of specific sarcolemmal ion channels in the spontaneous membrane potential of cardiac pacemaker cells has been firmly established by decades of research applying electrophysiological techniques.[6–12] The role of intracellular Ca^{2+} releases in cardiac arrhythmias had previously been most appreciated,[13] but earlier studies,[14–16] however, had failed to embrace or had not emphasized a specific role for local increases in Ca^{2+} inside the cell membrane with respect to modulation of the pacemaker potential and the spontaneous beating rate of SANC. The advent of Ca^{2+} sensitive indicators, coupled to fluorescence microscopy or confocal imaging, and simultaneous measurement of membrane potential or current, has permitted investigation into the role of Ca^{2+} release in normal cardiac pacemaker function.[17–32]

INTRACELLULAR CALCIUM CYCLING

Localized, Submembrane Ryanodine Receptor Ca^{2+} Release during Diastolic Depolarization in SANC

We studied single, spindle-shaped, spontaneously beating SANC isolated from the rabbit hearts. The beating rate of these SANC (n = 118) was between 2 Hz and 5 Hz and their capacitance between 20 pF and 95 pF. Both beating rate and cell size were normally distributed around the average values of 3 Hz and 50 pF, respectively (FIG. 1A). There was no correlation between the spontaneous beating rate and cell size (FIG 1A).

Like other excitable cells, SANC cycle Ca^{2+} into and out of their SR.[19,24,25,28,32] Activation of ryanodine receptors (RyRs), the SR Ca^{2+} release channel, like that of many other classically described sarcolemmal ion channels, is cyclically regulated by time- and Ca^{2+}-dependent gating mechanisms.[33,34] Upon activation, RyRs release

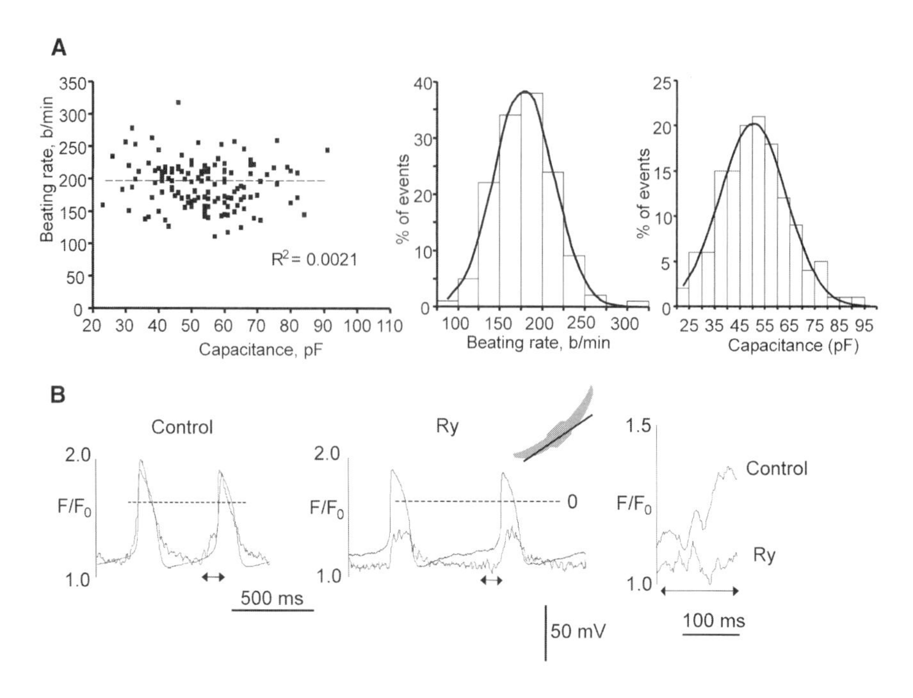

FIGURE 1. (**A**) Sinoatrial nodal pacemaker cells (SANC) size and beating rate: (*Top left*) Correlation between SANC (n = 118) size and beating rate; (*middle*) SANC beating rate histogram; (*right*) capacitance histogram. (**B**) Variation of subsarcolemmal Ca^{2+} release in the presence or absence of ryanodine (Ry). Simultaneous recordings of membrane potential and subsarcolemmal [Ca^{2+}] defined using confocal line-scan images of SANC (***inset***) in the presence or absence of 3 μmol/L Ry. F/F_0 is the instantaneous fluorescence of the Ca^{2+} indicator, fluo-3, relative to its minimal fluorescence (F_0), and other experimental conditions are as previously published.[19,25] Note that in the predrug control state, the Ca^{2+} waveform exhibits an increase during the later part of the spontaneous diastolic depolarization (DD) (*bracketed by arrows and shown at greater resolution in rightmost panels*). Thus, DD component of the Ca^{2+} waveform is abolished by Ry concurrent with the beating rate slowing.

Ca^{2+} into a narrow cleft beneath the SANC sarcolemma at two distinct times during each spontaneous SANC duty cycle (FIGS. 1B and 2A): in response to the AP rapid upstroke, as in ventricular myocytes, and during the later part of spontaneous diastolic depolarization (DD) before the next AP upstroke.

Local RyR Ca^{2+} Releases Do Not Require Membrane Depolarization

Studies in SANC that are voltage clamped at the maximum diastolic potential or permeabilized with saponin indicate that ignition of local Ca^{2+} releases (LCRs) during DD in spontaneously beating cells does not require the concomitant depolarization. LCRs occur in permeabilized cells and their characteristics vary with the bathing [Ca^{2+}] (FIG. 2B). Elevating free [Ca^{2+}] from 100 to 250 nmol/L results in

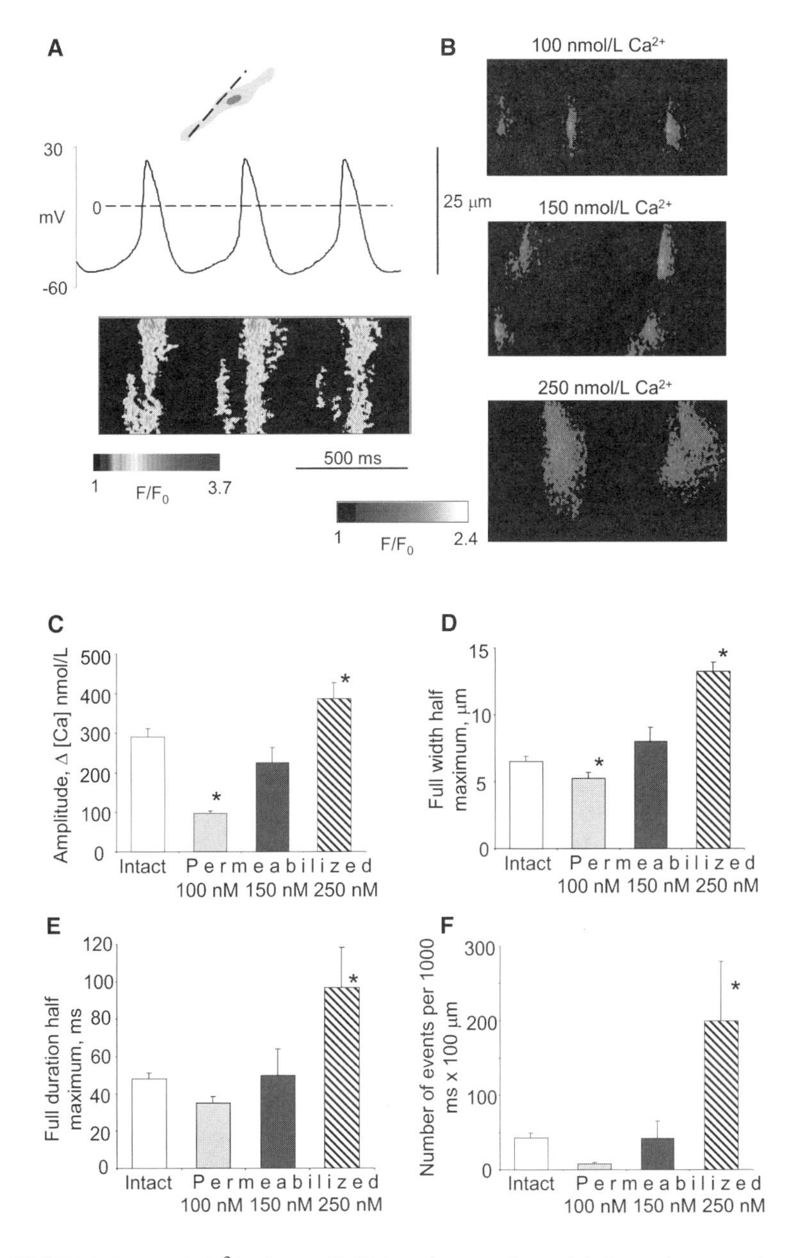

FIGURE 2. Local Ca^{2+} releases (LCRs) and comparison of their spatiotemporal properties in intact spontaneously beating and permeabilized sinoatrial nodal pacemaker cells (SANCs). (**A**) Recordings of action potentials (APs) and confocal line-scan images (see cell cartoon) in a representative, spontaneously beating SANC. (**B**) Confocal line-scan images in a "skinned" cell recorded with Fluo-4 K salt, in a representative "skinned" cell bathed in 100nmol/L, 150 nmol/L and 250 nmol/L [Ca^{2+}]. (**C–F**) Average characteristics of subsarcolemmal LCRs in intact, spontaneously beating cells and skinned cells.[32]

increases in the absolute LCR amplitude (FIG. 2C), spatial width (FIG. 2D), duration (FIG. 2E), and frequency (FIG. 2F). Calibration of fluo-3 fluorescence[32] estimated an average diastolic and systolic [Ca^{2+}] in intact cells of 160 ± 13 and 1150 ± 213 nmol/L, respectively. FIGURE 2C–2F also shows that spatiotemporal properties of LCRs in permeabilized cells at 150 nmol/L Ca^{2+} most closely resemble those in spontaneously beating cells.

Voltage clamping of SANC terminates spontaneous beating, which is restored on voltage clamp removal. FIGURE 3A illustrates a confocal line-scan image of subsarcolemmal Ca^{2+} in a representative, spontaneously beating SANC prior to and during acute voltage clamp at the maximum diastolic potential. The main point of the figure is that LCRs not only occur during DD during spontaneous beating, but also occur during voltage clamp, in the absence of a change in membrane potential. Similar LCRs were recorded when cells are paced under voltage clamp conditions with an AP waveform (to ensure SR Ca^{2+} loading) before voltage clamping of the membrane at –60 mV. The time during voltage clamp of each cell was divided into "would-be" cycles; that is, time intervals equal to the cycle length during spontaneous beating

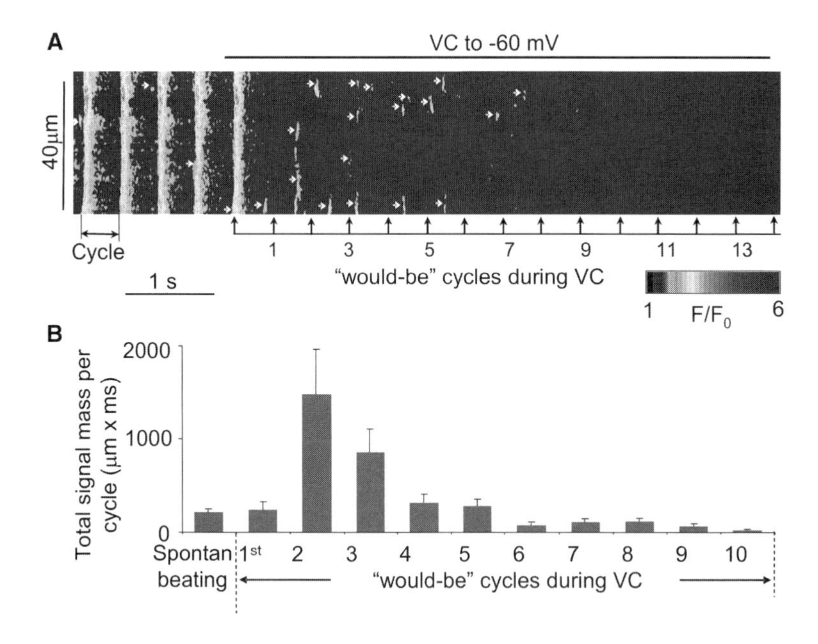

FIGURE 3. Local Ca^{2+} releases (LCRs) total signal mass within the cycle during spontaneous beating and during would-be inter-action-potential (AP) intervals during voltage clamp at the maximum diastolic potential. (**A**) Line-scan image of a representative sinoatrial nodal pacemaker cell (SANC) during spontaneous beating and during voltage clamp. *Horizontal arrowheads* indicate LCRs, and *vertical arrows* indicate would-be cycles during voltage clamp (i.e., times corresponding to the inter-AP interval if spontaneous firing were to have continued in the absence of voltage clamp). Local change in Ca^{2+} resulting from each LCR event during diastolic depolarization or during voltage clamp was characterized by its signal mass. (**B**) Average total LCR signal mass in 9 cells in control during spontaneous beating and during each would-be cycle of voltage clamp.[32]

(FIG. 3A). The first would-be cycle during voltage clamp is of particular significance, because at this time SR Ca^{2+} loading or RyR inactivation status would not be expected to differ from that during the prior spontaneous beating. FIGURE 3B demonstrates that the signal mass of local Ca^{2+} releases during the first would-be cycle does not differ from that during spontaneous beating.

Local RyR Ca^{2+} Releases Are Roughly Periodic

Since LCRs in intact SANC during voltage clamp or in SANC permeabilized with saponin do not require a depolarizing trigger, it is possible that they are a manifestation of an intracellular rhythmic or roughly periodic Ca^{2+} oscillator. FIGURE 4 shows that the fast Fourier transform (FFT) of the Ca^{2+} releases during voltage clamp (FIG. 4B) exhibits periodicity, with a dominant power at 2.9 Hz. Membrane current fluctuations accompanied Ca^{2+} fluctuations during the voltage clamp. The FFT of these current oscillations exhibits similar periodicity as that of LCRs, with the dominant power at 2.9 Hz (FIG. 4B). Importantly, the dominant period of 345 ms both in calcium and current oscillations during voltage clamp is about 20% shorter than the spontaneous cycle length of 435 ms in this cell before voltage clamp.

Periodicity of Local RyR Ca^{2+} Releases during Steady-State Spontaneous Beating is Linked to the Spontaneous Cycle Length

The spontaneous beating rate among isolated rabbit SANC varies over a substantial range of frequencies (FIG. 1A). If submembrane Ca^{2+} releases during spontaneous beating are rhythmic and their occurrence does indeed modulate the spontaneous beating frequency, then cells that beat faster ought to have shorter spontaneous LCR periods than those that beat at a slower rate. FIGURE 4D shows that the interval from the rapid upstroke of the Ca^{2+} transient caused by the prior AP to the upstroke of the subsequent LCR (FIG. 4C) directly correlates with the variation in the spontaneous cycle length among different cells. Importantly, regardless of the absolute spontaneous cycle length, which varies from cell to cell, LCRs predominantly occur at constant relative time (i.e., at 80% to 90% of that cycle length) (FIG. 4E). Most of the spontaneous local Ca^{2+} release occurs beneath the cell membrane during the later part of the DDs, producing an inward current to accelerate the terminal part of the DD and driving the membrane potential to reach threshold to fire on AP.

Na^{+}-Ca^{2+} Exchanger is a Crucial Element Linking LCRs to Membrane Depolarization

Although the experiments in FIGURES 2 and 3 demonstrate that LCRs in spontaneously beating cells do not require changes in membrane potential, there is still a possibility that they could be triggered by Ca^{2+} influx during spontaneous beating as well as during voltage clamp (e.g., by T-type Ca^{2+} current).[24] A comparison of LCRs during spontaneous beating before and after exposure to 50 μmol/L Ni^{2+} indicates no difference in the number of LCRs per cycle during spontaneous beating, and in all cells subjected to 50 μmol/L Ni^{2+} LCRs are not abolished during voltage clamp.[32]

Several time- and voltage-dependent ion currents become activated during the DD, and therefore may link RyR Ca^{2+} release from the SR to DD modulation. T-type and L-Type Ca^{2+} currents ($I_{Ca,T}$, $I_{Ca,L}$), hyperpolarization-activated inward current

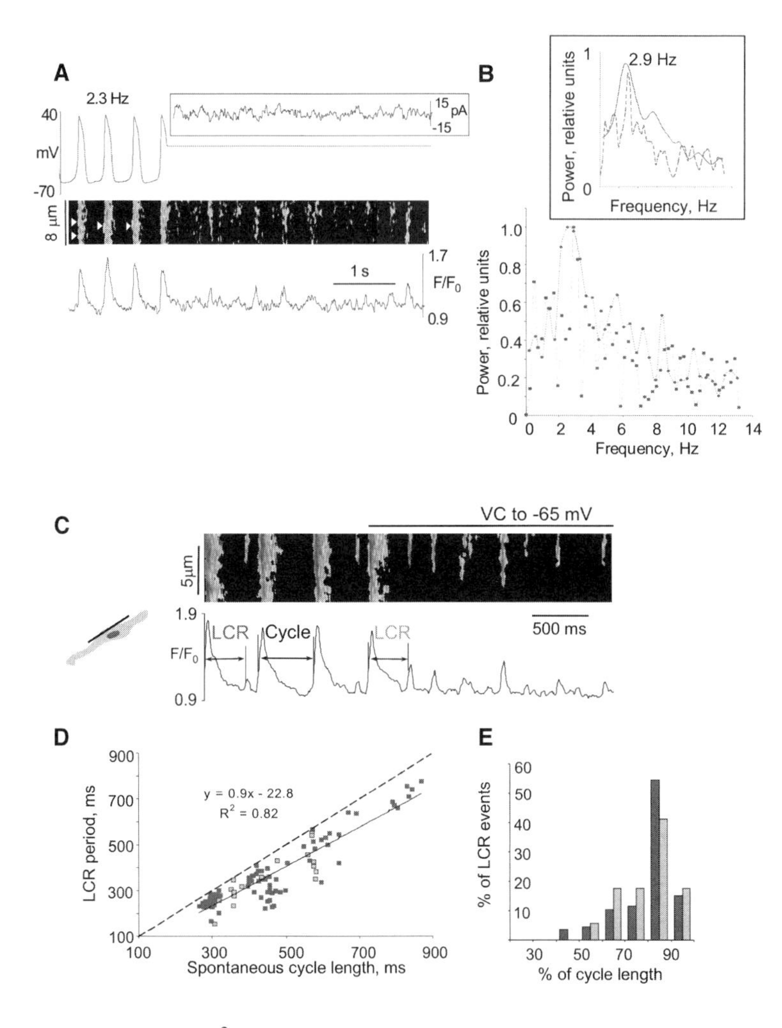

FIGURE 4. Local Ca^{2+} releases (LCRs) exhibit periodicity during voltage clamp and during spontaneous beating. (**A**) Simultaneous recordings of action potential (AP) (*top*), line-scan image (*middle*), and normalized fluorescence (*bottom*) averaged over the image width in a representative cell before and during voltage clamp to -10 mV. ***Inset*** shows current oscillations recorded during voltage clamp. (**B**) Fast Fourier transform of Ca^{2+} waveform (*solid line*) and current oscillations (*dotted line*) shown in **A**. ***Inset*** depicts smoothed curves obtained with Gaussian filter. (**C**) Confocal linescan image (see cell illustration) and normalized subsarcolemmal fluorescence averaged over the image width in a representative sinoatrial nodal pacemaker cell measured before and during voltage clamp. *Double-headed arrows* delineate the LCR period during spontaneous beating and during the first "would-be" cycle during voltage clamp. (**D**) Periodicity of LCRs during spontaneous beating and during the first "would-be" cycle of voltage clamp: (black symbols) relationship between cycle length and the LCR period during spontaneous beating (10 cells, 86 LCRs); (gray symbols) the relationship between LCR period and the first "would-be" cycle during voltage clamp (13 cells, 17 LCRs). *Dashed line* is the line of identity. (**E**) (*black bars*) Probability of LCR occurrence as a function of the relative cycle length in cells in **D** during spontaneous beating; (*gray bars*) during the first "would-be" cycle during voltage clamp.[32]

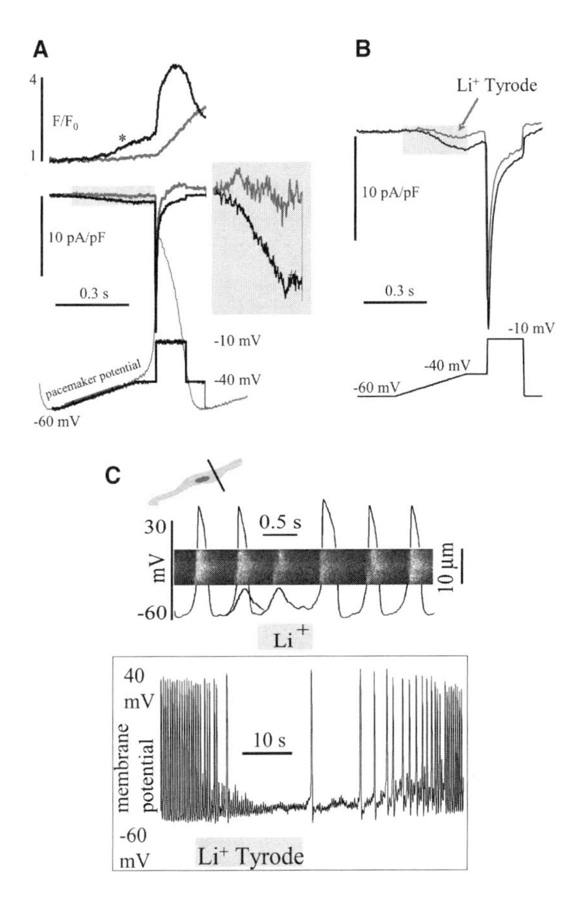

FIGURE 5. Effects of ryanodine (Ry) and Li^+ on inward current during diastolic depolarization (DD) and spontaneous beating in sinoatrial nodal pacemaker cells (SANCs). (**A**) Ry inhibits submembrane Ca^{2+} increase and inward current during the DD without affecting peak $I_{Ca,L}$. Normalized fluorescence (*top*) and membrane current (*middle*) under voltage clamp before (black) and 4 min after (gray) addition of 3 μmol/l ryanodine; voltage-clamp protocol is shown in *bottom panel*. *Middle right panel* displays the indicated part of the current record at greater magnification to more optimally show the inhibitory effect of Ry on inward current during the voltage ramp. (*) Local $[Ca^{2+}]_i$ increase in the absence of ryanodine and corresponding inward current. Both are inhibited in the presence of ryanodine. (**B**) Membrane current during voltage clamp before (black) and 10 s after (gray) superfusion with Na^+-free, Li^+-containing solution during voltage protocol (*bottom*). Highlighted area shows an inhibitory effect of Na^+-free Li^+ spritz on inward current during the voltage ramp. (**C**) (*Top*) Linescan image of Ca^{2+} release with superimposed action potential (AP) records during rapid and brief superfusion with a solution in which Na^+ was replaced by Li^+. Note that the maneuver blocked the subsequent AP firing. Scanned line was oriented perpendicular to long cell axis at half its depth (***inset***). Gray curve superimposed on the last AP preceding spritz of Na^+-free solution is a copy of the residual membrane potential oscillation observed during the Li^+ solution spritz. (*Bottom*) The inhibitory effect of Na^+ replacement by Li^+ on spontaneous beating in SANCs. AP recordings during onset and after washout of Na^+-free, Li^+-containing solution.[25]

(I_f), chloride current, the rapid component of delayed rectifier K^+ current (I_{Kr}), a time-independent background current carried by Na^+, and Na^+-Ca^{2+} exchanger (NCX) current all exhibit a Ca^{2+}-dependence. The effect of LCR occurrence during DD on the beating rate, however, is mediated by an inward current as ryanodine, which abolishes the diastolic Ca^{2+} release, suppresses the simultaneously measured inward current (FIG. 5A). Earlier studies had identified NCX current in cat latent pacemaker cells,[16] and this current has also been implicated in the spontaneous beating of toad pacemaker cells.[35] FIGURE 5B indicates that the electrogenic NCX (blocked by lithium) generates the Ca^{2+}-dependent inward current resulting from RyR Ca^{2+} release. Blockade of either the local Ca^{2+} release or the NCX should have potent effects on the spontaneous beating of these cells. An acute, rapid, transient application of lithium reduces the slope of the later DD and prevents the subsequent AP from firing without altering the diastolic RyR Ca^{2+} release (FIG. 5C, *top*). A longer exposure to Li^+ abolishes beating (FIG. 5C, *bottom*).

The magnitude of the ryanodine effect on adult pacemaker cell spontaneous beating rate has varied among studies (from 20% to 100%, review[28]). A critical dependence of cardiac pacemaker regulation on RyR Ca^{2+} release-NCX activation demonstrated in adult rabbit SANC becomes manifest early during development. The normal developmental increase in heart rate *in vitro* is markedly depressed in cardiac myocytes differentiated *in vitro* from embryonic stem (ES) cells with a functional knockout (KO) of the RyR2 (FIG. 6A), and these cells exhibit a markedly depressed late DD slope (FIG. 6C).[30] Exposure of wild-type ES-derived cardiomyocytes to ryanodine, which at low concentration locks RyRs in a subconductance state, imparts the characteristics of these cells to RyR KO cells (FIG. 6B). NCX KO mice die in embryo without evidence of the heart ever beating.[36]

Numerical Modeling of Submembrane Ca^{2+} Release Effect on SANC Beating Rate

Numerical modeling indicates that coupling of the RyR-generated Ca^{2+} waveform to the NCX current, when the latter is the only Ca^{2+}-dependent current operational during DD, is sufficient to increase the beating rate via DD acceleration.[28] A novel numerical model[31] that features local diastolic Ca^{2+} release component predicts that both the timing and amplitude of the RyR Ca^{2+} release during the DD determine its impact on SANC spontaneous beating rate (FIG. 7A–C). Thus, NCX is a partner of the RyR in amplification of DD and in modulation of the SANC beating rate.

We utilized a combination of experimental and numerical approaches to characterize how depolarizations produced by individual LCRs via NCX impact on the DD fine structure to control the spontaneous SANC firing rate.[37] Exponential LCR growth during the DD imparted a nonlinearity to its later part, which exhibited beat-to-beat potential (V_M) fluctuations, averaging about +2 mV. Numerical modeling, utilizing experimentally measured LCR amplitudes and delay times measured from the prior beat, simulated the stochastic ensemble of LCRs and respective NCX currents (FIG. 7D–F) that reproduced V_M fluctuations of the same magnitude and timing as those measured experimentally. The modeling also showed that the timing and amplitude of V_M fluctuations were determinants of the onset and amplitude of the nonlinear DD

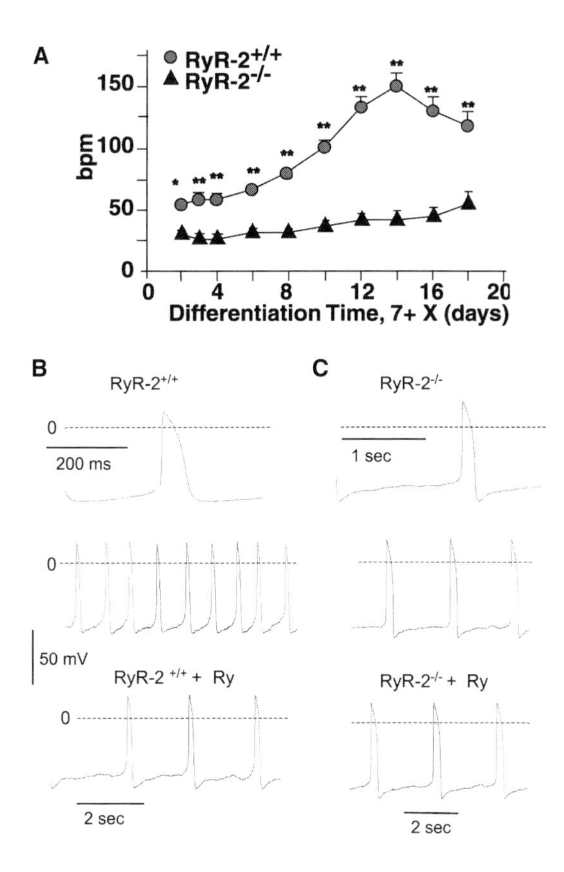

FIGURE 6. Spontaneous beating rates of ryanodine receptor (RyR)-$2^{+/+}$ and RyR-$2^{-/-}$ cardiomyocytes in embryoid bodies and single isolated cells in the absence and presence of ryanodine. (**A**) Beating rate in the RyR-$2^{-/-}$ cardiomyocytes within embryoid bodies and its increase with developmental time is significantly suppressed compared to RyR-$2^{+/+}$ cells. (**B, C**) Action potential recordings in representative RyR-$2^{+/+}$ and (**B**) RyR-$2^{-/-}$ cardiomyocytes. Firing rate of RyR-$2^{+/+}$ cell before (*middle*, 71 b/min) and after (*bottom*, 26 b/min) Ry (10 µmol/L). (**C**) The firing rate of RyR2 KO cardiomyocytes before (*middle*) and after (*bottom*) exposure to Ry did not change (26 b/min).[30]

component. Maneuvers that altered LCR timing or amplitude (i.e., ryanodine, BAPTA [1,2-Bis(2-aminophenoxy)ethane-N,N,N′,N′-tetraacetic acid tetrakis(acetoxymethyl ester)], nifedipine, or isoproterenol) similarly affected V_M fluctuations during the late DD, and the V_M fluctuation response evoked by these maneuvers was tightly correlated with the change in spontaneous beating rate. When changes in LCR properties and numbers are introduced into our new stochastic SANC model, the model predicts experimentally measured changes in the spontaneous beating rate induced by these maneuvers. Thus, ensemble of elementary NCX currents induced by LCRs links LCRs to the spontaneous V_M fluctuations and modulates the late, nonlinear DD fine structure, linking LCRs to the spontaneous SANC beating rate.[37]

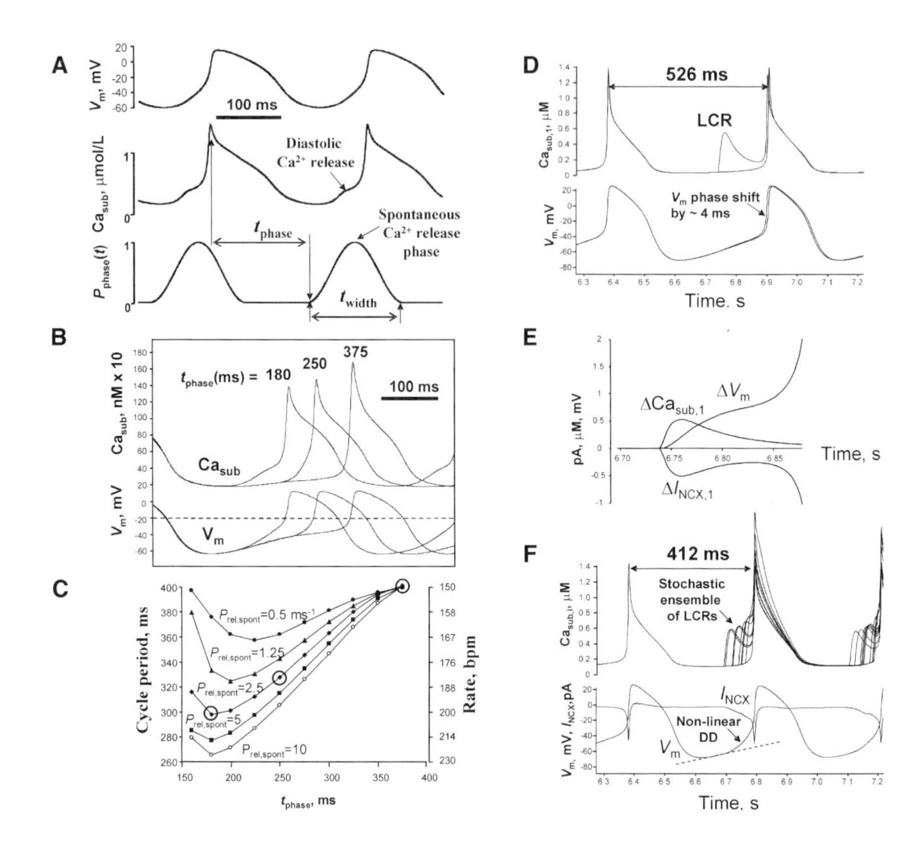

FIGURE 7. Diastolic calcium release controls the beating rate of rabbit sinoatrial nodal pacemaker cells (SANCs): numerical modeling of the coupling process. (**A**) Simulated spontaneous action potentials (APs), V_m and submembrane Ca^{2+}, Ca_{sub}. The spontaneous Ca^{2+} release flux was described as $j_{spont} = P_{phase}(t) \cdot P_{rel,spont} \cdot (Ca_{rel} - Ca_{sub})/[1 + (K_{rel}/Ca_{sub})^2]$, where $P_{phase}(t) = 0.5 - 0.5 \cdot \cos[2 \cdot \pi \cdot (t - t_{phase})/t_{width}]$ is the diastolic Ca^{2+} release phase function, t is time after the previous Ca^{2+} transient peak, $t_{width} = 150$ ms, $t_{phase} = 150$ ms, $P_{rel,spont} = 5\ ms^{-1}$ is the rate constant for the spontaneous Ca^{2+} release from the junctional sarcoplasmic reticulum. (**B**) Steady-state firing of APs, V_m, and respective Ca_{sub} transients simulated with varying phase of the diastolic release (t_{phase}). The simulated traces were shifted along the time axis to be in phase at maximum diastolic potential. (**C**) Both the phase and the amplitude of the diastolic Ca^{2+} release control the steady-state rate of AP firing. The family of plots shows relationships of cycle period (or rate) versus the phase of the diastolic release (t_{phase}) for various Ca^{2+} release amplitude determined by $P_{rel,spont}$ (in ms^{-1}). Circled data points correspond to simulations shown in panel **B**. (**D**) Introduction of only one local Ca^{2+} release (LCR) (*top*) into our recent stochastic sinoatrial nodal pacemaker cell model[31] results in AP upstroke phase shift towards earlier occurrence (*bottom*). (**E**) Unitary membrane depolarization (ΔV_m) and local unitary Na^+-Ca^{2+} exchanger (NCX) current ($\Delta I_{NCX,1}$) were determined as the difference (Δ) between simulations with and without an LCR occurrence shown in panel **D**. (**F**) Introduction of a stochastic ensemble of 40 LCRs results in significant shortening of the cycle length (426 ms vs. 526 ms, panel **D**) as the respective stochastic ensemble of inward-going unitary NCX currents (I_{NCX}) changes diastolic depolarization (DD) from linear to nonlinear fashion ("Nonlinear DD," *bottom*) reducing time to reach the threshold for L-type Ca^{2+} channel activation and thus the time at which the subsequent AP fires.[31,37]

Synchronization of Ryanodine Receptor Ca^{2+} Release in Sinoatrial Pacemaker Cells Underlies β-Adrenergic Receptor Modulation of Spontaneous Beating Rate

While it has long been recognized that β-adrenergic receptor (β-AR) stimulation increases the spontaneous beating rate of SANC, specific links between β-AR signaling and the resultant increase in firing rate have not been fully elucidated. Multiple ion channels are affected by β-ARs-cAMP signaling.[38–42] Given the more recently recognized role of diastolic RyR Ca^{2+} release in pacemaker cells spontaneous firing rates,[28,29] an important issue arises as to whether the increase in SANC beating rate effected by β-ARs also involves an effect on this Ca^{2+} release.

There is evidence to indicate that altered Ca^{2+} flux via RyR is indeed involved in the chronotropic effect of β-ARs in toad, guinea pig, and rabbit SANCs.[17–19,40] The total signal mass of Ca^{2+} released during DD reflects the total number of RyRs activated during a given DD. The signal mass of LCRs during diastole in SANC, and its phase, determine the impact of this Ca^{2+} release with respect to NCX activation on membrane potential depolarization and beating rate.[28,31] β-AR stimulation increases the number of local subsarcolemmal Ca^{2+} releases during a given DD (FIG. 8A and B) and also increases their amplitudes. Thus, synchronization of RyRs activation states (RyR recruitment) by β-AR stimulation in SANC leads to an increased number of RyRs participating in local diastolic Ca^{2+} release in SANC. This spatiotemporal synchronization of RyR Ca^{2+} release, due to synchronization RyR gating by β-AR stimulation, augments the inward current via NCX during the DD (FIG. 8C). Ryanodine prevents the effect of β-ARs to increase the diastolic Ca^{2+} release signal mass and the inward current it produces (FIG. 8C). The effect of β-AR stimulation to increase the spontaneous SANC firing rate is markedly blunted if normal RyR function is disabled (FIG. 8D), suggesting that modulation of Ca^{2+} release via RyR is an essential feature of optimal β-AR stimulation-induced acceleration of the SANC firing rate. In other terms, the RyR Ca^{2+} release during DD acts as a "switchboard" to link β-AR stimulation to an increase in SANC firing rate: recruitment of additional activated RyRs by β-AR stimulation, and partial synchronization of LCR occurrence leads to an increase in the heart rate. New evidence indicates that autonomic nervous signals to elicit tachycardia *in vivo* are decoded via ryanodine-sensitive, rhythmic Ca^{2+} cycling within pacemaker cells.[43]

Recent evidence indicates that constitutive protein kinase A (PKA) activation generates the rhythmic spontaneous Ca^{2+} oscillations in rabbit SANC and regulates intrinsic, spontaneous beating even in the absence of β-AR stimulation.[44] Constitutive β-AR activation is not a factor in the PKA activation dependence of beating rate, as neither the β_1-AR or β_2-AR blockers affect the beating rate. PKA inhibition substantially reduces phosphorylation of phospholamban (PLB) at serine 16, and confocal imaging of LCR shows that PKA inhibition also markedly increases the LCR period, and this effect is highly correlated with the concomitant reduction in the beating rate.[44] Stimulation of β-AR with isoproterenol produces changes opposite to PKA inhibition (i.e., PLB phosphorylation increases, LCR period decreases, and beating rate increases).

A modified version of our primary SANC model[31] simulated the effects of PKA inhibition.[45] When experimentally measured changes in both LCR (numbers and phase) due to PKA inhibition and sarcolemmal ion conductances (g_{CaL}, g_h, g_{Kr} decreased to 80% of control) were introduced into the model, the simulation faithfully

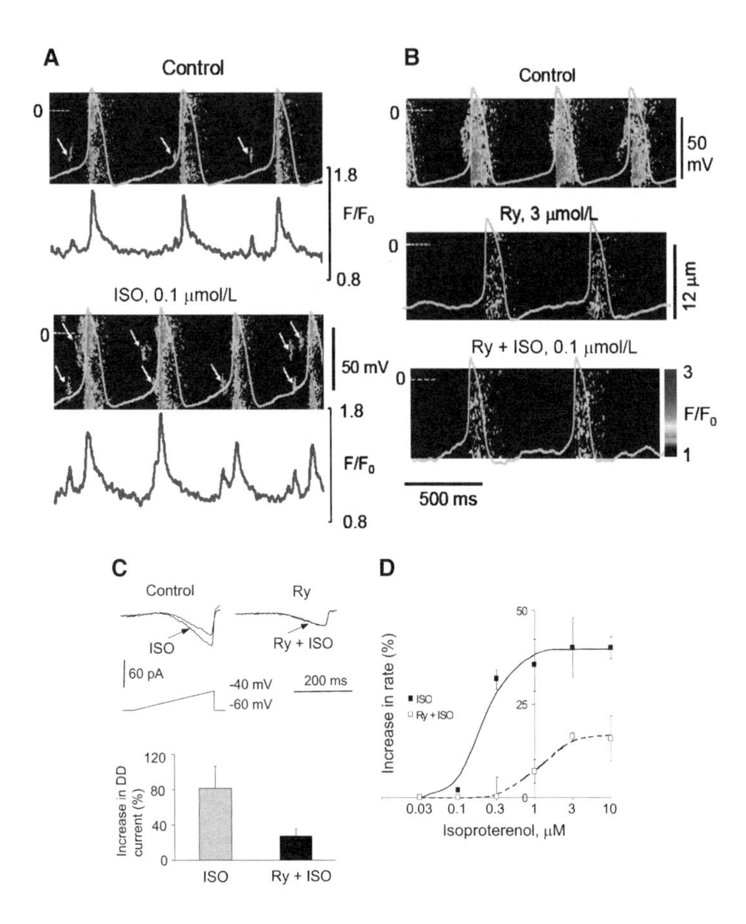

FIGURE 8. β-adrenergic stimulation effects on local ryanodine receptor (RyR) diastolic Ca^{2+} releases and diastolic depolarization (DD) current in sinoatrial nodal pacemaker cells (SANCs) with normal and disabled RyR function. (**A**) (*Top*) Simultaneous action potential (AP) recordings (gray curve) and linescan images of Ca^{2+} releases in the same cell, along the same scanline, before and during application of Isoproterenol. (*Bottom*) Normalized subsarcolemmal fluorescence. Isoproterenol (ISO) induced an increase in the number, brightness, and spatial width of local diastolic Ca^{2+} releases (*white arrows*), and this was accompanied by a 20% increase in the firing rate. (**B**) Simultaneous AP recordings and linescan images of Ca^{2+} releases along the same scanned line in control, after application of 3 μmol/L of ryanodine and of ryanodine + ISO. When RyRs are inhibited by ryanodine, local diastolic Ca^{2+} releases are suppressed, the firing SANC rate decreases, and β-adrenergic receptor (β-AR) stimulation fails to amplify local diastolic Ca^{2+} releases and fails to increase the SANC firing rate. (**C**) Effects of RyR inhibition on the ISO potentiation of inward current during a voltage clamp ramp simulating the DD in the SANC. *Top panels* show the voltage protocols and original current recordings for the simulated DD current. *Bottom panel* shows relative increase in mean DD current amplitude by 1 μmol/L ISO before (n = 5) and after block of RyR with 3 μmol/L. D, Effect of RyR inhibition on the concentration response of SANC firing rate to β-AR stimulation. ISO causes a dose-dependent increase in the firing rate with an apparent threshold concentration of 0.1 μmol/L. This effect is markedly decreased in the presence of ryanodine. The maximal effect of ISO occurred at 1 and 3 μmol/L in the absence and presence of ryanodine, respectively.[19]

reproduced the experimental effect on beating rate. Introduction into the model of changes induced in LCR by PKA inhibition alone produced a 23.2% effect, whereas introducing ion channel changes alone resulted in relatively small effect of 8.7%; normal rhythm was maintained in both cases. When sarcolemmal ion conductances were further decreased (i.e., to 66.7% of control without the LCR support), the beating rhythm became unstable and disrhythmic, similar to the experimental extremes during PKA inhibition. Introduction of LCRs into the model under the very same conditions restored the stable beating rhythm. Thus, the model predicts that LCRs occurring during the late phase of DD provide a powerful and timely prompt for the membrane to depolarize to the level of Ca^{2+}channel activation threshold, not only to regulate the beating rate but also to ensure a healthy rhythmic beat.[45]

Thus, constitutive PKA activation (i.e., not dependent on β-AR stimulation) generates intrinsic SANC rhythmicity; stimulation of β-AR extends the modulatory range of PKA activation.[44] The link between the spontaneous beating of SANC and synchronization of diastolic RyR Ca^{2+} release is in part attributable to cAMP-PKA-dependent phosphorylation of Ca^{2+} flux into the cell and Ca^{2+} pumping into the SR and release from SR, mediated by PKA-dependent phosphorylation of PLB, RyRs, and L-type Ca^{2+} channels. The potential role of a direct action of β-AR stimulation on NCX in pacemaker cells, to date, has not been determined.

Synchronization of RyR Ca^{2+} Release in Ventricular Cell Underlies β-ARs Modulation of Contractility

In ventricular myocytes, synchronization of L-type Ca^{2+} channel activation by cAMP-PKA signaling following depolarization, and synchronization of RyR Ca^{2+} releases generate a global cytosolic Ca_i transient of increased amplitude and accelerated kinetics. Thus, synchronization of RyR activation and synchronization of the ensuing Ca^{2+} release by cAMP-PKA signaling causes the heart to beat both stronger and faster. It follows, therefore, that such synchronization of RyR activation is a unified mechanism that links cardiac inotropy and chronotropy. In myocytes synchronization of RYR firing by cAMP-PKA signaling is manifest following AP (i.e., in response to activation of L-type Ca^{2+} current); whereas in SANC, it occurs prior to the AP and modulates activation of the NCX and the late DD rate, thus determining when the next AP will occur. An apparently similar or analogous mechanism that augments the beating rate of SANC leads to arrhythmias in ventricular cells.[46–49] Experimentally, however, β-AR stimulation-induced arrhythmias are usually observed when the same external pacing rate applied prior to β-AR stimulation also prevails during β-AR stimulation. In nature, however, β-AR stimulation modulates Ca^{2+} in both SANC and ventricular myocytes, and commensurate with the ensuing enhanced beating frequency (shorter diastole), the subsequent AP occurs at a sufficiently early time to "overdrive" the late diastolic release (i.e., to suppress its occurrence by activation of L-type channels which trigger RyRs to produce RyR Ca^{2+} release as a systolic event).[45,46]

Spontaneous SR Ca^{2+} Oscillations Generate Automaticity in Cardiac Cell Types other than Sinoatrial Nodal Pacemaker Cells

Spontaneous, RyR SR-generated oscillations of intracellular Ca^{2+} have previously been indirectly detected by laser light scattering technique; or recordings of

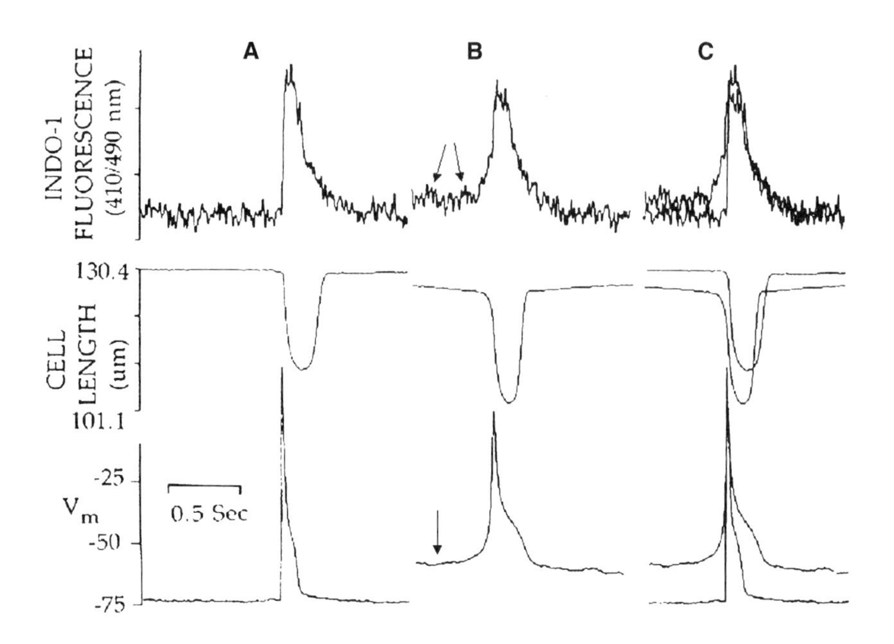

FIGURE 9. Spontaneous sarcoplasmic reticulum Ca^{2+} release triggers action potential (AP) in ventricular myocytes. (**A**) Electrical stimulator triggers an abrupt AP that generates a synchronized Ca^{2+}-triggered (INDO-1) fluorescence and contraction. (**B**) Spontaneous Ca^{2+} release (*arrow*) generates a slow change in membrane potential that reaches threshold to open Na^+ channels to trigger spontaneous AP. (**C**) Superimposition of A and B.[51]

voltage; or current, or myofilament motion or tension in unstimulated, multicellular cardiac muscle and Purkinje fibers (see Ref. 50 for review). With the advent of intracellular Ca^{2+} probes (see Refs. 50 and 51 for review) the underlying Ca^{2+} oscillations were subsequently directly demonstrated. These spontaneous Ca^{2+} oscillations originate locally within cells, and while stochastic they are roughly periodic and their periodicity (<0.1 Hz to ~7 Hz) varies with cell and SR Ca^{2+} loading. Further, during electrical stimulation, AP-triggered and spontaneous Ca^{2+} release are interactive with respect to their magnitude and occurrence (see Ref. 51 for review). In ventricular, atrial, and Purkinje fiber cells, spontaneous SR calcium oscillations produce delayed after-depolarizations[52] driven by a calcium-activated transient inward current that may initiate arrhythmias.[53] Spontaneous Ca^{2+} release of this sort leads to spontaneous depolarization of ventricular myocytes and is a cause of "abnormal automaticity" of these cells.[54] FIGURE 9 illustrates this phenomenon.

SUMMARY

On the basis of the evidence reviewed above, it can be envisioned that the spontaneous beating rate of pacemaker cells involves cyclic variation of subsarcolemmal Ca^{2+} produced by local releases of SR Ca^{2+} via RyRs, acting both in concert with

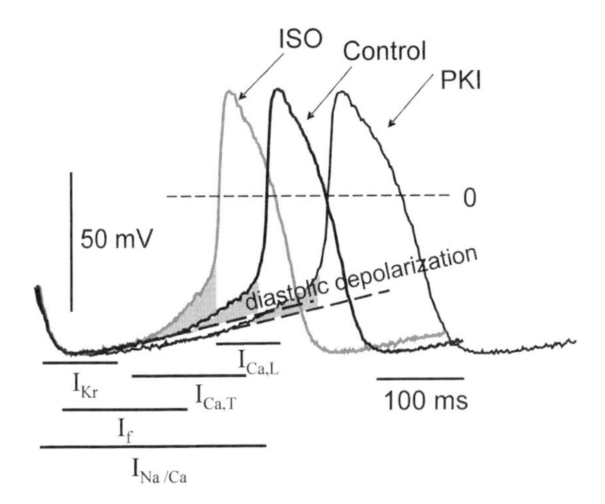

FIGURE 10. A schematic depicting the membrane potential and timing of the occurrence of the subsarcolemmal ryanodine receptor (RyR) Ca^{2+} release during the later part of diastolic depolarization (DD) (shaded area) of sinoatrial nodal pacemaker cells, before or after protein kinase A (PKA) activation by β-adrenergic receptor (β-AR) stimulation with isoproterenol (ISO) or PKA inhibition by a specific peptide inhibitor PKI (myristoylated protein kinase A inhibitor amide 14–22), and on the timing of activation of an ensemble of sarcolemmal ion channels as reported in the literature (in the absence of maneuvers that affect PKA activity). The *dashed line* is an extension of the rate of DD following the maximum diastolic potential. The deviation of the actual membrane potential from the *dashed line* occurs concomitantly with an increase in diastolic subsarcolemmal [Ca^{2+}], due to RyR release (shaded areas). Following PKA activation by β-AR stimulation, this deviation exhibits a phase shift; that is, its initial trajectory occurs earlier than in control, due largely to enhanced Ca^{2+} activation of the Na^+-Ca^{2+} exchanger current, and it drives the membrane potential to the threshold required for action potential (AP) firing at an earlier time. Critical influences of I_f and I_{Kr} mostly occur prior to the period of the shaded area; $I_{Ca,T}$ is activated prior to and during the earlier part of the shaded area, and $I_{Ca,L}$ begins to activate during the later part of the shaded area and explosively activates during the rapid AP upstroke. Following suppression of PKA by PKI, Local Ca^{2+} release (LCR) amplitude is suppressed and LCR occurrence is delayed in time, leading to a decrease in the beating rate.

the NCX and with an ensemble of sarcolemmal ionic channels[6–12,29] to regulate the SANC spontaneous diastolic membrane depolarizations (FIG. 10). While no single factor likely confers the status of dominance with respect to pacemaker function of SANC, the rhythmic SR subsarcolemmal Ca^{2+} releases during the DD impart physiological robustness (stability) to the oscillating sarcolemmal ion currents[28] and strengthen responses to hormonal regulation.

The number of local Ca^{2+} releases within a given epoch (e.g., during DD in SANC, or following an AP in ventricular cells, or an increase in local Ca^{2+} release amplitude) is indicative of a recruitment of more RyRs to fire within that epoch. Such recruitment reflects a synchronization of individual RyR gating and Ca^{2+} release. Synchronization of RyR activation and Ca^{2+} release is a common mechanism that underlies the cAMP-PKA signaling to increase both heart rate and myocardial contraction amplitude. In SANC, the integral of the total number of elementary RyR

Ca^{2+} releases during a given epoch (i.e., during DD) and the elementary inward currents via NCX that these generate, determine the magnitude (amplitude) of the effect of these releases on the change in membrane potential during DD.

ACKNOWLEDGMENT

The authors would like to thank Christina R. Link for her editorial assistance in preparing this document.

REFERENCES

1. KUNERTH, S., G.W. MAYR, F. KOCH-NOLTE & A.H. GUSE. 2003. Analysis of subcellular calcium signals in T- lymphocytes. Cell. Signal. **15:** 783–792.
2. DING, J.M., G.F. BUCHANAN, S.A. TISCHKAU, *et al.* 1998. A neuronal ryanodine receptor mediates light-induced phase delays of the circadian clock. Nature **394:** 381–384.
3. IKEDA, M., T. SUGIYAMA, C.S. WALLACE, *et al.* 2003. Circadian dynamics of cytosolic and nuclear Ca^{2+} in single suprachiasmatic nucleus neurons. Neuron **38:** 253–263.
4. KOSHIYA, N. & J.C. SMITH. 1999. Neuronal pacemaker for breathing visualized in vitro. Nature **400:** 360–363.
5. DEVOR, A. 2002. The great gate: control of sensory information flow to the cerebellum. Cerebellum **1:** 27– 34.
6. BROWN, H.F., J. KIMURA, S.J. NOBLE & A. TAUPIGNON. 1984. The ionic currents underlying pacemaker activity in rabbit sino-atrial node: experimental results and computer simulation. Proc. R. Soc. Lond. B. Biol. Sci. **222:** 329–347.
7. DEMIR, S.S., J.W. CLARK, C.R. MURPHEY & W.R. GILES. 1994. A mathematical model of a rabbit sinoatrial node cell. Am. J. Physiol. **266:** C832–C852.
8. KUROTA, Y., I. HISATOME, S. IMANISHI, *et al.* 2002. Dynamical description of sinoatrial node pacemaking: improved mathematical model for primary pacemaker cell. Am. J. Physiol. Heart Circ. Physiol. **238:** H2074–H2101.
9. ZHANG, H., A.V. HOLDEN, I. KODAMA, *et al.* 2000. Mathematical models of action potentials in the periphery and center of the rabbit sinoatrial node. Am. J. Physiol. **279:** H397–H421.
10. IRISAWA, H., H.F. BROWN & W. GILES. 1993. Cardiac pacemaking in the sinoatrial node. Annu. Rev. Physiol. **73:** 197–227.
11. DOKOS, S., B. CELLER & N. NOVELL. 1996. Ion currents underlying sinoatrial node pacemaker activity: a new single cell mathematical model. J. Theor. Biol. **181:** 245–272.
12. DIFRANCESCO, D. 1993. Pacemaker mechanisms in cardiac tissue. Annu. Rev. Physiol. **55:** 455–472.
13. CRANFIELD, P.F. 1977. Action potentials, afterpotentials, and arrhythmias. Circ. Res. **41:** 415–423.
14. RUBENSTEIN, D.S. & S.L. LIPSIUS. 1989. Mechanisms of automaticity in subsidiary pacemakers from cat right atrium. Circ. Res. **64:** 648–657.
15. ESCANDE, D., E. CORABOEUF & C. PLANCHE. 1987. Abnormal pacemaking is modulated by sarcoplasmic reticulum in partially-depolarized myocardium from dilated right atria in human. J. Mol. Cell. Cardiol. **19:** 231–241.
16. ZHOU, Z. & S.L. LIPSIUS. 1993. Na^{+}-Ca^{2+} exchange current in latent pacemaker cells isolated from cat right atrium. J. Physiol. **466:** 263–285.
17. JU, Y.-K. & D.G. ALLEN. 1999. How does β-adrenergic stimulation increase the heart rate? The role of intracellular Ca^{2+} release in amphibian pacemaker cells. J. Physiol. **516:** 793–804.
18. RIGG, L., B.M. HEATH, Y. CUI, *et al.* 2000. Localization and functional significance of ryanodine receptors during β-adrenergic stimulation in guinea-pig sino-atrial node. Cardiovasc. Res. **48:** 254–264.
19. VINOGRADOVA, T.M., K.Y. BOGDANOV & E.G. LAKATTA. 2002. β-Adrenergic stimulation modulates ryanodine receptor Ca^{2+} release during diastolic depolarization to accelerate pacemaker activity in rabbit sinoatrial nodal cells. Circ. Res. **90:** 73–79.

20. LI, J., J. QU & R.D. NATHAN. 1997. Ionic basis of ryanodine's negative chronotropic effect on pacemaker cells isolated from the sinoatrial node. Am. J. Physiol. **273:** H2481–H2489.
21. RIGG, L. & D.A. TERRAR. 1996. Possible role of calcium release from the sarcoplasmic reticulum in pacemaking in guinea-pig sino-atrial node. Exp. Physiol. **81:** 877–880.
22. HATA, T., T. NODA, M. NISHIMURA, *et al.* 1996. The role of Ca^{2+} release from sarcoplasmic reticulum in the regulation of sinoatrial node automaticity. Heart Vessels **11:** 234–241.
23. SATOH, H. 1997. Electrophysiological actions of ryanodine on single rabbit sinoatrial nodal cells. Comp. Gen. Pharmacol. **28:** 31–38.
24. HUSER, J., L.A. BLATTER & S.L. LIPSIUS. 2000. Intracellular Ca^{2+} release contributes to automaticity in cat atrial pacemaker cells. J. Physiol. **524:** 415–422.
25. BOGDANOV, K.Y., T.M. VINOGRADOVA & E.G. LAKATTA. 2001. Sinoatrial nodal cell ryanodine receptor and Na^+-Ca^{2+} exchanger: molecular partners in pacemaker regulation. Circ. Res. **88:** 1254–1258.
26. VINOGRADOVA, T.M., Y.-Y. ZHOU, K.Y. BOGDANOV, *et al.* 2000. Sinoatrial node pacemaker activity requires Ca^{2+}/calmodulin-dependent protein kinase II activation. Circ. Res. **87:** 760–767.
27. JU, Y.-K. & D.G. ALLEN. 1998. Intracellular calcium and Na^+-Ca^{2+} exchange current in isolated toad pacemaker cells. J. Physiol. **508:** 153–166.
28. LAKATTA, E.G., V.A. MALTSEV, K.Y. BOGDANOV, *et al.* 2003. Cyclic variation of intracellular calcium: a critical factor for cardiac pacemaker cell dominance. Circ. Res. **92:** e45–e50.
29. LIPSIUS, S.L., J. HUSER & L.A. BLATTER. 2001. Intracellular Ca^{2+} release sparks atrial pacemaker activity. News Physiol. Sci. **16:** 101–106.
30. YANG, H.-T., D. TWEEDIE, S. WANG, *et al.* 2002. The ryanodine receptor modulates the spontaneous beating rate of cardiomyocytes during development. Proc. Natl. Acad. Sci. U.S.A. **99:** 9225–9230.
31. MALTSEV, V.A., T.M. VINOGRADOVA, K.Y. BOGDANOV, *et al.* 2004. Diastolic calcium release controls the beating rate of rabbit sinoatrial node cells: numerical modeling of the coupling process. Biophys. J. **86:** 2596–2605.
32. VINOGRADOVA, T.M., Y.Y. ZHOU, V. MALTSEV, *et al.* 2004. Rhythmic ryanodine receptor Ca^{2+} releases during diastolic depolarization of sinoatrial pacemaker cells do not require membrane depolarization. Circ. Res. **94:** 802–809.
33. FABIATO, A. 1985. Time and calcium dependence of activation and inactivation of calcium-induced release of calcium from the sarcoplasmic reticulum of a skinned canine cardiac Purkinje cell. J. Gen. Physiol. **85:** 247–289.
34. GYORKE, S. & M. FILL. 1993. Ryanodine receptor adaptation: control mechanism of Ca^{2+} release in heart. Science. **260:** 807–809.
35. JU, Y.-K. & D.G. ALLEN. 1998. Intracellular calcium and Na^+Ca^{2+} exchange current in isolated toad pacemaker cells. J. Physiol. **508:** 153–166.
36. KOUSHIK, S.V., J. WANG, R. ROGERS, *et al.* 2001. Targeted inactivation of the sodium-calcium exchanger (NCX1) results in the lack of a heartbeat and abnormal myofibrillar organization. FASEB J. **15:** 1209–1211.
37. MALTSEV, V, KY. BOGDANOV, T.M. VINOGRADOVA, *et al.* 2004. Membrane depolarizations produced by individual local Ca^{2+} releases are elementary events in regulation of sinoatrial nodal cell beating rate. Circulation **110**(Suppl.): III-195.
38. DIFRANCESCO, D. & M. MANGONI. 1994. Modulation of single hyperpolarization-activated channels (i_f) by cAMP in the rabbit sino-atrial node. J. Physiol. **474:** 473–482.
39. LEI, M., H.F. BROWN & D.A. TERRAR. 2000. Modulation of delayed rectifier potassium current, i_K, by isoprenaline in rabbit isolated pacemaker cells. Exp. Physiol. **85:** 27–35.
40. BUCCHI, A, M. BARUSCOTTI, R.B. ROBINSON & D. DIFRANCESCO. 2003. I(f)-dependent modulation of pacemaker rate mediated by cAMP in the presence of ryanodine in rabbit sino-atrial node cells. J. Mol. Cell. Cardiol. **35:** 905–913.
41. TAKANO, M. & A. NOMA. 1992. Distribution of the isoprenaline-induced chloride current in rabbit heart. Pfluegers Arch. **420:** 223–226.
42. HOROWITZ, B., S.S. TSUNG, P. HART, *et al.* 1993. Alternative splicing of CFTR Cl- channels in heart. Am. J. Physiol. **264:** H2214–H2220.

43. Lakatta, E.G., S. Deo, M. Barlow, *et al.* 2003. Beta-adrenergic receptor stimulation induced cardioacceleration *in situ* requires intact ryanodine receptor function of sinoatrial nodal cells. Circ. Suppl. **108**(Suppl.): IV-35.
44. Vinogradova, T.M., A. Lyashkov, W. Zhu, *et al.* 2004. Constitutive protein kinase A activation generates rhythmic Ca^{2+} oscillations in rabbit sinoatrial nodal cells that modulate intrinsic, spontaneous beating. Circulation **110**(Suppl.): III-85.
45. Maltsev, V.A., T.M. Vinogradova, M.D. Stern & E.G. Lakatta. 2005. Local subsarcolemmal Ca^{2+} releases within rabbit sinoatrial nodal cells not only regulate beating rate but also ensure normal rhythm during protein kinase A inhibition. Abstract. Biophys. J. **88:** 89a.
46. Capogrossi, M.C. & E.G. Lakatta. 1985. Frequency modulation and synchronization of spontaneous oscillations in cardiac cells. Am. J. Physiol. **248:** H412–H418.
47. Capogrossi, M.C., B.A. Suarez-Isla & E.G. Lakatta. 1986. The interaction of electrically stimulated twitches and spontaneous contractile waves in single cardiac myocytes. J. Gen. Physiol. **88:** 615–633.
48. Kort, A.A. & E.G. Lakatta. 1988. Biomodal effect of stimulation on light fluctuation transients monitoring spontaneous sarcoplasmic reticulum calcium release in rat cardiac muscle. Circ. Res. **63:** 960–968.
49. Xiao, R.-P. & E.G. Lakatta. 1993. β1-Adrenoceptor stimulation and β2 adrenoceptor stimulation differ in their effects on contraction, cytosolic Ca^{2+}, and Ca^{2+} current in single rat ventricular cells. Circ. Res. **73:** 286–300.
50. Wier, W.G., M.B. Cannel, J.R. Berlin, *et al.* 1987. Cellular and subcellular heterogeneity of $[Ca^{2+}]_i$ in single heart cells revealed by fura-2. Science **235:** 325–328.
51. Lakatta, E.G. 1992. Functional implications of spontaneous sarcoplasmic reticulum Ca^{2+} release in the heart. Cardiovasc. Res. **26:** 193–214.
52. Ferrier, G.R., J.H. Saunders & C. Mendez. 1973. A cellular mechanism for the generation of ventricular arrhythmias by acetylstrophanthidin. Circ. Res. **32:** 600–609.
53. Kass, R.S. & R.W. Tsien. 1982. Fluctuations in membrane current driven by intracellular calcium in cardiac Purkinje fibers. Biophys. J. **38:** 259–269.
54. Capogrossi, M.C., S. Houser, A. Bahinski & E.G. Lakatta. 1987. Synchronous occurrence of spontaneous localized calcium release from the sarcoplasmic reticulum generates action potentials in rat cardiac ventricular myocytes at normal resting membrane potential. Circ. Res. **61:** 498–503.

Modification of Cellular Communication by Gene Transfer

J. KEVIN DONAHUE, ALEXANDER BAUER, KAN KIKUCHI, AND TETSUO SASANO

Institute of Molecular Cardiobiology and Division of Cardiology, Johns Hopkins University School of Medicine, Baltimore, Maryland 21205, USA

ABSTRACT: Hope has been expressed that gene and cell therapies will one day reduce the morbidity and mortality associated with cardiovascular diseases. Work in these fields has shown that the road from bench to bedside is filled with obstacles. Still, the possibility for treatment or even cure of cardiac disease is real. Continuing work will improve understanding of the underlying physiology and vector biology. The current review focuses on the potential use of gene therapy to affect cellular communication. Included is a review of communication effects on a transcellular level with angiogenesis, AV nodal conduction and sinus nodal automaticity, and effects on an intracellular level with cardiac myocyte repolarization. Challenges facing the field of gene therapy are also reviewed. If these problems can be solved, gene therapy will become a viable alternative for clinical use.

KEYWORDS: angiogenesis; arrhythmia; gene therapy; growth factor; ion channel

INTRODUCTION

Recent clinical trials have shown the utility of a variety of drugs and interventions for prevention and treatment of cardiac diseases, including antilipid therapy, adrenergic modulation, afterload reduction, antiplatelet agents, drug-eluting stents, implantable cardioverter-defibrillators, etc. Still, in spite of these innovations, the inevitable course for most patients is one of progressive disease. In part, this is due to the palliative nature of the available therapies. These limitations in currently available therapies have led to considerable speculation on the potential utility for gene or stem cell therapeutics for treatment of cardiovascular diseases. In the current review, we will focus on gene therapy as a means to alter intracellular or transcellular communication and discuss the use of genetic agents to achieve angiogenesis, to alter cardiac automaticity, and to modify conduction and repolarization. We will conclude with a discussion of the current problems preventing large-scale use of these therapies and potential answers to these problems.

Address for correspondence: A. Prof. Kevin Donahue, M.D., Division of Cardiology, Johns Hopkins University, Ross 844, 720 N. Rutland Avenue, Baltimore, MD 21205, USA. Voice: 410-955-2775; fax: 410-502-2096.
kdonahue@jhmi.edu

Ann. N.Y. Acad. Sci. 1047: 157–165 (2005).
doi: 10.1196/annals.1341.014

INDUCTION OF ANGIOGENESIS IN ISCHEMIC REGIONS

The first cardiovascular gene therapy clinical trials focused on the use of growth factors to induce angiogenesis as a treatment for coronary disease. The underlying hypothesis was that focal application of gene transfer vectors encoding growth factors would lead to localized blood vessel formation, increasing the network of collateral vessels from less-diseased coronary vessels to areas served by partially or completely occluded coronary arteries. Animal work centered on use of two models: limb ischemia, caused by abrupt ligation of the femoral artery, and gradual coronary occlusion days or weeks after implantation of ameroid constrictors around coronary arteries. Both models employed otherwise healthy, young animals. Numerous growth factors were tested in these models, but the majority of the work to date has centered on use of either vascular endothelial growth factor (VEGF) or fibroblast growth factor (FGF). The VEGF work utilized two splice variants of the factor: $VEGF_{121}$ tested with adenoviral vectors, and $VEGF_{165}$ evaluated with plasmid vectors. All FGF testing has been performed with adenoviral vectors.

Animal work in both limb and coronary ischemia models gave suggestive evidence that gene therapy with any of these vectors was capable of inducing sufficient angiogenesis to relieve the ischemic burden.[1,2] Endpoints of these studies included microscopic quantification of blood vessel number in treated and control groups, various analyses of regional blood flow, and indirect indicators of ischemia and organ function. All endpoints were positive, providing evidence for initiation of several phase 1 clinical trials for limb and coronary ischemia.

The human work has been less convincing. Phase 1 trials for both peripheral and coronary ischemia showed improvement. The peripheral vascular disease trials documented marginal increases in the ankle-brachial index (0.33 improving to 0.48) but improved healing of ischemic ulcers.[3,4] Coronary phase 1 trials documented reductions in nitroglycerin usage and improved flow in dobutamine or treadmill stress tests.[5,6] Problems with these trials included the absence of blinding or of controls. Phase 2 trials included both control populations and investigator blinding to the treatment regimen. A peripheral phase 2 clinical trial using Ad-$VEGF_{121}$ found no change in the primary endpoint of peak walk time or the composite endpoint of peak walk time, ankle-brachial index, and time to onset of claudication.[7] The phase 2 coronary trial using Ad-$VEGF_{121}$ found no significant change in functional class, work ability, or nitrate use when comparing between control and treatment groups.[8] The Ad-FGF4 clinical found a nonsignificant trend toward improvement in perfusion.[9] Encouraging news from all trials is that the side effect profile appears to be minimal for all agents tested. To date, however, it is fair to say that angiogenesis gene therapy has been a disappointment. Ongoing studies will evaluate the use of multiple splice variants of VEGF or the hypoxia inducibility factor 1α to test the idea that more generalized activation of ischemia-reactive pathways might give better results than those seen with application of specific single agents like individual VEGF splice variants or the single serotype of FGF.

RECREATION OR AMPLIFICATION OF CARDIAC PACEMAKER ACTIVITY

Efforts to improve or recreate the sinus node have centered around three genes: the β-adrenergic receptor (βAR), the hyperpolarization-activated, cyclic nucleotide-gated cation channel (HCN) gene encoding the I_f current, and the Kir 2.1 gene encoding the I_{K1} current.[10–12] The first reported manipulation of cardiac pacemaker activity involved amplification of the β-AR in the sinus node. Edelberg and co-workers mapped the earliest site of atrial activation, presumed to be the sinus node, and injected plasmid vectors containing the β-AR.[10] Two days after injection, the heart rate in the treatment group increased from a baseline of 108 beats per minute to a peak effect of 163 beats per minute. This effect was significantly reduced by day 3 and eliminated by day 4. Immunostaining results showed the presence of β-AR only in the sinus nodes of the treatment group. Comparing treatment to control groups showed a significant difference in heart rate, but no quantitation of gene transfer or comparison of expression level to effect was documented. Given the usual time course of gene expression after most gene transfer techniques, where peak effects occur 5 to 7 days after plasmid or adenovirus injection and gene expression is eliminated 2 to 3 weeks after injection, the results of these studies are difficult to explain and further investigation is warranted.

Subsequent to the initial report of amplified sinus node activity with β-AR gene transfer, Miake and colleagues showed that gene transfer of a dominant negative mutant of Kir 2.1 (GYG144-146AAA) caused automaticity in normally quiescent ventricular cells.[12] Kir 2.1 is the major subunit of the inward rectifier current I_{K1}, which is responsible for maintaining cellular resting membrane potential. In a guinea pig model, Miake *et al.* delivered adenoviruses throughout the ventricles using an aortic cross-clamping method initially reported by Hajjar *et al.*[13] They documented gene transfer to 20% of cells using the green fluorescent protein (GFP) reporter, and with the Kir 2.1 AAA dominant negative mutant they found increased automaticity on both a cellular and whole heart level (FIG. 1A). In a subsequent report, Miake *et al.* found that the I_{K1} current needed to be reduced by at least 80% in order to get cellular automaticity; lesser reductions caused a prolongation of the action potential (AP) duration with unstable repolarization.[14]

Recently, Qu *et al.* reported another method for increased myocyte automaticity.[11] In a dog model, they transferred the HCN2 gene, producing a protein thought to be responsible for I_f, the putative pacemaker current. Qu *et al.* injected solutions containing Ad-HCN2 into the left atria and noted spontaneous left atrial rhythms during vagally induced sinus arrest in four dogs (FIG. 1B). In contrast, three dogs injected with an adenovirus encoding GFP did not have any spontaneous rhythms. Subsequent work with Ad-HCN2 injection into the left bundle branch by the same group showed left bundle branch premature ventricular beats intermittently during sinus rhythm and left bundle escape rhythms in six of seven animals during vagal stimulation.[15] These data suggested that HCN2 overexpression could increase automaticity of either atrial or ventricular tissue.

Overall, gene transfer approaches to increase cardiac automaticity are in early stages of development. None of the published interventions completely recreates the sinus node, but all show promise. Ongoing work in this field has the potential for tremendous impact if the correct gene or combination of genes is identified to allow recreation of true pacemaker activity.

MODIFICATION OF ELECTRICAL CONDUCTION

Altering cardiac transcellular conduction could potentially be done using a number of strategies. The most obvious approach would be manipulation of gap junction

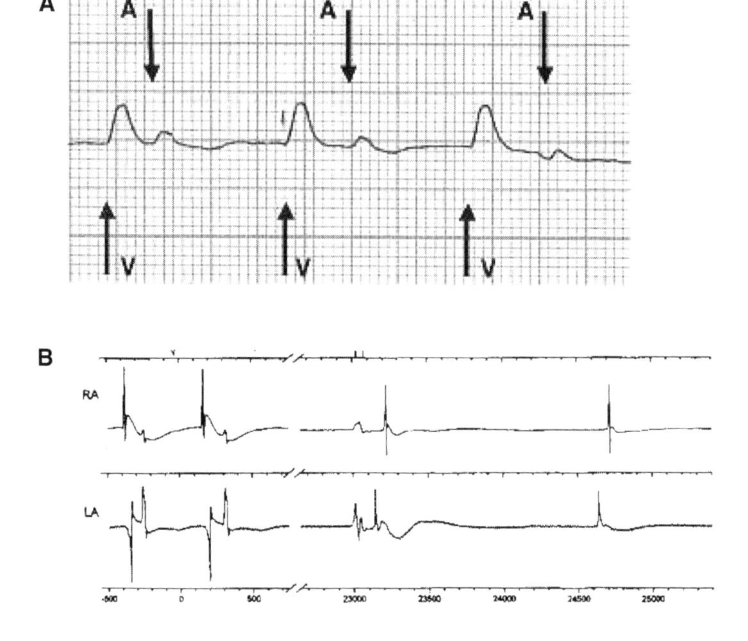

FIGURE 1. Gene transfer-induced alterations in cardiac automaticity. (**A**) Idioventricular rhythm originating in ventricular cells expressing a knock-out mutation for Kir2.1. (From Miake *et al.*[12] Reproduced by permission.) (**B**) Left atrial escape rhythm coming from cells expressing HCN2. The two beats on the *left* originate in the right atrium during normal sinus rhythm. The two beats on the *right* originate in the left atrium, from the gene transfer region, during vagally mediated sinus arrest. (From Qu *et al.*[11] Reproduced by permission.)

behavior by up- or down-regulation of connexins. In the specialized conducting regions of the heart, calcium channel alterations would affect conduction by slowing the rate of cellular activation. Similar alterations could be achieved in atrial or ventricular myocytes by manipulating sodium channel function. Manipulation of the extracellular matrix or of the distribution of gap junctions and ion channels also has potential for modification of electrical conduction, albeit at a much higher level of complexity than the other simpler manipulations.

To date, the primary strategy chosen for modifying intracardiac conduction has been manipulation of the G protein signaling cascade, which has a presumed downstream effect of slowing conduction through the L-type calcium channel. We initially achieved this effect by gene transfer of adenoviral vectors encoding wild-type $G\alpha_{i2}$ in a porcine model of acute atrial fibrillation. The viruses were delivered to the atrioventricular (AV) node by perfusion of the AV nodal artery after treatment with VEGF and nitroglycerin to increase vascular permeability to the viruses.[16,17] Control experiments with adenoviruses encoding the β-galactosidase reporter gene documented gene transfer to almost 50% of cells in the AV node.[18] AV nodal function was not affected by gene transfer with the control virus. After gene transfer of $G\alpha_{i2}$, conduction slowing and refractory period prolongation were observed in the AV node.

The ventricular response rate to acutely induced atrial fibrillation (AF) was decreased by 20% at baseline, with a persistent 15% reduction after administration of 1 mg of epinephrine (FIG. 2A). Subsequent work in a model of persistent AF and severe congestive heart failure[19] found that heart rate control was overdriven by the waking state when the wild type $G\alpha_{i2}$ was used, but a constitutively active mutant $G\alpha_{i2}$ Q205L was capable of sustained reductions in heart rate (FIG. 2B).[20] The per-

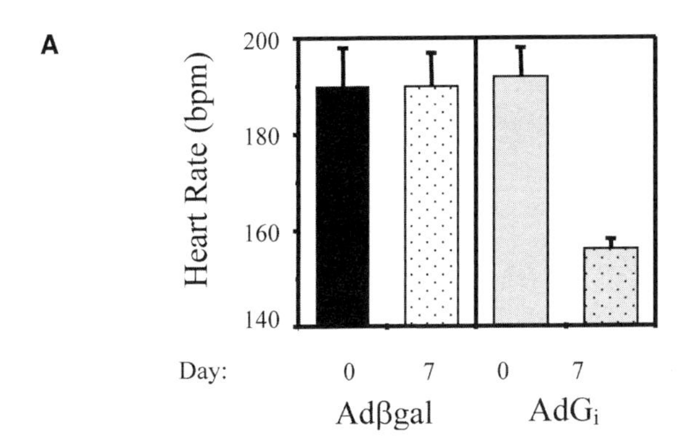

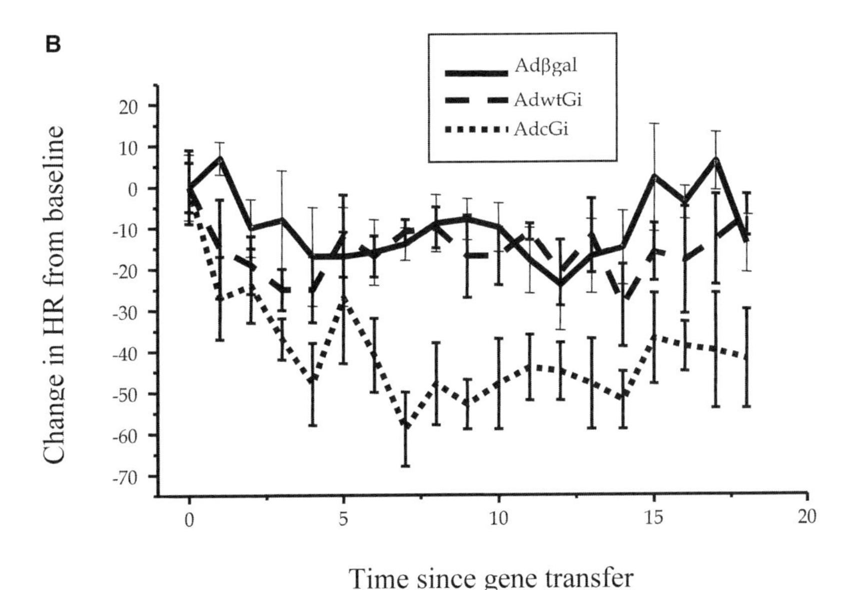

FIGURE 2. Effects of $G\alpha_{i2}$ overexpression in porcine AV node. (**A**) Percentage change in atrial fibrillation (AF) heart rate during acute atrial fibrillation. (Adapted with permission from Donahue *et al.*[18]) (**B**) Change in ventricular rate in persistent AF after gene transfer of β-galactosidase, wild type $G\alpha_{i2}$, and a constitutively active mutant $G\alpha_{i2}$ Q205L. (From Bauer *et al.*[20] Reproduced by permission.)

sistent AF model had an underlying tachycardiomyopathy caused by the uncontrolled heart rate in AF. The 20% reduction in heart rate achieved by gene transfer of the constitutively active mutant was sufficient to reverse the tachycardiomyopathy.

GENE TRANSFER EFFECTS ON CARDIAC REPOLARIZATION

Some of the earliest work in myocardial gene transfer evaluated the effects of potassium channel expression in primary cultures of ventricular myocytes. The first reported alteration of the action potential morphology by *in vitro* gene transfer used an adenovirus expressing the Shaker potassium channel, an ion channel expressed in Drosophila that causes an I_{to1}-like current.[21] Expression of Shaker in canine ventricular myocytes caused numerous changes in the AP, including amplification of phase 1, suppression of phase 2, and abbreviation of the overall AP duration. The level of expression correlated with the extent of changes in the AP, with more robustly expressing cells having the most bizarre AP morphologies.[22] Subsequent work by Nuss *et al.* showed that overexpression of HERG also shortened the AP duration but with a more physiological AP morphology.[23] HERG channels are the major component of the I_{Kr} current, and as such are activated later than the I_{to}-like current caused by the Shaker channel. This later activation allowed a normal AP notch and plateau but still abbreviated the AP duration.

Two *in vitro* reports have evaluated the therapeutic possibilities of cardiac myocyte gene transfer. Ennis *et al.* transferred a bicistronic construct encoding both a sarcoplasmic reticulum calcium adenosine triphosphate (SERCA-1) and the Kv2.1 channel responsible, in part, for the I_{K1} current.[24] They found that I_{K1} overexpression shortened AP duration and SERCA-1 overexpression increased the efficiency of intracellular calcium handling. This work suggested that combination gene therapy might be effective in heart failure, where the potassium channel overexpression could have an antiarrhythmic effect by shortening repolarization, and the SERCA overexpression could have a direct increase in contractility or in the least could obviate any negative effects on contractility from the AP shortening caused by the potassium channel overexpression. Kodirov *et al.* were able to regenerate potassium current in myocytes from Kv1.1 knock-out transgenic mice by overexpression of Kv1.5.[25] This gene-switching strategy illustrated the ability to overcome the dominant negative effects of one mutation by overexpressing a non-interacting protein of similar function.

In a recent publication, we reported a method for transmural atrial gene transfer that involved painting viral vectors complexed in poloxamer gels onto the atrial epicardium.[26] The inclusion of trypsin at low concentrations in the gel matrix allows penetration of the virus across the atrial myocardium without structural damage to the heart (FIG. 3A). Using this vector delivery method, we found that transfer of a dominant negative mutation of the HERG channel caused prolongation of atrial repolarization without affecting ventricular cells (FIG. 3B).

PROBLEMS LIMITING THE CLINICAL UTILITY OF GENE TRANSFER

Translation of arrhythmia gene therapy to clinical practice will require solutions to a number of problems that face the field, including those related to homogeneous

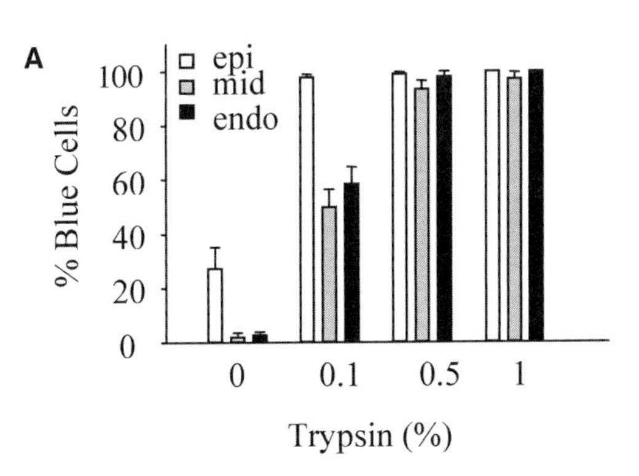

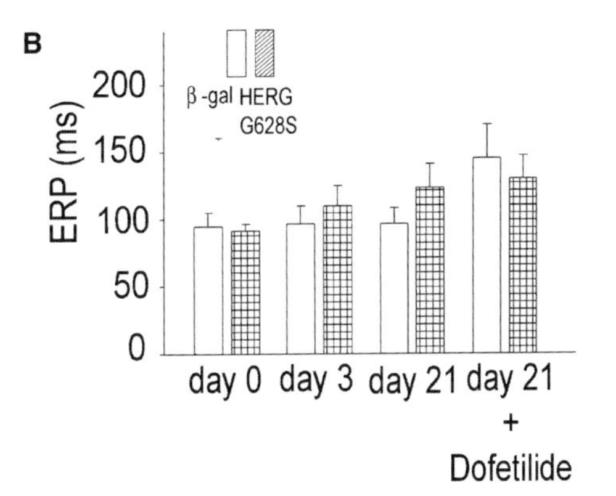

FIGURE 3. Atrial gene transfer using a newly described epicardial painting method. (**A**) Percentage of myocytes expressing the reporter gene after epicardial application of 5×10^9 pfu adenovirus in 5 ml of a gel-matrix with the given concentration of trypsin. (**B**) Effect of gene transfer on the effective refractory period (ERP) after painting atria with either Adβgal or AdHERG-G628S. βgal encodes the lacZ reporter gene. HERG-G628S is a dominant negative mutant of the HERG potassium channel. Day 21 effects are shown in the absence and presence of dofetilide. (From Kikuchi *et al.*[26] Reproduced by permission.)

delivery of the vector to the target tissue, control of gene expression, evaluation of potentially toxic effects of the vector or the transgene, and control of nontarget organ gene transfer and of host immune responses. Efforts to solve these problems are taking place on multiple fronts. Recent advances in vector design have documented the ability of adeno-associated virus and helper-dependent adenovirus vectors to sustain long-term gene expression and to reduce host immune responses.[27,28] Discovery of methods to increase microvascular permeability to vectors and to improve physical parameters relevant to gene transfer efficiency have improved efficacy and homogeneity of gene delivery.[16,17,29–31] New situation-specific promoters or response elements that activate in response to hypoxia, temperature, and steroid or drug exposure have been identified,[32] lending hope to the possibility that the timing and amount of gene expression can be controlled.

SUMMARY

This review has illustrated several ways that gene transfer techniques can be used to explore or manipulate cellular communication and ultimately cardiac function. Relatively straightforward techniques have been employed to test hypotheses regarding the role of growth factors, ion channels, connexins, and G proteins in cardiac myocytes. The ultimate goal of these studies is translation of animal study results to human therapies. Continuing effort to improve gene delivery, to control gene expression, and to investigate gene transfer effects in a variety of "near-human" animal models will be required to realize this goal.

ACKNOWLEDGMENTS

This work was supported by the National Institutes of Health grants HL-67148 and EB-2846, the American Heart Association grant 130350N, and the Donald W. Reynolds Center at Johns Hopkins University.

REFERENCES

1. GIORDANO, F.J., P.P. PING, M.D. MCKIRNAN, *et al.* 1996. Intracoronary gene transfer of fibroblast growth factor-5 increases blood flow and contractile function in an ischemic region of the heart. Nat. Med. **2:** 534–539.
2. TAKESHITA, S., Y. TSURUMI, T. COUFFINAHL, *et al.* 1996. Gene transfer of naked DNA encoding for three isoforms of vascular endothelial growth factor stimulates collateral development *in vivo*. Lab. Invest. **75:** 487–501.
3. BAUMGARTNER, I., A. PIECZEK, O. MANOR, *et al.* 1998. Constitutive expression of phVEGF165 after intramuscular gene transfer promotes collateral vessel development in patients with critical limb ischemia. Circulation **97:** 1114–1123.
4. KIM, H.J., S.Y. JANG, J.I. PARK, *et al.* 2004. Vascular endothelial growth factor-induced angiogenic gene therapy in patients with peripheral artery disease. Exp. Mol. Med. **36:** 336–344.
5. ROSENGART, T.K., L.Y. LEE, S.R. PATEL, *et al.* 1999. Six-month assessment of a phase I trial of angiogenic gene therapy for the treatment of coronary artery disease using direct intramyocardial administration of an adenovirus vector expressing the VEGF121 cDNA. Ann. Surg. **230:** 466–470.
6. LOSORDO, D.W., P.R. VALE, J.F. SYMES, *et al.* 1998. Gene therapy for myocardial angiogenesis: initial clinical results with direct myocardial injection of phVEGF165 as sole therapy for myocardial ischemia. Circulation **98:** 2800–2804.
7. RAJAGOPALAN, S., E.R. MOHLER, III, R.J. LEDERMAN, *et al.* 2003. Regional angiogenesis with vascular endothelial growth factor in peripheral arterial disease: a phase II randomized, double-blind, controlled study of adenoviral delivery of vascular endothelial growth factor 121 in patients with disabling intermittent claudication. Circulation **108:** 1933–1938.
8. HEDMAN, M., J. HARTIKAINEN, M. SYVANNE, *et al.* 2003. Safety and feasibility of catheter-based local intracoronary vascular endothelial growth factor gene transfer in the prevention of postangioplasty and in-stent restenosis and in the treatment of chronic myocardial ischemia: phase II results of the Kuopio Angioogenesis Trial (KAT). Circulation **107:** 2677–2683.
9. GRINES, C.L., M.W. WATKINS, J.J. MAHMARIAN, *et al.* 2003. A randomized, double-blind, placebo-controlled trial of Ad5FGF-4 gene therapy and its effect on myocardial perfusion in patients with stable angina. J. Am. Coll. Cardiol. **42:** 1339–1347.
10. EDELBERG, J., D. HUANG, M. JOSEPHSON & R. ROSENBERG. 2001. Molecular enhancement of porcine cardiac chronotropy. Heart **86:** 559–562.
11. QU, J., A.N. PLOTNIKOV, P. DANILO, JR., *et al.* 2003. Expression and function of a biological pacemaker in canine heart. Circulation **107:** 1106–1109.
12. MIAKE, J., E. MARBAN & H. NUSS. 2002. Biological pacemaker created by gene transfer. Nature **419:** 132–133.
13. HAJJAR, R.J., U. SCHMIDT, T. MATSUI, *et al.* 1998. Modulation of ventricular function through gene transfer *in vivo*. Proc. Natl. Acad. Sci. U.S.A. **95:** 5251–5256.

14. MIAKE, J., E. MARBAN & H. NUSS. 2003. Functional role of inward rectifier current in heart probed by Kir2.1 overexpression and dominant-negative suppression. J. Clin. Invest. **111:** 1529–1536.
15. PLOTNIKOV, A.N., E.A. SOSUNOV, J. QU, *et al.* 2004. Biological pacemaker implanted in canine left bundle branch provides ventricular escape rhythms that have physiologically acceptable rates. Circulation **109:** 506–512.
16. DONAHUE, J.K., K. KIKKAWA, A.D. THOMAS, *et al.* 1998. Acceleration of widespread adenoviral gene transfer to intact rabbit hearts by coronary perfusion with low calcium and serotonin. Gene Ther. **5:** 630–634.
17. NAGATA, K., E. MARBAN, J.H. LAWRENCE & J.K. DONAHUE. 2001. Phosphodiesterase inhibitor-mediated potentiation of adenovirus delivery to myocardium. J. Mol. Cell Cardiol. **33:** 575–580.
18. DONAHUE, J.K., A.W. HELDMAN, H. FRASER, *et al.* 2000. Focal modification of electrical conduction in the heart by viral gene transfer. Nat. Med. **6:** 1395–1398.
19. BAUER, A., A.D. MCDONALD & J.K. DONAHUE. 2004. Pathophysiological findings in a model of atrial fibrillation and severe congestive heart failure. Cardiovasc. Res. **61:** 764–770.
20. BAUER, A., A.D. MCDONALD, K. NASIR, *et al.* 2004. Inhibitory G protein overexpression provides physiologically relevant heart rate control in persistent atrial fibrillation Circulation. **110:** 3115–3120.
21. JOHNS, D.C., H.B. NUSS, N. CHIAMVIMONVAT, *et al.* 1995. Adenovirus-mediated expression of a voltage-gated potassium channel *in vitro* (rat cardiac myocytes) and *in vivo* (rat liver). J. Clin. Invest. **96:** 1152–1158.
22. NUSS, H.B., D. JOHNS, S. KAAB, *et al.* 1996. Reversal of potassium channel deficiency in cells from failing hearts by adenoviral gene transfer: a prototype for gene therapy for disorders of cardiac excitability and contractility. Gene Ther. **3:** 900–912.
23. NUSS, B., E. MARBAN & D. JOHNS. 1999. Overexpression of a human potassium channel suppresses cardiac hyperexcitability in rabbit ventricular myocytes. J. Clin. Invest. **103:** 889–896.
24. ENNIS, I., R. LI, A. MURPHY, *et al.* 2002. Dual gene therapy with SERCA1 and Kir 2.1 abbreviates excitation without suppressing contractility. J. Clin. Invest. **109:** 393–400.
25. KODIROV, S., M. BRUNNER, L. BUSCONI & G. KOREN. 2003. Long-term restitution of 4-aminopyridine-sensitive currents in Kv1DN ventricular myocytes using adeno-associated virus-mediated delivery of Kv1.5. FEBS Lett. **550:** 74–78.
26. KIKUCHI, K., A. MCDONALD, T. SASANO & J. DONAHUE. 2005. Targeted modification of atrial electrophysiology by homogeneous transmural atrial gene transfer. Circulation **111:** 264–270.
27. MONAHAN, P. & R. SAMULSKI. 2000. AAV vectors: is clinical success on the horizon? Gene Ther. **7:** 24–30.
28. CHEN, H.H., L.M. MACK, R. KELLY, *et al.* 1997. Persistence in muscle of an adenoviral vector that lacks all viral genes. Proc. Natl. Acad. Sci. U.S.A. **94:** 1645–1650.
29. DONAHUE, J., K. KIKKAWA, D.C. JOHNS, *et al.* 1997. Ultrarapid, highly efficient viral gene transfer to the heart. Proc. Natl. Acad. Sci. U.S.A. **94:** 4664–4668.
30. HOSHIJIMA, M., Y. IKEDA, Y. IWANAGA, *et al.* 2002. Chronic suppression of heart-failure progression by a pseudophosphorylated mutant of phospholamban via *in vivo* cardiac rAAV gene delivery. Nat. Med. **8:** 864–871.
31. IKEDA, Y., Y. GU, Y. IWANAGA, *et al.* 2002. Restoration of deficient membrane proteins in the cardiomyopathic hamster by *in vivo* cardiac gene transfer. Circulation **105:** 502–508.
32. FUSSENEGGER, M. 2001. The impact of mammalian gene regulation concepts on functional genomic research, metabolic engineering, and advanced gene therapies. Biotechnol. Prog. **17:** 1–51.

Caveolae and Lipid Rafts

G Protein–Coupled Receptor Signaling Microdomains in Cardiac Myocytes

PAUL A. INSEL,[a] BRIAN P. HEAD,[a] RENNOLDS S. OSTROM,[b] HEMAL H. PATEL,[a] JAMES S. SWANEY,[a] CHIH-MIN TANG,[c] AND DAVID M. ROTH[c]

[a] *Department of Pharmacology, University of California, San Diego, La Jolla, California 92093, USA*

[b] *Department of Pharmacology, University of Tennessee Health Science Center, Memphis, Tennessee 38163, USA*

[c] *Department of Anesthesiology, University of California, San Diego, La Jolla, California 92093, USA*

ABSTRACT: A growing body of data indicates that multiple signal transduction events in the heart occur via plasma membrane receptors located in signaling microdomains. Lipid rafts, enriched in cholesterol and sphingolipids, form one such microdomain along with a subset of lipid rafts, caveolae, enriched in the protein caveolin. In the heart, a key caveolin is caveolin-3, whose scaffolding domain is thought to serve as an anchor for other proteins. In spite of the original morphologic definition of caveolae ("little caves"), most work related to their role in compartmenting signal transduction molecules has involved subcellular fractionation or immunoprecipitation with anti-caveolin antibodies. Use of such approaches has documented that several G protein–coupled receptors (GPCR), and their cognate heterotrimeric G proteins and effectors, localize to lipid rafts/caveolae in neonatal cardiac myocytes. Our recent findings support the view that adult cardiac myocytes appear to have different patterns of localization of such components compared to neonatal myocytes and cardiac fibroblasts. Such results imply the existence of multiple subcellular microdomains for GPCR-mediated signal transduction in cardiac myocytes, in particular adult myocytes, and raise a major unanswered question: what are the precise mechanism(s) that determine co-localization of GPCR and post-receptor components with lipid rafts/caveolae in cardiac myocytes and other cell types?

KEYWORDS: adenylyl cyclase; cardiac fibroblast; cardiac myocyte; caveolin; G proteins; GPCR

INTRODUCTION

The transfer of information between the hydrophilic extracellular and intracellular environments across the hydrophobic barrier of the plasma membrane is a funda-

Address for correspondence: Prof. Paul A. Insel, M.D., Departments of Pharmacology and Medicine, University of California, San Diego, 9500 Gilman Drive, Mail code 0636, La Jolla, CA 92093, USA. Voice: 858-534-2295; fax: 858-822-1007.
pinsel@ucsd.edu

Ann. N.Y. Acad. Sci. 1047: 166–172 (2005). © 2005 New York Academy of Sciences.
doi: 10.1196/annals.1341.015

mental cellular property. Nature has developed multiple ways to achieve this information transfer, perhaps most importantly by utilizing various types of membrane-bound receptors. Based on data available from genomic analyses, GPCR (also termed heptahelical or 7-transmembrane receptors) are the largest such receptor family with ~1,000 unique GPCR genes, that is, ~4% of the genome, with perhaps half that number as GPCR for endogenous ligands (see reviews).[1,2]

Particular types of cells, including cardiac myocytes and fibroblasts, may express >100 different GPCR, many of which are "orphans," that is, receptors without currently defined endogenous agonists.[3] This raises the possibility that widely held notions regarding the key physiologic agonists and GPCRs will require modification as the identity and functional role of such orphans are revealed.

Aside from the presence or absence of GPCR in a given cell type, a further key issue is that the overall level of expression of particular receptors and post-receptor components, for example, heterotrimeric G proteins and G protein–regulated effector molecules cannot readily explain the speed, fidelity, and extent of response (measured as second messenger generation) to GPCR agonists. These observations, including the selectivity of activation with distal effectors, have provided a rationale in support of the concept of compartmentation of GPCR together with key "downstream" cognate components.[4,5] This concept was proposed many years ago, primarily on the basis of theoretical grounds and poorly explainable results, for example, selective activation of distal events by different classes of agonists that were thought to work in similar ways.[5] It has recently received considerable experimental support with the identification of plasma membrane subdomains, in particular, lipid rafts, cholesterol/sphingolipid-enriched membrane domains and caveolae, a subset of lipid rafts that are 50 to 100 nm invaginations of the plasma membrane and include the protein caveolin. Caveolae were originally discovered 50 years ago when electron microscopy was used to examine plasma membranes,[6,7] but the precise functional role of caveolae has taken much longer to define and is still not completely understood. In this article, we briefly review recent findings for cardiac GPCR–G protein signaling in lipid raft/caveolin-enriched microdomains of cardiac myocytes by addressing the following three questions:

(1) Which GPCR and G protein signaling components localize in lipid raft/caveolin-enriched microdomains of cardiac myocytes?
(2) Does localization of such components in those microdomains differ for neonatal and adult cardiac myocytes?
(3) What determines the lipid raft/caveolae localization of GPCR and G protein signaling components?

RESULTS AND DISCUSSION

Which GPCR and G Protein Signaling Components Localize in Lipid Raft/Caveolin-Enriched Microdomains of Cardiac Myocytes?

A definitive answer to this question is currently not possible. Work in progress that involves proteomic analyses of lipid raft and caveolin-rich domains will likely prove crucial for defining the full complement of proteins, including GPCR and

G protein signaling components in those domains in cardiac myocytes and other cell types. However, a detailed understanding of not only the proteins themselves but of their covalent modifications will involve substantial effort. Such efforts will likely require the development and use of new separation and analytic techniques, especially because most methods employed for proteomic analysis are not well suited for membrane-associated proteins that must be solubilized with detergents prior to characterization.

In spite of this current gap in knowledge, evidence has accrued, primarily based on studies with neonatal rat cardiac myocytes, that a number of functionally important proteins involved in GPCR–G protein–mediated signal transduction and cell regulation are found in caveolin-rich domains of cardiac cells. TABLE 1 summarizes examples from studies (e.g., Refs. 4 and 5, and Refs. 8 through 10) showing that GPCR, including β_1-, β_2-, and α_1-adrenergic, muscarinic cholinergic (M2), and adenosine (A1) receptors, as well as $G\alpha_s$, $G\alpha_i$, $G\alpha_q$, $G_{\beta\gamma}$ adenylyl cyclase types 5 and 6, certain phospholipase C isoforms, and other G protein–regulated effectors can be identified in lipid raft/caveolar domains of cardiac myocytes or cardiac membranes. In addition, exclusion of certain G_s-linked GPCR from caveolin-enriched microdomains appears to help determine preferential linkage of such receptors to adenylyl cyclase (reviewed in Refs. 4 and 9).

Conclusions regarding expression of GPCR and post-receptor signaling components in caveolae derive almost exclusively from use of two experimental approaches: (1) subcellular fractionation techniques that exploit the lipid composition and resultant buoyancy (because of the lipid content) of lipid raft/caveolar fractions[11] and (2) co-immunoprecipitation with antibodies to caveolins, in particular caveolin-3, a muscle-specific caveolin. In spite of the original identification of caveolae as morphologic entities, little information has been available from the use of microscopic approaches to visually document the localization of receptors and post-receptor components in such domains. This is likely to be an active area of study in the future.

In some cases, GPCR and post-receptor components appear to localize to lipid raft/caveolar domains prior to treatment with agonists (e.g., β-adrenergic receptor sub-types), whereas in others (e.g., certain muscarinic cholinergic and bradykinin receptors), such localization preferentially occurs only after interaction with agonists. The latter findings suggest a role for lipid raft/caveolar domains in receptor desensitization and internalization.[12–14] Indeed, it has been suggested that certain GPCR internalize in response to agonist activation following phosphorylation by G protein–receptor kinase, interaction with beta-arrestins, and internalization via clathrin-coated pits while other GPCR are phosphorylated by protein kinase A and then internalized

TABLE 1. Examples of G protein–coupled receptors, G proteins, and G protein–regulated enzymes/effectors that have been localized in lipid raft/caveolin-rich regions of cardiac myocytes and membranes

Receptors	G protein subunits	G protein–regulated effectors
β-adrenergic, α_1-adrenergic	$G\alpha_s$, $G\alpha_i$	Adenylyl cyclase 5/6
Muscarinic cholinergic	$G\alpha_q$	Phospholipase C_βs
Adenosine A1	$G_{\beta\gamma}$ subunits	RhoA, Rac1, Src, ERK 1/2

from caveolae.[15] It will be of interest to learn of the generality of such an idea by studying a much wider number of GPCR than have been evaluated thus far.

Does Localization of Components in the Microdomains Differ for Neonatal and Adult Cardiac Myocytes?

For a variety of reasons, including ease of isolation of viable cells and ability to be maintained in culture, almost all studies of lipid raft/caveolar localization of GPCR and post-receptor membrane components in the heart have involved use of neonatal cardiac myocytes, more specifically, neonatal rat myocytes. Neonatal cardiac myocytes differ from adult myocytes in a number of properties.[16] Accordingly, we have recently begun to undertake studies of localization of GPCR and post-receptor components in cardiac myocytes acutely isolated from adult rats and then studied within several hours[17] (see also B.P. Head, H.H. Patel, D.M. Roth, N.C. Lai, I.R. Neisman, M.G. Farquhar, and P.A. Insel. 2005, unpublished data).

We have used several different approaches to assess compartmentation of GPCR and post-receptor components in lipid raft/caveolin-rich microdomains: subcellular fractionation, immunoprecipitation, fluorescent microscopy with deconvolution analysis, and immunoelectron microscopy. Fractionation studies (FIG. 1) indicate that the adult cardiac myocytes show a much broader distribution of caveolin expression than do other cardiac cell types, such as coronary artery smooth muscle cells and cardiac fibroblasts. Unlike the latter two cell types, which yield only "light" buoyant fractions enriched in caveolin, adult cardiac myocytes also have enrichment for caveolin in "heavy" fractions that co-migrate with markers for cellular membranes other than just the sarcolemma.

The data imply that caveolins may be located in intracellular membranes in addition to the plasma membrane, a notion that is consistent with results showing that caveolar antibodies are able to immunoprecipitate several GPCR and post-receptor

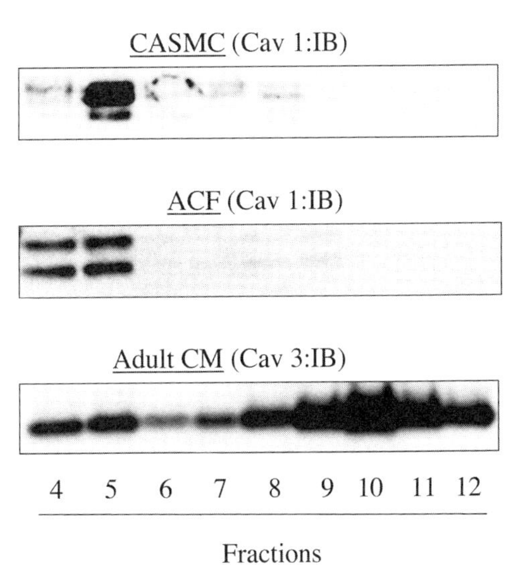

FIGURE 1. Distribution of caveolin-1 (Cav-1) and caveolin-3 (Cav-3) in human coronary artery smooth muscle cells (CASMC), adult rat cardiac fibroblasts (ACF), and adult rat cardiac myocytes (Adult CM). Sodium carbonate extraction followed by sucrose density fractionation was undertaken with CASMC, ACF, and Adult CM. Cav-1 was detected in buoyant cav-enriched fractions (4 & 5) in both CASMC and ACF, while Cav-3 was detected in both buoyant fractions and heavier caveolin-enriched fractions (8–12) in Adult CM by immunoblotting (IB) with antibodies for Cav-1 and Cav-3, as indicated.

components from both "light" and "heavy" fractions. Thus, GPCR and post-receptor agonists appear to localize in caveolin-rich fraction in adult cardiac myocytes in a manner different from that of cardiac fibroblasts, smooth muscle cells, or neonatal cardiac myocytes (FIG. 1; see also Refs. 8 and 17, as well as B.P. Head, H.H. Patel, D.M. Roth, N.C. Lai, I.R. Neisman, M.G. Farquhar, and P.A. Insel. 2005, unpublished data). "Heavy" fractions prepared from adult cardiac myocytes, and that contain caveolin, are likely heterogeneous but almost certainly include cellular membranes from regions such as the sarcoplasmic reticulum and T-tubules, thus raising the possibility that caveolae contribute to the functional activity of those regions.

What Determines the Lipid Raft/Caveolae Localization of GPCR and G Protein Signaling Components?

In spite of the evidence that certain GPCR and post-receptor signaling components reside (either natively or following agonist treatment) in lipid rafts/caveolae of cardiac myocytes, the precise determinants of this localization are not known. This is a vexing problem, in part because GPCR and post-receptor signaling components can show cell-type selective patterns of localization in those microdomains (as recently reviewed).[8] There are several possible explanations, only some of which have been tested.

(1) Protein–protein interaction: Interactions of particular proteins might be favored because of charge, size, and/or steric factors, but why such factors should differ in a cell type-selective manner is not clear. Recent data related to localization in caveolae of different isoforms of adenylyl cyclase imply a role for extramembrane portions of the proteins,[18] but clearly additional data are needed to address the nature of the protein–protein interactions that contribute to co-localization of components in lipid rafts/caveolae. For GPCR, perhaps the ability to form various oligomers with different composition contributes to such localization.[19] The availability of caveolin knockout mice should prove useful in helping to define the role of different caveolins in the localization of GPCR signaling components.[20]

(2) Lipid–protein interaction: Membrane lipids are critical to the composition of lipid rafts and caveolae; differences have been observed in the nature of membrane lipids in different portions of the plasma membrane and among different cell types.[11,21] Although such differences have the potential to contribute to changes in localization during states of altered lipid composition and may help explain cell-specific patterns of localization, direct evidence to test this possibility has not been provided. This has been especially difficult for studies of lipid rafts, whose biology has numerous controversial aspects.[22] Lipid modification of proteins (e.g., palmitoylation, myristoylation) almost certainly contributes to the localization of G protein signaling components in lipid raft/caveolin-enriched domains.[23–25]

(3) Caveolin-associated proteins: The nature of caveolin-associated proteins might also contribute to the differences in the ability of different signaling molecules to localize in caveolae, especially among different cell types. There are likely to be numerous potential candidates: recent data suggest that ~150 proteins can be identified in caveolin-enriched fractions (J.

Schnitzer, personal communication). The role of the scaffolding domain of caveolin as a means for localization of signaling proteins has been an attractive hypothesis (as recently reviewed),[26,27] but it is difficult to imagine that such a domain can accommodate such a wide array of proteins and thus be the only means for co-localization. Perhaps there is a supramolecular assembly of multiple proteins that create a "caveolin signaling particle," akin to complexes involved in other cellular events (transcription, translation, secretion, etc.). The application of proteomic strategies with different cell types prepared under different experimental conditions may offer an attractive way to define the existence of such "particles."

SUMMARY AND CONCLUSIONS

Co-localization of GPCR and post-GPCR signaling components most likely contributes to both the activation and deactivation of heterotrimeric G protein–mediated signal transduction events in cardiac myocytes. Although much has been learned, especially from studies with neonatal rat cardiac myocytes, understanding of localization of GPCR signaling components in adult myocytes is just beginning. Further data are needed to define the nature of the components that co-localize together both before and following treatment of myocytes with agonists. In addition, several major, unresolved questions remain regarding mechanisms responsible for co-localization and impact on functional activities of myocytes. Our view is that the answers are likely to reveal unexpected mechanisms with important consequences for health and disease.

ACKNOWLEDGMENTS

We thank Marilyn Farquhar, James Feramisco, and their colleagues for assistance in the conduct and interpretation of microscopic studies. Work in our laboratory on this topic is supported by NIH Grants HL-66941, HL-53773, HL-63885, and HL-71781.

REFERENCES

1. ARMBRUSTER, B.N. & B.L. ROTH. 2005. Mining the receptorome. J. Biol. Chem. **280:** 5129–5132.
2. WISE, A., S.C. JUPE & S. REES. 2004. The identification of ligands at orphan G-protein coupled receptors. Annu. Rev. Pharmacol. Toxicol. **44:** 43–66.
3. TANG, C.M. & P.A. INSEL. 2004. GPCR expression in the heart; "new" receptors in myocytes and fibroblasts. Trends Cardiovasc. Med. **14:** 94–99.
4. OSTROM, R.S., S.R. POST & P.A. INSEL. 2000. Stoichiometry and compartmentation in G protein-coupled receptor signaling: implications for therapeutic interventions involving G(s). J. Pharmacol. Exp. Ther. **294:** 407–412.
5. STEINBERG, S.F. & L.L. BRUNTON. 2001. Compartmentation of G protein-coupled signaling pathways in cardiac myocytes. Annu. Rev. Pharmacol. Toxicol. **41:** 751–773.
6. PALADE, G.E. 1953. Fine structure of blood capillaries. J. Appl. Physics. **24:** 1424.
7. YAMADA, E. 1955. The fine structure of the gall bladder epithelium of the mouse. J. Biophys. Biochem. Cytol. **1:** 445-458.

8. OSTROM, R.S. & P.A. INSEL. 2004. The evolving role of lipid rafts and caveolae in G protein-coupled receptor signaling: implications for molecular pharmacology. Br. J. Pharmacol. **143**: 235–245.
9. STEINBERG, S.F. 2004. Beta(2)-adrenergic receptor signaling complexes in cardiomyocyte caveolae/lipid rafts. J. Mol. Cell. Cardiol. **37:** 407–415.
10. FUJITA, T., Y. TOYA, K. IWATSUBO, *et al.* 2001. Accumulation of molecules involved in alpha1-adrenergic signal within caveolae: caveolin expression and the development of cardiac hypertrophy. Cardiovasc. Res. **51:** 709–716.
11. PIKE, L.J. 2003. Lipid rafts: bringing order to chaos. J. Lipid Res. **44:** 655–667.
12. DESSY, C., R.A. KELLY, J.L. BALLIGAND & O. FERON. 2000. Dynamin mediates caveolar sequestration of muscarinic cholinergic receptors and alteration in NO signaling. EMBO J. **19:** 4272–4280.
13. TOEWS, M.L., S.C. PRINSTER & N.A. SCHULTE. 2003. Regulation of alpha-1B adrenergic receptor localization, trafficking, function, and stability. Life Sci. **74:** 379–389.
14. HAASEMANN, M., J. CARTAUD, W. MULLER-ESTERL & I. DUNIA. 1998. Agonist-induced redistribution of bradykinin B2 receptor in caveolae. J. Cell Sci. **111:** 917–928.
15. RAPACCIUOLO, A., S. SUVARNA, L. BARKI-HARRINGTON, *et al.* 2003. Protein kinase A and G protein-coupled receptor kinase phosphorylation mediates beta-1 adrenergic receptor endocytosis through different pathways. J. Biol. Chem. **278:** 35403–35411.
16. MITCHESON, J.S., J.C. HANCOX & A.J. LEVI. 1998. Cultured adult cardiac myocytes: future applications, culture methods, morphological and electrophysiological properties. Cardiovasc. Res. **39:** 280–300.
17. HEAD, B.P., R.S. OSTROM, A.N. ANDER, *et al.* 2002. Caveolar microdomains concentrate signal transduction of β_1-adrenergic receptors, but not β_2-adrenergic receptors in adult rat cardiac myocytes. Circulation **106**(Suppl. 2): 48.
18. CROSSTHWAITE, A.J., T. SEEBACHER, N. MASADA, *et al.* 2005. The cytosolic domains of Ca2+-sensitive adenylyl cyclases dictate their targeting to plasma membrane lipid rafts. J. Biol. Chem. **280:** 6380–6391.
19. PARK, P.S., S. FILIPEK, J.W. WELLS & K. PALCZEWSKI. 2004. Oligomerization of G protein-coupled receptors: past, present, and future. Biochemistry **43:** 15643–15656.
20. HNASKO, R. & M.P. LISANTI. 2003. The biology of caveolae: lessons from caveolin knockout mice and implications for human disease. Mol. Interv. **3:** 445–464.
21. ANDERSON, R.G. & K. JACOBSON. 2002. A role for lipid shells in targeting proteins to caveolae, rafts, and other lipid domains. Science **296:** 1821–1825.
22. LAI, E.C. 2003. Lipid rafts make for slippery platforms. J. Cell Biol. **162:** 365–370.
23. MOFFETT, S., D.A. BROWN & M.E. LINDNER. 2000. Lipid-dependent targeting of G-proteins into rafts. J. Biol. Chem. **275:** 2191–2198.
24. OH, P. & J.E. SCHNITZER. 2001. Segregation of heterotrimeric G proteins in cell surface microdomains. G(q) binds caveolin to concentrate in caveolae, whereas G(i) and G(s) target lipid rafts by default. Mol. Biol. Cell **12:** 685–698.
25. RESH, M.D. 2004. Membrane targeting of lipid modified signal transduction proteins. Subcell. Biochem. **37:** 217–232.
26. RAZANI, B., S.E. WOODMAN & M.P. LISANTI. 2002. Caveolae: from cell biology to animal physiology. Pharmacol. Rev. **54:** 431–467.
27. COHEN, A.W., R. HNASKO, W. SCHUBERT & M.P. LISANTI. 2004. Role of caveolae and caveolins in health and disease. Physiol. Rev. **84:** 1341–1379.

Nitric Oxide and the Heart

Update on New Paradigms

C. BELGE, PAUL B. MASSION, M. PELAT, AND J.L. BALLIGAND

Department of Pharmacology and Therapeutics, Université Catholique de Louvain, Brussels 1200, Belgium

ABSTRACT: The role of nitric oxide (NO) as a regulator of cardiac contraction was suggested in the early nineties, but a consensual view of its main functions in cardiac physiology has only recently emerged with the help of experiments using genetic deletion or overexpression of the three nitric oxide synthase (NOS) isoforms in cardiomyocytes. Contrary to the effects of exogenous, pharmacologic NO donors, signaling by endogenous NO is restricted to intracellular effectors co-localized with NOS in specific subcellular compartments. This both ensures coordinate signaling by the three NOS isoforms on different aspects of the cardiomyocyte function and helps to reconcile previous apparently contradictory observations based on the use of non-isoform–specific NOS inhibitors. This review will emphasize the role of NOS on excitation–contraction coupling in the normal and diseased heart. Endothelial NOS and neuronal NOS contribute to maintain an adequate balance between adrenergic and vagal input to the myocardium and participate in the early and late phases of the Frank-Starling adaptation of the heart. At the early phases of cardiac diseases, inducible NOS reinforces these effects, which may become maladaptive as disease progresses.

KEYWORDS: excitation-contraction coupling; inhibitors; NO; NOS

INTRODUCTION

As an endothelial vasorelaxing factor modulating coronary reserve, nitric oxide (NO) is an important indirect regulator of cardiac function. The focus of more recent research on cardiac NO, however, is its direct modulation of myocardial contractility, as well as other aspects of cardiac cellular biology, such as oxygen consumption, hypertrophic remodeling, apoptosis, or myocardial regeneration (for a review, see Ref. 1). The apparent redundancy of NO synthase (NOS) isoforms within the myocardium (including in cardiomyocytes themselves) supports the multitude of its biological effects in the heart, but at the same time poses the fundamental problem of specificity in NO signaling, given the gaseous nature of NO and its theoretical diffusibility within the range of multiple putative molecular targets. This review will emphasize

Address for correspondence: Prof. J.L. Balligand, M.D., Ph.D., Unit of Pharmacology and Therapeutics, FATH 5349, Faculty of Medicine, Université Catholique de Louvain (UCL), 53 Avenue Mounier, 1200 Brussels, Belgium. Voice: 32-2-764 5349; fax: 32-2-764 9322.
balligand@mint.ucl.ac.be

Ann. N.Y. Acad. Sci. 1047: 173–182 (2005).
doi: 10.1196/annals.1341.016

the importance of the specific isoforms of NOS and their compartmentation as critical determinants of NO's effect on cardiac function in response to specific stimuli.

THREE NO SYNTHASES ARE WIDELY DISTRIBUTED IN HEART TISSUE

The endothelial nitric oxide synthase (eNOS; encoded by the *NOS3* gene) and the neuronal nitric oxide synthase (nNOS; encoded by *NOS1*) are constitutively expressed in cardiac myocytes. In these cells, a subset of eNOS is enriched in plasmalemmal and T-tubular caveolae, where it is associated with caveolin-3 (the myocyte-specific structural protein of caveolae).[2] In normal heart extracts, the canonical nNOS is co-immunoprecipitated with the cardiac ryanodine receptor (RyR2), indirect evidence of its localization in the sarcoplasmic reticulum, and nNOS-alpha is expressed in mitochondria.[3] Neuronal NOS is also expressed in both adrenergic and cholinergic nervous fibers, and eNOS is expressed in both endothelial and endocardial cells. Under stress conditions (ischemia, sepsis), inflammatory cytokines induce the expression of the calcium-independent NOS (inducible nitric oxide synthase [iNOS], encoded by *NOS2*) in cardiomyocytes[4] and inflammatory cells infiltrating the myocardium.

Despite Their Apparent Promiscuity, the Cardiac NOS Coordinately Signal to Limited Co-Localized Effectors

Like many signaling molecules, the NOS are part of a complex network of proteins mutually interacting in a dynamic mode within "hubs" or signaling modules. High-throughput proteomic techniques coupled with bioinformatics analysis have started to map these interactions, producing "interactomes" for organisms such as yeast, *Drosophila* or *C. elegans*. These are also being prospectively stored in databases, such as BIND (http://bind.ca/), which produces more than 4,100 signaling interactions only for eNOS. However, because of their dynamic nature, only a few are likely to be operative in response to specific stimuli in cardiac cells. From the perspective of cell physiology, this justifies the relevance of the dissection of NOS protein–protein interactions using more conventional pharmacologic or mutational approaches that yield important insights into functionally relevant interactions for NOS signaling in the heart. This approach also provides key answers to the question of specificity of NO signaling by resolving at least some determinants of NOS posttranslational regulation or spatial confinement restricting NO signaling to nearby effectors, such as we identified through eNOS-hsp90 or eNOS-caveolin interactions in caveolae.[5,6] This compartmentation of NO signaling is further supported because the theoretic diffusion radius for NO (free radical) is limited in cardiac myocytes by at least two facts: (1) actively contracting myocytes abundantly produce superoxide anions that combine with NO to produce other nitrogen derivatives, thereby limiting NO bioavailability, and (2) myocytes contain high concentrations of cytoplasmic myoglobin, a heme protein with a high affinity for NO, that acts as a natural scavenger for this radical, as recently identified by experiments in which myoglobin was either inhibited or genetically invalidated.[7,8] Therefore, the source (i.e., specific NOS isoform) of NO is critical in order to predict its functional impact, in part because subcellular compartmentation of each NOS restricts NO's effects on closely localized targets. One should also bear in mind the dynamic pattern of expression of the NOS isoforms with aging and cardiac disease, such as the induction and transient expression of iNOS in response to sepsis or nNOS in the postinfarcted myocardium,[9] thereby modulating the relative influence of each isoform at any disease stage. Finally, the three NOS differ in their intrinsic catalytic activities (V_{max}, K_m for oxygen) and posttranscriptional regulation (calcium sensitivity, phosphorylations, cofactor availability, protein–protein interactions with caveolin, hsp90, endothelial nitric oxide synthase interacting protein [NOSIP], endothelial nitric oxide synthase traffic inducer [NOSTRIN], etc.), which adds another layer of complexity to predicting the final effect of NOS on cardiomyocyte biology.

NO PHYSIOLOGICALLY REGULATES NORMAL HEART FUNCTION

We will focus on the direct regulation of cardiac contraction by the endogenous NOS, both in the basal (unstimulated) state and under classical inotropic interventions, such as cardiac muscle distension (Frank-Starling response), increase in contraction frequency (force-frequency response), and β-adrenergic stimulation (FIG. 1).

Systolic Properties

Nitric oxide synthase inhibition or NOS genetic deletion (of eNOS and iNOS—virtually absent in the normal cardiomyocyte) have little effect on the basal contractile function in the mammalian heart and in isolated cardiomyocytes. Conversely, acute inhibition of nNOS, or its genetic ablation, results in an increase in L-type calcium current (albeit not uniformly) and amplitude of contractile shortening in isolated mouse cardiomyocytes (at low catecholamines concentrations), as well as contractility *in vivo*. This would indicate a role for nNOS in reducing calcium entry (and excitation–contraction [E-C] coupling) on a beat-to-beat basis.[10]

Diastolic Properties

Basal relaxation properties of isolated myocytes or intact hearts in mice are unaffected by chronic eNOS deletion, contrary to the previous proposition of a facilitatory role of paracrine NO produced from endothelial eNOS on cardiac relaxation in larger animals and man, but this may be explained by compensatory mechanisms (such as upregulation of ANP).[11] Conversely, inhibition or genetic ablation of nNOS results in a slowing of calcium transient decay and myocyte relengthening in one study,[12] but unaltered basal relaxation *in vivo* in another.[13] This is consistent with a facilitatory effect of nNOS on calcium reuptake into the sarcoplasmic reticulum (SR) by sarco- and endoplasmic reticulum calcium-ATPase (SERCA), perhaps through tonic phospholamban (PLB) phosphorylation (B. Casadei, American Heart Association, Scientific Sessions 2004, New Orleans) by as yet uncharacterized mechanisms.

Inotropic Response

The dual influence of nNOS on transsarcolemmal Ca entry and SR calcium reuptake also influences SR calcium loading and contractile reserve after inotropic interventions. In one study, $nNOS^{-/-}$ mice have an attenuated positive force-frequency

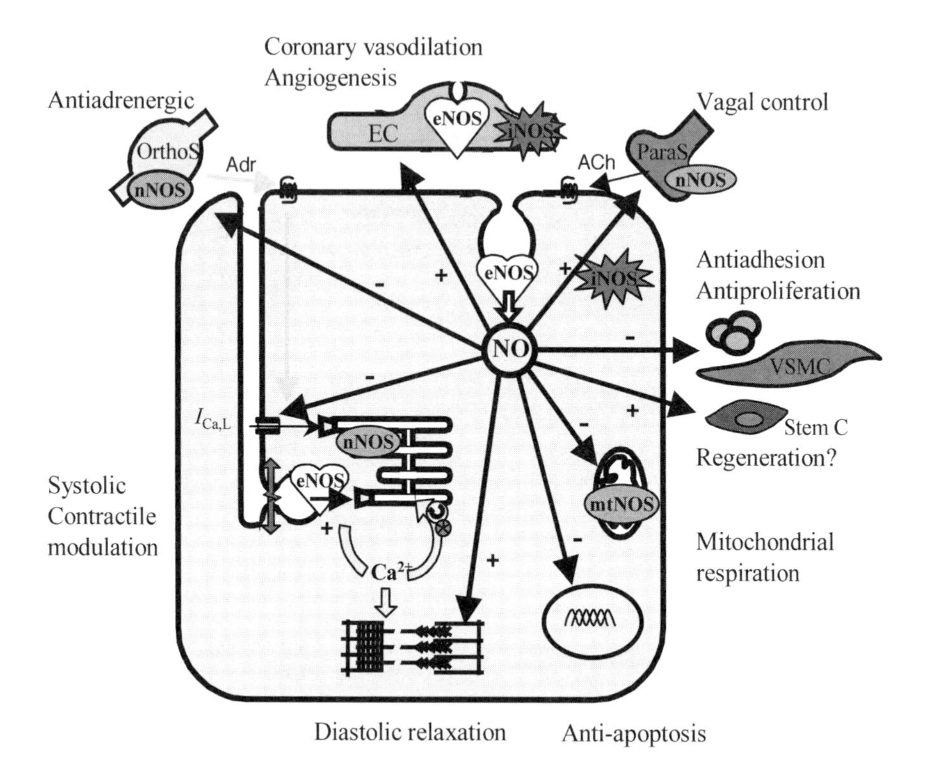

FIGURE 1. Main functional roles of endogenous nitric oxide (NO) in cardiomyocytes and coronary endothelium. Endothelial nitric oxide synthase (eNOS)–derived NO stimulates coronary vasodilation and angiogenesis while it inhibits platelet adhesion and vascular smooth muscle cell (VSMC) proliferation. Neuronal nitric oxide synthase (nNOS) and/or eNOS-derived NO also modulates the sympathovagal balance, systolic and diastolic function, as well as apoptosis and mitochondrial respiration. A possible positive role of eNOS on stem cell recruitment and/or transdifferentiation is also suggested. OrthoS, sympathetic nerve terminals; ParaS, parasympathetic nerve terminals; mtNOS, mitochondrial nitric oxide synthase.

response in conjunction with impaired increase in SR calcium load,[14] which is consistent with a facilitatory effect of nNOS on SERCA activity. Conversely, another group in Oxford found an increased SR calcium load in myocytes from nNOS$^{-/-}$ mice (at a fixed –3 Hz stimulation rate), possibly attributable to a compensatory increase in transsarcolemmal calcium entry in their hands[12] (but not in the Baltimore group).[13] Neuronal NOS$^{-/-}$ hearts also have a decreased PLB/SERCA ratio, which may explain part of the diminished contractile reserve at increased contraction frequency and may be a compensatory chronic adaptation to decreased SERCA function.

Such effects on E-C coupling would also be expected to impact the contractile response to β-adrenergic stimulation, the inotropic intervention most consistently found to be modulated by the three cardiac NOS isoforms.[1] Historically, this was first observed for a constitutive NOS by demonstrating that the contractile effect of isoproterenol was potentiated by treatment of isolated rat cardiomyocytes with a NOS inhibitor, L-N-monomethyl arginine (L-NMMA).[15] This paradigm has subsequently been confirmed in mice genetically deficient in eNOS *in vivo*, further confirming the role of eNOS-derived NO to limit the β-adrenergic effect of catecholamines, just like an "endogenous beta-blocker." Several groups[16–20] have now linked this effect to the stimulation, by catecholamines, of a distinct β3-adrenoceptor isoform on cardiac myocytes activating eNOS through $G_{\alpha i}$ coupling and leading to increases in cyclic-guanosine mono-phosphate (cGMP), with its well-known antagonistic effects on classic β1- and β2-adrenergic signaling in cardiomyocytes. Neuronal NOS may similarly attenuate the inotropic response to β-adrenergic stimulation, at least at low concentrations of isoproterenol, through the inhibitory effect of nNOS-derived NO on L-type calcium current,[21] although the implication of cGMP or any other molecular mechanism to this effect remains unresolved, as is the mechanism of nNOS activation after β-adrenergic stimulation. Of note, although an increase in β-adrenergic response after nNOS inhibition/deletion is consistently observed at low catecholamine concentrations[13,21] (and perhaps at higher concentrations as well by the Oxford group) in isolated cells, this effect is lost and even reversed *in vivo* in $nNOS^{-/-}$ animals,[13] suggesting that the increase in L-type calcium current may not offset the more profound alteration of calcium cycling (i.e., deficient SR calcium loading due to decreased SERCA function, as discussed above) in the stressed heart. Altogether, this would point to a more important role of nNOS in sustaining SERCA function and SR calcium load to increase inotropy (and promote diastoly) in catecholamine-stressed hearts *in vivo*, whereas eNOS (and possibly iNOS under specific circumstances) act as a countervailing "brake" against over stimulation. The latter is in keeping with recent observations where overexpression of eNOS in cardiomyocytes (by adenoviral infection[22] or transgenesis,[23] under the cardiomyocyte-specific alpha-myosin heavy chain [MHC] promoter) decreases peak inotropic (and chronotropic) responses to catecholamines *in vitro* and *in vivo* (e.g., in non-anesthetized mice studied by implanted telemetry.)[23] Cardiac iNOS overexpressors exhibit a similar right-shift of dose-response curve to catecholamines.[24,25] Neuronal NOS overexpressors have not yet been studied.

Although all NOS can attenuate the β-adrenergic response, eNOS distinctively facilitates E-C coupling in response to sarcomere stretch.[26] Indeed, although 10% to 14% sarcomere stretch increased calcium sparks rate (indicative of SR calcium release) in cells and muscles from wild-type mice, this effect was totally abrogated in tissues from eNOS-deficient mice. Unlike the attenuation of the β-adrenergic inotropic effect, the mechanism for eNOS' increase in E-C coupling gain was independent of cGMP and possibly involved an activating S-nitrosylation of the RyR2 by eNOS-derived NO facilitated by the co-localization of both proteins in the diads (see below). This clearly implicates eNOS in the slow rise in calcium transient and force, referred to as the Anrep effect that participates in the length-dependent recruitment of contractile reserve. Overall, the constitutive NOS likely contribute to different phases of the Frank-Starling response of the cardiac muscle, with nNOS and perhaps paracrine NO from surrounding endothelial cells facilitating myocyte relaxation in the early phase, while eNOS sustains the late length-dependent increase in calcium transient and force generation in stretched fibers. Again, these multifaceted influences of the different NOS on E-C coupling would be facilitated by their subcellular localization, eNOS being enriched more than three-fold in caveolar and T-tubular structures close to the diads, as recently shown by us,[23] while nNOS is in SR membranes close to SERCA.

MODULATION OF AUTONOMIC TRANSMISSION

In addition to directly opposing the β1- and β2-adrenergic effect of catecholamines, eNOS was shown to potentiate the postsynaptic effect of acetylcholine (i.e., reinforce vagal effects) in cardiomyocytes from different mammalian species.[1,15] This again requires proper targeting of eNOS to cardiomyocyte caveolae, where muscarinic cholinergic receptors translocate after ligand binding.[27] Endothelial NOS also potentiates the "accentuated antagonism" mediated by acetylcholine in catecholamine-prestimulated cardiac muscles,[22] as confirmed recently by us in cardiomyocytes from eNOS overexpressing mice.[23] This postsynaptic potentiation of vagal signaling by eNOS coordinately reinforces the presynaptic effect of nNOS in parasympathetic fibers, where nNOS potentiates the release of acetylcholine in the synaptic cleft.[28,29] Neuronal NOS also centrally attenuates the adrenergic output as well as the peripheral release of catecholamines in orthosympathetic fibres.[30] As a result, nNOS and eNOS cooperate to maintain an optimal sympathovagal balance for the regulation of cardiac function at the pre- and postsynaptic levels, respectively.

DRAMATIC CHANGES IN NOS EXPRESSION/REGULATION CONTRIBUTE TO CARDIAC DYSFUNCTION IN THE DISEASED HEART

Cardiac diseases are characterized by changes in the relative abundance of each NOS, with nNOS and iNOS being upregulated, whereas eNOS is downregulated.[1] In addition, their specific catalytic activity may be differentially affected by concurrent changes in the abundance/interaction of allosteric modulators (caveolin-3, hsp90),[31,32] availability of cofactors (tetrahydrobiopterin [THB4]) and substrate (L-arginine), concentration of endogenous inhibitors (asymetric dimethyl arginine [ADMA]), or biochemical uncoupling. Cell ultrastructure remodeling such as loss of T-tubular structure in the diseased cardiomyocytes will affect the NOS compartmentation. Increased circulating catecholamines, relative changes in β-adrenergic isoforms expression (e.g., upregulation of β3-adrenoceptors), cytosolic calcium overload, stimulation by inflammatory cytokines, or increased mechanical strain, as commonly observed in cardiac failure, will profoundly alter the nature and intensity of stimuli affecting each NOS, potentially leading to deregulated NO production. Some of these dynamic alterations may be adaptive in the early development of a specific disease (e.g., reinforcement of anti-adrenergic effects) but may become maladaptive at a later stage.

Systolic and Diastolic Properties

Despite the well-documented upregulation of iNOS and nNOS with cardiac disease, non-specific NOS inhibitors have little effect on basal function and force-frequency relationships in heart failure.[1] Increased biodegradation of the NO produced, for example, by oxidant radicals, could explain some insensitivity of the failing myocardium. Of note, some components of the NOS-sensitive cardiac function may be preferentially affected by this "desensitization," as exemplified by the phenotype of iNOS overexpressors. These mice have no alteration of mitochondrial respiration or lusitropic properties, but exquisite sensitivity of their β-adrenergic response.[8,33]

Endothelial NOS downregulation may impair the Frank-Starling adaptation of the terminally failing heart, both through defective stretch-dependent contractile reserve and altered relaxation. The latter may not be fully compensated by nNOS or iNOS upregulation, as illustrated by the lack of correction of altered relaxation properties of the postinfarcted heart in transgenic mice overexpressing iNOS (A. Godecke, Society for Experimental Biology meeting, Capri, 2004). This strongly argues against a role for this isoform in adaptive preservation of the diastolic reserve,[34] at least in the mouse.

β-*Adrenergic Response*

Conversely, the NO-dependent attenuation of the inotropic response to β1-/β2-adrenergic stimulation is again consistently observed in isolated cardiomyocytes, whole heart animal or *in vivo* preparations, and in human cardiac diseases. Originally, this was attributed to the induction of iNOS in cardiomyocytes (as well as infiltrating inflammatory cells) and, recently, was confirmed in transgenic mice overexpressing iNOS or tumor necrosis factor (TNF)-alpha.[25] Neuronal NOS is also upregulated in animal models of cardiac aging and disease, as well as in infarcted patients,[35] and through preferential expression at the sarcolemmal membrane in these circumstances, may further inhibit the inotropic effect of catecholamines (e.g., through inhibition of L-type calcium currents), although this remains to be tested. Albeit downregulated in the failing heart, eNOS remains coupled to β3-adrenoceptors, which are upregulated in heart failure and mediate a prevailing negative inotropic effect in the human ventricle.[19]

Autonomic Transmission

After myocardial infarction, nNOS expression increases in preganglionic cardiac nerves,[36] where it may contribute to the restoration of the sympathovagal balance and perhaps compensate for the downregulation of eNOS in cardiomyocytes. Indeed, adenoviral infection of the sinoatrial node with nNOS encoding plasmids potentiated the presynaptic vagal control of heart rate,[29] which would help to contain the adrenergically induced tachycardia in the failing heart. Transgenic overexpression of eNOS in cardiomyocytes induced similar protective effects in the infarcted heart.[37]

CONCLUSIONS

Despite the apparent redundancy of NOS isoforms in the myocardium, subcellular compartmentation dictates specific NO signaling from each isoform to co-localized effectors in response to physical (e.g., stretch) or receptor-mediated stimuli. Genetic deletion or overexpression experiments helped to characterize each isoform's respective role in the normal or diseased heart. Endothelial NOS and nNOS both contribute to sustaining normal E-C coupling and contribute to the early and late phases of the Frank-Starling mechanism of the heart. They also negatively modulate the β1-/β2-

adrenergic increase in inotropy and chronotropy, and reinforce the (pre- and postsynaptic) vagal control of cardiac contraction, thereby protecting the heart against excessive stimulation by catecholamines. In the ischemic and failing myocardium, iNOS expression is induced and further contributes to attenuate the inotropic effect of catecholamines, as does eNOS coupled to overexpressed β3-adrenoceptors. Neuronal NOS expression also increases in the aging and ischemic heart, but its role (compensatory or deleterious) remains to be defined. Many drugs currently used for the treatment of ischemic or failing cardiac diseases also activate and/or upregulate eNOS in the myocardium, which supports its proposed protective role (e.g., as "endogenous beta-blocker"). Future pharmacologic modulation of the cardiac NOS will have to take into account their specific modulation of the various aspects of cardiac function, if one hopes to deliver more targeted and efficient therapy than currently achieved with exogenous NO donors.

ACKNOWLEDGMENTS

This work was supported by the Fonds National de la Recherche Scientifique, Belgium; the Communauté Française de Belgique (Action de Recherche Concertée); and the Politique Scientifique Fédérale de Belgique (Pôle d'Attraction Interuniversitaire). M.P. was supported by a Prix du Fonds Spécial de Recherche, Université Catholique de Louvain; P.M. was supported by the Fondation Damman.

REFERENCES

1. MASSION, P.B., O. FERON, C. DESSY & J.L. BALLIGAND. 2003. Nitric oxide and cardiac function: ten years after, and continuing. Circ. Res. **93:** 388–398.
2. FERON, O., L. BELHASSEN, L. KOBZIK, *et al.* 1996. Endothelial nitric oxide synthase targeting to caveolae. Specific interactions with caveolin isoforms in cardiac myocytes and endothelial cells. J. Biol. Chem. **271:** 22810–22814.
3. ELFERING, S.L., T.M. SARKELA & C. GIULIVI. 2002. Biochemistry of mitochondrial nitric-oxide synthase. J. Biol. Chem. **277:** 38079–38086.
4. BALLIGAND, J.L., D. UNGUREANU, R.A. KELLY, *et al.* 1993. Abnormal contractile function due to induction of nitric oxide synthesis in rat cardiac myocytes follows exposure to activated macrophage-conditioned medium. J. Clin. Invest. **91:** 2314–2319.
5. BALLIGAND, J.L. 2002. Heat shock protein 90 in endothelial nitric oxide synthase signaling: following the lead(er)? Circ. Res. **90:** 838–841.
6. FERON, O., C. DESSY, D.J. OPEL, *et al.* 1998. Modulation of the endothelial nitric-oxide synthase-caveolin interaction in cardiac myocytes. Implications for the autonomic regulation of heart rate. J. Biol. Chem. **273:** 30249–30254.
7. FLOGEL, U., M.W. MERX, A. GODECKE, *et al.* 2001. Myoglobin: a scavenger of bioactive NO. Proc. Natl. Acad. Sci. U.S.A. **98:** 735–740.
8. GODECKE, A., A. MOLOJAVYI, J HEGER, *et al.* 2003. Myoglobin protects the heart from inducible nitric-oxide synthase (iNOS)-mediated nitrosative stress. J. Biol. Chem. **278:** 21761–21766.
9. DAMY, T., P. RATAJCZAK, E. ROBIDEL, *et al.* 2003. Up-regulation of cardiac nitric oxide synthase 1-derived nitric oxide after myocardial infarction in senescent rats. FASEB J. **17:** 1934–1936.
10. SEARS, C.E., E.A. ASHLEY & B. CASADEI. 2004. Nitric oxide control of cardiac function: is neuronal nitric oxide synthase a key component? Philos. Trans. R. Soc. Lond. B. Biol. Sci. **359:** 1021–1044.
11. MASSION, P.B. & J.L. BALLIGAND. 2003. Modulation of cardiac contraction, relaxation

and rate by the endothelial nitric oxide synthase (eNOS): lessons from genetically modified mice. J. Physiol. **546:** 63–75.
12. Sears, C.E., S.M. Bryant, E.A. Ashley, *et al.* 2003. Cardiac neuronal nitric oxide synthase isoform regulates myocardial contraction and calcium handling. Circ. Res. **92:** e52–e59.
13. Barouch, L.A., R.W. Harrison, M.W. Skaf, *et al.* 2002. Nitric oxide regulates the heart by spatial confinement of nitric oxide synthase isoforms. Nature **416:** 337–339.
14. Khan, S.A., M.W. Skaf, R.W. Harrison, *et al.* 2003. Nitric oxide regulation of myocardial contractility and calcium cycling: independent impact of neuronal and endothelial nitric oxide synthases. Circ. Res. **92:** 1322–1329.
15. Balligand, J.L., R.A. Kelly, P. Marsden, *et al.* 1993. Control of cardiac muscle cell function by an endogenous nitric oxide signaling system. Proc. Natl. Acad. Sci. U.S.A. **90:** 347–351.
16. Gauthier, C., V. Leblais, L. Kobzik, *et al.* 1998. The negative inotropic effect of β3-adrenoceptor stimulation is mediated by activation of a nitric oxide synthase pathway in human ventricle. J. Clin. Invest **102:** 1377–1384.
17. Cheng, H.J., Z.S. Zhang, K. Onishi, *et al.* 2001. Upregulation of functional β(3)-adrenergic receptor in the failing canine myocardium. Circ. Res. **89:** 599–606.
18. Morimoto, A., H. Hasegawa, H.J. Cheng, *et al.* 2004. Endogenous β3-adrenoreceptor activation contributes to left ventricular and cardiomyocyte dysfunction in heart failure. Am. J. Physiol. Heart. Circ. Physiol. **286:** H2425–H2433.
19. Moniotte, S., L. Kobzik, O. Feron, *et al.* 2001. Upregulation of β(3)-adrenoceptors and altered contractile response to inotropic amines in human failing myocardium. Circulation **103:** 1649–1655.
20. Varghese, P., R.W. Harrison, R.A. Lofthouse, *et al.* 2000. β(3)-adrenoceptor deficiency blocks nitric oxide-dependent inhibition of myocardial contractility. J. Clin. Invest. **106:** 697–703.
21. Ashley, E.A., C.E. Sears, S.M. Bryant, *et al.* 2002. Cardiac nitric oxide synthase 1 regulates basal and β-adrenergic contractility in murine ventricular myocytes. Circulation **105:** 3011–3016.
22. Champion, H.C., D. Georgakopoulos, E. Takimoto, *et al.* 2004. Modulation of in vivo cardiac function by myocyte-specific nitric oxide synthase-3. Circ. Res. **94:** 657–663.
23. Massion, P.B., C. Dessy, F. Desjardins, *et al.* 2004. Cardiomyocyte-restricted overexpression of endothelial nitric oxide synthase (NOS3) attenuates β-adrenergic stimulation and reinforces vagal inhibition of cardiac contraction. Circulation **110:** 2666–2672.
24. Heger, J., A. Godecke, U. Flogel, *et al.* 2002. Cardiac-specific overexpression of inducible nitric oxide synthase does not result in severe cardiac dysfunction. Circ. Res. **90:** 93–99.
25. Funakoshi, H., T. Kubota, N. Kawamura, *et al.* 2002. Disruption of inducible nitric oxide synthase improves β-adrenergic inotropic responsiveness but not the survival of mice with cytokine-induced cardiomyopathy. Circ. Res. **90:** 959–965.
26. Petroff, M.G., S.H. Kim, S. Pepe, *et al.* 2001. Endogenous nitric oxide mechanisms mediate the stretch dependence of Ca2+ release in cardiomyocytes. Nat. Cell Biol. **3:** 867–873.
27. Feron, O., T.W. Smith, T. Michel, R.A. Kelly. 1997. Dynamic targeting of the agonist-stimulated m2 muscarinic acetylcholine receptor to caveolae in cardiac myocytes. J. Biol. Chem. **272:** 17744–17748.
28. Herring, N. & D.J. Paterson. 2001. Nitric oxide-cGMP pathway facilitates acetylcholine release and bradycardia during vagal nerve stimulation in the guinea-pig in vitro. J. Physiol. **535:** 507–518.
29. Mohan, R.M., D.A. Heaton, E.J. Danson, *et al.* 2002. Neuronal nitric oxide synthase gene transfer promotes cardiac vagal gain of function. Circ. Res. **91:** 1089–1091.
30. Paton, J.F., S. Kasparov & D.J. Paterson. 2002. Nitric oxide and autonomic control of heart rate: a question of specificity. Trends Neurosci. **25:** 626–631.
31. Piech, A., P.E. Massart, C. Dessy, *et al.* 2002. Decreased expression of myocardial eNOS and caveolin in dogs with hypertrophic cardiomyopathy. Am. J. Physiol. Heart. Circ. Physiol. **282:** H219–H231.

32. Piech, A., C. Dessy, X. Havaux, *et al.* 2003. Differential regulation of nitric oxide synthases and their allosteric regulators in heart and vessels of hypertensive rats. Cardiovasc. Res. **57:** 456–467.
33. Wunderlich, C., U. Flogel, A. Godecke, *et al.* 2003. Acute inhibition of myoglobin impairs contractility and energy state of iNOS-overexpressing hearts. Circ. Res. **92:** 1352–1358.
34. Paulus, W.J., S. Frantz & R.A. Kelly. 2001. Nitric oxide and cardiac contractility in human heart failure: time for reappraisal. Circulation **104:** 2260–2262.
35. Damy, T., P. Ratajczak, A.M. Shah, *et al.* 2004. Increased neuronal nitric oxide synthase-derived NO production in the failing human heart. Lancet **363:** 1365–1367.
36. Takimoto, Y., T. Aoyama, K. Tanaka, *et al.* 2002. Augmented expression of neuronal nitric oxide synthase in the atria parasympathetically decreases heart rate during acute myocardial infarction in rats. Circulation **105:** 490–496.
37. Janssens, S., P. Pokreisz, L. Schoonjans, *et al.* 2004. Cardiomyocyte-specific overexpression of nitric oxide synthase 3 improves left ventricular performance and reduces compensatory hypertrophy after myocardial infarction. Circ. Res. **94:** 1256–1262.

Cardiac Neurobiology of Nitric Oxide Synthases

EDWARD J. DANSON AND DAVID J. PATERSON

Laboratory of Physiology, University of Oxford, Oxford OX1 3PT, UK

ABSTRACT: Nitric oxide (NO) is a potent modulator of cardiac and vascular regulation. Its role in cardiac-autonomic neural signaling has received much attention over the last decade because of the ability of NO to alter cardiac sympathovagal balance to favor more anti-arrhythmic states. Complexity and controversy have arisen, however, because of the numerous sources of NO in the brain, peripheral nerves, and cardiomyocytes, all of which are potential regulators of cardiac excitability and calcium signaling. This review addresses the integrative role of NO as a relatively ubiquitous signaling molecule with respect to cardiac neurobiology. The present idea, that divergent NO-signaling pathways from multiple sources within the heart and nervous system converge to modulate cardiac excitability and impact on morbidity and mortality in health and disease, is discussed.

KEYWORDS: autonomic; Ca^{2+} signaling; heart regulation; nitric oxide

INTRODUCTION

The autonomic nervous system has become recognized as a reliable and accurate means of assessing cardiac health and morbidity from fetal to geriatric ages. It is well established that impaired parasympathetic neural signaling and enhanced sympathetic nerve activity characterize cardiac disease states (e.g., heart failure, myocardial ischemia/infarction), whereas enhanced vagal control over heart rate (HR) is a feature of a healthy, nondiseased or athlete's heart.[1–4] Furthermore, indices of cardiac-parasympathetic neural control (HR variability, baroreflex sensitivity[3]) or the rate of heart rate recovery following exertion,[5] are uniquely strong correlates of mortality and morbidity in the general population and of the development of ventricular arrhythmia in heart failure patients independently of ventricular ejection fraction.[2,6] Therefore, clinical interest in cardiac neurobiology has mainly focused on its apparent ability to stratify prognosis in cardiac disease; although the underlying causal role of autonomic dysfunction in mortality is not fully understood.

The idea that sympathetic nerve stimulation can acutely induce ventricular arrhythmias dates back as far as 1859.[7] In addition, human survivors of sudden cardiac death often describe shock, fear, or excitement[8] (associated with sympathetic nervous activity) as being temporally associated with the cardiac event. In support of this

Address for correspondence: E.J.F. Danson, M.D., Ph.D., University Laboratory of Physiology, University of Oxford, Parks Road, Oxford OX1 3PT, UK. Voice: 44 1865 282503; fax: 44 1865 272453.

edward.danson@physiol.ox.ac.uk

Ann. N.Y. Acad. Sci. 1047: 183–196 (2005). © 2005 New York Academy of Sciences.
doi: 10.1196/annals.1341.017

idea, it has been repeatedly shown that behaviorally mediated arrhythmias induced following coronary occlusion can be prevented by β-adrenoceptor blockade.[9,10] In addition to this, it is evident that parasympathetic neural activation is anti-arrhythmic under conditions of heightened sympathetic discharge.[11,12] Low baroreflex sensitivity also identifies conscious dogs with a healed myocardial infarction that are susceptible to ventricular fibrillation induced by treadmill exercise coupled with a transient occlusion of the circumflex artery.[13] Similarly, vagus nerve stimulation reduced the incidence of ventricular arrhythmia in these dogs,[14] suggesting that vagal impairment is causal in the development of arrhythmia. This vagal cardioprotection cannot be explained by changes in heart rate[7,11,14] and is in keeping with vagal prolongation of ventricular effective refractory period.[15] There is also data to suggest that chronically vagal nerve stimulation improves cardiac remodeling and mortality after myocardial infarction.[16]

Establishing which mediators are responsible for regulating cardiac-autonomic signaling in health and disease has proved a difficult task. Until recently, it was thought that the marked alteration in sympathovagal influence on HR after exercise-training could be ascribed to changes on central baroreflex load and the expression of adrenoceptors.[17] However, it is now known that cardiac-autonomic signaling is determined by powerful modulatory signals at numerous levels, both peripheral and central, which involve both limbs of the autonomic nervous system.

ROLE OF NITRIC OXIDE IN CARDIAC NEUROBIOLOGY

Over the last ten years, substantial work has gone into investigating the roles of nitric oxide (NO) in regulating cardiac function. NO has a therapeutically recognized role in coronary vaso-relaxation and angina. However, many of the more recent experimentally documented roles of NO involve facilitating cardiac-parasympathetic signaling or limiting sympathetically enhanced cardiac excitability (see Refs. 18–22 for examples). As evidence within this field has grown, this idea has also become more controversial, with many reports casting doubts on the magnitude of effects produced by NO (see Refs. 23–25 for examples). Notwithstanding the controversy, there are multiple sources of NO within the heart and the autonomic nervous system (and even within cardiomyocytes themselves); a fact not all studies have fully appreciated. Both constitutive isoforms of NO synthase (endothelial, eNOS and neuronal, nNOS) are expressed in cardiac tissues and may activate discrete biochemical pathways. The question of how these various signaling pathways interact is complex, especially since NO can act in both an autocrine and paracrine fashion, and because the regulatory mechanisms governing NO bioavailability are numerous and highly sensitive to changes in physical and biochemical factors.

Autocrine vs. Paracrine Actions of NO

Although NO was first characterized in relation to vascular function, it soon became clear that it could also modulate myocardial function. Experiments carried out in the early 1990s[26] took as their premise the vascular paracrine model: since NO derived from the endothelium could modulate the function of nearby smooth muscle cells, so NO released by coronary vessels or endocardium was thought to similarly influence cardiac myocytes. Such thinking was reinforced by the finding that NO could diffuse several hundred micrometers in aqueous solutions,[27] albeit without the endogenous buffers present (mainly iron-containing heme residues).

To this day it is still thought that many of the effects of endogenous NO on the heart in fact come from NO synthesized from endothelial/endocardial cells, which express large amounts of eNOS protein. This is particularly true for many of the lusitropic effects of NO. Pinsky *et al.*[28] showed in 1997 that intermittent compressive or distending forces applied to *ex vivo* nonbeating hearts were associated with bursts of NO synthesis, with peak NO linearly related to ventricular transmural pressure. This signal was dependent on intact endothelial and endocardial cells. Functionally this effect is thought to accelerate diastolic ventricular relaxation[26] and improve ventricular filling during exercise.[29] Indeed, such endothelium-dependent lusitropic effects can be augmented by substance P[30] or bradykinin,[31] suggesting that either exercise or sympathetic nerve activation—the physiological stimuli for these mediators—may be fundamental triggers for paracrine activation of endothelial eNOS in the heart, as they are in the vasculature. Many now also believe that failure of this mechanism may represent an important feature of diastolic cardiomyopathy.

Despite the contribution of functionally important NO synthesis from endothelial cells, it has also become abundantly clear that other sources of endogenous NO evoke entirely different responses in cardiac tissue. In experiments where the autocrine effect of NO was isolated by the use of single, left ventricular myocytes, a recurrent finding was potentiation of contraction following NOS inhibition (i.e., a negative inotropic effect of endogenous NO production on basal contractile function[32,33]). This idea is supported by expression of eNOS in ventricular myocytes,[32] but effects are often not observed in eNOS knockout animals[23] and are highly dependent on β-adrenergic stimulation, the contribution from paracrine endothelial NO, and the redox state of the microenvironment at the time. Recently, it has also emerged that the level of eNOS protein expression may also play a role in the effects observed.[34,35] However, the triggers and promoters for eNOS expression in cardiomyocytes are not fully understood.

NO as a Regulator of Basal Myocardial Function versus Cardiac-Autonomic Signaling

Most of the literature on NO in cardiac regulation places great emphasis on β-adrenergic stimulation as a crucial determinant of the functional effect observed. Although Klabunde *et al.*[36] were first to report an effect of NOS inhibition on the β-adrenergic pathway, it was Balligand *et al.*[18] who characterized the recurrent finding of potentiation of the β-adrenergic response following non-isoform-specific pharmacological blockade. In general, smaller order magnitude or absent effects are seen more commonly in *in vitro* cardiac preparations without β-adrenergic influence (see Ref. 37 for review). This has led many to suspect that elevated levels of intracellular calcium are required for intracardiac effects of NO,[38] or that β-adrenergic activation itself facilitates NO-synthesis by a calcium-dependent activation. Furthermore, autonomic activation is critical to the ability of the heart to adapt to volume or pressure stress, and an entirely unstimulated state is not necessarily a physiologically relevant representation of *in vivo* cardiac function.

eNOS vs. nNOS

The first isoform of nitric oxide synthase (NOS) to be identified in non-endothelial tissue within the heart was eNOS, which resides in plasmalemmal caveolae in ventricular[39] and atrial myocytes.[40] The functional roles ascribed to this at the time were in attenuating the inotropic effects of β-agonists and potentiating the chronotropic effects of acetylcholine on neonatal rat ventricular myocytes,[18] with notably no effects on basal function. Subsequently this idea has become highly controversial within myocytes themselves, which are sensitive to temperature and biochemical conditions that modulate NO-bioavailability. Augmented eNOS-signaling in ventricular myocytes leads to similar effects that were originally observed in neonatal rat myoyctes;[35] however, whether this is achieved through an inhibition of calcium influx through the L-type calcium channel,[41] inhibition of calcium release from the sarcoplasmic reticulum,[42] or inhibition of myofilament sensitivity[34] has not been firmly established.

Aside from the differences in experimental preparations used, it is also possible that some of the earlier studies into the role of eNOS in myocytes were confounded by the yet undiscovered existence of nNOS in ventricular myocytes, which is associated with both sarcolemmal proteins[43] and the sarcoplasmic reticulum.[44]

Recent evidence from nNOS knockout mice suggests that in unstimulated ventricular myocytes, nNOS may play a role in inhibiting myocardial contractility[22] by inhibiting the L-type calcium current.[45] However, others have not observed this effect.[46] Interestingly, the inotropic effect of nNOS knockout in unstimulated myocytes may also be dependent on the rate of beating, since some have observed depressed force-frequency responses in these cells.[47] The effect of myocardial nNOS on β-adrenergic signaling in myocytes may also depend on the concentration of β-agonist used. In particular, nanomolar doses of isoprenaline produce augmented inotropic effects in nNOS knockout myocytes,[22] while micromolar doses have depressed effects.[46] Because much of this data remains controversial, and to date no preparation that isolates the role of nNOS in one tissue or cellular microdomain has been made available, the true nature of the integrative role of nNOS in normal cardiomyocytes remains elusive.

NO in Peripheral Neurons

Intracardiac cholinergic neurones express nNOS protein in most species[48–50] (see FIG. 1). The effects of NO on peripheral parasympathetic nerves are still somewhat controversial however, and are largely extrapolated from experiments by vagotomising animals *in vivo*, or isolating the heart with intact vagal innervation for study *in vitro*. Peripherally, inhibition of nNOS causes a dramatic reduction in the vagally mediated bradycardia in the ferret and guinea pig *in vivo*,[51] and a modest effect in the dog,[52] whereas others report no significant effect in the young guinea pig *in vitro*[53] or rabbit *in vivo*.[53,54] In humans, NO donors can increase the high-frequency component of heart rate variability, an index of cardiac vagal tone.[55] Additionally, NO donors and the cGMP analogue 8-bromo-cGMP enhance the HR response to vagal nerve stimulation in rabbits and guinea pigs.[56] However, the bradycardic response to carbamylcholine (the stable analogue of acetylcholine) is unaffected by NOS or guanylate cyclase inhibitors, consistent with the idea that NO is

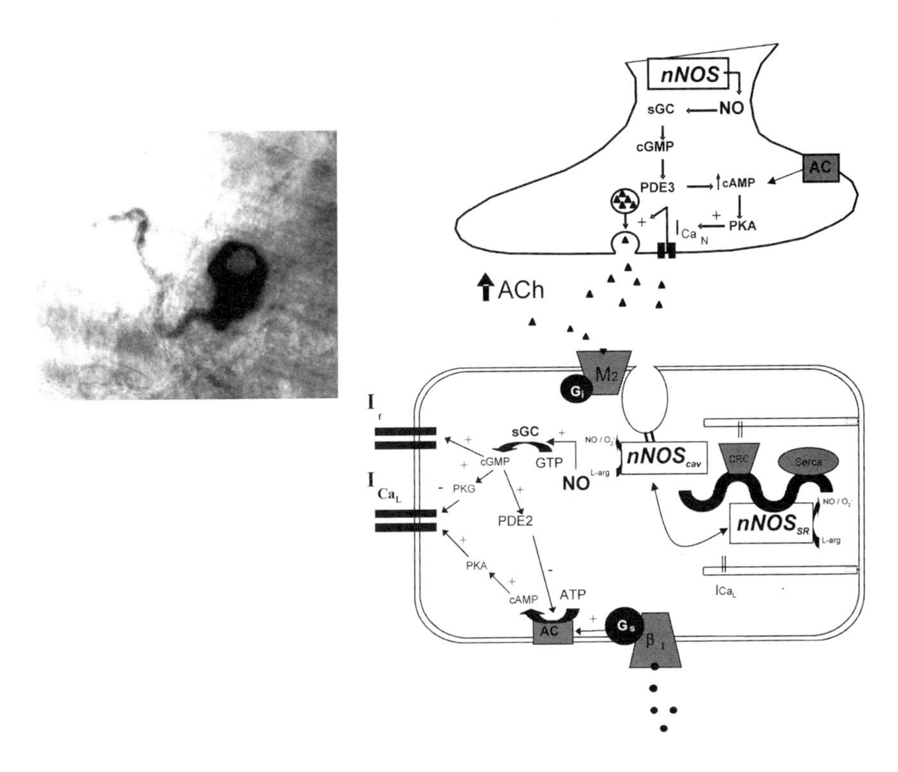

FIGURE 1. Putative roles of neuronal nitric oxide synthase (nNOS) in cardiac cholinergic neurons and myocytes. In neurons, nNOS synthesizes nitric oxide (NO), which acts to increase cyclic guanosine monophosphate (cGMP) via soluble guanylyl cyclase (GC). This may inhibit PDE3 to increase cAMP- and protein kinase A (PKA)-dependent phosphorylation of N-type calcium channels (I_{CaN}) and facilitate acetylcholine (ACh) release and vagal bradycardia. In myocytes, nNOS may reside in microdomains associated with the sarcoplasmic reticulum (SR) and caveolae. At the SR, NO may modulate calcium cycling by attenuating calcium release at the calcium release channel (CRC) or by modulating sarcoplasmic reticulum calcium-ATPase pump (SERCA) activity. Within caveolae, NO may inhibit calcium influx by cGMP-dependent inhibition of the L-type calcium channel (I_{CaL}). Micrograph shows immunolocalization of nNOS in intrinsic cardiac parasympathetic neuron. (From Choate *et al.*[50] Adapted by permission.)

a presynaptic modulator of vagal neurotransmission.[56,57] Furthermore, NO or cGMP can cause an increase in acetylcholine release during field stimulation of the right atrium.[58] This is thought to result from increased cGMP-mediated inhibition of PDE3, leading to increased PKA-dependent phosphorylation of N-type calcium channels and calcium mediated exocytosis. Whether this neuromodulation takes place at the pre-post ganglionic junction or at the neuroeffector junction is controversial, since the HR response to electrical ganglionic stimulation[59] or activation of cholinergic ganglia with nicotine[60] are not attenuated by nNOS inhibition in the dog heart. In contrast, activation of mouse cholinergic ganglia with anatoxin fumarate is sensitive to nNOS inhibition (see FIG. 2).

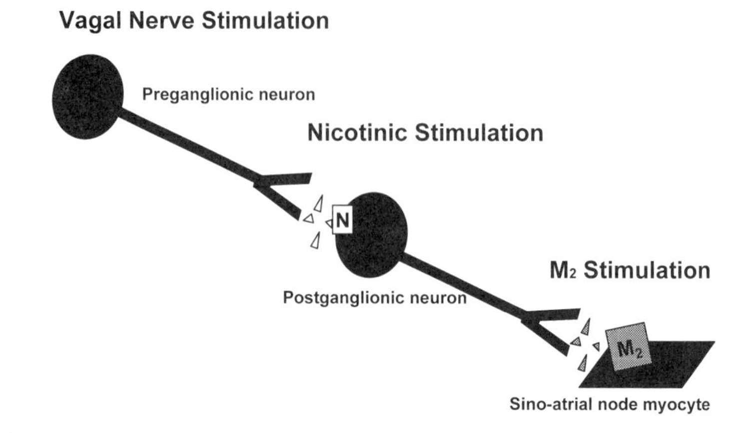

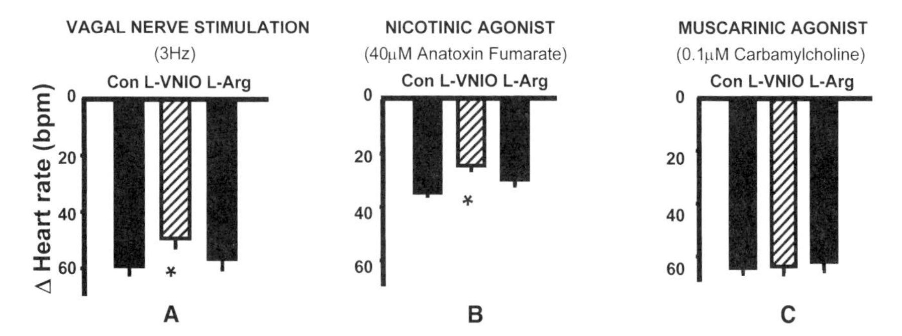

FIGURE 2. (**A**) Neuronal nitric oxide synthase (nNOS) inhibition with 100 μM L-VNIO attenuates bradycardia in response to 3 Hz vagal nerve stimulation in the *in vitro* mouse double atrial preparation ($P < 0.05$). Addition of excess L-arginine (1 mM) reverses this effect. (**B**) The nicotinic agonist Anatoxin Fumarate (40 μM) was used to selectively stimulate postganglionic vagal neurons. Again, the resulting bradycardia was attenuated by 100 μM L-VNIO ($P < 0.05$), suggesting that nNOS-derived NO acts to facilitate vagal neurotransmission at the post-ganglionic–neuroeffector junction. Reversal of the effect of nNOS inhibition was again seen on addition of L-arginine (1 mM). (**C**) Bradycardia in response to 0.1 μM carbamylcholine is unaffected by nNOS inhibition, indicating that the facilitatory effect of NO on vagal neurotransmission occurs at the pre-synaptic level. (From Mohan *et al.*[89] Adapted by permission.)

Moreover, there is no convincing evidence for significant nNOS expression in peripheral pre-ganglionic cholinergic neurons.[50] Nevertheless, artificial upregulation of nNOS in these neurons following targeted gene delivery of nNOS with adenovirus significantly increases vagal responsiveness.[61] When all data are taken together, it highlights a role for both pre- and post-ganglionic sites for nNOS-derived NO facilitating cholinergic transmission. Functionally, and perhaps more importantly, conscious nNOS knockout mice have a higher baseline HR and blunted HR response to atropine compared to wild-type controls.[19]

It is more difficult to determine the magnitude of any effect of NO on peripheral sympathetic nerve terminals, since evidence for localization of NOS isoforms in these tissues is more scarce. Some evidence exists nonetheless that NO acts presynaptically to inhibit cardiac sympathetic neurotransmission, both *in vivo* and *in vitro*. NOS inhibition enhances overflow of noradrenaline during cardiac sympathetic nerve stimulation in the isolated perfused rat heart,[62] and increases the inotropic effect of stellate ganglion stimulation[63] and β-agonists[52] in vagotomised dogs. Inhibition of NOS also increases chronotropic responsiveness to stimulation of the stellate ganglion, both in rabbits *in vivo*[53] and guinea pigs *in vitro*,[64] although chronotropic responsiveness to β-agonists is unaffected. It is unclear, therefore, whether endogenous NO may exert differential effects on cardiac contraction and rate under conditions of β-adrenergic stimulation; acting pre-junctionally to inhibit release of noradrenaline from the sympathetic varicosity, but also post-junctionally to inhibit the inotropic responsiveness of target myocytes.

NO in the Brain

In the central nervous system, NO is synthesized (1) by spatially distributed nNOS in neurons, and (2) by eNOS in the vasculature; although eNOS may also be present in neurons.[65,66] Cardiac autonomic function is regulated at the level of the nucleus tractus solarius (NTS) by central integration of afferent neurons from baroreceptors and decending pathways from the cerebellum. Microinjection of L-arginine or NO donors into the nucleus tractus solitarius (NTS) elicits both a bradycardia and hypotension.[67] Moreover, similar effects can be produced by adenoviral vector-mediated overexpression of eNOS within the NTS.[68] However, low concentrations of L-arginine or NO donors powerfully depress activity of cardiac vagal afferent neurons.[65] This suggests that a more sensitive mechanism responding to lower levels of NO may operate to activate inhibitory transmission in the NTS and depress the afferent side of the baroreflex. The current proposal is that eNOS-derived NO increases soluble guanylate cyclase via a paracrine mechanism to enhance GABAergic-mediated inhibition of baroreflex afferents.[65,69] In contrast, on the efferent side of the baroreflex, microinjection of NO donors facilitates activity in the nucleus ambiguous to cause vagal bradycardia.[70] Therefore, specificity of action of NO in the brainstem may depend on the NOS isoforms involved and their spatial localisation to the target autonomic cells.

FUNCTIONAL SIGNIFICANCE OF CARDIAC-NOS SIGNALING

Unraveling the integrative role of NO in modulating autonomic signaling at various different levels of the neuro-cardiac axis is clearly complex. This problem is even further complicated where at each location numerous roles may exist for NO under different conditions. However, the roles of NO may also be sufficiently variable to account for the differences in cardiac-autonomic signaling in exercise-trained or diseased hearts. Understanding the triggers and mechanisms that underpin regulation of NO bioavailability in different tissues therefore may be key to promoting artificial intervention and modification of cardiac sympathovagal balance.

Exercise

Exercise is a promoter of NO bioavailability in a number of cardiac tissues. There are at least two important ways through which this occurs. Sessa *et al.*[71] first showed that levels of eNOS protein in coronary vascular endothelium were markedly increased in pigs subjected to treadmill exercise. This has subsequently been shown to be a global response in numerous vascular beds, and a hemodynamic shear-stress responsive element has been identified on the eNOS gene.[72] This effect is known to improve vascular relaxation and therefore myocardial perfusion; however, any other paracrine role of this response is not known. A similar pattern is seen with nNOS after exercise in neural tissue innervating the heart both peripherally[73,74] and centrally.[54] Increases in nNOS expression have been detected in both peripheral sympathetic[74] and parasympathetic intrinsic neurons.[73] The facilitation of parasympathetic neural signaling by nNOS is notable for being entirely presynaptic (i.e., the response of the cardiac preparation to acetylcholine is not altered) and because it can be reversibly abolished by nNOS inhibition or single-allele gene knockout[75] (see FIG. 3).

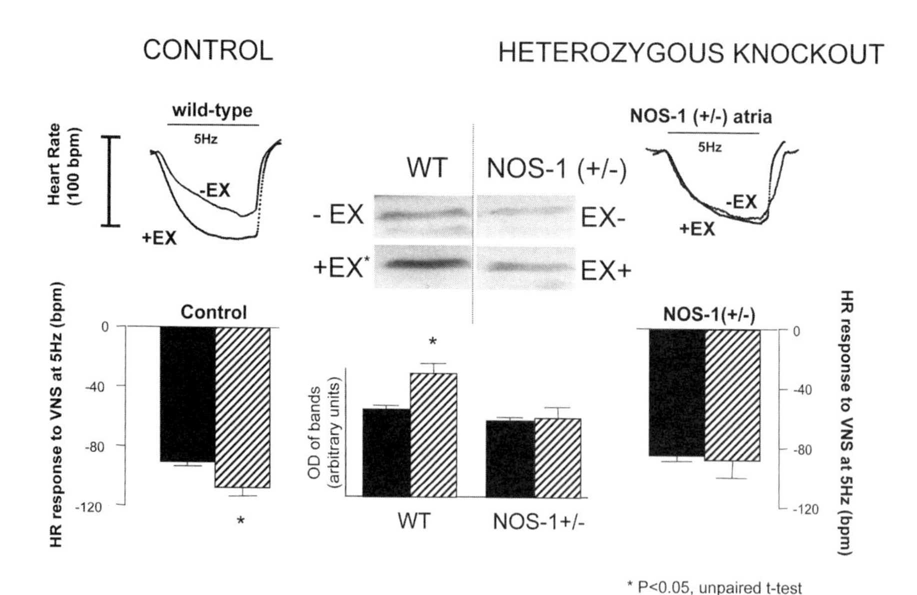

FIGURE 3. Raw heart rate (HR) traces to the *far left* and *right*: the decrease in HR associated with vagal nerve stimulation (5 Hz) in isolated atria *in vitro* was enhanced in trained (+EX) wild-type atria with respect to untrained (-EX). (Grouped data *below traces*; *P < 0.05, unpaired t-test). This effect was not evident in neuronal nitric oxide synthase (*nNOS*) +/- atria. This effect was consistent with data from Western blot analysis of nNOS expression in +EX and –EX atria from wild-type (WT) and *nNOS* +/- mice (*center*), since WT mice showed increased nNOS protein in response to training (*P < 0.01), whereas this effect was absent in *nNOS* +/- mice. (From Danson *et al.*[75] Adapted by permission.)

Sympathetic signaling is attenuated by nNOS-derived NO and exercise. However, the site of action of this effect may be more complicated, since exercise-training is known to more commonly depress the effects of β-adrenergic agonists, suggesting that either paracrine actions of NO or other signaling pathways may be operating. Whether NOS enzyme systems within cardiomyocytes are upregulated during exercise-training has never been specifically tested, and the resulting consequences if this were the case can only be guessed or inferred from data obtained from cardiomyocyte-specific overexpression of eNOS.[34,35] The consequences of physiological overexpression of nNOS in cardiomyocytes leaves even more unanswered questions, since this effect has only ever been observed in diseased hearts.[43,76,77]

The second effect of exercise on NOS-signaling is to augment NO bioavailability and signaling through dual upregulation of antioxidant enzyme systems, such as superoxide dismutase (SOD),[78] catalase, and glutathione peroxidase.[79] This is liable to enhance NO bioavailability, since superoxide inhibits NOS enzyme activity, favoring uncoupled NOS activity and leading to less viable NO synthesis. Although the ways in which this might affect cardiac-autonomic signaling haven't been shown directly, improvements in baroreflex sensitivity have been observed after the intravenous administration of some antioxidants in heart failure patients.[80,81]

Heart Failure

The relationship between heart disease and NO is not simple since most evidence is representative of only one time-point in a multifactorial pathological process. Increased myocardial eNOS expression has been observed in human failing hearts[82,83] and in an animal model of dilated cardiomyopathy.[84] However, others observe no changes[85] or decreases[76] in total eNOS expression. This is further complicated by concomitant changes in the expression of known eNOS protein regulators[85] and other NOS isoforms[43,76,77] such that protein expression itself may not be a good indicator of resultant change in NO synthesis and bioavailability.

On the other hand, evidence reporting decreased nNOS protein expression in the brainstem appears to demonstrate a simple relationship with heart failure and a resultant increase in sympathetic neural outflow to the heart.[86] This may also be augmented by the accompanying effect of increased brain tissue oxidative stress during heart failure, since microinjection of SOD into the rostral ventrolateral medulla of pigs can reduce sympathetic outflow to the heart by up to 70%.[87] Within the heart itself, however, this relationship between nNOS and cardiac disease appears to be less clear cut. Within intrinsic parasympathetic ganglia, limited evidence suggests that nNOS expression may acutely increase following myocardial infarct,[88] an effect accompanied by a parasympathetically mediated decrease in HR within three days of the ischemic event. In addition, within ventricular myocytes, recent reports have suggested that nNOS protein expression increases along with an increase in its coimmunoprecipitation with the plasmalemmal-associated NOS regulator caveolin-3.[76] Functionally, this effect is accompanied by a nNOS-sensitive decrease in inotropic responses to isoprenaline, suggesting that augmented membrane-bound nNOS-derived NO synthesis may play a role in decreasing myocardial β-adrenergic responsiveness in heart failure.[77] Whether these effects are sustained or only a transient compensatory response remains to be established.

CONCLUSION

The various roles ascribed to NO in modulating cardiac function suggest that it is principally a regulator of cardiac-autonomic signaling. These roles are often relatively isolated in terms of anatomical site, although NO is a modulator in this respect at a number of different sites. Emerging evidence suggests that most of these NOS enzyme systems synthesize NO to cope with physiological and pathophysiological stresses and may be responsible for some of the changes in cardiac-autonomic signaling seen in athletes or those with heart disease.

REFERENCES

1. Cohn, J.N., T.B. Levine, M.T. Olivari, *et al.* 1984. Plasma norepinephrine as a guide to prognosis in patients with chronic congestive heart failure. N. Engl. J. Med. **311:** 819–823.
2. Kleiger, R.E., J.P. Miller, J.T. Bigger, *et al.* 1987. Decreased heart rate variability and its association with increased mortality after acute myocardial infarction. Am. J. Cardiol. **59:** 256–262.
3. La Rovere, M.T., J.T. Bigger, Jr., F.I. Marcus, *et al.* 1998. Baroreflex sensitivity and heart rate variability in prediction of total cardiac mortality after myocardial infarction. ATRAMI (Autonomic Tone and Reflexes after Myocardial Infarction) Investigators. J. Lancet **351:** 478–484.
4. Nolan, J., P.D. Batin, R. Andrews, *et al.* 1998. Prospective study of heart rate variability and mortality in chronic heart failure: results of the United Kingdom heart failure evaluation and assessment of risk trial (UK-heart). Circulation **98:** 1510–1516.
5. Cole, C.R., E.H. Blackstone, F.J. Pashkow, *et al.* 1999. Heart-rate recovery immediately after exercise as a predictor of mortality. N. Engl. J. Med. **341:** 1351–1357.
6. Bigger, J.T., J.L. Fleiss, R.C. Steinman, *et al.* 1992. Frequency domain measures of heart period variability and mortality after myocardial infarction. Circulation **85:** 164–171.
7. Lown, B. & R.L. Verrier. 1976. Neural activity and ventricular fibrillation. N. Engl. J. Med. **294:** 1165–1170.
8. Lown, B. 1987. Sudden cardiac death: biobehavioral perspective. Circulation **76:** I186–I196.
9. Rosenfeld, J., M.R. Rosen & B.F. Hoffman. 1978. Pharmacologic and behavioral effects on arrhythmias that immediately follow abrupt coronary occlusion: a canine model of sudden coronary death. Am. J. Cardiol. **41:** 1075–1082.
10. Anderson, J.L., H.E. Rodier & L.S. Green. 1983. Comparative effects of beta-adrenergic blocking drugs on experimental ventricular fibrillation threshold. Am. J. Cardiol. **51:** 1196–1202.
11. Kent, K.M., E.R. Smith, D.R. Redwood, *et al.* 1973. Electrical stability of acutely ischemic myocardium. Influences of heart rate and vagal stimulation. Circulation **47:** 291–298.
12. Corr, P.B. & R.A. Gillis. 1974. Role of the vagus nerves in the cardiovascular changes induced by coronary occlusion. Circulation **49:** 86–97.
13. Schwartz, P.J., E. Vanoli, M. Stramba-Badiale, *et al.* 1988. Autonomic mechanisms and sudden death. New insights from analysis of baroreceptor reflexes in conscious dogs with and without a myocardial infarction. Circulation **78:** 969–979.
14. Vanoli, E., G.M. De Ferrari, M. Stramba-Badiale, *et al.* 1991. Vagal stimulation and prevention of sudden death in conscious dogs with a healed myocardial infarction. Circ. Res. **68:** 1471–1481.
15. Morady, F., W.H. Kou, S.D. Nelson, *et al.* 1988. Accentuated antagonism between beta-adrenergic and vagal effects on ventricular refractoriness in humans. Circulation **77:** 289–297.

16. LI, M., C. ZHENG, T. SATO, *et al.* 2004. Vagal nerve stimulation markedly improves long-term survival after chronic heart failure in rats. Circulation **109:** 120–124.
17. RAVEN, P.B. & X. SHI. 1995. Cardiovascular adaptations following detraining. J. Gravit. Physiol. **2:** P17–P18.
18. BALLIGAND, J.L., R.A. KELLY, P.A. MARSDEN, *et al.* 1993. Control of cardiac muscle cell function by an endogenous nitric oxide signaling system. Proc. Natl. Acad. Sci. U.S.A. **90:** 347–351.
19. JUMRUSSIRIKUL, P., J. DINERMAN, T.M. DAWSON, *et al.* 1998. Interaction between neuronal nitric oxide synthase and inhibitory G protein activity in heart rate regulation in conscious mice. J. Clin. Invest. **102:** 1279–1285.
20. HERRING, N., E.J. DANSON & D.J. PATERSON. 2002. Cholinergic control of heart rate by nitric oxide is site specific. News. Physiol. Sci. **17:** 202–206.
21. PATON, J.F., S. KASPAROV & D.J. PATERSON. 2002. Nitric oxide and autonomic control of heart rate: a question of specificity. Trends Neurosci. **25:** 626–631.
22. ASHLEY, E.A., C.E. SEARS, S.M. BRYANT, *et al.* 2002. Cardiac nitric oxide synthase 1 regulates basal and beta-adrenergic contractility in murine ventricular myocytes. Circulation **105:** 3011–3016.
23. VANDECASTEELE, G., T. ESCHENHAGEN, H. SCHOLZ, *et al.* 1999. Muscarinic and beta-adrenergic regulation of heart rate, force of contraction and calcium current is preserved in mice lacking endothelial nitric oxide synthase. Nat. Med. **5:** 331–334.
24. BELEVYCH, A.E. & R.D. HARVEY. 2000. Muscarinic inhibitory and stimulatory regulation of the L-type Ca2+ current is not altered in cardiac ventricular myocytes from mice lacking endothelial nitric oxide synthase. J. Physiol. **528:** 279–289.
25. GODECKE, A., T. HEINICKE, A. KAMKIN, *et al.* 2001. Inotropic response to beta-adrenergic receptor stimulation and anti-adrenergic effect of ACh in endothelial NO synthase-deficient mouse hearts. J. Physiol. **532:** 195–204.
26. SMITH, J.A., A.M. SHAH & M.J. LEWIS. 1991. Factors released from endocardium of the ferret and pig modulate myocardial contraction. J. Physiol. **439:** 1–14.
27. KNOWLES, R.G. & S. MONCADA. 1992. Nitric oxide as a signal in blood vessels. Trends Biochem. Sci. **17:** 399–402.
28. PINSKY, D.J., S. PATTON, S. MESAROS, *et al.* 1997. Mechanical transduction of nitric oxide synthesis in the beating heart. Circ. Res. **81:** 372–379.
29. SHAH, A.M. 1996. Paracrine modulation of heart cell function by endothelial cells. Cardiovasc. Res. **31:** 847–867.
30. PAULUS, W.J., P.J. VANTRIMPONT & A.M. SHAH. 1995. Paracrine coronary endothelial control of left ventricular function in humans. Circulation **92:** 2119–2126.
31. BRADY, A.J., J.B. WARREN, P.A. POOLE-WILSON, *et al.* 1993. Nitric oxide attenuates cardiac myocyte contraction. Am. J. Physiol. **265:** H176–H182.
32. BALLIGAND, J.L., L. KOBZIK, X. HAN, *et al.* 1995. Nitric oxide-dependent parasympathetic signaling is due to activation of constitutive endothelial (type III) nitric oxide synthase in cardiac myocytes. J. Biol. Chem. **270:** 14582–14586.
33. KAYE, D.M., S.D. WIVIOTT, J.L. BALLIGAND, *et al.* 1996. Frequency-dependent activation of a constitutive nitric oxide synthase and regulation of contractile function in adult rat ventricular myocytes. Circ. Res. **78:** 217–224.
34. BRUNNER, F., P. ANDREW, G. WOLKART, *et al.* 2001. Myocardial contractile function and heart rate in mice with myocyte-specific overexpression of endothelial nitric oxide synthase. Circulation **104:** 3097–3102.
35. MASSION, P.B., C. DESSY, F. DESJARDINS, *et al.* 2004. Cardiomyocyte-restricted overexpression of endothelial nitric oxide synthase (NOS3) attenuates beta-adrenergic stimulation and reinforces vagal inhibition of cardiac contraction. Circulation **110:** 2666–2672.
36. KLABUNDE, R.E., R.C. RITGER & M.C. HELGREN. 1991. Cardiovascular actions of inhibitors of endothelium-derived relaxing factor (nitric oxide) formation/release in anesthetized dogs. Eur. J. Pharmacol. **199:** 51–59.
37. MASSION, P.B., O. FERON, C. DESSY & J.L. BALLIGAND. 2003. Nitric oxide and cardiac function: ten years after, and continuing. Circ. Res. **93:** 388–398.
38. MOHAN, P., D.L. BRUTSAERT, W.J. PAULUS & S.U. SYS. 1996. Myocardial contractile response to nitric oxide and cGMP. Circulation **93:** 1223–1229.

39. Feron, O., T.W. Smith, T. Michel & R.A. Kelly. 1997. Dynamic targeting of the agonist-stimulated m2 muscarinic acetylcholine receptor to caveolae in cardiac myocytes. J. Biol. Chem. **272:** 17744–17748.
40. Wei, C., S. Jiang, J.A. Lust, *et al.* 1996. Genetic expression of endothelial nitric oxide synthase in human atrial myocardium. Mayo. Clin. Proc. **71:** 346–350.
41. Han, X., I. Kubota, O. Feron, *et al.* 1998. Muscarinic cholinergic regulation of cardiac myocyte ICa-L is absent in mice with targeted disruption of endothelial nitric oxide synthase. Proc. Natl. Acad. Sci. U.S.A. **95:** 6510–6515.
42. Petroff, M.G., S.H. Kim, S. Pepe, *et al.* 2001. Endogenous nitric oxide mechanisms mediate the stretch dependence of Ca2+ release in cardiomyocytes. Nat. Cell Bio. **3:** 867–873.
43. Damy, T., P. Ratajczak, E. Robidel, *et al.* 2003. Up-regulation of cardiac nitric oxide synthase 1-derived nitric oxide after myocardial infarction in senescent rats. FASEB J. **17:** 1934–1936.
44. Xu, K.Y., D.L. Huso, T.M. Dawson, *et al.* 1999. Nitric oxide synthase in cardiac sarcoplasmic reticulum. Proc. Natl. Acad. Sci. U.S.A. **96:** 657–662.
45. Sears, C.E., S.M. Bryant, E.A. Ashley, *et al.* 2003. Cardiac neuronal nitric oxide synthase isoform regulates myocardial contraction and calcium handling. Circ. Res. **92:** e52–e59.
46. Barouch, L.A., R.W. Harrison, M.W. Skaf, *et al.* 2002. Nitric oxide regulates the heart by spatial confinement of nitric oxide synthase isoforms. Nature **416:** 337–339.
47. Khan, S.A., M.W. Skaf, R.W. Harrison, *et al.* 2003. Nitric oxide regulation of myocardial contractility and calcium cycling: independent impact of neuronal and endothelial nitric oxide synthases. Circ. Res. **92:** 1322–1329.
48. Klimaschewski, L., W. Kummer, B. Mayer, *et al.* 1992. Nitric oxide synthase in cardiac nerve fibers and neurons of rat and guinea pig heart. Circ. Res. **71:** 1533–1537.
49. Mawe, G.M., E.K. Talmage, K.P. Lee & R.L. Parsons. 1996. Expression of choline acetyltransferase immunoreactivity in guinea pig cardiac ganglia. Cell Tissue Res. **285:** 281–286.
50. Choate, J.K., E.J. Danson, J.F. Morris, *et al.* 2001. Peripheral vagal control of heart rate is impaired in neuronal NOS knockout mice. Am. J. Physiol. Heart Circ. Physiol. **281:** H2310–H2317.
51. Conlon, K. & C. Kidd. 1999. Neuronal nitric oxide facilitates vagal chronotropic and dromotropic actions on the heart. J. Auton. Nerv. Syst. **75:** 136–146.
52. Elvan, A., M. Rubart & D.P. Zipes. 1997. NO modulates autonomic effects on sinus discharge rate and AV nodal conduction in open-chest dogs. Am. J. Physiol. **272:** H263–H271.
53. Sears, C.E., J.K. Choate & D.J. Paterson. 1998. Effect of nitric oxide synthase inhibition on the sympatho-vagal contol of heart rate. J. Auton. Nerv. Syst. **73:** 63–73.
54. Liu, J.L., H. Murakami & I.H. Zucker. 1996. Effects of NO on baroreflex control of heart rate and renal nerve activity in conscious rabbits. Am. J. Physiol. **270:** R1361–R1370.
55. Chowdhary, S. & J.N. Townend. 1999. Role of nitric oxide in the regulation of cardiovascular autonomic control. Clin. Sci. (Lond). **97:** 5–17.
56. Sears, C.E., J.K. Choate & D.J. Paterson. 1999. NO-cGMP pathway accentuates the decrease in heart rate caused by cardiac vagal nerve stimulation. J. Appl. Physiol. **86:** 510–516.
57. Herring, N., S. Golding & D.J. Paterson. 2000. Pre-synaptic NO-cGMP pathway modulates vagal control of heart rate in isolated adult guinea pig atria. J. Mol. Cell. Cardiol. **32:** 1795–1804.
58. Herring, N. & D.J. Paterson. 2001. Nitric oxide-cGMP pathway facilitates acetylcholine release and bradycardia during vagal nerve stimulation in the guinea-pig *in vitro*. J. Physiol. **535:** 507–518.
59. Markos, F., H.M. Snow, C. Kidd, *et al.* 2002. Nitric oxide facilitates vagal control of heart rate via actions in the cardiac parasympathetic ganglia of the anaesthetised dog. Exp. Physiol. **87:** 49–52.
60. Sweeney, C. & F. Markos. 2004. The role of neuronal nitric oxide in the vagal control of cardiac interval of the rat heart *in vitro*. Auton. Neurosci. **111:** 110–115.

61. PATERSON, D.J., S. GOLDING, D.A. HEATON, *et al.* 2004. Adenoviral nNOS gene transfer into the cardiac vagus increases protein expression within 9 hours and enhances vagal responsiveness in the pig. Circulation **110:** S957.
62. SCHWARZ, P., R. DIEM, N.J. DUN, *et al.* 1995. Endogenous and exogenous nitric oxide inhibits norepinephrine release from rat heart sympathetic nerves. Circ. Res. **77:** 841–848.
63. TAKITA, T., J. IKEDA, Y. SEKIGUCHI, *et al.* 1998. Nitric oxide modulates sympathetic control of left ventricular contraction *in vivo* in the dog. J. Auton. Nerv. Syst. **71:** 69–74.
64. CHOATE, J.K. & D.J. PATERSON. 1999. Nitric oxide inhibits the positive chronotropic and inotropic responses to sympathetic nerve stimulation in the isolated guinea-pig atria. J. Auton. Nerv. Syst. **75:** 100–108.
65. PATON, J.F., J. DEUCHARS, Z. AHMAD, *et al.* 2001. Adenoviral vector demonstrates that angiotensin II-induced depression of the cardiac baroreflex is mediated by endothelial nitric oxide synthase in the nucleus tractus solitarii of the rat. J Physiol. **531:** 445–458.
66. TEICHERT, A.M., T.L. MILLER, S.C. TAI, *et al.* 2000. *In vivo* expression profile of an endothelial nitric oxide synthase promoter-reporter transgene. Am. J. Physiol. Heart Circ. Physiol. **278:** H1352–H1361.
67. LIN, H.C., F.J. WAN, K.K. CHENG, *et al.* 1999. Nitric oxide signaling pathway mediates the L-arginine-induced cardiovascular effects in the nucleus tractus solitarii of rats. Life Sci. **65:** 2439–2451.
68. SAKAI, K., Y. HIROOKA, I. MATSUO, *et al.* 2000. Overexpression of eNOS in NTS causes hypotension and bradycardia *in vivo*. Hypertension **36:** 1023–1028.
69. KASPAROV, S. & J.F. PATON. 1999. Differential effects of angiotensin II in the nucleus tractus solitarii of the rat-plausible neuronal mechanism. J. Physiol. **521:** 227–238.
70. RUGGERI, P., A. BATTAGLIA, R. ERMIRIO, *et al.* 2000. Role of nitric oxide in the control of the heart rate within the nucleus ambiguus of rats. Neuroreport **11:** 481–485.
71. SESSA, W.C., K. PRITCHARD, N. SEYEDI, *et al.* 1994. Chronic exercise in dogs increases coronary vascular nitric oxide production and endothelial cell nitric oxide synthase gene expression. Circ. Res. **74:** 349–353.
72. ZIEGLER, T., P. SILACCI, V.J. HARRISON, *et al.* 1998. Nitric oxide synthase expression in endothelial cells exposed to mechanical forces. Hypertension **32:** 351–355.
73. DANSON, E.J. & D.J. PATERSON. 2003. Enhanced neuronal nitric oxide synthase expression is central to cardiac vagal phenotype in exercise-trained mice. J. Physiol. **546:** 225–232.
74. MOHAN, R.M., J.K. CHOATE, S. GOLDING, *et al.* 2000. Peripheral pre-synaptic pathway reduces the heart rate response to sympathetic activation following exercise training: role of NO. Cardiovasc. Res. **47:** 90–98.
75. DANSON, E.J., K.S. MANKIA, S. GOLDING, *et al.* 2004. Impaired regulation of neuronal nitric oxide synthase and heart rate during exercise in mice lacking one nNOS allele. J. Physiol. **558:** 963–974.
76. DAMY, T., P. RATAJCZAK, A.M. SHAH, *et al.* 2004. Increased neuronal nitric oxide synthase-derived NO production in the failing human heart. Lancet **363:** 1365–1367.
77. BENDALL, J.K., T. DAMY, P. RATAJCZAK, *et al.* 2004. Role of myocardial neuronal nitric oxide synthase-derived nitric oxide in beta-adrenergic hyporesponsiveness after myocardial infarction-induced heart failure in rat. Circulation **110:** 2368–2375.
78. POWERS, S.K., D. CRISWELL, J. LAWLER, *et al.* 1993. Rigorous exercise training increases superoxide dismutase activity in ventricular myocardium. Am. J. Physiol. **265:** H2094–H2098.
79. GUNDUZ, F., U.K. SENTURK, O. KURU, *et al.* 2004. The effect of one year's swimming exercise on oxidant stress and antioxidant capacity in aged rats. Physiol. Res. **53:** 171–176.
80. NIGHTINGALE, A.K., D.J. BLACKMAN, R. FIELD, *et al.* 2003. Role of nitric oxide and oxidative stress in baroreceptor dysfunction in patients with chronic heart failure. Clin. Sci. (Lond). **104:** 529–535.
81. PICCIRILLO, G., M. NOCCO, A. MOISE, *et al.* 2003. Influence of vitamin C on baroreflex sensitivity in chronic heart failure. Hypertension **41:** 1240–1245.

82. FUKUCHI, M., S.N. HUSSAIN & A. GIAID 1998. Heterogeneous expression and activity of endothelial and inducible nitric oxide synthases in end-stage human heart failure: their relation to lesion site and beta-adrenergic receptor therapy. Circulation **98:** 132–139.
83. STEIN, B., T. ESCHENHAGEN, J. RUDIGER, *et al.* 1998. Increased expression of constitutive nitric oxide synthase III, but not inducible nitric oxide synthase II, in human heart failure. J. Am. Coll. Cardiol. **32:** 1179–1186.
84. KHADOUR, F.H., D.W. O'BRIEN, Y. FU, *et al.* 1998. Endothelial nitric oxide synthase increases in left atria of dogs with pacing-induced heart failure. Am. J. Physiol. **275:** H1971–H1978.
85. HARE, J.M., R.A. LOFTHOUSE, G.J. JUANG, *et al.* 2000. Contribution of caveolin protein abundance to augmented nitric oxide signaling in conscious dogs with pacing-induced heart failure. Circ. Res. **86:** 1085–1092.
86. PATEL, K.P., K. ZHANG, I.H. ZUCKER, *et al.* 1996. Decreased gene expression of neuronal nitric oxide synthase in hypothalamus and brainstem of rats in heart failure. Brain Res. **734:** 109–115.
87. ZANZINGER, J. & J. CZACHURSKI. 2000. Chronic oxidative stress in the RVLM modulates sympathetic control of circulation in pigs. Pfleugers Arch. Eur. J. Physiol. **439:** 489–494.
88. TAKIMOTO, Y., T. AOYAMA, K. TANAKA, *et al.* 2002. Augmented expression of neuronal nitric oxide synthase in the atria parasympathetically decreases heart rate during acute myocardial infarction in rats. Circulation **105:** 490–496.
89. MOHAN, R.M., S. GOLDING, D.A. HEATON, *et al.* 2004. Targeting neuronal nitric oxide synthase with gene transfer to modulate cardiac autonomic function. Prog. Biophys. Mol. Biol. **84:** 321–344.

The Murine Cardiac 26S Proteasome

An Organelle Awaiting Exploration

ALDRIN V. GOMES,[a] CHENGGONG ZONG,[a,d] RICKY D. EDMONDSON,[c] BENIAM T. BERHANE,[a,b] GUANG-WU WANG,[a] STEVEN LE,[a] GLEN YOUNG,[a] JUN ZHANG,[a] THOMAS M. VONDRISKA,[a] JULIAN P. WHITELEGGE,[a,e] RICHARD C. JONES,[c] IRVING G. JOSHUA,[d] SHEENO THYPARAMBIL,[c] DAWN PANTALEON,[a] JOE QIAO,[a] JOSEPH LOO,[b] AND PEIPEI PING[a,d]

[a] *Department of Physiology and Medicine, and Cardiac Proteomics and Signaling Lab at the Cardiovascular Research Laboratory, University of California at Los Angeles, School of Medicine, Los Angeles, California 90095, USA*

[b] *Department of Chemistry, University of California at Los Angeles, Los Angeles, California 90095, USA*

[c] *Food and Drug Administration/National Center for Toxicology Research, Jefferson, Arkansas 72079, USA*

[d] *Department of Physiology and Biophysics, University of Louisville, Louisville, Kentucky 40202, USA*

[e] *Department of Psychiatry and Behavior Sciences, Pasarow Mass Spectrometry Lab, University of California at Los Angeles, Los Angeles, California 90095,USA*

Abstract: Multiprotein complexes have been increasingly recognized as essential functional units for a variety of cellular processes, including the protein degradation system. Selective degradation of proteins in eukaryotes is primarily conducted by the ubiquitin proteasome system. The current knowledge base, pertaining to the proteasome complexes in mammalian cells, relies largely upon information gained in the yeast system, where the 26S proteasome is hypothesized to contain a 20S multiprotein core complex and one or two 19S regulatory complexes. To date, the molecular structure of the proteasome system, the proteomic composition of the entire 26S multiprotein complexes, and the specific designated function of individual components within this essential protein degradation system in the heart remain virtually unknown. A functional proteomic approach, employing multidimensional chromatography purification combined with liquid chromatography tandem mass spectrometry and protein chemistry, was utilized to explore the murine cardiac 26S proteasome system. This article presents an overview on the subject of protein degradation in mammalian cells. In addition, this review shares the limited information that has been garnered thus far pertaining to the molecular composition, function, and regulation of this important organelle in the cardiac cells.

Keywords: protein degradation; proteomics; 26S proteasome; ubiquitination

Address for correspondence: Prof. Peipei Ping, Ph.D., 675 CE Young Drive, CVRL at UCLA, School of Medicine, MRL Building, Suite 1609, Los Angeles, CA 90095, USA. Voice: 310-267-5624; fax: 310-267-5623.
peipeiping@earthlink.net

**Ann. N.Y. Acad. Sci. 1047: 197–207 (2005). © 2005 New York Academy of Sciences.
doi: 10.1196/annals.1341.018**

INTRODUCTION

Great advancement has been made in the past several decades in our understanding of protein synthesis and construction of molecular networks in the cardiac system. In contrast, little effort has been devoted to determine how proteins are degraded and how signaling pathways are dissembled. Recently, the proteasome system has been increasingly recognized as the major proteolytic machinery responsible for degrading many rate-limiting enzymes and essential regulatory proteins in yeast and in mammalian cells.[1] Therefore, despite the tremendous lack of information concerning its role in the heart, the potential significance of proteasomes in modulating cardiac cell function is deemed to be high.

The proteasomes are distinct cellular organelles composed of various multiprotein complexes, and they are known to be located in the cytosol, nucleus, and microsomes of eukaryotic cells.[2,3] Most of our current information regarding the cytosolic 26S proteasome (~2000kDa) has come from molecular characterization of the yeast proteasome.[4,5] The 20S proteasome (~700kDa) is reported to selectively degrade oxidized proteins in an adenosine triphosphate (ATP)- and ubiquitin-independent manner. The large 26S proteasome is formed when the 20S complex and one or two 19S regulatory complexes associate with each other. Thus, the classically denoted "26S proteasome" is misleading with respect to its molecular weight since it may consist of either a 30S complex, which contains a 20S complex capped at both ends by 19S complexes, or a 26S complex, which includes a 20S complex capped at one end with a 19S complex.[6,7]

Most proteins are not degraded by purified 20S proteasomes because substrates are unable to access the proteolytic sites. As the catalytic center of proteasome complexes, the 20S proteasome is a barrel-shaped complex of 28 protein subunits with αββα rings made of fourteen distinct subunits, including 7 α subunits and 7 β subunits. The two β rings possess three different proteolytic active sites in three β subunits (β1, β2, and β5), each with its own distinct substrate specificity. The 19S regulatory complex contains a lid and a base, the latter conferring the ability to recognize poly-ubiquitinated proteins. The 19S complex hydrolyzes ATP to provide the energy for unfolding and inserting protein substrates into the core of the 20S proteasome. This 19S regulatory complex of yeast[8,9] contains six ATPase subunits and a variable number of non-ATPase subunits. The 20S proteasome is activated by the 19S complexes that bind to either or both the α subunit end rings[10,11] of the 20S. The 19S proteasome mediates most of the biological effects of the proteasome by facilitating ATP- and ubiquitinylation-dependent substrate degradation.[1,12] Despite recent advances in our knowledge of the structure and function of the proteasome, the enzymatic activities and specific functions displayed by many regulatory subunits remain completely unknown.

ISOLATION AND PURIFICATION OF THE MURINE CARDIAC PROTEASOME

To the best of our knowledge, the molecular composition, the function, and the regulation of the cardiac 26S proteasome is virtually unknown. We began to attempt the isolation and purification of the murine cardiac proteasome nearly three years ago. Here we report many of the challenges we were presented with and the lessons we learned during this process.

The 26S proteasome, as well as the 20S proteasome, exists in a form of multiprotein complexes where protein-protein interactions are thought to be primarily ionic in nature.[13,14] Several strategies have been previously applied for the purification of either yeast or mammalian 20S proteasome complexes.[15–17] In particular, Claverol *et al.*[17] reports the purification of the 20S complexes from the human platelets using a monoclonal antibody to the α2 subunit. Others have used chromatography based approaches to separate these particles.[18,19] Although the reported yield varies significantly among methods,[13–19] the proteolytic activities were retained through these separation processes.

We encountered tremendous difficulty in applying the previously reported protocols for the purification of the cardiac 20S and 26S proteasome complexes. Encouraged by our successful experience in using the monoclonal antibody based immunoprecipitation to isolate PKCε multiprotein complexes in previous investigations,[20] we initially attempted a similar procedure to purify the cardiac 20S proteasome complexes, using antibodies against several subunits including the mouse monoclonal antibodies, α2 (MCP236), α3 (M. CP257), α6 (MCP106), β2 (MCP165), and β3 (MCP102) (obtained from Biomol, Plymouth Meeting, Pennsylvania). Unfortunately, we were unable to obtain sufficient purity and quantity of samples for subsequent proteomic analyses. In addition, we performed immunoprecipitation experiments, using the rabbit antibodies against the "core" 20S subunits (PW8155; Biomol), against the β5 subunit (PW8895; Biomol), and against the bovine pituitary 20S proteasome (36-3801; Zymed, San Francisco, CA). However, we were unsuccessful with these approaches as well. In summary, we were unable to obtain adequate purified proteasome for mass spectrometry analyses using immunoprecipitation, and therefore we chose to pursue a chromatography based approach. The protocols reported by French and Bardag-Gorce[21,22] were used as templates, which, along with months of trial-and-error pilot experiments, culminated in the final set of protocols optimized for purification of the 20S and 26S proteasome complexes from the heart.

The critical steps for the purification of murine cardiac 20S proteasome include ammonium sulfate precipitation of the cytosolic fraction from hearts of male mice, Q-fast flow ion-exchange chromatography, ultra-centrifugation at 200,000 xg for 19 h, and Mono Q ion-exchange chromatography. The steps for the purification of murine cardiac 26S proteasome include centrifugation of homogenized male mice hearts at 100,000 xg for 1 h to remove non-cytosolic components, centrifugation of the 100,000 xg supernatant at 70,100 xg for 6 h to pellet the proteasome, and separation of the proteasome from other high molecular weight protein complexes on a 10% to 40% glycerol gradient at 100,000 xg for 22 h. The purified cardiac proteasome samples were characterized by a comprehensive biochemical and proteomic analyses including proteolytic activity assays, western immunoblotting, and mass spectrometry. The technology platform, which depicts the purification and characterization process, is documented in FIG. 1. An example of chromatography analyses of the 26S proteasome is shown in FIG. 2. The detailed map of the cardiac 20S proteasome complexes and the cardiac 26S proteasome complexes will be published elsewhere.

REGULATION OF PROTEIN DEGRADATION

There is a consensus that selective protein degradation in mammalian cells is primarily conducted via the 26S proteasome system, a notion supported by most of the

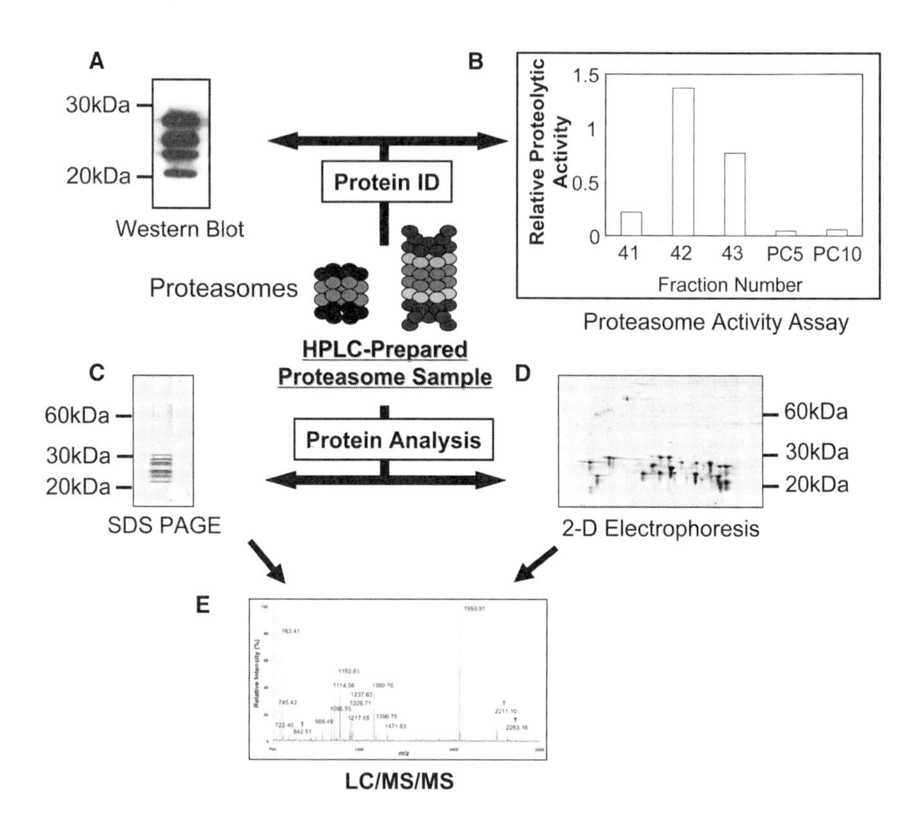

FIGURE 1. Schematic depiction of experimental procedures involved in the identification and analysis of the murine cardiac proteasome. After the murine cardiac proteasome fraction is purified, both the proteasome subunits and the proteasome activity are verified by multiple techniques. (**A**) Western blotting confirms the presence of the various proteasome subunits. (**B**) Proteasome activity assays using fluorescently labeled substrates are used to determine which of the high pressure liquid chromatography (HPLC) fractions contain viable proteolytic activity for proteasome subunits. PC5 and PC10 are negative controls. (**C**) Purified proteasome complexes separated by large format sodium dodecyl sulphate-polyacrylamide gel electrophoresis (SDS-PAGE) with Sypro Ruby-stained protein species. (**D**) Two dimensional electrophoretic separation of the purified proteasome complexes labeled by Sypro Ruby fluorescent staining. (**E**) Protein gel plugs or spots obtained from **C** or **D** were excised and analyzed by liquid chromatography tandem mass spectrometry (LC/MS/MS) to determine the protein identity as well as the post-translational modifications.

reported experimental findings. Nevertheless, our current knowledge of the proteasome is very limited; to date, little is known pertaining to the regulatory mechanisms of protein degradation, the governing principles to selectivity and specificity, and the specific molecular machineries involved. If protein substrates are to be degraded into peptides by the proteasome system (FIG. 3), then there are two fundamental components of this biological process subject to regulation: the protein to be degraded and the proteasome system itself.

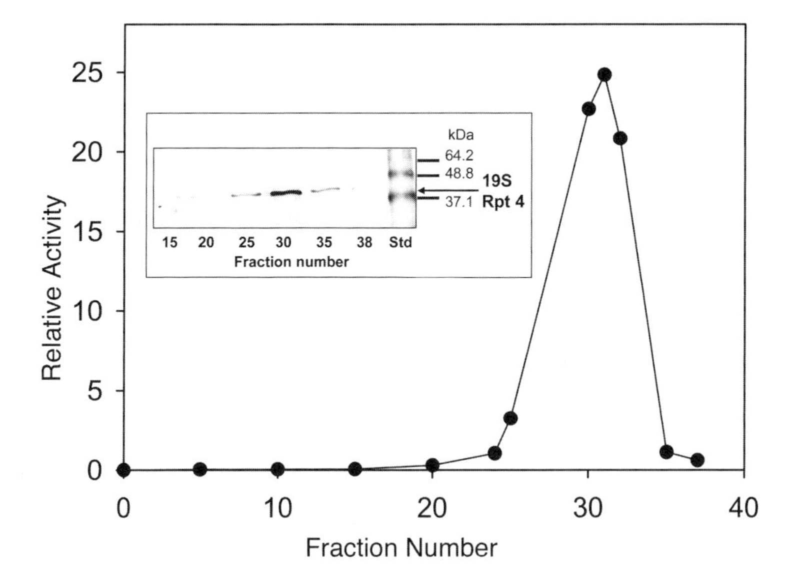

FIGURE 2. Purification of the murine cardiac 26S proteasome by glycerol density gradient. Fractions collected from the glycerol gradient were assayed for 26S proteasome activity using a fluorescent-labeled substrate in the presence of 2mM adenosine triphosphate (ATP). (**Inset**) Western immunoblotting shows the 19S subunit Rpt4 is present predominantly in those glycerol fractions that exhibited high proteolytic activities, indicating the presence of high amounts of proteasome complexes in these fractions (i.e., fractions 25–35).

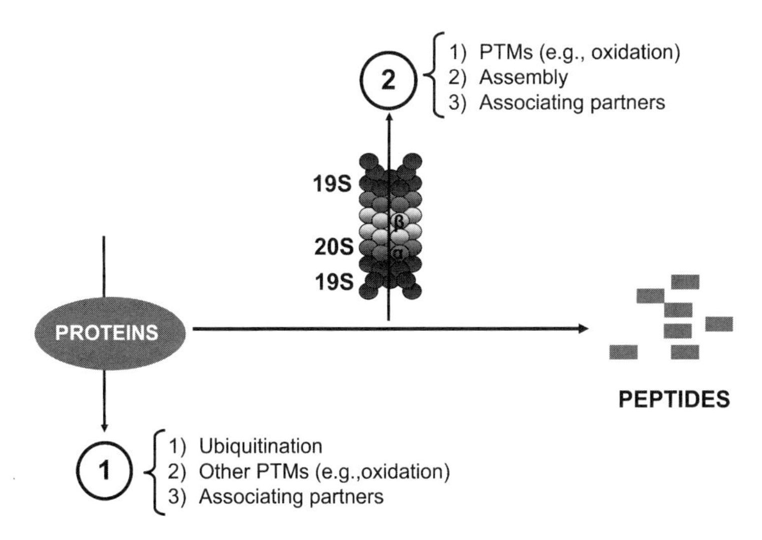

FIGURE 3. Proposed regulatory mechanisms of protein degradation. Two principle regulatory sites are proposed in the depicted process of protein degradation. *Site 1* is the substrate protein to be degraded. Modulation of the substrate protein through ubiquitination, post-translational modifications (PTMs), or associating partners will regulate the degradation process of this protein. *Site 2* is the proteasome system, which offers the proteolytic activity to degrade the substrate protein. Modulation of the proteasome system through PTMs, distinct assembly of the proteasome complexes, or associating partners will impact either the magnitude or the specificity of proteolytic activity.

Proteins to Be Degraded

Significant advances have been made in the past decade to enhance our understanding of modulation of the substrate proteins to be degraded. In particular, ubiquitination as an essential step to initiate protein degradation process is now widely appreciated, based on the experimental finding first made by Avram Hershko, Aaron Ciechanover,[23,24] and Irwin Rose, who shared the 2004 Nobel Prize in Chemistry for their extraordinary contributions to the field. The tagging of protein substrates and the subsequent degradation of these substrates have been shown to be closely coupled *in vivo*.[1,25] In contrast, the delivery of ubiquitinylated proteins to the proteasome is not well understood, but a recent report by Richly *et al.*[26] suggests possibilities on how this may occur. Their investigation demonstrates that an artificial substrate was first ubiquitinated with only one or two ubiquitin moieties in the presence of E1, E2, and E3. The ubiquitin chain is further extended in the presence of another specialized E3 enzyme (often called the E4 enzyme).

A main polyubiquitin recognition feature intrinsic to the 26S proteasome is the ubiquitin interacting motif (UIM) present in the 19S subunit Rpn10.[27,28] Although the precise mechanisms involved remain elusive, there is strong evidence[27,29] indicating the participation of several proteasome cofactors (such as Rad23 and Dsk2), that are thought to bind ubiquitinylated proteins and promote degradation of certain substrates. In other cases, proteins are able to form complexes which promote the degradation of a limited number of substrates; examples include the Cdc48 and its cofactor.[26] The significance of each of the above mechanisms remains to be determined, but it is highly likely that multiple independent ways exist in parallel for targeting ubiquitylated proteins to the proteasome.

In addition to ubiquitination, the oxidation-dependent modification of substrate proteins is also considered to significantly affect the protein degradation rate. Oxidation-modified proteins are preferentially degraded *in vitro* by the 20S proteasome in an ATP-*independent* fashion, while covalent attachment of ubiquitin directs the protein for ATP-*dependent* degradation by the 26S proteasome. Both forms of the proteasome cleave substrate proteins at the carboxyl end of basic (trypsin-like), hydrophobic (chymotrypsin-like), and acidic (peptidylglutamyl-peptide hydrolase) amino acids. Protein degradation facilitated by oxidation may have unique clinical significance to the cardiovascular field, because elevated levels of free radicals and oxidative stress occur concomitant with ischemia-reperfusion injury and have been implicated as key molecular perturbations underlying the pathogenesis of a variety of cardiovascular diseases. In fact, proteasomes were previously found to be targets of ischemia-reperfusion injury of liver, brain, kidney, and heart.[30–33] Intriguingly, ischemic insult has been reported to have distinct effects on proteasome activity in various organs. The liver model of ischemia-reperfusion was accompanied by an increased proteasome activity,[32] whereas pre-inhibition of the proteasome attenuated ischemia-reperfusion injury of the brain, heart, and kidney. A global attenuation of the proteasome function by proteasomal inhibitors was found to protect cardiomyocytes against hyperthermic and oxidative injury.[34] However, the causative role of proteasome activity during ischemia reperfusion, and the specific mechanisms of regulation, has not been established.

The Proteasome Complexes

In addition to modulation of the protein substrate to be degraded, the other obvious subject important in the protein degradation process is the proteasome complexes (FIG. 3). Mechanisms regarding how these multiprotein complexes may be modified and assembled to regulate proteolytic activity and to impart specificity await to be elucidated. Post-translational modification of the proteasome subunits themselves has also not been explored in detail; however, recent investigations indicate that this may also be a powerful mechanism to regulate the protein degradation process. For example, the 26S proteasome was recently shown to be inhibited by modification with the enzyme, O-GlcNAc transferase (OGT).[35] Increased glycosylation of the 19S regulatory subcomplex was correlated with decreased proteasomal activity, suggesting a new model of proteasomal regulation.[36] This modification is reversible and may serve as a mechanism to couple the proteasome to the general metabolic state of the cell.

A key challenge in modulation of the proteasome activity is selectivity. To date, all commonly used proteasome inhibitors block proteolytic activity in a global fashion. An example is lactacystin, which is also capable of inhibiting other cysteine proteases (such as cathepsin A and tripeptidyl peptidase) in addition to the proteasome. Another commonly used proteasome inhibitor, MG132 (N-Cbz-Leu-Leu-norleucinal), also inhibits other cysteine proteases (such as cathepsin B and other Ca^{2+}-activated proteases). Unfortunately, these inhibitors were widely used in many investigations, including studies on homogenized human brain tissue of patients with and without Parkinson's disease, investigations on changes in "proteasomal" activity in the cytosolic fractions from rat liver and kidney in rats with experimentally-induced diabetes,[37] and the studies reporting altered proteasome activities in the ischemic-injured tissues.[30–33] The specificity issues of these inhibitors impact data collected using crude fractions, and they highlight the need for more discerning techniques for assaying proteasome activity. The recent availability of a more specific inhibitor of the tryptic activity of the proteasome, epoxomycin,[38] will hopefully help to resolve this issue.

THE CARDIAC PROTEASOMES: WORKING MODELS

Recent investigations have emphasized modulation of protein substrates as means for regulating protein degradation. Evidence obtained from a number of laboratories support the concept of E1-E2-E3 ligase cascade.[1,12,14] In the proposed models, substrate proteins will be modified via both protein-protein interactions and chemical modifications, with a number of associating partners designated as E1, E2, and E3 (E4). Moreover, ubiquitination and/or other post-translational modifications of the substrate proteins are proposed to provide selectivity and specificity and ultimately facilitate the entry of these proteins to the various proteasome complexes.[13,14,39,40]

In contrast, limited information is available regarding the roles of various proteasome complexes in the regulation of protein degradation. Many important questions await answers. Are these multiprotein complexes exclusive for proteolytic activity? Are they contributing actively to selectivity and specificity of the protein degradation process? And if so, how are selectivity and specificity achieved? Recent reports indicate that distinct assembly of proteasome subunits may interfere with proteolytic

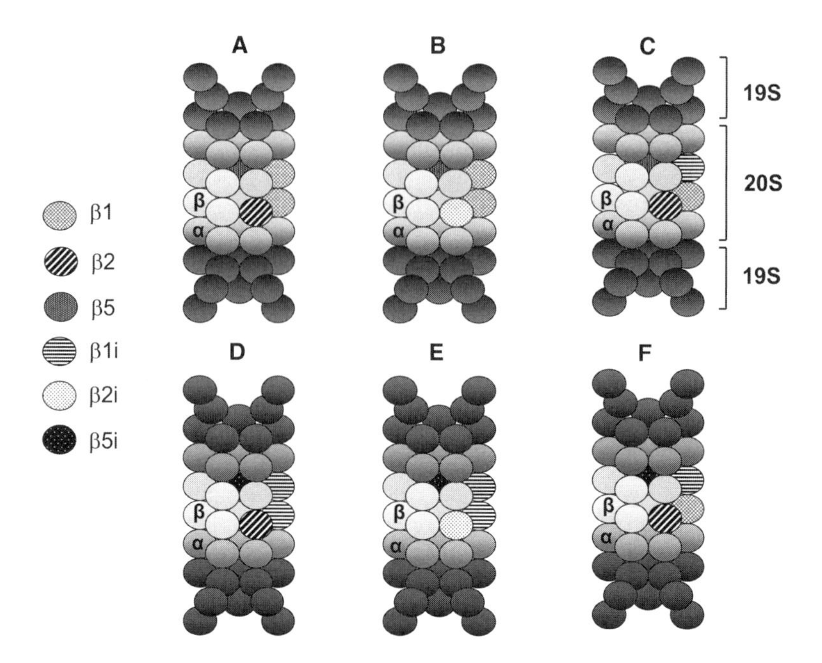

FIGURE 4. Diagrammatic representation of the 26S proteasome assembly. The 26S proteasome is made up of 19S and 20S complexes. The 20S complex consists of a duplicate of 14 distinct subunits (7α and 7β). Each 20S subunit occurs in duplicate with the possible exception of three subunits, β1, β2, and β5, which may be replaced by three inducible isoforms (β1i, β2i, and β5i). Based on our preliminary data, we propose that the cardiac proteasome system contains multiple complexes with distinct molecular compositions. Several of the proposed multiprotein complexes are modeled in this figure. (**A**) The 20S proteasome complex contains two β1 subunits, two β2 subunits, and two β5 subunits. Please note that the two β rings are aligned in counter orientation, thus except for the two β1 subunits, others are not visible in pairs. (**B**) In contrast to the 20S proteasome shown in **A**, this 20S proteasome contains one β2 subunit and one β2i subunit. (**C**) In contrast to the 20S proteasome shown in **A**, this 20S proteasome contains one β1i subunit and one β1 subunit. (**D**) In contrast to the 20S proteasome shown in **A**, this 20S proteasome contains two β1i subunits and one β5i subunit (β5i is shown in the back of the β ring), respectively. (**E**) In contrast to the 20S proteasome shown in **A**, this 26S proteasome contains two β1i, one β2i, and one β5i subunit, respectively. (**F**) In contrast to the 20S proteasome shown in **A**, this 20S proteasome contains one β1i subunit, one β2i subunit, and one β5i subunit, respectively.

specificity,[19,41,42] providing exciting evidence that regulation of the proteasome complexes themselves may be of critical importance.

The major challenges we face in cardiac proteasome research stem from a lack of fundamental information. The molecular compositions of all cardiac proteasome complexes are completely unknown. An anatomical delineation of the various proteasome complexes in cardiac cells is desperately needed, as it is the indispensable first step toward understanding the regulatory mechanisms as well as physical structure of this central protein degradation machinery.

We have recently organized a joint effort to characterize the murine cardiac proteasome complexes. Our investigations have yielded considerable information pertaining to the proteomic basis of this organelle. Our preliminary results support the co-expression of multiple proteasome complexes with distinct molecular compositions. In the light of this evidence, we propose the following working model for the cardiac proteasomes: the cardiac 26S proteasome system consists of various 20S and 19S complexes, each assembled with different proteasome subunits. This diversity in cardiac proteasome complex formation affords specificity and selectivity in proteolytic activity and serves as a major mechanism for the regulation of protein degradation. Examples of the modeled assembly of the cardiac 26S complexes with distinct proteasome subunits are illustrated in FIG. 4.

CONCLUSION

Protein degradation is an exciting area of research that we have only begun to explore in the cardiovascular field. The implementation of functional proteomic technologies will facilitate complete delineation of the proteomic basis of the cardiac proteasome system. The anatomical information of these proteasome complexes will facilitate discoveries regarding mechanisms governing the specificity and selectivity of protein degradation. We anticipate that significant information about this essential protein degradation system in the heart, and the impact of its function on cardiac diseases, will be forthcoming.

REFERENCES

1. GLICKMAN, M.H. & A. CIECHANOVER. 2002. The ubiquitin-proteasome proteolytic pathway: destruction for the sake of construction. Physiol. Rev. **82:** 373–428.
2. BROOKS, P., G. FUERTES & R.Z. MURRAY. 2000. Subcellular localization of proteasomes and their regulatory complexes in mammalian cells. Biochem. J. **346:** 155–161.
3. TANAKA, K., A. Kumatori, A. KUMATORI, II, K. ICHIHARA & A. ICHIHARA. 1989. Direct evidence for nuclear and cytoplasmic colocalization of proteasomes (multiprotease complexes) in liver. J. Cell Physiol. **139:** 34–41.
4. FISCHER, M., W. HILT, B. RICHTER-RUOFF, *et al.* 1994. The 26S proteasome of the yeast *Saccharomyces cerevisiae*. FEBS Lett. **355:** 69–75.
5. CHEN, P. & M. HOCHSTRASSER. 1995. Biogenesis, structure and function of the yeast 20S proteasome. EMBO J. **14:** 2620–2630.
6. YOSHIMURA, T., K. AMEYAMA, T. TAKAGI, *et al.* 1993. Molecular characterization of the "26S" proteasome complex from rat liver. J. Struct. Biol. **111:** 200–211.
7. VOGES, D., P. ZWICKL & W. BAUMEISTER. 1999. The 26S proteasome: a molecular machine designed for controlled proteolysis. Annu. Rev. Biochem. **68:** 1015–1068.
8. GORBEA, C., D. TAILLANDIER & M. RECHSTEINER. 1999. Assembly of the regulatory complex of the 26S proteasome. Mol. Biol. Rep. **26:** 15–19.
9. KIMURA, Y., Y. SAEKI, H. YOKOSAWA, *et al.* 2003. N-Terminal modifications of the 19S regulatory particle subunits of the yeast proteasome. Arch. Biochem. Biophys. **409:** 341–348.
10. OGURA, T. & K. TANAKA. 2003. Dissecting various ATP-dependent steps involved in proteasomal degradation. Mol. Cell. **11:** 3–5.
11. GROLL, M. & R. HUBER. 2003. Substrate access and processing by the 20S proteasome core particle. Int. J. Biochem. Cell Biol. **35:** 606–616.
12. DOHERTY, F.J., S. DAWSON & R.J. MAYER. 2002. The ubiquitin-proteasome pathway of intracellular proteolysis. Essays Biochem. **38:** 51–63.

13. PICKART, C.M. & R.E. COHEN. 2004. Proteasomes and their kin: proteases in the machine age. Nat. Rev. Mol. Cell Biol. **5:** 177–187.
14. ADAMS, J. 2003. The proteasome: structure, function, and role in the cell. Cancer Treat. Rev. **29**(Suppl. 1): 3–9.
15. SAEKI, Y., A. TOH-E & H. YOKOSAWA. 2000. Rapid isolation and characterization of the yeast proteasome regulatory complex. Biochem. Biophys. Res. Commun. **273:** 509–515.
16. GORBEA, C., D. TAILLANDIER & M. RECHSTEINER. 2000. Mapping subunit contacts in the regulatory complex of the 26S proteasome. S2 and S5b form a tetramer with ATPase subunits S4 and S7. J. Biol. Chem. **275:** 875–882.
17. CLAVEROL, S., O. BURLET-SCHILTZ, E. GIRBAL-NEUHAUSER, *et al.* 2002. Mapping and structural dissection of human 20 S proteasome using proteomic approaches. Mol. Cell. Proteomics. **1:** 567–578.
18. AKAISHI, T., H. YOKOSAWA & H. SAWADA. 1995. Regulatory subunit complex dissociated from 26S proteasome: isolation and characterization. Biochim. Biophys. Acta **1245:** 331–338.
19. FERRINGTON, D.A., A.D. HUSOM & L.V. THOMPSON. 2005. Altered proteasome structure, function, and oxidation in aged muscle. FASEB J. **19:** 644–646.
20. EDMONDSON, R.D., T.M. VONDRISKA, K.J. BIEDERMAN, *et al.* 2002. Protein kinase C epsilon signaling complexes include metabolism- and transcription/translation-related proteins: complimentary separation techniques with LC/MS/MS. Mol. Cell. Proteomics. **1:** 421–433.
21. FRENCH, S.W., R.J. MAYER, F. BARDAG-GORCE, *et al.* 2001. The ubiquitin-proteasome 26s pathway in liver cell protein turnover: effect of ethanol and drugs. Alcohol Clin. Exp. Res. **25:** 225S–229S.
22. BARDAG-GORCE, F., F.W. VAN LEEUWEN, V. NGUYEN. *et al.* 2002. The role of the ubiquitin-proteasome pathway in the formation of mallory bodies. Exp. Mol. Pathol. **73:** 75–83.
23. HERSHKO, A. 1983. Ubiquitin: roles in protein modification and breakdown. Cell **34:** 11–12.
24. HERSHKO, A., H. HELLER, S. ELIAS & A. CIECHANOVER. 1983. Components of ubiquitin-protein ligase system. Resolution, affinity purification, and role in protein breakdown. J. Biol. Chem. **258:** 8206–8214.
25. CIECHANOVER, A. 2005. Proteolysis: from the lysosome to ubiquitin and the proteasome. Nat. Rev. Mol. Cell Biol. **6:** 79–87.
26. RICHLY, H., M. RAPE, S. BRAUN, *et al.* 2005. A series of ubiquitin binding factors connects CDC48/p97 to substrate multiubiquitylation and proteasomal targeting. Cell **120:** 73–84.
27. VERMA, R., R. OANIA, J. GRAUMANN & R.J. DESHAIES. 2004. Multiubiquitin chain receptors define a layer of substrate selectivity in the ubiquitin-proteasome system. Cell **118:** 99–110.
28. WALTERS, K.J., M.F. KLEIJNEN, A.M. GOH, *et al.* 2002. Structural studies of the interaction between ubiquitin family proteins and proteasome subunit S5a. Biochemistry **41:** 1767–1777.
29. MADURA, K. 2004. Rad23 and Rpn10: perennial wallflowers join the melee. Trends Biochem. Sci. **29:** 637–640.
30. KUKAN, M. 2004. Emerging roles of proteasomes in ischemia-reperfusion injury of organs. J. Physiol. Pharmacol. **55:** 3–15.
31. KELLER, J.N., F.F. HUANG, H. ZHU, *et al.* 2000. Oxidative stress-associated impairment of proteasome activity during ischemia-reperfusion injury. J. Cereb. Blood Flow Metab. **20:** 1467–1473.
32. WILLMORE, W.G. & K.B. STOREY. 1996. Multicatalytic proteinase activity in turtle liver: responses to anoxia stress and recovery. Biochem. Mol. Biol. Int. **38:** 445–451.
33. TAKAOKA, M., M. ITOH, S. HAYASHI, *et al.* 1999. Proteasome participates in the pathogenesis of ischemic acute renal failure in rats. Eur. J. Pharmacol. **384:** 43–46.
34. LUSS, H., W. SCHMITZ & J. NEUMANN. 2002. A proteasome inhibitor confers cardioprotection. Cardiovasc. Res. **54:** 140–151.

35. Zhang, F., K. Su, X. Yang, *et al.* 2003. O-GlcNAc modification is an endogenous inhibitor of the proteasome. Cell **115:** 715–725.
36. Zachara, N.E. & G.W. Hart. 2004. O-GlcNAc modification: a nutritional sensor that modulates proteasome function. Trends Cell Biol. **14:** 218–221.
37. Portero-Otin, M., R. Pamplona, M.C. Ruiz, *et al.* 1999. Diabetes induces an impairment in the proteolytic activity against oxidized proteins and a heterogeneous effect in nonenzymatic protein modifications in the cytosol of rat liver and kidney. Diabetes **48:** 2215–2220.
38. Meng, L., R. Mohan, B.H. Kwok, *et al.* 1999. Epoxomicin, a potent and selective proteasome inhibitor, exhibits *in vivo* antiinflammatory activity. Proc. Natl. Acad. Sci. U.S.A. **96:** 10403–10408.
39. Feng, J., R. Tamaskovic, Z. Yang, *et al.* 2004. Stabilization of Mdm2 via decreased ubiquitination is mediated by protein kinase B/Akt-dependent phosphorylation. J. Biol. Chem. **279:** 35510–35517.
40. Pickart, C.M. 2000. Ubiquitin in chains. Trends Biochem. Sci. **25:** 544–548.
41. Kingsbury, D.J., T.A. Griffin & R.A. Colbert. 2000. Novel propeptide function in 20 S proteasome assembly influences beta subunit composition. J. Biol. Chem. **275:** 24156–24162.
42. Khan, S., M. van den Broek, K. Schwarz, *et al.* 2001. Immunoproteasomes largely replace constitutive proteasomes during an antiviral and antibacterial immune response in the liver. J. Immunol. **167:** 6859–6868.

Metabolic Energetics and Genetics in the Heart

HEINRICH TAEGTMEYER, CHRISTOPHER R. WILSON, PETER RAZEGHI, AND SAUMYA SHARMA

University of Texas Health Science Center, Department of Internal Medicine, Division of Cardiology, Houston, Texas 77030, USA

ABSTRACT: From the first stages of differentiation in the embryo to the end of life, energy substrate metabolism and function are inextricably linked features of the heart. The principle of energy substrate metabolism is simple. For a given developmental stage and for a given environment, the heart oxidizes the most efficient fuel on the path to ATP. The "multitasking" of energy substrate metabolism in the heart entails more than the generation of reducing equivalents for oxidative phosphorylation of ADP in the respiratory chain. In the postnatal heart, substrate switching and metabolic flexibility are features of normal function. In the stressed heart, metabolic remodeling precedes, triggers, and sustains functional and structural remodeling. This manuscript reviews the pleiotropic actions of metabolism in energy transfer, signal transduction, cardiac growth, gene expression, and viability. Examples are presented to illustrate that metabolic signals of stressed and failing heart are the product of complex cellular processes. An early feature of the maladapted heart is a loss of metabolic flexibility. The example of lipotoxic heart failure illustrates the concept of sustained metabolic dysregulation as a cause of contractile dysfunction of the heart. Thus, a paradigm emerges in which metabolic signals not only regulate fluxes through enzyme catalyzed reactions in existing metabolic pathways, but also regulate transcriptional, translational, and post-translational signaling in the heart. As new insights are gained into metabolic adaptation and maladaptation of the heart, metabolic modulation may become an effective strategy for the treatment of heart failure.

KEYWORDS: cardiac growth; gene expression; substrate switching

INTRODUCTION

The year 2004 marked the centenary of two important discoveries in the field of metabolism: the 1904 discoveries of beta-oxidation of fatty acids by Franz Knoop,[1] and of the oxygen dependence on normal pump function of the adult heart by Hans Winterstein.[2] The year 2004 also marked the fiftieth anniversary of the discovery by Richard Bing and his colleagues that the human heart prefers fatty acids for respiration.[3] Our views of cardiac metabolism are grounded in the nineteenth century and are currently undergoing rapid evolutionary changes.

Address for correspondence: Prof. Heinrich Taegtmeyer, M.D., D.Phil., University of Texas Medical School at Houston, Division of Cardiology, 6431 Fannin, MSB 1.246, Houston, TX 77030, USA. Voice: 713-500-6569; fax: 713-500-0637.
Heinrich.Taegtmeyer@uth.tmc.edu

**Ann. N.Y. Acad. Sci. 1047: 208–218 (2005). © 2005 New York Academy of Sciences.
doi: 10.1196/annals.1341.019**

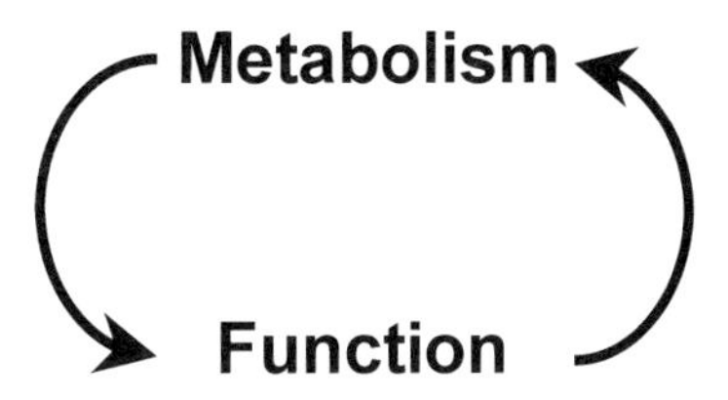

FIGURE 1. The coupling of metabolism and function.

The early studies support the concept that the postnatal heart is well designed, both anatomically and biochemically, for uninterrupted, rhythmic aerobic work. Although heart muscle has certain distinctive biochemical characteristics, many of the basic biochemical reaction patterns are similar to those of other tissues. Because the heart works even when the body is at rest, myocardial O_2 consumption is greater than O_2 consumption of any other organ of the body. In short, metabolism and function of the heart are inextricably linked (FIG. 1).

Heart muscle is an efficient converter of energy. The enzymatic catabolism of substrates results in the production of free energy, which is then used for cell work and for various biosynthetic activities including the synthesis of glycogen, triglycerides, proteins, membranes, and enzymes. The aim of this review is to highlight the actions of cardiac metabolism in energy transfer, cardiac growth, gene expression, and viability.

THE LOGIC OF METABOLISM

The heart "makes a living" by liberating energy from different oxidizable substrates. The logic of metabolism is grounded in the First Law of Thermodynamics—the Law of Energy Conservation—which states that energy can neither be created nor destroyed. In his early experiments on the chemistry of muscle contraction, Helmholtz observed that "a chemical transformation of the compounds contained in them takes place during the action of muscles."[4] The work culminated in the famous treatise, "On the Conservation of Force." The first law of thermodynamics forms the basis for the stoichiometry of metabolism and the calculation of the efficiency of cardiac performance.[4] Efficiency is, as we recall, the ratio of useful work over energy input, and the latter is measured in the heart either as the rate of substrate utilization or the rate of O_2 consumption.

Substrate Switching and Metabolic Flexibility

Heart muscle is a metabolic omnivore with the capacity to oxidize fatty acids, carbohydrates, and also, in certain circumstances, amino acids (either simultaneously or vicariously). Much work has been done in the isolated perfused rat heart to elucidate the mechanisms by which substrates compete for the fuel of respiration. In their celebrated studies in the 1960s, Philip Randle and his group established that, when present in sufficiently high concentrations, fatty acids suppress glucose oxidation to a greater extent than glycolysis, and glycolysis to a greater extent than glucose uptake. These observations gave rise to the concept of a "glucose-fatty acid-cycle."[5]

However, Randle's cycle is not a true metabolic cycle like Krebs' urea cycle or citric acid cycle. Rather, the phenomenon describes the complex interactions between two classes of energy producing substrates, carbohydrates, and fatty acids. What is, in fact, described is the inhibition of glucose oxidation at the level of the pyruvate dehydrogenase complex (PDC) by fatty acids, which, in turn, is greater than the inhibition of glycolysis and the inhibition of glucose uptake by the heart.

The concept of substrate metabolism was later modified with the discovery of the suppression of fatty acid oxidation by glucose[6] through inhibition of the enzyme carnitine-palmitoyl transferase I (CPTI) by malonyl-CoA.[7] This phenomenon is appropriately termed the "reverse Randle effect." Malonyl-CoA is, in turn, regulated by its rate of synthesis (by acetyl-CoA carboxylase [ACC]) and its rate of degradation (by malonyl-CoA decarboxylase [MCD]). Of the two enzymes, MCD is transcriptionally regulated by the nuclear receptor peroxisome proliferator activated receptor α (PPARα),[8] while ACCβ, the isoform that predominates in cardiac and skeletal muscle, is regulated both allosterically and covalently.[9,10] High-fat feeding, fasting, and diabetes increase MCD mRNA and activity in the heart muscle. Conversely, cardiac hypertrophy, which is associated with decreased PPARα expression[11] and a switch from fatty acid to glucose oxidation,[12,13] results in decreased MCD expression and activity, an effect that is independent of fatty acids. Thus, MCD is regulated both transcriptionally and post-transcriptionally, and, in a feed-forward mechanism, fatty acids induce MCD gene expression. The same principle applies to the regulation of other enzymes governing fatty acid metabolism in the heart, including the expression of uncoupling protein 3 (UCP3).[14] Here, fatty acids up-regulate UCP3 expression, while UCP3 is down-regulated in the hypertrophied heart that has switched to glucose for its main fuel of respiration. In short, these are examples for an effective interplay of short- and long-term mechanisms that offer the heart metabolic flexibility to support and maintain its required contractile function.

The above concept has far-reaching implications. For a given physiologic environment, the heart selects the most efficient substrate for energy production. A fitting example is the switch from fatty acid to carbohydrate oxidation with an acute "work jump" or increase in workload.[15] The transient increase in rates of glycogen oxidation is followed by a sustained increase in rates of glucose and lactate oxidation (FIG. 2). Because oleate oxidation remains unaffected by the work jump, the increase in O_2 consumption and cardiac work are entirely accounted for by the increase in carbohydrate oxidation. In short, the heart possesses a built-in metabolic machinery to switch to the most energy efficient fuel when it is stressed. How does the heart mobilize its "back-up fuel?"

The enzymes of glucose and glycogen metabolism are highly regulated by either allosteric activation or covalent modification. The regulation of glycogen phosphorylase by the metabolic signals AMP and glucose is a case in point (FIG. 3). The acute increase in workload of the heart is accompanied by increases in [AMP] and intracellular free [glucose]. While AMP activates phosphorylase and promotes glycogen breakdown, free glucose inhibits phosphorylase and promotes glycogen synthesis via an increase in glucose 6-phosphate and inhibition of glycogen synthase kinase. Thus, the metabolite measurements depicted in FIGURE 3 serve as illustrations for metabolic signals regulating fluxes through metabolic pathways and, hence, determining metabolic flexibility.

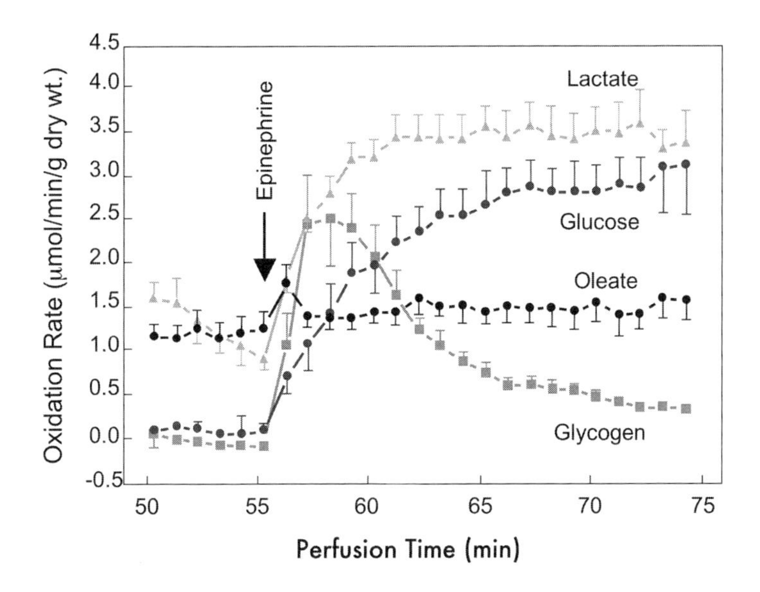

FIGURE 2. Substrate oxidation rates before and after contractile stimulation. (From Goodwin *et al.*[15] Adapted by permission.)

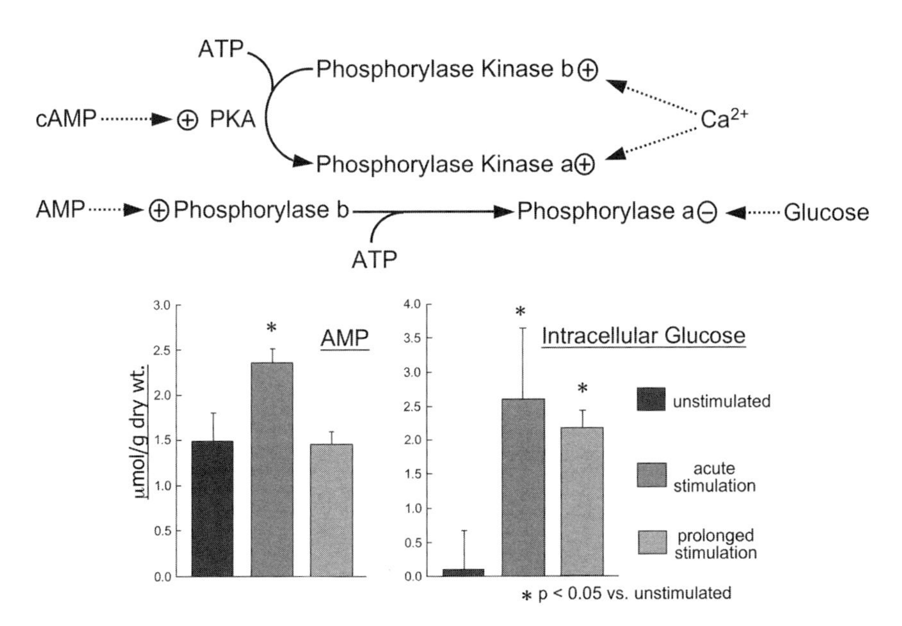

FIGURE 3. Metabolic regulating phosphorylase activity. (From Goodwin *et al.*[15] Adapted by permission.)

However, adaptations to sustained or chronic changes in the environment induce changes of the metabolic machinery at a transcriptional and/or translational level of the enzymes of metabolic pathways. We proposed earlier that metabolic remodeling precedes, triggers, and maintains structural and functional remodeling of the heart.[16] Here the nuclear receptor PPARα (and its coactivator PGC-1) need to be mentioned again, because they have been identified as master switches for the metabolic remodeling of the heart.[16–18] For example, pressure overload[19] and unloading of the heart,[20] hypoxia,[21] and, unexpectedly, also insulin-deficient diabetes[22] all result in the downregulation of genes controlling fatty acid oxidation and in reactivation of the fetal gene program (FIG. 4). These recent observations are in line with earlier work showing increased glucose metabolic activity in the pressure overloaded heart,[23] even before the onset of hypertrophy.[24] They are also in line with work showing impaired fatty acid oxidation by failing heart muscle *in vitro*[25] and *in vivo*.[26,27] We have recently proposed that metabolic flexibility is lost in diseased heart.[28] The loss of metabolic flexibility is in line with development of myocardial insulin resistance in conscious dogs with advanced dilated cardiomyopathy induced by rapid ventricular pacing.[29] These early observations point to an important new concept: metabolism and hormones affecting metabolic regulation give rise to signals that affect the biologic properties of the heart.

	Hypertrophy	Atrophy	Diabetes
Contractile proteins			
α-MHC	↓	↓	↓
β-MHC	↑	↑	↑
Cardiac α-actin	↓	↓	↓
Skeletal α-actin	↑	↑	↑
Ion pumps			
α_2 Na/K-ATPase	↓	↓	↓
SERCA 2a	↓	↓	↓
Metabolic proteins			
GLUT4	↓	↓	↓
GLUT1	=	=	↓
Muscle CPT-1	↓	↓	=
Liver CPT-1	=	=	=
mCK	↓	↓	↓
PPARα	↓	↓	↓
PDK4	↓	↓	↑
MCD	↓	↓	↑
UCP2	↓	↓	=
UCP3	↓	↓	↓
Protooncogenes			
c-fos	↑	↑	↓

FIGURE 4. Coordinated transcriptional responses of the heart in hypertrophy, atrophy, and diabetes.

PLEIOTROPIC ACTIONS OF METABOLISM

It is apparent from the previous discussion that the actions of metabolism are more diverse than those found in the network of energy transfer and function of the heart. In addition to function, metabolism provides signals for growth, gene expression, and viability (FIG. 5).

Metabolic Signals for Cardiac Growth

A case in point is the mammalian target of rapamycin (mTOR), an evolutionary conserved kinase and regulator of cell growth that serves as a point of convergence for nutrient sensing and growth factor signaling. In preliminary studies with the isolated working rat heart, we found that both glucose and amino acids are required for the activation of mTOR by insulin (Sharma *et al.*, unpublished observations). In the same model we observed an unexpected dissociation between insulin stimulated Akt and mTOR activity, suggesting that Akt is not an upstream regulator of mTOR. We found that, irrespective of the stimulus, nutrients are critical for the activation of mTOR in the heart (Sharma *et al.*, unpublished observations). The studies are ongoing.

Metabolic Signals of Cardiac Gene Expression

We propose that a single factor linking myosin heavy chain (MHC) isoform expression in the fetal, hypertrophied, and diabetic heart may be intracellular free glucose or one of its metabolites. Compared to fatty acids, relatively little is known about the effects of glucose metabolism on cardiac gene expression.[30] The mechanisms by which glucose availability affects the DNA binding of transcription factors are not known precisely, although glucose availability and/or insulin affect the expression of specific genes in the liver.[31] A number of candidate transcription factors have been identified that are believed to be involved in glucose-mediated gene expression, mainly through investigations on the glucose/carbohydrate responsive

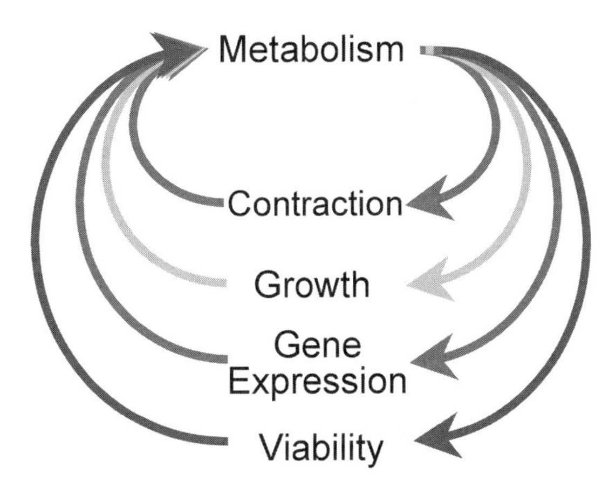

FIGURE 5. The pleiotropic actions of metabolism.

elements. Carbohydrate responsive element-binding protein (ChREBP),[32] sterol regulatory element binding proteins (SREBPs), stimulatory protein 1 (Sp1), and upstream stimulatory factor 1 (USF1)[33] have all been implicated in glucose sensing by non-muscle tissues.[30]

We have begun to investigate whether glucose-sensing mechanisms exist in heart muscle.[30] In preliminary experiments, we observed that altered glucose homeostasis through feeding of an isocaloric low-carbohydrate, high-fat diet completely abolishes MHC isoform switching in the hypertrophied heart (Young *et al.*, unpublished observations). One mechanism by which glucose affects gene expression is through O-linked glycosylation of transcription factors. Glutamine:fructose-6-phosphate amidotransferase (*gfat*) catalyzes the flux-generating step in UDP-N-acetylglucosamine biosynthesis, the rate determining metabolite in protein glycosylation (FIG. 6). In preliminary studies we observed that overload increases the intracellular levels of UDP-N-acetylglucosamine and the expression of *gfat2*, but not *gfat1*, in the heart (McClain *et al.*, unpublished work). Thus, there is early evidence for glucose-regulated gene expression in the heart and, more specifically, for the involvement of glucose metabolites in isoform switching of sarcomeric proteins. This work is ongoing.

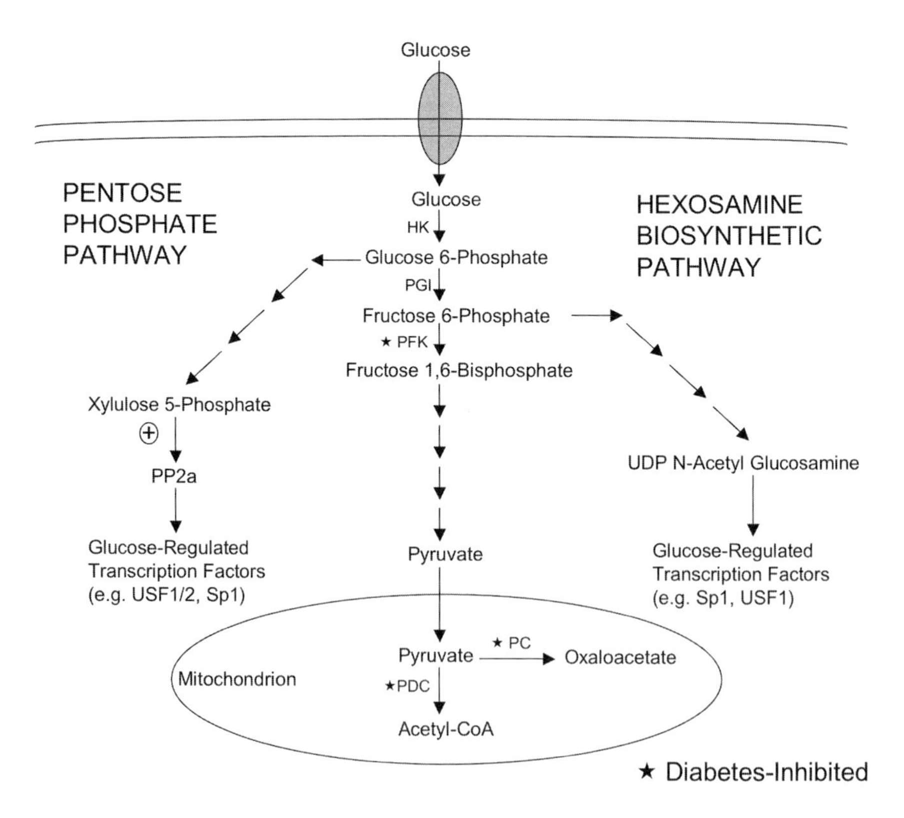

FIGURE 6. Mechanisms for glucose sensing in heart. (From Young *et al.*[30] Adapted by permission.)

Viability and Programmed Cell Survival

Perhaps the most dramatic example of chronic metabolic adaptation is the hibernating myocardium. Hibernating myocardium represents a chronically dysfunctional myocardium, which is most likely the result of extensive cellular reprogramming due to repetitive episodes of ischemia. The adaptation to reduced oxygen delivery results in the prevention of irreversible tissue damage. A functional characteristic of hibernating myocardium is improved contractile function with inotropic stimulation or reperfusion. A metabolic characteristic of hibernating myocardium is the switch from fat to glucose metabolism, accompanied by reactivation of the fetal gene program. Because glucose transport and phosphorylation is readily traced by the uptake and retention of [^{18}F] 2-deoxy, 2-fluoroglucose (FDG), hibernating myocardium is readily detected by enhanced glucose uptake and glycogen accumulation in the same regions.[34,35] Like in fetal heart, the glycogen content of hibernating myocardium is dramatically increased. There is a direct correlation between glycogen content and myocardial levels of ATP,[36] and one is tempted to speculate that improved "energetics" may be the result of improved glycogen metabolism in hibernating myocardium. The true mechanism for "viability remodeling" of ischemic myocardium encompasses antiapoptotic signaling pathways and is likely to be very complex. The point is, however, that metabolic adaptation is a key feature of this process.

The vast literature on programmed cell death, or apoptosis,[37,38] and our own observations on programmed cell survival[36] support the idea of a direct link between metabolic pathways and the pathways of cell survival and destruction. Striking evidence for a link between cell survival and metabolism is found in cancer cells. Cancer cells not only possess an increased rate of glucose metabolism,[39] but they are also less likely to "commit suicide" when stressed.[40] The same general principle appears to apply to the hibernating myocardium, where the downregulation of function and oxygen consumption is viewed as an adaptive response when coronary flow is impaired.[41] In other words, metabolic reprogramming initiates and sustains the functional and structural feature of hibernating myocardium.

Recently, the hypothesis has been advanced that insulin promotes tolerance against ischemic cell death via the activation of innate cell-survival pathways in the heart.[42] Specifically, activation of PI3 kinase, a downstream target of the insulin receptor substrate (IRS), and activation of protein kinase B/Akt are mediators of antiapoptotic, cardioprotective signaling through activation of p70s6 kinase and inactivation of proapoptotic peptides. The major actor is Akt (pun not intended). Akt is located at the center of insulin and insulin-like growth factor 1 (IGF1) signaling. As the downstream serine-threonine kinase effector of PI3 kinase, Akt plays a key role in regulating cardiomyocyte growth and survival.[43] Overexpression of constitutively active Akt raises myocardial glycogen levels and protects against ischemic damage *in vivo* and *in vitro*.[44] Akt is also a modulator of metabolic substrate utilization.[45] Phosphorylation of GLUT4 by Akt promotes its translocation and increases glucose uptake. Although the "insulin hypothesis" is attractive, there is good evidence showing that the signaling cascade is dependent on the first committed step of glycolysis and translocation of hexokinase to the outer mitochondrial membrane.[46,47] These few examples illustrate the fact that signals detected by metabolic imaging of stressed or failing heart are the product of complex cellular reactions—truly only the tip of an iceberg.

FIGURE 7. Metabolic adaptation and maladaptation of the heart. (From Young *et al.*[30] Adapted by permission.)

FUTURE DIRECTIONS OF RESEARCH IN CARDIAC METABOLISM

The future of research in cardiac metabolism will require some form of reorientation according to a new and more refined understanding of pathways of metabolic adaptation and maladaptation in the heart (FIG. 7). In this paradigm, metabolic signals alter metabolic fluxes and give rise to specific metabolic signals that, in turn, lead to changes in translational and/or transcriptional activities in the cardiac myocyte. In other words, metabolism provides the link between environmental stimuli and signaling pathways of cardiac growth and function. Although we chose the title of this review for its phonetic alliteration ("metabolic energetics and genetics in the heart"), the message is clear. As new insights are gained into metabolic adaptation and maladaptation of the heart, metabolic modulation may become an effective strategy for the treatment of heart failure. The concepts are worth exploring.

ACKNOWLEDGMENTS

We thank Stacey Vigil and Roxy Ann Tate for their help with the preparation of the manuscript, and the past and present members of the laboratory for their many contributions to the ideas discussed in this review. Work in our laboratory is supported by Grants from the National Institutes of Health of the U.S. Public Health Service.

REFERENCES

1. KNOOP, F. 1904. Der Abbau aromatischer Fettsaeuren im Tierkoerper. Beitr. chem. Physiol. Pathol. **6:** 150–162.
2. WINTERSTEIN, H. 1904. Ueber die Sauerstoffatmung des isolierten Saeugetierherzens. Z. Allg. Physiol. **4:** 339–359.
3. BING, R.J., A. SIEGEL, I. UNGAR & M. GILBERT. 1954. Metabolism of the human heart. II. Studies on fat, ketone and amino acid metabolism. Am. J. Med. **16:** 504–515.
4. HOLMES;, F.L. 1992. Between Biology and Medicine: The Formation of Intermediary Metabolism. University of California at Berkeley. Berkeley, CA.
5. RANDLE, P.J., P.B. GARLAND, G.N. HALES & E.A. NEWSHOLME. 1963. The glucose fatty-acid cycle. Its role in insulin sensitivity and the metabolic disturbances of diabetes mellitus. Lancet **1:** 785–789.
6. TAEGTMEYER, H., R. HEMS & H.A. KREBS. 1980. Utilization of energy providing substrates in the isolated working rat heart. Biochem. J. **186:** 701–711.
7. MCGARRY, J.D., S.E. MILLS, C.S. LONG & D.W. FOSTER. 1983. Observations on the affinity for carnitine and malonyl-CoA sensitivity of carnitine palmitoyl transferase I in animal and human tissues. Demonstration of the presence of malonyl-CoA in non-hepatic tissues of the rat. Biochem. J. **214:** 21–28.
8. YOUNG, M.E., G.W. GOODWIN, J. YING, *et al.* 2001. Regulation of cardiac and skeletal muscle malonyl-CoA decarboxylase by fatty acids. Am. J. Physiol. Endocrinol. Metab. **280:** E471–E479.
9. HARDIE, D.G. & D. CARLING. 1997. The AMP-activated protein kinase—fuel gauge of the mammalian cell? Eur. J. Biochem. **246:** 259–273.
10. RUDERMAN, N.B., A.K. SAHA, D. VAVVAS & L.A. WITTERS. 1999. Malonyl-CoA, fuel sensing, and insulin resistance. Am. J. Physiol. **276:** E1–E18.
11. BARGER, P.M., J.M. BRANDT, T.C. LEONE, *et al.* 2000. Deactivation of peroxisome proliferator-activated receptor-alpha during cardiac hypertrophic growth. J. Clin. Invest. **105:** 1723–1730.
12. ALLARD, M.F., B.O. SCHONEKESS, S.L. HENNING, *et al.* 1994. Contribution of oxidative metabolism and glycolysis to ATP production in hypertrophied hearts. Am. J. Physiol. **267:** H742–H750.
13. DOENST, T., G.W. GOODWIN, A.M. CEDARS, *et al.* 2001. Load-induced changes *in vivo* alter substrate fluxes and insulin responsiveness of rat heart *in vitro*. Metabolism **50:** 1083–1090.
14. YOUNG, M.E., S. PATIL, J. YING, *et al.* 2001. Uncoupling protein 3 transcription is regulated by peroxisome proliferator-activated receptor (alpha) in the adult rodent heart. FASEB J. **15:** 833–845.
15. GOODWIN, G.W., C.S. TAYLOR & H. TAEGTMEYER. 1998. Regulation of energy metabolism of the heart during acute increase in heart work. J. Biol. Chem. **273:** 29530–29539.
16. TAEGTMEYER, H. 2000. Genetics of energetics: transcriptional responses in cardiac metabolism. Ann. Biomed. Eng. **28:** 871–876.
17. BARGER, P.M. & D.P. KELLY. 2000. PPAR signaling in the control of cardiac energy metabolism. Trends Cardiovasc. Med. **10:** 238–245.
18. KELLY, D.P. 2003. PPARs of the heart: three is a crowd. Circ. Res. **92:** 482–484.
19. LEHMAN, J.J. & D.P. KELLY. 2002. Gene regulatory mechanisms governing energy metabolism during cardiac hypertrophic growth. Heart Fail. Rev. **7:** 175–185.
20. DEPRE, C., G.L. SHIPLEY, W. CHEN, *et al.* 1998. Unloaded heart *in vivo* replicates fetal gene expression of cardiac hypertrophy. Nat. Med. **4:** 1269–1275.
21. RAZEGHI, P., M.E. YOUNG, S. ABBASI & H. TAEGTMEYER. 2001. Hypoxia *in vivo* decreases peroxisome proliferator-activated receptor alpha-regulated gene expression in rat heart. Biochem. Biophys. Res. Commun. **287:** 5–10.
22. YOUNG, M.E., P. GUTHRIE, S. STEPKOWSKI & H. TAEGTMEYER. 2001. Glucose regulation of sarcomeric protein gene expression in the rat heart. J. Mol. Cell Cardiol. **33:** A181-(Abstr.).
23. BISHOP, S. & R. ALTSCHULD. 1970. Increased glycolytic metabolism in cardiac hypertrophy and congestive heart failure. Am. J. Physiol. **218:** 153–159.

24. TAEGTMEYER, H. & M.L. OVERTURF. 1988. Effects of moderate hypertension on cardiac function and metabolism in the rabbit. Hypertension **11:** 416–426.
25. WITTELS, B. & J.F. SPANN. 1968. Defective lipid metabolism in the failing heart. J. Clin. Invest. **47:** 1787–1794.
26. SACK, M.N., T.A. RADER, S. PARK, *et al.* 1996. Fatty acid oxidation enzyme gene expression is downregulated in the failing heart. Circulation **94:** 2837–2842.
27. DAVILA-ROMAN, V.G., G. VEDALA, P. HERRERO, *et al.* 2002. Altered myocardial fatty acid and glucose metabolism in idiopathic dilated cardiomyopathy. J. Am. Coll. Cardiol. **40:** 271–277.
28. TAEGTMEYER, H., S. SHARMA, L. GOLFMAN, *et al.* 2004. Linking gene expression to function: metabolic flexibility in normal and diseased heart. Ann. N.Y. Acad. Sci. **1015:** 1–12.
29. NIKOLAIDIS, L.A., A. STURZU, C. STOLARSKI, *et al.* 2004. The development of myocardial insulin resistance in conscious dogs with advanced dilated cardiomyopathy. Cardiovasc. Res. **61:** 297–306.
30. YOUNG, M.E., P. MCNULTY & H. TAEGTMEYER. 2002. Adaptation and maladaptation of the heart in diabetes. II. Potential mechanisms. Circulation **105:** 1861–1870.
31. FERRE, P. 1999 Regulation of gene expression by glucose. Proc. Nutr. Soc. **58:** 621–623.
32. UYEDA, K., H. YAMASHITA & T. KAWAGUCHI. 2002. Carbohydrate responsive element-binding protein (ChREBP): a key regulator of glucose metabolism and fat storage. Biochem. Pharmacol. **63:** 2075–2080.
33. GIRARD, J., P. FERRE & F. FOUFELLE. 1997. Mechanisms by which carbohydrates regulate expression of genes for glycolytic and lipogenic enzymes. Annu. Rev. Nutr. **17:** 325–352.
34. MÄKI, M., M. LUOTOLAHTI, P. NUUTILA, *et al.* 1996. Glucose uptake in the chronically dysfunctional but viable myocardium. Circulation **93:** 1658–1666.
35. DEPRE, C., J.L. VAN OVERSCHELDE, B. GERBER, *et al.* 1997. Correlation of functional recovery with myocardial blood flow, glucose uptake, and morphologic features in patients with chronic left ventricular ischemic dysfunction undergoing coronary artery bypass grafting. J. Thorac. Cardiovasc. Surg. **113:** 82–87.
36. DEPRE, C. & H. TAEGTMEYER. 2000. Metabolic aspects of programmed cell survival and cell death in the heart. Cardiovasc. Res. **45:** 538–548.
37. GOTTLIEB, R.A. 1999. Mitochondria: ignition chamber for apoptosis. Mol. Genet. Metab. **68:** 227–231.
38. DOWNWARD, J. 2003. Metabolism meets death. Nature **424:** 896–897.
39. WARBURG, O. 1956. On the origin of cancer cells. Science **123:** 309–314.
40. HANAHAN, D. & R.A. WEINBERG. 2000. The hallmarks of cancer. Cell **100:** 57–70.
41. FALLAVOLLITA, J.A., B.J. MALM & J.M.J. CANTY. 2003. Hibernating myocardiam retains metabolic and contractile reserve despite regional reductions in flow, function, and oxygen consumption at rest. Circ. Res. **92:** 48–55.
42. SACK, M.N. & D.M. YELLON. 2003. Insulin therapy as an adjunct to reperfusion after acute coronary ischemia. A proposed direct myocardial cell survival effect independent of metabolic modulation. J. Am. Coll. Cardiol. **41:** 1404–1407.
43. MATSUI, T., T. NAGOSHI & A. ROSENZWEIG. 2003. Akt and PI 3-kinase signaling in cardiomyocyte hypertrophy and survival. Cell Cycle **2:** 220–223.
44. MATSUI, T., L. LI, J. WU, *et al.* 2002. Phenotypic spectrum caused by transgenic overexpression of activated Akt in the heart. J. Biol. Chem. **277:** 22896–22901.
45. WHITEMAN, E., H. CHO & M. BIRNBAUM. 2002. Role of Akt/protein kinase B in metabolism. Trends Endocrinol. Metab. **13:** 444–451.
46. GOTTLOB, K., N. MAJEWSKI, S. KENNEDY, *et al.* 2001. Inhibition of early apoptotic events by Akt/PKB is dependent on the first committed step of glycolysis and mitochondrial hexokinase. Genes Dev. **15:** 1406–1418.
47. MAJEWSKI, N., V. NOGUEIRA, R.B. ROBEY & N. HAY. 2004. Akt inhibits apoptosis downstream of BID cleavage via a glucose-dependent mechanism involving mitochondrial hexokinases. Mol. Cell. Biol. **24:** 730–740.

The Sarcomeric Control of Energy Conversion

CARMIT LEVY,[a] HENK E.D.J. TER KEURS,[b] YAEL YANIV,[a]
AND AMIR LANDESBERG[a]

[a] *Faculty of Biomedical Engineering, Technion, Israel Institute of Technology, Haifa, Israel*

[b] *Faculty of Medicine, Health Sciences Centre, University of Calgary, Calgary, Canada*

ABSTRACT: The Frank-Starling Law, Fenn Effect, and Suga's suggestions of cardiac muscle constant contractile efficiency establish the dependence of cardiac mechanics and energetics on the loading conditions. Consistent with these observations, this review suggests that the sarcomere control of contraction consists of two dominant feedbacks: (1) a cooperativity mechanism (positive feedback), whereby the number of force-generating cross-bridges (XBs) determines the affinity of calcium binding to the troponin regulatory protein; and (2) a mechanical (negative) feedback, whereby the filament shortening velocity affects the rate of XB turnover from the force to the non-force generating conformation. The study explains the roles of these feedbacks in providing the adaptive control of energy consumption by the loading conditions and validates the dependence of the cooperativity mechanism on the number of strong XBs. The cooperativity mechanism regulates XB recruitment. It explains the cardiac force-length calcium relationship, the related Frank-Starling Law of the heart, and the adaptive control of new XB recruitment and the associated adenosine triphosphate (ATP) consumption. The mechanical feedback explains the force-velocity relationship and the constant and high-contractile efficiency. These mechanisms were validated by testing the force responses to large amplitude (100 nm/sarcomere) sarcomere length (SL) oscillations, in intact tetanized trabeculae (utilizing 30 μM cyclopiazonic). The force responses to large-length oscillations lag behind the imposed oscillations at low extracellular calcium concentration ($[Ca^{2+}]_0$) and slow frequencies (<4 Hz, 25°C), yielding counterclockwise hystereses in the force-length plane. The force was higher during shortening than during lengthening. The area within these hystereses corresponds to the external work generated from new XB recruitment during each oscillation, and it is determined by the delay in the force response. Characterization of the delayed response and its dependence on the SL, force, and calcium allows identification of the regulation of XB recruitment. The direct dependence of the phase on force indicates that XB recruitment is determined directly by the force (i.e., the number of strong XBs) and indirectly by SL or calcium. The suggested feedbacks determine cardiac energetics: 1) the constant and high contractile efficiency is an intrinsic property of the single XB, due to the mechanical feedback; and 2) the XBs are the myocyte sensors that modulate XB recruitment in response to length and load changes through the cooperativity mechanism.

KEYWORDS: cardiac mechanics; cooperativity; cross-bridge; excitation-contraction coupling; Fenn Effect; Frank-Starling Law; myosin

Address for correspondence: Amir Landesberg, M.D., Ph.D., Faculty of Biomedical Engineering, Technion, Israel Institute of Technology, Haifa 32000, Israel. Voice: 972 4829 4143; fax: 972 4829 4599.

Ann. N.Y. Acad. Sci. 1047: 219–231 (2005).
doi: 10.1196/annals.1341.020

INTRODUCTION

The existence and importance of an adaptive control of the cardiac energy consumption by the loading conditions emerge from three fundamental characteristics of the cardiac muscle: the Frank-Starling Law,[1] the Fenn Effect,[2,3] and Suga's[4] linear relationship between energy consumption and the generated contractile mechanical energy.

The Frank-Starling Law of the heart describes the dependence of the cardiac maximal isovolumic pressure and the cardiac stroke work on the preload. This law allows the heart to cope with changes in the body's physiological demands, and to increase the cardiac output in response to an increase in the venous return. The Frank-Starling Law relates to the steep force-length relationship (FLR) of the isolated cardiac fiber, which is a basic feature of the sarcomere control of contraction.[1]

The Fenn Effect is one of the most intriguing features of the skeletal and cardiac muscles.[2,3] Fenn has shown that energy liberation by muscle contraction is not determined solely by the initial conditions, as is the case in the Frank-Starling Law, but is also modified by the load that the muscle encounters during the contraction. Fenn was the first to identify the dependence of energy consumption on the load and muscle shortening.[3] This phenomenon suggests the existence of feedback control mechanisms, as was suggested by Rall,[2] although the underlying mechanisms were not identified.

Suga[4] has obtained a linear relationship between the cardiac muscle oxygen consumption (MV_{O_2}) and the contractile mechanical energy. Suga[4] has suggested that the cardiac muscle contractile mechanical energy is comprised of two elements: the external work (EW) done during the contraction, and the potential energy (PE) stored in the cardiac elastic elements.

$$MV_{O_2} = a(EW + PE) + b. \tag{1}$$

The potential energy is determined mainly by the cardiac end-systolic pressure–volume relationship that is assumed to represent the cardiac maximal elastance and the end-systolic volume. The slope of this linear relationship, a, describes the efficiency of biochemical to mechanical energy conversion by the cardiac muscle. Suga has found that, for a wide spectrum of physiological conditions and loading conditions in the presence of various positive and negative ionotropic drugs, as well as in the presence of ischemia, the efficiency ($100/a$) is about 40% (i.e., 40% of the energy available from oxygen consumption is converted into contractile energy). As is well known, the efficiency of the mitochondrial oxidation phosphorylation, where adenosine triphosphate (ATP) is produced from metabolites oxidation, is about 70%. Therefore, the efficiency of the cardiac muscle contractile filaments that utilize ATP to generate the mechanical contractile energy is 60% to 70% (0.40/0.70) under a wide spectrum of physiological conditions. How can we explain such high and constant energy efficiency, which is independent of the loading conditions?

The relationship described by EQUATION 1 may seem trivial based on the first law of the thermodynamics, which literally states that the energy consumption determines the ability to generate mechanical output. However, cardiac muscle physiology presents the reverse signaling, as it is the energy demand that determines the energy consumption, rather than the availability of metabolites, ATP, or oxygen. In fact, the availability of the ATP is unlimited under most physiological conditions, and muscle activation and loading conditions determine the rate of energy consumption.

The present article addresses these basic phenomena and aims to resolve the following questions: How can we explain the adaptive control of energy consumption and the dependence of energy consumption on the loading conditions? What determines energy consumption: the load or the shortening velocity? And, finally, how can we explain the load independence of the contractile efficiency of the cardiac muscle?

Our theory of the sarcomere control of contraction[5–7] provides the basis for explaining these observations. Referring to our previous studies, we present here new, experimental validation for our earlier suggested[4,6] (cooperativity) mechanism that modulates energy consumption and cross-bridge (XB) recruitment. We suggest that the adaptive control of energy consumption and the contractile constant efficiency are maintained by two separate feedback mechanisms.[6,7]

Energy consumption by the sarcomere is determined by two main kinetics: (1) the kinetics of XB recruitment, which determines the number of cycling XBs; and (2) XB dynamics (i.e., the rate of the single XB turnover between the weak and strong states that determines the XB duty cycle). The generated force is determined by these two kinetics and by additional faster kinetics that determines the unitary force per XB. All these three kinetics are modulated by the filaments' sliding velocity.

The number of cycling XBs and the rate of XB recruitment is determined by the kinetics of calcium binding to troponin and by the sarcomere length (SL). Calcium binding to troponin turns the regulated actin from the "turned off" state to the "turned on" state and enables ATP hydrolysis by the adjacent heads of the myosin. The SL determines the length of the single overlap region between the thin and thick filaments, where the myosin heads can attached to actin-binding sites to form the strong, force-generating XBs.

The Frank-Starling Law of the heart, where the isovolumic pressure increases with the preload, implies that the cardiac muscle must work along the ascending limb of the FLR. Moreover, the cardiac muscle is characterized by a steep FLR, where a 10% change in the sarcomere length yields a 30% to 70% change in the generated force,[1,8] whereas according to the sliding filament concept[9] only a modest change in the force is expected. The length-dependent calcium sensitivity is an additional related basic feature of the cardiac muscle,[1] wherein sarcomere lengthening is associated with a shift to the left (toward lower calcium concentration) of the force-calcium (F-Ca) relationship, with an increase in the steepness of the F-Ca relationship.

Various mechanisms were suggested in the literature[1,8,10] to explain the steep cardiac FLR and the length-dependent calcium sensitivity, but there is still a dispute as to the dominant underlying mechanism. Some researchers strongly support the existence of a direct dependence of calcium affinity on the SL or on the interfilament spacing. Alternatively, three types of cooperativity mechanisms were postulated.[11,12] Some suggest[11,12] the existence of cooperativity between neighbor troponin complexes along the thin filaments (Ca-Ca type cooperativity), whereby calcium binding to troponin affects the affinity of calcium binding to the adjacent troponin complex. Another type assumes cooperativity between adjacent myosin heads (XB-XB cooperativity), whereby XB attachment facilitates the attachment of adjacent myosin head to other actin binding site, as is well established in the skeletal muscle. A third type of cooperativity, between myosin head and calcium binding to troponin (XB-Ca cooperativity), suggests that an increase in the number of strong XBs increases the affinity of troponin for calcium.[5,10,13] This mechanism was established by measuring the effect on the number of strong XBs (isometric force) on the amount of bound calcium to troponin[10] and by demonstrating the tight correlation in the time between the changes in the number of strong XB and the free calcium transients, following length perturbations.[14]

Our previous theoretical studies have suggested that the dominant cooperativity mechanism is the XB-Ca cooperativity.[5,6] This conclusion was derived from the analysis of a skinned fiber force-length calcium relationship[5] using our four-state model. We have also suggested[6,7] that this mechanism modulates XB recruitment rate and can explain the adaptive control of energy consumption by the loading conditions.

To further validate the relative role of this mechanism, we have recently characterized the dependence of the delay in the force response to length perturbation on the SL,[15] calcium concentration, and the number of strong XBs (isometric stress). These studies are based on the predicted existence of an inherent delay in the force response to SL perturbation in all the previous mechanisms. Indeed, this delay accommodates for the kinetics of calcium binding to troponin and XB cycling in our suggested XB-Ca cooperativity. SL perturbations affect the number of strong XBs, due to the changes in the filament overlap length and the effect of the shortening velocity on XB cycling rate, thereby modulating calcium affinity. Changes in calcium affinity affect the bound calcium, with a time constant that depends on calcium kinetics. Thereafter, changes in bound calcium alter the number of strong XBs with a time constant that depends on the rate of new XB recruitment. This prediction suggests that there is no constitutive law that describes a unique and time-independent FLR, and that the cardiac FLR is history dependent: the instantaneous force depends on the history of contraction. Note that a unique FLR is expected to exist if the rate kinetics of calcium binding to troponin and XB cycling are constants. Since a delay in the force response is expected to exist in the a.m. mechanisms, quantification and characterization of this time delay should allow to characterize the mechanism underlying XB recruitment.

The delay in the force response is quantified by imposing large sinusoidal length oscillations during tetanic contractions[15] (FIG. 1A). Our earlier study has established the existence of a delay in the force response to slow-length oscillation,[15] whereas the present study utilizes this phenomenon in order to characterize the underlying cooperativity mechanism.

METHODS

Thin, long, and unbranched trabeculae were isolated from the right ventricle of rats (Sprague Dawley, 250 to 300 gr). The fibers were perfused with modified Krebs-Henseleit (K-H) solution, containing (in mM): Na^+ 143.7; K^+ 5.0; Cl^- 130.4; Mg^{2+} 1.2; PO_4^- 2.26; SO_4^- 1.2; HCO_3^- 19.0; and glucose 10. The pH was set at 7.40 by adjusting the flow of 95% O_2 and 5% CO_2 (25°C, $[Ca^{2+}]_0$ = 1.5 mM). The muscles were stimulated at 0.5 Hz. Force was measured with a silicon strain gauge (SensoNor 801). SL was measured by a laser diffraction technique using HeNe laser (λ = 633 nm). Muscle length was controlled by a fast servomotor (model 308B, Aurora Scientific, Inc., Ontario, Canada) with a rise time of 250 μsec. Passive and twitch

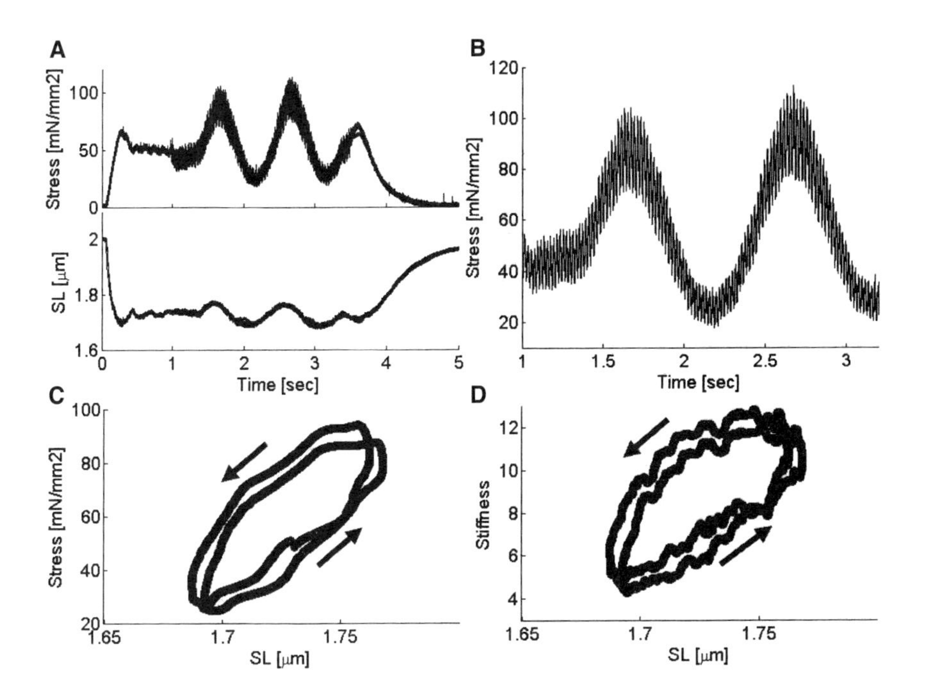

FIGURE 1. The effects of 1 Hz length oscillations on stress and dynamic stiffness. (**A**) The length oscillations (1 Hz) were imposed during tetanic contraction. Additional high-frequency (50 Hz) and small-amplitude (2.25 nm) oscillations were imposed on top of the large oscillations to measure the dynamic stiffness. (**B**) Magnification of the stress response to slow- and high-frequency oscillations. Note that the amplitude of the force response to high-frequency oscillations is modulated by the slow oscillations. Counterclockwise hystereses are obtained in the phase plots of stress (**C**) and stiffness (**D**) versus sarcomere length (SL) ($[Ca^{2+}]_0 = 6$ mM).

FLRs were measured after one hour of stabilization and were used as a control for evaluating the viability of the trabeculae.

The delay in the force response to SL oscillation was studied at a constant activation level by utilizing tetanus contractions.[15] Stable fused tetanii were achieved at different extra-cellular Ca^{2+} concentrations ($[Ca^{2+}]_0 = 1.5, 3, 4.5$, and 6 mM) by using 30 μM cyclopiazonic acid (CPA) (Sigma), which blocked the sarcoplasmic reticulum calcium ATPase, and 8 or 12 Hz stimulation. The twitch frequency was reduced to 0.2 Hz and tetanii were elicited every ten twitches. The tetanus duration was 3.5 s.

The delay in the force response to length perturbation was measured by imposing large length oscillations after a steady tetanic contraction was reached (FIG. 1A). The tetanic SL was determined during the first second of the tetanic contraction, and the length oscillations thereafter were imposed around the desired SL. Large-length oscillations with peak-to-peak amplitude of around 100 nm were used to impose perturbations well above the single XB stroke steps (assumed to be around 10 nm). These large oscillations are at least an order of magnitude larger than the small oscillations used for measuring the dynamic stiffness, where the oscillation amplitude is smaller than the XB stroke step. Tetanic contractions at different extracellular calcium concentrations (form 1.5 to 6 mM) were used to achieve constant calcium levels during the length oscillations and to eliminate the complexity of the time-dependent activation in the cardiac twitch contraction.

The variation in the number of strong XBs during the slow oscillation was evaluated by measuring the muscle dynamic stiffness.[15] The dynamic stiffness was measured by superimposing additional very small-amplitude (2.25 nm) and high-frequency (50 or 200 Hz) oscillations on top of the large and slow oscillations. The measurements of both the force and the dynamic stiffness during the large oscillations allowed to determine whether the oscillations in the force response resulted from variations in the number of XBs or from the effect of the slow oscillation on the average force generated by each XB. The dynamic stiffness depends on the oscillation frequency and on the number of strong XBs. At constant oscillation frequency (50 or 200 Hz), the changes in the dynamic stiffness are proportional only to the variations in the number of strong XBs.[16] The dynamic stiffness is defined as the ratio between the amplitude of the high frequency oscillation in the force and the high frequency component of the imposed oscillation in the sarcomere length. Since the oscillations in the SL were at constant frequency, the observed variation in the amplitude of the high-frequency oscillations in the force represents the variation in the number of XBs. The slow-frequency oscillations in the force (FIG. 1B), which were used to identify the dominant cooperativity, modulated the high-frequency oscillations used for measuring the dynamic stiffness. Note the increase in the amplitude of the high-frequency oscillations as the amplitude of the slow oscillations increases (FIG 1B). This observation implies that the oscillations in the force result mainly from changes in the number of strong XBs, and that the changes in the force per XB have only a secondary significance.

Our earlier study[15] explored the effects of the frequency of stimulation and the SL on the force responses, at a single extra cellular calcium concentration ($[Ca]_0 =$ 6 mM). More recently we have imposed large 1 Hz-length oscillations at various tetanic SLs and different Ca^{2+} concentrations.

The muscle length, force, and SL were sampled at 5,000 Hz. Each oscillation was repeated five times, and the results were averaged to reduce random noise. FIGURE 2 presents the observed delay between the force response and the SL, when the signals were normalized by the oscillation amplitudes. The phase delays between the force and the SL oscillations were calculated from the shift in the time required to obtain the least mean square error (LMSE) between the normalized force and the normalized SL oscillations. A trabecula was discarded when the twitch force dropped by more than 15% at the same $[Ca^{2+}]_0$ and SL.

RESULTS AND DISCUSSION

Our recent study[15] has established that the force responses lag behind the large-length oscillations at small-oscillation frequency and constant calcium concentration. The stress response lagged behind the SL oscillations by 111.9 ± 41 msec (40.3 ± 14.7°, mean ± SD, n = 6). The oscillations in the stress and the stiffness were closely aligned.[15] The phase delays between the stress or the stiffness and the SL are better demonstrated by plotting stress (FIG. 1C) and stiffness (FIG. 1D) against SL. The

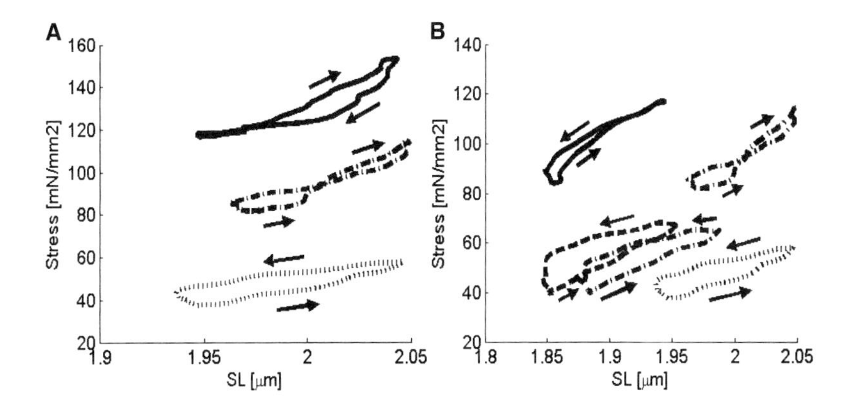

FIGURE 2. The phase of the hystereses depends directly on the stress level and indirectly on calcium concentration and sarcomere length (SL). (**A**) Raising $[Ca^{2+}]_0$ from 1.5 mM (*dotted*) to 3.0 mM (*dashed-dotted*) and 6.0 mM (*solid line*) at constant SL (1.99 μm), decreased the phase from +28° to −13°. (**B**) Similar phases were observed at identical mean stresses (27°± 1° for 50 mN/mm),[2] from one trabecula obtained by utilizing different pairs of SL and $[Ca^{2+}]_0$ (4.5 mM (*dashed*). Sarcomere lengthening (from 1.89 μm and 1.98 μm) decreases the phase at the same $[Ca^{2+}]_0$.

force or the stiffness responses to large-length oscillations at a constant calcium concentration lag behind the length oscillations, yielding a counterclockwise (CCW) hysteresis in the force-length (FIG. 1C) and stiffness-length (FIG. 1D) planes. The force is greater during shortening than during lengthening at the same SL during CCW hysteresis.[15] Note the similarity between the hysteresis in the force and the stiffness that strengthen the notion that the delay in the force response is due to the delay in XB recruitment.

The area inside the force-length loop is equal to the magnitude of the work generated. Hysteresis in the CCW direction implies that energy is generated by the muscle upon the system. Force is needed to stretch the muscle, but the muscle generates more force during shortening, and a net external work is generated during each cycle. This external work can be generated only by new XBs recruited during each cycle.

The mean SL, around which the length oscillations were imposed, had a significant effect on the phase of force response.[15] The largest delay in the force response was observed at the shortest SL (131.16 ± 31.75 s at SL = 1.78 ± 0.03 μm). There was a monotonic decrease in the delay with an increase in the SL. At SL of 1.99 ± 0.015 μm, the delay in the force response to 1 Hz length oscillation was reversed, and the force preceded the SL oscillation (14.73 ± 18.5 ms) leading to clockwise (CW) hysteresis.[15] This phenomenon can be simply interpreted according to the hypothesis that XB recruitment is a direct function of the SL, as was suggested by others.[1,8,17] However, this observation[15] was obtained at a constant calcium level; thus, sarcomere lengthening was associated with an increase in the generated force. Therefore, it remains to resolve whether the length per say or the number of strong XBs (isometric force) modulate the rate of new XB recruitment during the oscillations.

The area within the CCW hysteresis and the related external work are determined by the phase delay between the force and the SL. The area within the hysteresis increases as the phase increases from 0° to 90°, so that the phase delay can serve as an index to the number of new XBs recruited during a single hysteresis loop. Consequently, characterizing the effects of SL, calcium, and force level on the phase delay allows to determine whether the dominant cooperativity mechanism that regulates XB recruitment is force, length, or calcium dependent.

The hypothesis that the regulation of XB recruitment is directly regulated by the SL was tested by imposing SL oscillations at the same tetanic SL but at different $[Ca^{2+}]_0$ concentrations (3, 4.5, and 6 mM), and consequently at different mean tetanic stresses. If XB recruitment is determined by SL, then the hystereses obtained at the same SL should have the same phase. FIGURE 2A presents three stress-SL hystereses obtained at the same SL (1.99 μm) with different $[Ca^{2+}]_0$ (1.5, 3.0, and 6.0 mM). As seen, an increase in $[Ca^{2+}]_0$ decreased the phase of the stress-SL hysteresis (from 28° to –13°). A CCW hysteresis was observed at low activation levels and a clockwise (CW) hysteresis was obtained at high activation level ($[Ca^{2+}]_0$ = 6.0 mM).

At constant SL, the phase decreased with the elevation of $[Ca^{2+}]_0$ levels (FIG. 2A). At constant $[Ca^{2+}]_0$ the phase decreased with sarcomere lengthening.[15] Thus, the phase is both SL and calcium dependent. These observations are incongruent with the hypotheses that either Ca or the SL have a direct and dominant role in the regulation of XB recruitment. Note that both the increase in the free calcium concentration at constant SL (FIG. 2A) or sarcomere lengthening at constant calcium[15] decrease the phase of the hysteresis simultaneous with the increase in the stress. These observations insinuate that the force is the immediate modulator of XB recruitment, which determines the related phase of hysteresis.

The hypothesis that XB recruitment is determined by the number of strong XBs (stress) was further tested by generating hystereses at identical mean tetanic stress with different pairs of SL and $[Ca]_0$. If XB recruitment is determined by the number of strong XBs (the isometric tetanic stress), then the hystereses obtained at the same mean tetanic stress should be similar in shape. Stable tetanic stress was initially obtained with high $[Ca^{2+}]_0$ (6 mM) and shortest SL. Then $[Ca^{2+}]_0$ was decreased and the trabecula was stretched until identical stress level was achieved. FIGURE 2B presents hystereses observed at the same mean tetanic stress and with the same amplitude of length oscillations, but at different mean SLs utilizing different $[Ca^{2+}]_0$ concentrations. The following are the $[Ca^{2+}]_0$ concentrations that were studied: 1.5, 3, 4.5, and 6 mM, and the SL range was from 1.85 to 2.05 μm. Similar shapes and phases of hystereses were obtained at constant stress levels for the different pairs of SL and $[Ca^{2+}]_0$.

These results validate the hypothesis that the dominant cooperativity mechanism that regulates XB recruitment and the cardiac force-length-calcium relationships depends on the force (load) (i.e., it is determined by the number of strong XBs). The study negates the alternative hypotheses that the dominant cooperativity mechanism is directly determined either by the SL (and interfilament spacing) or the free calcium, based on analyzing the phase between force responses and SL oscillations. If the affinity is function of the SL, similar shape of hystereses should be obtained at the same SL at different calcium levels. Furthermore, these hystereses should have the same phase and should only be scaled in the amplitude of the force oscillations in proportion to the amplitude of the mean tetanic stress. However, the results invalidate this hypothesis by showing that the phase decreases and even reverses its direction with the elevation of calcium concentration, at the same SL.

The data suggest that the phase of the hysteresis depends directly on the stress level. Sarcomere lengthening and raising $[Ca^{2+}]_0$ independently decrease the phase of the hysteresis due to the associated increase in the generated force. However, for each fiber, similar phases were observed at the same stress level for different pairs of SL and $[Ca^{2+}]_0$ (FIG. 2B), suggesting that the effects of calcium and SL are through the changes in the force. Only XB dependent mechanisms (XB-XB cooperativity or XB-Ca cooperativity) can explain all the experimental observations; the same phase of hysteresis is obtained when the oscillations are imposed at the same stress level. An increase in the mean stress, either by an increase in calcium or sarcomere lengthening, decreases the phase. Hence, both the sarcomere length and calcium level have indirect effects on the phase of the hysteresis and on the rate of XB recruitment.

The observed hystereses in the isolated cardiac trabeculae represent the integrated effects of all different regulatory mechanisms between the contractile proteins and the regulatory proteins. The strength of the method used here—utilizing intact cardiac fiber—is that it does not alter the cellular control mechanism by any intervention (such as skinning or protein substitution) and allows us to quantify and differentiate between the relative roles of the various parameters. The method used here tests separately the roles of the SL, $[Ca^{2+}]_0$, and stress, and thus allows identification of the input of the dominant cooperativity mechanism, although several mechanisms may coexist.

The observed hystereses of the force response to length changes is consistent with earlier described hysteresis in the force-calcium relation at constant SL in intact and skinned cardiac muscle[17] and skinned skeletal fibers.[18] In both of these studies, the muscles generated more force at the same SL and calcium after the fibers had previously been subjected to a higher force level. Moreover, the two phenomena have a similar dependence on the SL: the phase of the force-length hysteresis[15] and the phase of force-calcium hysteresis[17] decreased with sarcomere lengthening. The advantage of the present study is in its ability to differentiate between the roles of SL, calcium, and the number of strong XBs (stress level). The findings define the afferent limb of the cooperativity mechanism and establish that force modulates the rate of XB recruitment (i.e., the XBs induce XB recruitment).

The dominant XB-Ca cooperativity has three significant features:

(1) It provides the explanation for the steep cardiac FLR with the length-dependent calcium sensitivity.[15]
(2) A positive feedback allows fast switching of muscle activation and steep force-calcium relationship where small changes in the free calcium concentration allow huge variation in the number of cycling XBs.
(3) It provides an adaptive control of energy consumption,[6,7] whereby changes in the loading conditions modulate calcium affinity and the bound calcium. The bound calcium determines the rate of XB recruitment and energy consumption. Thus, a decrease in the load decreases energy consumption and vice versa.

The cooperativity mechanism explains the adaptive control of XB recruitment and the adaptive control of ATP consumption, but this by itself does not explain the

Fenn Effect and Suga's observations. Only integration of the cooperativity with the mechanical feedback provides the complete explanation of the regulation of energy consumption.[7] The cooperativity regulates XB recruitment and the rate of energy consumption, while the mechanical feedback determines the XB duty cycle and the conversion of the consumed energy into external work and potential energy.

Woledge[3] has concluded from the Fenn Effect that muscle energy consumption is regulated by the load or the shortening. The classical Huxley model[9] suggests that XB kinetics are a function of the XB displacement (strain). Our theory of the XB dynamics[7] suggests that XB kinetics are determined by the filament sliding velocity.

To our best knowledge, there is also no analytical expression for energy conversion and for the Fenn Effect or Suga's observations based on Huxley's model. The energy consumption (E) based on Huxley's model, for constant shortening velocity (V) and under full activation, is given in the equation

$$E = C_1\left[1 - C_2\frac{V}{\varphi}\left(1 - e^{\varphi/V}\right)\right], \tag{2}$$

where C_1, C_2, and φ are constant. Huxley's model description of energy consumption does not allow to derive the generated external work and does not reflect Suga's empirical observations (EQUATION 1).

Our model suggests that XB cycling between the biochemical weak (non-force generating) and strong (force generating) conformations is modulated by the sarcomere velocity and not by the displacement. The kinetic of XB cycling between these conformations relate to nucleotides binding and release. ATP is bound to the myosin head in the weak state, while adenosine diphosphate (ADP) is bound to myosin in the strong state. ATP hydrolysis and phosphate release allow myosin head turnover to the strong state, and the corresponding rate constant (f) is constant and independent of the strain or strain rate (velocity). Note that according to the biochemical scheme of XB turnover, the hydrolysis may occur when the head is detached, since ATP binding to myosin facilitates myosin dissociation from actin. The rate of XB turnover back to the weak conformation (g) is a linear function of the sliding velocity, V,

$$g = g_0 + g_1 V. \tag{3}$$

We denote this mechanism as the mechanical feedback,[7] where g_1 is the mechanical feedback coefficient, since it describes the effect of the mechanical shortening velocity on the biochemical rate of XB cycling.

We have shown in our earlier studies that the model allows us to describe a wide spectrum of phenomena: it provides analytical expression to the well empirically established Hill equation for the force-velocity relationship. It describes the dependence of the shortening velocity on the activation level and the SL in accordance with experimental data. It also explains the regulation of relaxation. The model also provides analytical expressions for the force stiffness relationship and for the dependence of the rate of energy consumption on the shortening velocity. Most intriguing, it allows derivation of the equation for the energy conversion by the sarcomere[7]

$$\rho E = \mathrm{EW} + \mathrm{EP_P} + \mathrm{Q}, \tag{4}$$

where the energy consumption (E) is converted into three mechanical components: external work (EW), pseudo potential energy (EP_P), and heat dissipation due to the XB viscous property (Q).

This equation resembles Suga's observation[4] and Fenn Effect, since the heat dissipation due to the XB viscous property (Q) is negligible at reasonable physiological shortening velocity.

A most significant conclusion of this model is that the contractile efficiency (ρ) of biochemical to mechanical energy conversion can be quantitatively determined only by the single XB parameters[7]

$$\rho = \frac{\bar{F}U}{E_{ATP}} \frac{1}{g_0 + g_1 U}, \tag{5}$$

where $\bar{F}$ represents the unitary force per XB; U represents the maximal unloading shortening velocity; E_{ATP} represents the free energy liberated from the hydrolysis of ATP; and $g_0 + g_1 U$ represents the maximal rate of XB weakening at unloading conditions. Consequently, the contractile efficiency is an intrinsic property of the single XB and is independent of the loading conditions. The high-, constant-, and load-independent efficiency observed at the whole heart level[4] is not a unique feature of the whole ventricle but is derived from an intrinsic property of the XBs (i.e., the load-independent constant efficiency of the XBs).

Note that our EQUATION 4 for energy conversion is not limited to a constant activation or a steady shortening velocity, but applies to any activation level, varying shortening velocity, and any time interval.

SUMMARY AND CONCLUSION

The study explains the regulation of energy consumption in the cardiac muscle and the mechanisms underlying the Frank-Starling Law, the Fenn Effect, and Suga's observations. These phenomena are explained here by relating to the two main kinetics that determine energy consumption by the sarcomere: the kinetics of XB recruitment determines the number of available cycling XBs, and the kinetics of the single XB cycling between weak and strong conformations determines the XB duty cycle and muscle efficiency. Each kinetic is modulated by a single dominant feedback mechanism.

The efficiency of the biochemical to mechanical energy conversion is load-independent due to the mechanical feedback, where the rate of XB weakening is a function of the shortening velocity. XB recruitment is determined by the cooperativity mechanism. The present study integrates these mechanisms to explain the regulation of energy conversion and presents new experimental data that further substantiate the suggested XB-Ca cooperativity mechanism. The mechanisms underlying the regulation of XB recruitment were investigated by analyzing the phase delay of the force responses to SL oscillations imposed during tetanic contraction and characterizing its dependence on the SL, free calcium level, and the mean tetanic force. The phase decreases with sarcomere lengthening[15] and with raising $[Ca^{2+}]_0$ concentration, but an identical phase is obtained at the same stress level with different pairs of SL and $[Ca^{2+}]_0$. Direct length-dependent mechanisms cannot explain the decrease in the phase with rising $[Ca^{2+}]_0$ at constant SL. The suggested cooperativity mechanism, where XB recruitment is determined by the number of strong XBs, is the dominant mechanism that regulates the cardiac FLR. SL and calcium concentration have indirect effects on the rate constant of XB recruitment through the modulation of the number of strong XBs.

We conclude that the XBs play two key roles in the regulation of energy conversion:

(1) The high and constant contractile efficiency of the cardiac muscle is an intrinsic property of the single XB,[7] and it relates specifically to the mechanical feedback, where the XB weakening rate is function of the shortening velocity.

(2) The cycling XBs are the cardiac myocyte sensors for the prevailing mechanical loading condition (preload and afterload) that modulate XB recruitment and sarcomere energy consumption. The adaptive control of XB recruitment is attributed to the cooperativity mechanism that is determined by the number of strong XBs.

ACKNOWLEDGMENTS

This study was supported by the Fund for the Promotion of Research at the Technion (A.L.) and a grant from the United States–Israel Binational Science Foundation (BSF Research Project No. 2003399).

REFERENCES

1. ALLEN, D.G. & J.C. KENTISH. 1985. The cellular basis of the length-tension relation in cardiac muscle. J. Mol. Cell. Cardiol. **17:** 821–840.
2. RALL, J.A. 1982. Sense and nonsense about the Fenn Effect. Am. J. Physiol. **242:** H1–H6.
3. WOLEDGE, R.C., N.A. CURTIN & E. HOMSHER. 1985. Energetic Aspects of Muscle Contraction. Monographs of the Physiological Society. No. 41. Harcourt Brace Jovanovich. New York.
4. SUGA, H. 1990. Ventricular energetics. Physiol. Rev. **70:** 247–277.
5. LANDESBERG, A. & S. SIDEMAN. 1994. Coupling calcium binding to troponin-C and cross-bridge cycling in skinned cardiac cells. Am. J. Physiol. **266:** H1260–H1266.
6. LANDESBERG, A. & S. SIDEMAN. 1999. Regulation of energy consumption in cardiac muscle: analysis of isometric contraction. Am. J. Physiol. **276:** H998–H1011.
7. LANDESBERG, A. & S. SIDEMAN. 2000. Force-velocity relationship and biochemical to mechanical energy conversion by the sarcomere. Am. J. Physiol. **278:** H1274–H1284.
8. HIBBERD, M.G. & B.R. JEWELL. 1982. Calcium and length-dependent force production in rat ventricle muscle. J. Physiol. **329:** 527–540.
9. HUXLEY, A.F. 1957. Muscle structure and theories of contraction. Prog. Biophys. Byophys. Chem. **7:** 255–318.
10. HOFMANN, P.A. & F. FUCHS. 1987. Effect of length and cross-bridge attachment on Ca^{2+} binding to cardiac troponin C. Am. J. Physiol. **253:** C90–C96.
11. GRABAREK, Z., J. GRABAREK, P.C. LEAVIS & J. GERGELY. 1983. Cooperative binding to the Ca^{2+}-specific sites of troponin C in regulated actin and actomyosin. J. Biol. Chem. **258:** 14098–14102.
12. SHNIER, J.S. & J. SOLARO. 1982. Activation of thin-filament-regulated muscle by calcium ion: considerations based on nearest-neighbor lattice statistics. Proc. Natl. Acad. Sci. U.S.A. **79:** 4637–4641.

13. GUTH, K. & J.D. POTTER. 1987. Effect of rigor and cycling cross-bridges on the structure of troponin-C and on the Ca^{2+} affinity of the Ca^{2+}-specific regulatory sites in skinned rabbit psoas fibers. J. Biol. Chem. **262:** 13627–13635.
14. KENTISH, J.C. & A. WRZOSEK. 1998. Changes in force and cytosolic calcium concentration after length changes in isolated rat ventricular trabeculae. J. Physiol. **506:** 431–444.
15. LEVY, C. & A. LANDESBERG. 2004. Hystereses in the force-length relation and regulation of cross-bridge recruitment in tetanized rat trabeculae. Am. J. Physiol. **286:** H434–H441.
16. CAMPBELL, K.B., M.V. RAZUMOVA, R.D. KIRKPATRICK & B.K. SLINKER. 2001. Nonlinear myofilament regulatory processes affect frequency-dependent muscle fiber stiffness. Biophys. J. **81:** 2278–2296.
17. HARRISON, S.M., C. LAMONT & D.J. MILLER. 1988. Hysteresis and the length dependence calcium sensitivity in chemically skinned rat cardiac muscle. J. Physiol. **401:** 115–143.
18. RIDGWAY, E.B., A.M. GORDON & D.A. MARTYN. 1983. Hysteresis in the force-calcium relation in muscle. Science **219:** 1075–1077.

Structure-Function Relation of the Myosin Motor in Striated Muscle

MASSIMO RECONDITI,[a,b] MARCO LINARI,[a,b] LEONARDO LUCII,[a,b] ALEX STEWART,[c] YIN-BIAO SUN,[d] THEYENCHERI NARAYANAN,[e] TOM IRVING,[f] GABRIELLA PIAZZESI,[a,b] MALCOLM IRVING,[d] AND VINCENZO LOMBARDI[a,b]

[a] *Laboratorio di Fisiologia, DBAG, Università di Firenze, 50019 Sesto Fiorentino, Italy*

[b] *Istituto Nazionale di Fisica della Materia, Operative Group in Grenoble, F-38043 Grenoble Cedex, France*

[c] *Rosenstiel Center, Brandeis University, Waltham, Massachusetts 02545, USA*

[d] *Randall Division of Cell and Molecular Biophysics, King's College London, London SE1 1UL, UK*

[e] *European Synchrotron Radiation Facility, F-38043 Grenoble Cedex, France*

[f] *BioCAT Advanced Photon Source, Argonne National Laboratory, Argonne, Illinois 60439, USA*

ABSTRACT: **Force and shortening in striated muscle are driven by a structural working stroke in the globular portion of the myosin molecules—the myosin head—that cross-links the myosin-containing filaments and the actin-containing filaments. We use time-resolved X-ray diffraction in single fibers from frog skeletal muscle to link the conformational changes in the myosin head determined at atomic resolution in crystallographic studies with the kinetic and mechanical features of the molecular motor in the preserved sarcomeric structure. Our approach exploits the improved brightness and collimation of the X-ray beams of the third generation synchrotrons by using X-ray interference between the two arrays of myosin heads in each bipolar myosin filament to measure with Å sensitivity the axial motions of myosin heads *in situ* during the synchronous execution of the working stroke elicited by rapid decreases in length or load imposed during an active isometric contraction. Changes in the intensity and interference-fine structure of the axial X-ray reflections following the mechanical perturbation allowed to establish the average conformation of the myosin heads during the active isometric contraction and the extent of tilt during the elastic response and during the subsequent working stroke. The myosin working stroke is 12 nm at low loads, which is consistent with crystallographic studies, while it is smaller and slower at higher loads. The load dependence of the size and speed of the myosin working stroke is the molecular determinant of the macroscopic performance and efficiency of muscle.**

KEYWORDS: **myosin motor, striated muscle; working stroke; X-ray diffraction**

Address for correspondence: Prof. Vincenzo Lombardi, Laboratorio di Fisiologia, DBAG, Via G. Sansone 1, 50019 Sesto Fiorentino (FI), Italy. Voice: 39 055 457 2388; fax: 39 055 457 2387. vincenzo.lombardi@unifi.it

Ann. N.Y. Acad. Sci. 1047: 232–247 (2005).
doi: 10.1196/annals.1341.021

INTRODUCTION

Structure-Function Relation at Sarcomere Level

Skeletal and cardiac muscles share the structural feature that the contractile machinery is organized in well-ordered and parallel arrays of myosin and actin filaments called sarcomeres (FIG. 1A), which provide the striated appearance to the muscle. In each sarcomere, myosin-containing thick filaments (1.6 μm long) overlap with actin-containing thin filaments (~1 μm long) that attach to the Z line with their barbed end. During the contraction, the globular portions of the myosin molecules extending from the thick filaments—the "myosin heads"—attach to the actin monomers in the thin filaments. The attached myosin head (the cross-bridge) undergoes a conformational change (the working stroke) that generates piconewton scale force and nanometer scale shortening by pulling the actin filaments toward the center of the myosin filament. The polymeric arrangement of both the molecular motor (the myosin II) and its track (the actin filament) gives peculiar properties to this system in comparison with any other myosin driven motility system:

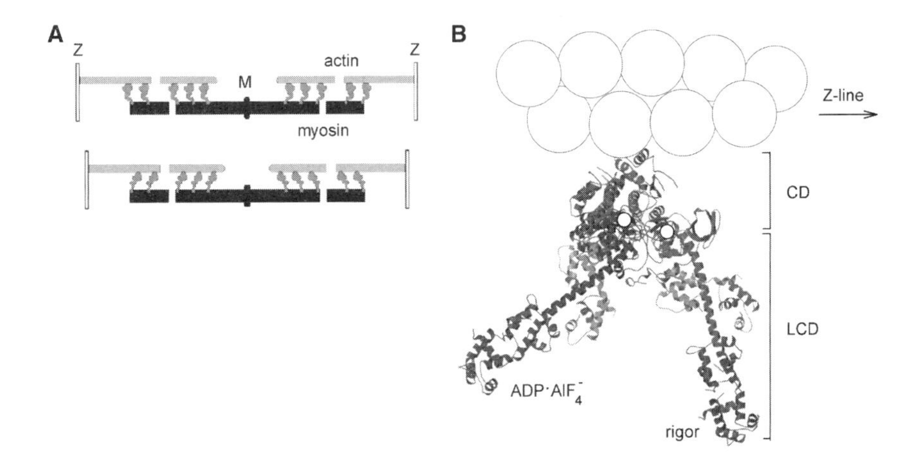

FIGURE 1. Structural model of the myosin motor in the sarcomere. (**A**) Diagram showing the overlapping actin (*light gray*) and myosin (*black*) filaments in the sarcomere. Myosin heads (*dark gray*) extend from the myosin filament in two arrays with antiparallel orientation, separated by the M line and the bare zone free from heads. Sarcomere shortening (*transition from upper to lower panel*) is associated with tilting of myosin heads so that their mass and the actin filaments move toward the center of the sarcomere. (**B**) Atomic model of the working stroke for one of the myosin heads of the right half-sarcomere in **A**. *White spheres* represent monomers of the actin filament; myosin head is shown in *ribbon* representation in both the nucleotide-free or rigor conformation (*right*) and in the adenosine diphosphate (ADP)·AlF$_4^-$ conformation (*left*). In the two conformations the catalytic domains (CD, residues 1–707) are made to coincide. The orientation of the light chain domain in the ADP·AlF$_4^-$ structure was determined by assuming that the converter–light chain domain (LCD, residues 711–843) moves as a rigid body about residue 707. The *white circles* between CD and LCD identify the position of the converter domain in the two conformations. Residue numbers refer to chicken skeletal myosin.[9] (From Irving *et al.*[17] Adapted by permission.)

(1) If the length of the sarcomere is maintained constant, the active force developed by cross-bridges acting in parallel in the half-sarcomere attains a maximum steady value T_0 (~300 kPa for frog skeletal muscle at 4°C); the corresponding strain (~5 nm per half-sarcomere) is distributed between the elastic components of the half-sarcomere, mainly cross-bridges and myofilaments.
(2) An external load on the sarcomere lower than T_0 results in sliding of each thin filament toward the center of the myosin filament and in micrometer scale shortening of the sarcomere sustained by cyclic cross-bridge interactions.[1,2] The arrangement in series of the two half-sarcomeres in the sarcomere (FIG. 1A), and of the thousand sarcomeres in the muscle fiber, provides a corresponding amplification factor for the macroscopic shortening of the muscle.
(3) The sliding velocity between track and motor depends not only on the action of a single motor, but also on the relation between the force of the ensemble of motors and the external load; in turn, the actin-myosin interaction may have quite different kinetics depending on the sliding velocity.

These features are the bases for the remarkable macroscopic outcomes of this polymeric machine: (1) the energy converted into work (i.e., the rate of adenosine triphosphate, ATP, hydrolysed by the myosin heads) at constant concentration of substrate increases with the increase in sliding velocity, but reduces again near unloaded shortening velocity; and (2) the efficiency of energy conversion (the ratio of the work produced over the energy used) is maximum during high load shortening.[3,4]

The sarcomeric organization of muscle means that its molecular structure-function relation cannot be elucidated by studies at the single molecule level. The necessity to preserve the structural and functional integrity at sarcomere level explains why both spectroscopic techniques on demembranated muscle cells and laser trap techniques on single molecules fail to produce definitive figures about mechanics, kinetics, and structural features of this motor. So far the only preparation of striated muscle in which both structural and functional features of the sarcomere can be reliably determined is the single fiber dissected intact from frog skeletal muscle. In this preparation, information on the size and the kinetics of a single acto-myosin interaction can be obtained by synchronization of the action of the motors following length or force steps controlled at the sarcomere level.[5,6]

A Crystallographic Model of the Working Stroke of the Myosin Motor

Biochemical and energetic studies[4,7] provided evidence that the work produced in the interaction of myosin with actin is accounted for by the free energy liberated by the hydrolysis of ATP on the catalytic site of the myosin head and more specifically with the release of the ATP hydrolysis products, phosphate and adenosine diphosphate (ADP). In the absence of ATP, the nucleotide-free myosin head strongly bound to actin is responsible for the rigor state of the muscle. Comparison of the crystallographic structure of a fragment of the myosin head complexed with an analogue of the ATP hydrolysis products ($ADP{\cdot}AlF_4^-$)[8], with that of the nucleotide-free head[9] (FIG. 1B) suggests that the release of hydrolysis products from the active site induces inter-domain molecular rearrangements resulting in a ~70° tilting of the converter region of the light chain domain (LCD) about the catalytic domain (CD), firmly attached to actin.[10] Due to the length of LCD acting as a lever arm, the tilting is amplified into ~11 nm axial motion of the head-rod junction (the tip of the lever arm) toward the Z line. In the native system (the sarcomere), the tip of the lever arm is connected to the myosin filament, and this motion would cause a corresponding relative sliding of the myosin and actin filaments in the direction of sarcomere shortening. Shortening of 11 nm is twice the size of the length step measured for one myosin-actin interaction in most single molecule measurements,[11,12] but is similar to the value estimated with the rapid force transients that follow step changes in sarcomere length superimposed on isometrically contracting fibers.[5,13]

STRUCTURAL STUDIES OF THE MYOSIN MOTOR *IN SITU*

X-Ray Diffraction from Muscle Fibers

The function of myosin in the sarcomere depends on the interaction between conformational changes in the motor and force or motion relative to the actin track, and this cannot be reproduced in crystallographic studies. Due to the quasi-crystalline

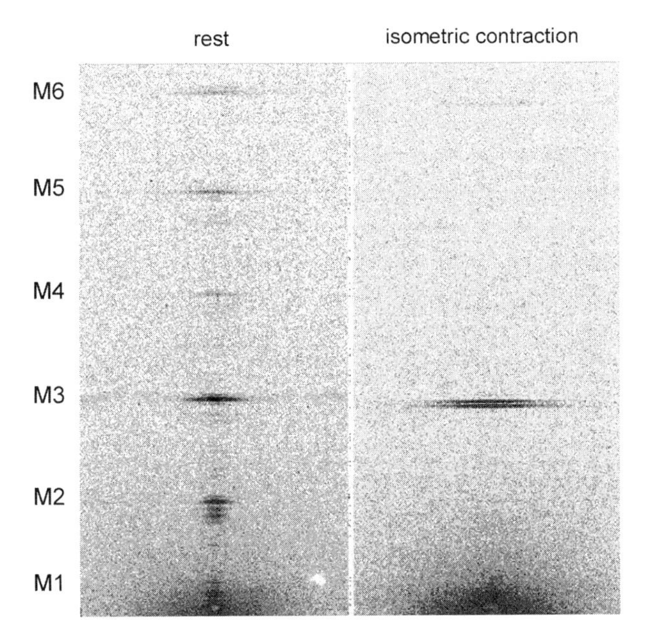

FIGURE 2. Axial X-ray diffraction patterns from a single fiber from tibialis anterior muscle of *Rana temporaria* at rest (*left panel*) and during an isometric tetanus (*right panel*), collected on an image plate detector at ID2, ESRF (European Synchrotron Radiation Facility, Grenoble, France). Average sarcomere length, 2.07 μm, temperature 4°C. The axial pattern at rest shows a series of reflections that index on the ~43 nm quasi-helical periodicity of the myosin heads in the thick filaments (called M1). Only the M3 and M6 remain evident in the active pattern.

arrangement of the motor proteins in the three-dimensional lattice of a muscle, X-ray diffraction can be used to investigate the structural working stroke at a lower spatial resolution than that of crystallographic studies but in the native environment for the motor function and with the adequate time resolution. Because only in single fibers it is possible to apply the high resolution mechanics that allow synchronization of cross-bridge action[5,6] necessary for collecting the corresponding structural change, and because muscle fibers diffract X-rays very weakly, it is necessary that these experiments are carried out with the very bright X-ray source available at the synchrotrons. In resting muscle the myosin heads have a helical arrangement on the surface of the filament, giving rise to a series of X-ray reflections called the M1, M2, M3, etc., which are orders of the ca 43 nm helical repeat of the myosin filament (FIG. 2). In actively contracting muscle the helical order is lost, but the M3 reflection from the 14.5 nm axial repeat of the myosin heads along the filaments (FIG. 1A) remains strong. The intensity of this reflection (I_{M3}) is sensitive to the changes in mass density projections of the myosin heads associated with force generation and filament sliding. However, changes in I_{M3} lack phase information and, to be interpreted in terms of conformational changes, they need crystallographic models.

X-Ray Diffraction Measurement of the Conformation of the Myosin Heads in the Isometric Contraction

FIGURE 3A shows I_{M3} changes expected from the structural model of the myosin head of FIGURE 1B for any bend between CD and LCD domains. I_{M3} is calculated as the square of the 1/14.5 nm^{-1} component of the Fourier transform of the axial mass distribution of the myosin heads. The horizontal axis shows the displacement (z) of the tip of the lever arm produced by the tilting, thus it is related to the sliding between the thick and thin filament. z is 0 for the nucleotide-free (rigor) structure. At $z =$ 10.6 nm, corresponding to the ADP·A1F_4^- structure (FIG. 3A), I_{M3} has a value similar to that at $z = 0$. I_{M3} is higher for intermediate tilt and is maximum for $z = 5.4$ nm. This occurs when the density distribution of the myosin head along the filament axis is the narrowest, because the axial coordinates of the centroids of the LCD and of the CD coincide (FIG. 3B). Thus, changes in I_{M3} can be used to define the conformational changes during the myosin working stroke. In a normal contraction, working strokes occur asynchronously, since the ensemble of myosin heads is spread out through the various steps of the chemo-mechanical transduction cycle. However, in single fibers from frog muscle, the working stroke in the attached myosin heads can be synchronized by step in sarcomere length imposed during an isometric contraction.[5,13] The drop in force from the isometric value T_0 to T_1 (phase 1 of Huxley and Simmons force transient; see also FIG. 6A), due to the discharge of elastic strain in the attached myosin heads and myofilaments, is followed by a quick recovery of force to T_2 (complete within 1 to 2 ms, phase 2), which represents the mechanical manifestation of the working stroke. Since intensities of axial X-ray reflections depend also on other parameters, such as the number of heads and their orientational dispersion, the critical assumption, supported by mechanical evidence,[14] is that the myosin heads that were attached to actin before the length step was applied remain attached during the elastic response and the subsequent execution of the working stroke. Repetitive synchronization of the working stroke by trains of steps separated by intervals necessary to reprime the response[14–16] allows averaging adequate to

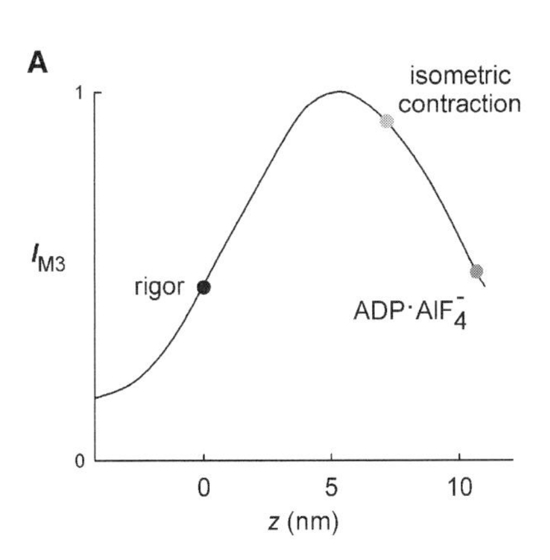

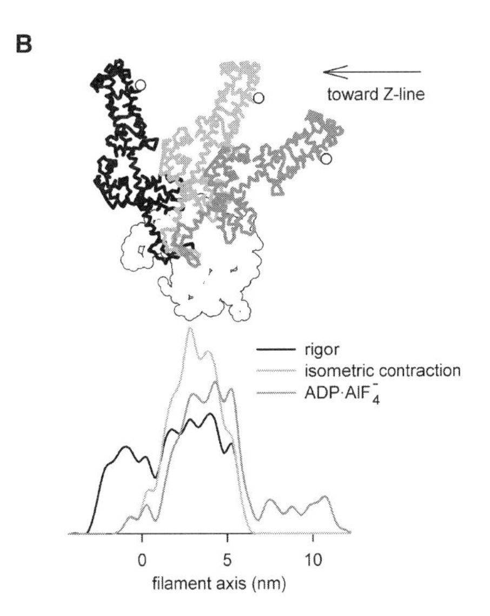

FIGURE 3. (**A**) Dependence of the intensity of the M3 X-ray reflection (I_{M3}) on z, the axial displacement of the tip of the lever arm from the position in the rigor conformation (*black circle*). *Dark gray circle* and *light gray circle* indicate the adenosine diphosphate (ADP)·AlF_4^- and the isometric contraction structures, respectively. (From Irving *et al.*[17] Adapted by permission.) (**B**) Mass projections along the filament axis of the myosin head with the catalytic domain (*white*) in a fixed position and the light chain domain oriented according to the three different conformations: rigor (*black*); isometric contraction (*light gray*); ADP·AlF_4^- (*dark gray*). The *white circle* identifies the tip of the lever arm (residue 843). The origin of the horizontal axis corresponds to the position of the tip in the rigor conformation.

reduce the time per X-ray frame to the 100 μs range necessary to resolve the rapid changes of I_{M3}.[15–17] As shown in FIGURE 4, a step release of ~6 nm per half-sarcomere produces a small increase of I_{M3} during the elastic drop in force in phase 1. I_{M3} shows a large decrease (to ~0.5 the isometric value) during the quick recovery in phase 2. The small change of I_{M3} during the elastic response suggests that z moves across the peak of the I_{M3}-z relation in FIGURE 3A. The synchronous execution of the working stroke moves the experimental point downhill along the ascending limb of the I_{M3}-z relation. Following a step stretch of the same size as the release, applied 1 ms after the shortening step, the force rises to 1.5 T_0 in phase 1 and then recovers toward its isometric value in phase 2. I_{M3} increases during the stretch due to the elastic response of the heads that brings z back toward the isometric value (FIG. 3A), and then it undergoes a small decrease during the force recovery associated with the reversal of the working stroke, due to the movement of z across the peak of the I_{M3}-z

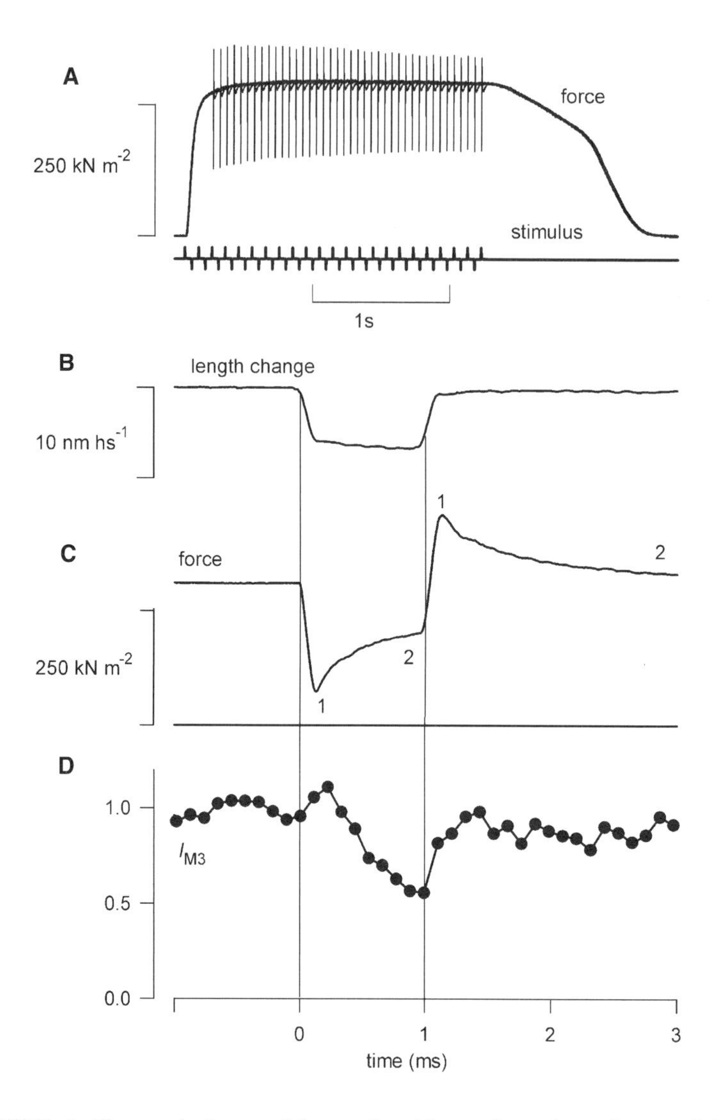

FIGURE 4. Changes in force and I_{M3} produced by a ~6 nm shortening step followed after 1 ms by a stretch during active contraction. (**A**) Superimposed slow time base force records in the presence and absence of 40 length change cycles imposed at 50 ms intervals. Before each shortening-stretch cycle, the force returned to the isometric tetanic value. Average sarcomere length 2.09 μm, temperature 4°C. Changes in length (**B**) and force (**C**), in the same fiber as panel **A**, sampled at 10 μs intervals, and I_{M3} (**D**, *filled circles*) in 100 μs time bins from 402 tetani in 14 fibers. X-ray data collected on a gas-filled detector at beamline 16, SRS (Synchrotron Radiation Source, Daresbury, UK). (From Irving *et al.*[17] Adapted by permission.)

relation. The best fit of the data with the I_{M3}-z curve obtained by the structural model shows that during isometric contraction, the LCD is tilted by 40° or ~6 nm (FIG. 3A and B) from the nucleotide-free (rigor) conformation.[17]

It must be noted that the I_{M3} versus z relation in FIGURE 3A was calculated assuming that all the myosin heads in the fiber have the same orientation, but the relationship is almost identical for a constant z dispersion of a few nanometers.[17] Thus the critical assumption is that, as mentioned before, the heads remain attached within the time of the response.[14]

The X-Ray Interference Effect

With a highly collimated beam—such as that of ID2/SAXS beamline at ESRF (European Synchrotron Radiation Facility, Grenoble, France), or a finely focused beam such as that of BioCAT beamline at APS (Advanced Photon Source, Argonne, IL)—it is possible to see that the myosin-based meridional reflections are split into several peaks (FIG. 2). This effect, which was first described in the whole muscle at rest by Huxley and Brown[18] and Haselgrove,[19] is the result of the interference between the two arrays of myosin heads in each thick filament. Each half of the thick filament (FIG. 5B) contains an array of forty-nine layers of heads with regular spacing d (~14.5 nm). This array generates the M3 X-ray reflection (FIG. 5C). The two arrays in each thick filament generate a sinusoidal interference modulation with a spatial frequency equal to the reciprocal of the center-to-center distance between the arrays (the interference distance, L, ~900 nm). The Fourier transform of the whole structure is the product of the Fourier transforms of the two separate structures. When, as in the case of the myosin filament, L is much larger than d ($L \sim 900$ nm $\approx 60\, d$), the product of the two intensity distributions gives as a result the fine sampling of the M3 reflection by the interference function (FIG. 5C).

In an isometrically contracting fiber at 4°C, the M3 reflection is split into two closely spaced peaks, at 14.47 nm and 14.66 nm (FIG. 5A). The ratio of intensity of the high angle peak to that of the low angle peak (R_{M3}) is ~0.8,[20] showing that the interference distance is close to an odd multiple of half the M3 spacing.

Due to the bipolar arrangement of the myosin heads in the two halves of the thick filament, when actin-attached myosin heads execute the working stroke that pulls the actin filaments toward the center of the sarcomere, the center of mass of the heads, and hence the center of mass of each array, also moves toward the center of the sarcomere (FIG. 5E). Consequently, the interference distance L reduces and the sinusoidal intensity distribution shifts to a higher reciprocal spacing (rightward in FIG. 5F), decreasing R_{M3}. If the fiber is stretched so that the actin filament and the catalytic domains attached to it are pulled away from the center of the sarcomere, L and R_{M3} increase accordingly. Thus, the interference method is able to resolve the direction of the movement, independent of a structural model for the head. The high order (~60th) by which the interference function samples the M3 reflection increases the sensitivity of the method to Å scale movements of the centroid of myosin heads. The structural information obtained from R_{M3} is independent of that from the total intensity of the reflection, I_{M3}. In fact, changes in R_{M3} depend only on the axial displacement of the center of mass of the myosin heads C (FIG. 5B), with respect to their attachment to the myosin filament (the head-rod junction) and, for a given center of mass position, are almost independent of the conformation and conformational dispersion.

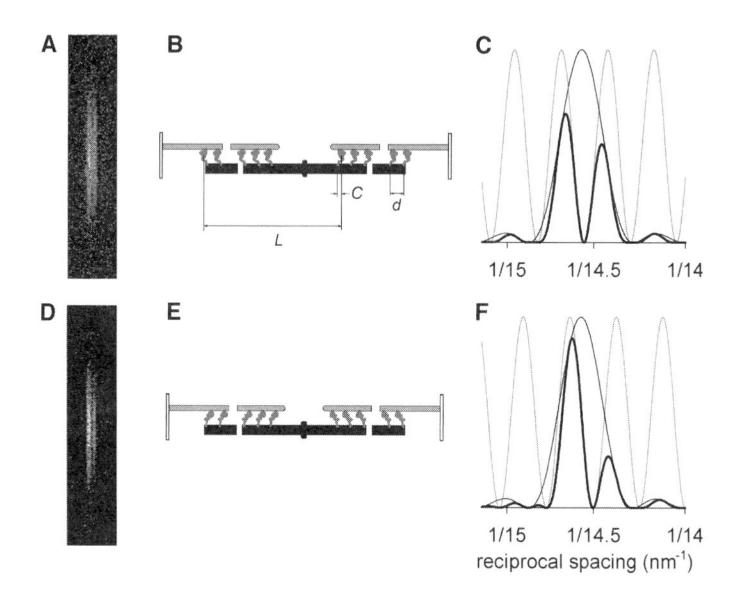

FIGURE 5. Interference fine structure of the M3 reflection and its dependence on axial motion of the myosin heads, defined by the displacement of the center of mass of the heads (**C**) with respect to their attachment to the myosin filament. (**A**, **D**) X-ray diffraction patterns in the region of the M3 reflection, collected on the image plate detector, at the isometric plateau (**A**), and 1 ms after a shortening step of ~6 nm (**D**). (**B**, **E**) Scheme of the two arrays of myosin heads (*dark gray*) in each myosin filament (*black*) with the LCD tilted as during the isometric contraction (**B**) and following the shortening step (**E**). Actin filament in *light gray*. (**C**, **F**) Axial X-ray intensity distribution (*thick black line*) generated by the sampling of the M3 reflection (*thin black line*) by the interference function (*gray line*).

Measurement of the Working Stroke Following Shortening Steps

We used the X-ray interference technique to measure the axial motions of myosin heads produced by applying 120 μs shortening steps during an isometric contraction.[21] As shown in FIG. 6B for a step of 5 nm, R_{M3} reduced following the shortening step as expected for displacement of the two arrays of actin-attached myosin heads toward the center of the sarcomere. However, most of the change occurred at the end of the shortening step that reduced the force from the isometric value T_0 to T_1 (FIG. 6A), and only very small further reduction occurred during the subsequent rapid force recovery to T_2.

The observed R_{M3} for different step sizes (range 2 to 9 nm per half-sarcomere) can be used to calculate the interference distance L, using the theoretical intensity distributions described above (FIG. 5). Then the change in L responsible for the change in R_{M3} can be used as an estimate of change in C. Changes in C are about five times smaller than the imposed filament sliding.[21] To make a quantitative interpretation of the observed axial motions of the myosin heads during the elastic response, we must take into account the compliances of the actin and the myosin filaments.[22–26] While the estimate of the instantaneous compliance of the actin filament by different

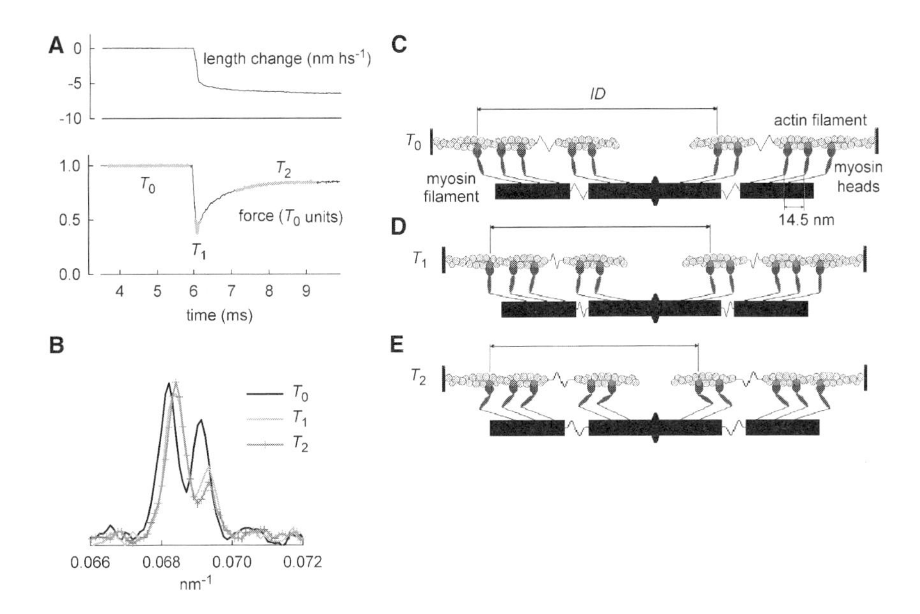

FIGURE 6. Change in interference fine structure of the M3 reflection following a shortening step, and explanation in terms of axial motion of myosin heads. (**A**) Length change (nm per half-sarcomere) and force normalized by isometric force T_0. The *thick gray* segments in the force trace mark the periods of X-ray exposure: T_0, before the length step; T_1, at the end of the elastic response; T_2, at the end of quick recovery. (**B**) Axial intensity distribution in the region of M3 reflection normalized to that of the lower angle peak recorded at the periods corresponding to the three X-ray exposure times shown in **A**. (From Piazzesi *et al.*[21] Adapted by permission.) (**C–E**) Motion of the myosin heads at the three times of the force transient shown in **A**.

methods converges to indicate ~0.26%/T_0, that of the myosin filament is less clearly defined. This is both because there are changes in spacing of axial myosin based reflections not related to mechanical compliance,[18] and because the most intense M3 reflection is dominated by the heads attached to actin and its spacing change may be influenced by the actin filament compliance. The spacing of the higher order myosin-based reflection, M6, is considered a better indicator of the change in extension of the myosin filament and indicates a compliance of ~0.26%/ T_0.[26] From these estimates it results that the change in extension of the myofilaments with force represents 70% of the total half-sarcomere compliance: of the 5.1 nm, which are the change in strain of the half-sarcomere for a change in force equal to T_0, 3.7 nm are accounted for by myofilaments and 1.4 nm by cross-bridges. The compliances of actin and myosin filaments can be used to calculate the axial displacements of each layer of myosin heads along the filament due to change in strain during the elastic response to length steps (FIG. 6C–D). The displacements can then be used to calculate the changes in R_{M3} with the structural model, where each layer of myosin heads is represented by the crystallographic model of the head with a variable tilt between the CD and the LCD (FIG. 1B). The observed changes in R_{M3} were reproduced if only one half of myosin heads that contribute to M3 reflection move in response to the step.

The absence of axial motion of the myosin heads during the quick recovery appears to contradict our previous interpretation of the decrease in I_{M3} as a tilting of the LCD of the heads (FIG. 3), since the tilting should displace the actin-bound CD toward the center of the sarcomere. The contradiction is resolved if we consider that the phase 2 recovery following a shortening step occurs at constant sarcomere length and then tilting of the heads is taken up by increase in strain of the filaments (FIG. 6D–E). Tilting of the LCD displaces the head-rod junction attached to the myosin filament away from the M-line and the catalytic domain attached to the actin filament toward the M-line, with the result that there is little net movement of the center of mass of the heads with respect to the M line.

In conclusion, the X-ray interference changes measured during the elastic response, and the subsequent rapid force recovery following shortening steps of different size (2–9 nm per half-sarcomere), do not allow a direct measurement of the axial motion of myosin heads but provide definitive evidence for a mechanism of force generation based on a ~10 nm working stroke in the myosin heads attached to actin and against the idea of rapid detachment of myosin heads followed by rapid reattachment to new actin monomers.[21]

The Size and Speed of the Working Stroke at Different Loads

Since the working stroke cannot be observed directly in X-ray interference experiments with length steps, we turned to a different type of protocol, in which the working stroke in the myosin heads is synchronised by a step (150 μs) reduction in load imposed on the fiber during the isometric contraction. The elastic or phase 1 response to the load step is the same as that to a length step, but the ensuing velocity transient[6,26] occurs at constant load (FIG. 7A), and therefore at constant filament strain. Thus, with a load-step protocol, the structural basis of the working stroke can be investigated by X-ray interference in the absence of the effects of filament compliance. Especially at low loads, shortening in phase 2 of the velocity transient is very fast, and that is why it has been difficult to isolate this phase from phase 1 response until recent improvements in mechanical techniques.[6] The velocity transient, induced by a step in force from T_0 (length L_0) to 0.5 T_0 (FIG. 7A), is made by the elastic shortening (phase 1) followed by a rapid shortening that lasts for a few milliseconds after the load step (phase 2, associated with the working stroke in the myosin heads); the shortening velocity then decreases (phase 3), before increasing once more to attain the steady state value characteristic of the force-velocity relationship (phase 4).[6]

The interference fine structure of the M3 reflection (FIG. 7B) was recorded during isometric contraction (L_0,), at the start and end of phase 2 (L_{2s} and L_{2e}, respectively), and at the end of phase 3 (L_3,). R_{M3} decreased during phase 1, as seen previously with the length step protocol, then, in contrast with the length step protocol, R_{M3} continued to decrease during phase 2 of the velocity transient, before recovering in phase 3. These results show that myosin heads remain attached to actin and move toward the center of the sarcomere during both phase 1 and phase 2 of the response to a load step (FIG. 7C–E). The recovery of the interference fine structure of the M3 reflection during phase 3 of the velocity transient shows that myosin heads detach

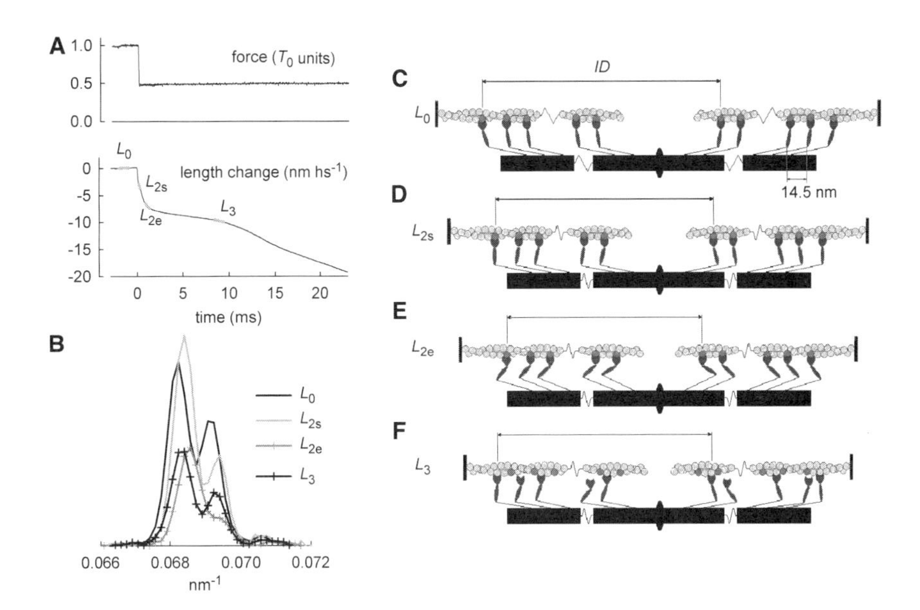

FIGURE 7. Change in the interference fine structure of the M3 reflection following a load step, and explanation in terms of the axial motions of the myosin heads. (**A**) Force step normalized by the isometric force (T_0), and length change in nm per half-sarcomere. The *thick gray* segments in the length trace mark the periods of X-ray exposure: L_0, before the load step; L_{2s}, start of phase 2; L_{2e}, end of phase 2; L_3, end of phase 3. (**B**) Axial intensity distribution in the region of the M3 reflection at the periods corresponding to the four exposure times shown in **A**. (**C–F**) Motions of the myosin heads at the four times of the velocity transient shown in **A**. (From Reconditi *et al.*[26] Adapted by permission.)

from actin in this phase and reattach to new actin monomers farther from the center of the sarcomere (FIG. 7E–F).

The change of the load in the range 0.75 to 0.25 T_0 (FIG. 8A and B) shows that the speeds of both the working stroke and the subsequent detachment from actin depend on the load and are slower at higher load.[26] The effect of the load on the size of the working stroke is made evident in FIGURE 8C, where R_{M3} is plotted against the extent of filament sliding in each half-sarcomere: data for phase 2 of the velocity transient fall on the same line, independent of the load. In fact, the extent of axial motion of the heads in phase 2 depends only on the filament sliding and not on the load. The recovery of R_{M3} during phase 3 of the velocity transient is revealed as a deviation from the phase 2 relationship when the heads start to detach from actin. This deviation occurs after different extents of filament sliding at different loads: at the lowest load (0.25 T_0), there was about 12 nm filament sliding before R_{M3} recovery, but at the highest load (0.75 T_0), there was only about 6 nm sliding before R_{M3} recovered. These results show directly that the working stroke of the myosin motor is smaller at higher load. At low load, the working stroke is about 12 nm, a value which is consistent with the atomic model for the working stroke derived from X-ray crystallography (FIG. 1B). At higher loads, the myosin head detaches from actin before reaching the struc-

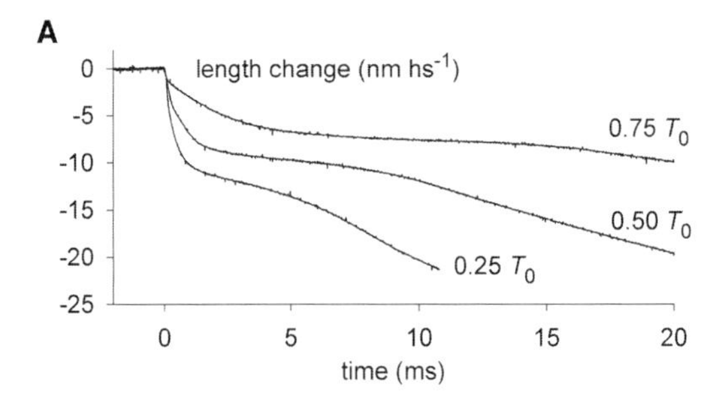

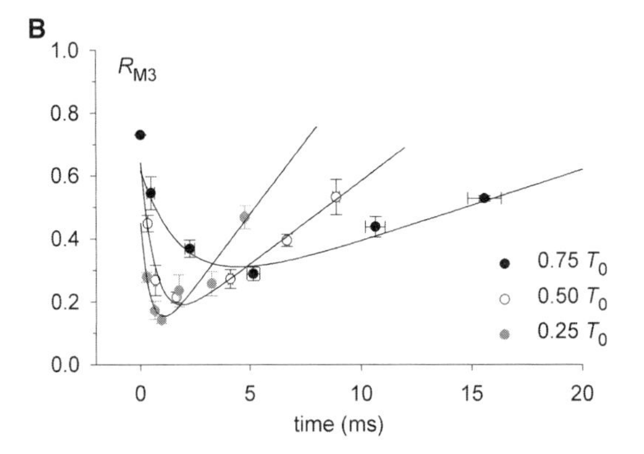

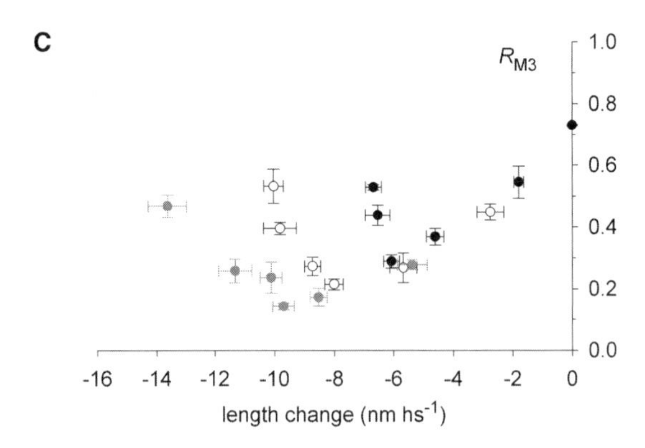

FIGURE 8. Load dependence of the axial motions of myosin heads. (**A**) Velocity transients following step reductions in force to 0.75, 0.5, and 0.25 T_0. (**B**) R_{M3} plotted against time for the three different loads: 0.75 T_0 (*black*); 0.5 T_0 (*white*); 0.25 T_0 (*gray*). (**C**) Relationship between R_{M3} and the extent of filament sliding in nm per half-sarcomere, from the same experiments as **B**. The lines are drawn for visual help to follow the time courses. (From Reconditi *et al.*[26] Adapted by permission.)

tural limit. The load dependence of the size and the speed of the working stroke are basic molecular parameters that determine the macroscopic performance and efficiency of muscle contraction.

CONCLUSIONS AND PERSPECTIVES

This article focuses on the investigation of the working stroke of the myosin motor in striated muscle made by a combination of X-ray diffraction and X-ray interference with high resolution mechanical methods in intact single fibers from frog muscle.

Changes in axial myosin-based X-ray reflections following step changes in sarcomere length were interpreted by using the crystallographic model of the myosin head and made it possible to establish that the average orientation of the myosin heads during an isometric contraction is near the perpendicular to the filament axis and that the heads tilt both during the elastic response to a change in strain and during the subsequent working stroke.

The application of the X-ray interference method allowed us to directly measure the axial movement of the heads in the sarcomere with Å sensitivity. We found that during the working stroke elicited by a shortening step, the axial movement of the heads is absorbed by the increase in myofilament strain, with little change in the interference fine structure. Interference measurements during the working stroke elicited by a step reduction in load revealed that the size and the speed of the working stroke of the myosin heads depend on the load: at 0.75 T_0 the stroke is ~6 nm and takes ~5 ms, while at 0.25 T_0 it increases to ~12 nm and takes 0.5 ms. The load-dependent reduction in the size of the working stroke is less than the increase in load, so that the work per stroke is maximum at high loads.[6] This result provides the molecular determinant of the efficiency of muscle.

The size of myosin working stroke at low loads agrees with the 11 nm movement expected from the crystallographic model. Whether the reduced size of the stroke at high load is the result of a lower proportion of motors going through a switch-like conformational change, as postulated by crystallographers, or it is the result of a reduced movement in myosin motor, is a question for the future. A challenge to the crystallographers' view is the increasing evidence that the myosin motor has a very small compliance: the strain in the myosin head to generate the isometric force T_0 is 1.4 nm,[26] which is one order of magnitude smaller than the 11 nm working stroke. Models of the working stroke where force in the motor is generated by stretching a distinct elastic element[15] become less plausible when the working stroke is much larger than the compliance. The high stiffness of the motor rather suggests a series of structural transitions within the myosin head.

Another fundamental question that will drive future experiments is related to the peculiar polymeric organization of the myosin motor in the sarcomere. How are the size of the stroke and the nature of myosin interaction with actin modified by the load at the steady state of isotonic shortening? Is there a structural explanation for the reduced rate of energy used at high shortening speed? The combination of X-ray interference and high resolution mechanics in single muscle fibers can address these questions and provide the molecular and supramolecular bases of the energetics of the relation between force and shortening speed in striated muscle.

ACKNOWLEDGMENTS

The work described in this article was supported by the Ministero dell'Istruzione, dell'Università e della Ricerca, Telethon-945, Italy; the National Institutes of Health, USA; the Medical Research Council, UK; the European Molecular Biology Laboratory; and the European Union and European Synchrotron Radiation Facility. Use of the Advanced Photon Source was supported by the U.S. Department of Energy. BioCAT is a National Institutes of Health-supported Research Center. We thank A. Aiazzi, M. Dolfi, R. Fischetti, and J. Gorini for technical support.

REFERENCES

1. Huxley, A.F. & R. Niedergerke. 1954. Structural changes in muscle during contraction. Nature **173:** 971–973.
2. Huxley, H. & J. Hanson. 1954. Changes in the cross-striations of muscle during contraction and stretch and their structural interpretation. Nature **173:** 973–976.
3. Kushmerick, M.J. & R.E. Davies. 1969. The chemical energetics of muscle contraction. II. The chemistry, efficiency and power of maximally working sartorius muscles. Proc. R. Soc. Lond. B. Biol. Sci. **174:** 315–353.
4. Woledge, R.C., N.A. Curtin & E. Homsher. 1985. Energetic Aspects of Muscle Contraction. Academic Press. London.
5. Huxley, A.F. & R.M. Simmons. 1971. Proposed mechanism of force generation in striated muscle. Nature **233:** 533–538.
6. Piazzesi, G., L. Lucii & V. Lombardi. 2002. The size and the speed of the working stroke of muscle myosin and its dependence on the force. J. Physiol. **545:** 145–151.
7. Lymn, R.W. & E.W. Taylor. 1971. Mechanism of adenosine triphosphate hydrolysis by actomyosin. Biochemistry **10:** 4617–4624.
8. Dominguez, R., R. Freyzon, K.M. Trybus & C. Cohen. 1998. Crystal structure of a vertebrate smooth muscle myosin motor domain and its complex with the essential light chain: visualization of the pre-power stroke state. Cell **94:** 559–571.
9. Rayment, I., W.R. Rypniewski, K. Schmidt-Base, *et al.* 1993. Three-Dimensional structure of myosin subfragment-1: a molecular motor. Science **261:** 50–58.
10. Geeves, M.A. & K.C. Holmes. 1999. Structural mechanism of muscle contraction. Annu. Rev. Biochem. **68:** 687–728.
11. Molloy, J.E., J.E. Burns, J. Kendrick-Jones, *et al.* 1995. Movement and force produced by a single myosin head. Nature **378:** 209–212.
12. Mehta, A.D., J.T. Finer & J.A. Spudich. 1997. Detection of single-molecule interactions using correlated thermal diffusion. Proc. Natl. Acad. Sci. U.S.A. **94:** 7927–7931.
13. Piazzesi, G. & V. Lombardi. 1995. A cross-bridge model that is able to explain mechanical and energetic properties of shortening muscle. Biophys. J. **68:** 1966–1979.
14. Lombardi, V., G. Piazzesi & M. Linari. 1992. Rapid regeneration of the actin-myosin power stroke in contracting muscle. Nature **355:** 638–641.
15. Irving, M., V. Lombardi, G. Piazzesi & M.A. Ferenczi. 1992. Myosin head movements are synchronous with the elementary force-generating process in muscle. Nature **357:** 156–158.
16. Lombardi, V., G. Piazzesi, M.A. Ferenczi, *et al.* 1995. Elastic distortion of myosin heads and repriming of the working stroke in muscle. Nature **374:** 553–555.
17. Irving, M., G. Piazzesi, L. Lucii, *et al.* 2000. Conformation of the myosin motor during force generation in skeletal muscle. Nat. Struct. Biol. **7:** 482–485.
18. Huxley, H.E. & W. Brown. 1967. The low-angle x-ray diagram of vertebrate striated muscle and its behaviour during contraction and rigor. J. Mol. Biol. **30:** 383–434.
19. Haselgrove, J.C. 1975. X-ray evidence for conformational changes in the myosin filaments of vertebrate striated muscle. J. Mol. Biol. **92:** 113–143.

20. LINARI, M., G. Piazzesi, I. Dobbie, *et al.* 2000. Interference fine structure and sarcomere length dependence of the axial x-ray pattern from active single muscle fibers. Proc. Natl. Acad. Sci. U.S.A. **97:** 7226–7231.
21. PIAZZESI, G., M. Reconditi, M. Linari, *et al.* 2002. Mechanism of force generation by myosin heads in skeletal muscle. Nature **415:** 659–662.
22. WAKABAYASHI, K., Y. Sugimoto, H. Tanaka, *et al.* 1994. X-ray diffraction evidence for the extensibility of actin and myosin filaments during muscle contraction. Biophys. J. **67:** 2422–2435.
23. HUXLEY, H.E., A. Stewart, H. Sosa & T. Irving. 1994. X-ray diffraction measurements of the extensibility of actin and myosin filaments in contracting muscle. Biophys. J. **67:** 2411–2421.
24. LINARI, M., I. Dobbie, M. Reconditi, *et al.* 1998. The stiffness of skeletal muscle in isometric contraction and rigor: the fraction of myosin heads bound to actin. Biophys. J. **74:** 2459–2473.
25. DOBBIE, I., M. Linari, G. Piazzesi, *et al.* 1998. Elastic bending and active tilting of myosin heads during muscle contraction. Nature **396:** 383–387.
26. RECONDITI, M., M. Linari, L. Lucii, *et al.* 2004. The myosin motor in muscle generates a smaller and slower working stroke at higher load. Nature **428:** 578–581.

Mitochondria and Ischemia/Reperfusion Injury

HENRY M. HONDA, PAAVO KORGE, AND JAMES N. WEISS

Departments of Medicine (Division of Cardiology) and Physiology, David Geffen School of Medicine at the University of California Los Angeles, Los Angeles, California 90095, USA

Abstract: Cardiac ischemia/reperfusion injury results in a variable mixture of apoptotic, necrotic, and normal tissue that depends on both the duration and severity of ischemia. Injury can be abrogated by activation of protective pathways via ischemic and pharmacologic preconditioning. Mitochondria serve as final arbiters of life and death of the cell as these organelles not only are required to generate ATP but also can trigger apoptosis or necrosis. A key mechanism of mitochondrial injury is by the mitochondrial permeability transition (MPT) that has been shown to occur at reperfusion. The article hypothesizes that ischemia/reperfusion promotes MPT in two phases: (1) MPT priming during ischemia occurs as progressive inner mitochondrial membrane leak is accompanied by depressed electron transport in the setting of fatty acid accumulation and loss of cytochrome *c* and antioxidants; and (2) Triggering of MPT at reperfusion is determined by the interplay of mitochondrial membrane potential ($\Delta\Psi_m$) with mitochondrial matrix Ca, reactive oxygen species, and pH. It has been found that strategies that promote mitochondrial recovery such as pharmacologic preconditioning by diazoxide are mediated by K^+-dependent regulation of matrix volume and $\Delta\Psi_m$, resulting in improved efficiency of ATP synthesis as well as prevention of cytochrome *c* loss. If mitochondria fail to recover, then MPT and hypercontracture can result as $\Delta\Psi_m$ depolarization waves regeneratively cross the cell (0.1 to 0.2 μm/s).

Keywords: apoptosis; cytochrome *c*; mitochondrial injury; permeability transition; proton leakage; reperfusion

INTRODUCTION

Each year over 3.8 million men and 3.4 million women die from myocardial infarctions (MI) throughout the world.[1] Thrombolytic therapy and revascularization by percutaneous coronary intervention or coronary artery bypass graft surgery are effective means of treating MI, whereas medical therapy with statins, angiotensin-converting enzyme inhibitors, beta blockers, aspirin, and fish oil, as well as alteration of lifestyle by smoking cessation and control of hypertension and diabetes, can help prevent or delay MI. However, given the health, economic, and personal burden

Address for correspondence: Prof. James N. Weiss, M.D., Chief, Division of Cardiology, 3645 MRL Bldg., David Geffen School of Medicine at UCLA, 675 Charles Young Drive South, Los Angeles, CA 90095-1760, USA. Voice: 310-825-9029; fax: 310-206-5777.
jweiss@mednet.ucla.edu

Ann. N.Y. Acad. Sci. 1047: 248–258 (2005).
doi: 10.1196/annals.1341.022

caused by ischemic heart disease, research into additional treatment modalities is imperative.

Discovery of Ischemic Preconditioning

In 1986, using a canine model, Murry *et al.*[2] pioneered the concept of ischemic preconditioning (IPC), whereby short periods of ischemia decreased the size of a subsequent (index) infarction by 75%. The finding that brief periods of ischemia can protect against a subsequent infarction has since been reproduced in all species from mice to man.[3] The mechanism has been intensively studied and is likely to involve multiple signaling pathways including PKC-ε[4] and GSK-3β.[5] Although the targets of ischemic preconditioning are likely multiple, the key signaling pathways ultimately must converge on recovery of mitochondrial function after infarction and, specifically, on prevention of the mitochondrial permeability transition (MPT). Recently, similar protective effects have been demonstrated with ischemic postconditioning,[6] in which brief periods of ischemia are delivered during reperfusion after the index ischemia. Cardioprotection by postconditioning appears to involve activation of many of the same pathways as in IPC, and suggests that the reperfusion period is a critical period during which cell fate is decided.

MITOCHONDRIA AND ISCHEMIC/REPERFUSION INJURY

Mitochondria are essential for cell survival, both because of their roles as metabolic energy producers and as regulators of programmed cell death. Cardioprotection by IPC must ultimately involve protection of mitochondria from injury, which is supported by the observation that certain drugs acting primarily on mitochondria, such as diazoxide or 5-hydroxydecanoate (5-HD), can mimic or block, respectively, IPC.[7] One of the major factors believed to mediate mitochondrial injury is permeability transition pore (PTP) opening. PTPs are multiprotein complexes that form nonselective pores in the inner mitochondrial membrane (IMM). PTP opening causes immediate depolarization of mitochondrial membrane potential ($\Delta\psi_m$) and transforms mitochondria from ATP producers to ATP consumers by reversing ATP synthase in a futile attempt to maintain $\Delta\psi_m$. In addition, long-lasting PTP opening can lead to excessive water entry into the matrix, matrix swelling, and outer mitochondrial membrane (OMM) rupture. The last causes release of proapoptotic molecules within the intermembrane space (IMS), including cytochrome *c*, apoptosis-inducing factor, Smac/DIABLO, and APAF-1,[8] leading to cell death via both caspase-dependent and caspase-independent mechanisms.

The exact molecular composition of PTP is unknown, but the primary components appear to be the adenine nucleotide transporter in the IMM, voltage-dependent anion channel (VDAC) in the OMM, and cyclophilin D in the matrix, in addition to regulatory components such as the benzodiazepine receptor, hexokinase, and creatine kinase.[8,9] PTPs are Ca^{2+}, redox, voltage, and pH-sensitive such that their open probability is increased by matrix free [Ca], reactive oxygen species (ROS), $\Delta\psi_m$ depolarization, and alkalosis. Other promoters of PTP opening include inorganic phosphate (P_i), various oxidizing and SH-group crosslinking agents, fatty acids (FA), and proapoptotic members of the Bcl family of proteins such as Bax, Bak, Bad, Noxa, Puma, and Bid.[10]

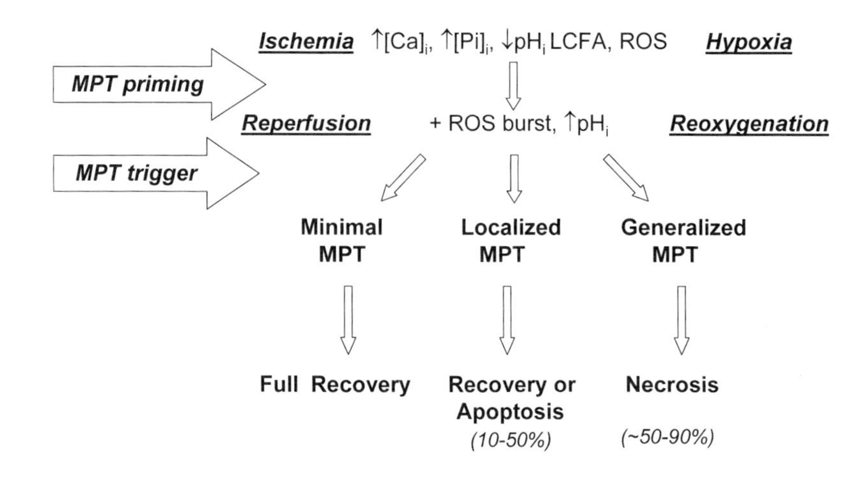

FIGURE 1. Role of the mitochondrial permeability transition (MPT) in determining cell fate after ischemia/reperfusion. Hypothetical role of MPT in cell death during ischemia/reperfusion of the heart. LCFA, long-chain fatty acids; ROS, reactive oxygen species. See text for details. (From Weiss *et al.*[9] Reproduced by permission.)

Crompton *et al.*[11] were the first to propose a pivotal role of MPT in cardiac ischemic/reperfusion injury, and Griffiths and Halestrap subsequently demonstrated that MPT occurs upon reperfusion of the ischemic heart.[12] With reperfusion, mitochondria with depressed electron transport power due to ischemia are placed at high risk for PTP opening due to elevated Ca and P_i, the ROS burst, and clearing of acidosis. The fate of the cell is then determined by the extent of PTP opening. If minimal, the cell may recover and live; if moderate, the cell may undergo programmed cell death even though energy production is adequate; if severe, the cell will die from necrosis due to inadequate energy production (FIG. 1).

We hypothesize that ischemia and reperfusion each make important and separate contributions to the probability that PTP opening will occur during reperfusion. Events during ischemia prime mitochondria for PTP opening by increasing IMM proton leak and depressing electron transport power, compromising their ability to maintain $\Delta\psi_m$. Matrix remodeling can lead to cytochrome *c* release, further weakening electron transport power. The conditions of reperfusion then act as the trigger for PTP opening, depending on the interplay between $\Delta\psi_m$, matrix Ca levels, ROS, and pH.

The MPT Priming Phase

Mitochondria with reduced ability to support $\Delta\psi_m$, due to reduced electron transport power and/or increased IMM leak, are more susceptible to MPT because $\Delta\psi_m$ depolarization directly increases PTP open probability. During ischemia, electron transport power of mitochondria becomes compromised, and IMM leakiness increases. Both free FA and activated FA (e.g., arachidonyl-CoA) accumulate in ischemic myocardium. To investigate their potential role in promoting PTP opening, we performed studies using *in vitro* and *in situ* mitochondria.[13] We found that activated long-chain FA in particular increased susceptibility to MPT. This effect was mediated by a combination of increased IMM leakiness and cytochrome *c* mobilization and release that was independent of MPT. Accompanying cytochrome *c* loss, activated long-chain FA depressed the rate of ATP synthesis and impaired the ability of mitochondria to resist Ca-induced MPT.[13] Cytochrome *c* resides in the IMS, where it is normally bound to the outer leaflet of the IMM and is required by cytochrome oxidase in Complex IV for electron transport. However, only about 15% of cytochrome *c* is in close proximity to the OMM. Approximately 85% is sequestered within the infoldings of the IMM called crista, and is accessible to OMM only via ~20 nm tubular necks of the crista[14] (FIG. 2). After an apoptotic stimulus, the crista remodel, widening their tubular necks to ~60 nm to give the bulk of cytochrome *c* access to the OMM.[14] However, the release of cytochrome *c* into the cytosol is a two-step process. First, it must dissociate from the outer leaflet of the IMM, which is facilitated by oxidation of cardiolipin by ROS.[15] Second, it must pass through the OMM, which can occur via formation of OMM megachannels formed by a combination of Bid, Bax, and cardiolipin,[16] or rupture of the OMM if the matrix volume expands excessively, (e.g., due to PTP opening[17] or closure of VDAC).[18] In the isolated mitochondria in our study, activated long-chain FA induced cytochrome *c* mobilization and release without PTP opening and OMM rupture, since normal mitochondrial function could be restored by adding exogenous cytochrome *c*.[13] The effects of long-chain activated FA were prevented by ROS scavengers, consistent with the finding that long-chain activated FA enhance ROS production, and that ROS are involved in the mobilization and release of cytochrome *c*. However, whether the mechanism involved matrix remodeling and OMM megachannel formation, similar to tBid-mediated cytochrome *c* release,[14] remains to be investigated.

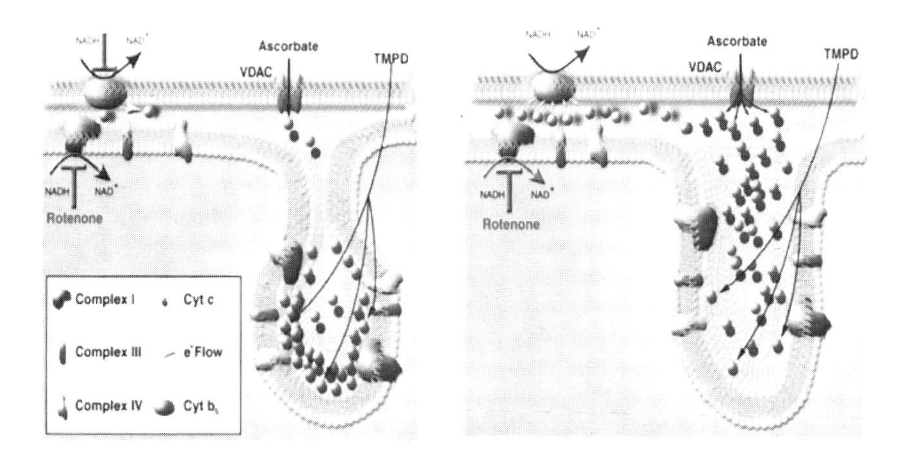

FIGURE 2. Crista remodeling. Cartoon illustrating remodeling of cristae in response to the apoptotic stimuli tBID, allowing cytochrome *c* (red spheres) access to the OMM and subsequent release through OMM megachannels [in color in Annals Online]. (From Scorrano and Korsmeyer.[14] Reproduced by permission.)

The MPT Trigger Phase

During ischemia, acidosis, elevated Mg^{2+}, and depressed electron transport tend to keep PTP closed. However, at reperfusion, the open probability for PTP is increased by elevated Ca^{2+}, P_i, ROS, as well as by clearing of acidosis and decreasing intracellular free Mg^{2+}. Recovery of $\Delta\psi_m$ during reperfusion represents a vulnerable period for mitochondria: although regeneration of $\Delta\psi_m$ decreases the PTP open probability, additional Ca^{2+} can be driven by $\Delta\psi_m$ into the matrix via the uniporter. Thus, if mitochondria are depolarized before they have accumulated significant amounts of Ca^{2+}, then they are protected from subsequent Ca^{2+} uptake and MPT. On the other hand, if mitochondria have already accumulated Ca^{2+}, then $\Delta\psi_m$ depolarization triggers MPT. Therefore, whether MPT occurs during reperfusion is determined by the interplay of the electron transport capacity for regenerating $\Delta\psi_m$ and the conditions present at reperfusion.[19] FIGURE 3 illustrates schematically how these various factors (FA, ROS, matrix volume, $\Delta\psi_m$) may all interact to promote PTP opening.

ROLE OF MITOCHONDRIAL ATP-SENSITIVE K CHANNELS IN CARDIOPROTECTION

The importance of mitochondrial recovery in IPC was emphasized when it was demonstrated early on that ATP generation after ischemia/reperfusion was better in protected than unprotected hearts.[20] Interest in the role of mitochondria in IPC was rekindled when it was demonstrated that drugs that open mitochondrial ATP-sensitive K (mitoK_{ATP}) channels mimic the protective effects of IPC, and drugs that block mitoK_{ATP} channels also blocked IPC. Although the role of nonselective effects of these drugs in mimicking and blocking IPC remains a controversial topic, mitoK_{ATP} channels recently have been successfully reconstituted in lipid bilayers[21] as part of a multiprotein signaling complex.[22]

At least three mechanisms have been postulated to explain the cardioprotective effects of opening mitoK_{ATP} channels: (1) Opening of mitoK_{ATP} channels leads to a depolarization of $\Delta\psi_m$, decreasing the electrochemical gradient for Ca^{2+} entry during reperfusion;[23,24] (2) opening of mitoK_{ATP} channels promotes matrix swelling and helps to maintain OMM-IMM contact sites and matrix integrity required for efficient electron transport;[25] and (3) opening of mitoK_{ATP} channels during preconditioning induces low levels of ROS production that triggers cardioprotection, and during ischemia/reperfusion limits subsequent higher levels of ROS production, thus helping to prevent MPT.[19,26]

In our investigation of these mechanisms, we have found that all three have merit, and all three influence both the MPT priming and trigger phases (FIG. 3). Diazoxide attenuated cytochrome *c* loss in response to activated long-chain FA, probably by attenuating ROS production through mild $\Delta\psi_m$ depolarization,[19] as ROS production is highly sensitive to $\Delta\psi_m$.[27,28] We also found that diazoxide attenuated severe matrix condensation and $\Delta\psi_m$ dissipation during ADP phosphorylation in mitochondria subjected to anoxia-reoxygenation.[29] Both of these effects are very relevant to the MPT priming phase, by helping to preserve electron transport power and the ability to maintain $\Delta\psi_m$ so that PTP opening can be avoided during reperfusion. We also found that diazoxide protected against the MPT trigger phase. In isolated mitochondria

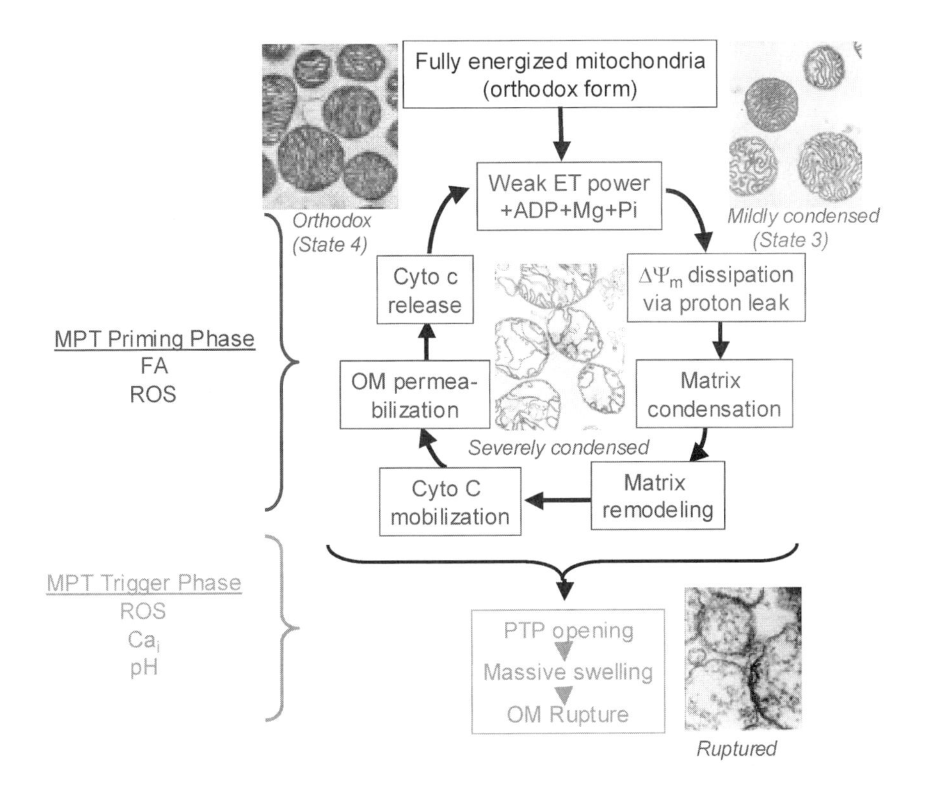

FIGURE 3. Hypothetical schema of events promoting mitochondrial permeability transition (MPT) during ischemia/reperfusion. With weakened electron transport power and increased inner mitochondrial membrane (IMM) leak during ischemia, mitochondria are progressively less able to resist severe matrix shrinkage, which promotes matrix remodeling leading to cytochrome *c* release, further weakening electron transport power in a vicious cycle. These events during the MPT priming phase leave mitochondria progressively more susceptible to MPT during reperfusion (the MPT trigger phase). Representative electron micrographs of isolated mitochondria during these phases are shown as inserts. Diazoxide is protective by preventing severe matrix shrinkage and resisting matrix remodeling leading to cytochrome *c* loss and loss of electron transport power during the MPT priming phase. Diazoxide also protects mitochondria against the MPT trigger components during reperfusion by limiting matrix Ca^{2+} accumulation. FA, fatty acid; OM; outer membrane; ROS, reactive oxygen species [in color in Annals Online].

with reduced electron transport power due to either anoxia/reoxygenation or low substrate availability, diazoxide partially dissipated $\Delta\psi_m$, preventing matrix Ca^{2+} loading and thereby averting PTP opening.[19] In contrast, diazoxide had very little effect on $\Delta\psi_m$ in fully energized mitochondria, because electron transport power readily compensated for the K^+ leak induced by diazoxide. Thus, diazoxide has an ideal profile for therapeutic efficacy—very little effect on normal, well-oxygenated, fully energized mitochondria, but prominent protective effects when mitochondria are weakened by ischemia conditions.

MITOCHONDRIAL $\Delta\psi_m$ WAVES AND ISCHEMIA/REPERFUSION INJURY

Under conditions in which the matrix is Ca^{2+} overloaded, PTP can act as Ca^{2+} induced Ca^{2+} release channels. When PTP opening occurs locally and dissipates $\Delta\psi_m$, the accumulated matrix Ca^{2+} efflux flows down its electrochemical gradient from matrix to cytoplasm, where it is available to be taken up by adjacent mitochondria, potentially triggering PTP opening in a regenerative fashion, as has been demonstrated *in vitro* gel-immobilized mitochondria,[30] as well as in isolated cardiac myocytes.[26] We found that after reoxygenation of anoxic isolated intact rabbit ventricular myocytes loaded with tetramethylrhodamine methyl ester (TMRM) to monitor $\Delta\psi_m$, reversible $\Delta\psi_m$ depolarization waves, both focal and global, were observed. This was ultimately followed by a final irreversible $\Delta\psi_m$ depolarization wave that crossed from one end to the other at a velocity of approximately 0.1-0.2 μm/s and was associated with hypercontracture of the cell (FIG. 4). Both the reversible and irreversible waves were prevented by the cell permeant, free-radical scavengers Trolox.

The reversible $\Delta\psi_m$ depolarization waves were also stimulated by increased laser intensity, which is known to generate ROS, and cause ROS-induced ROS waves.[31,32]

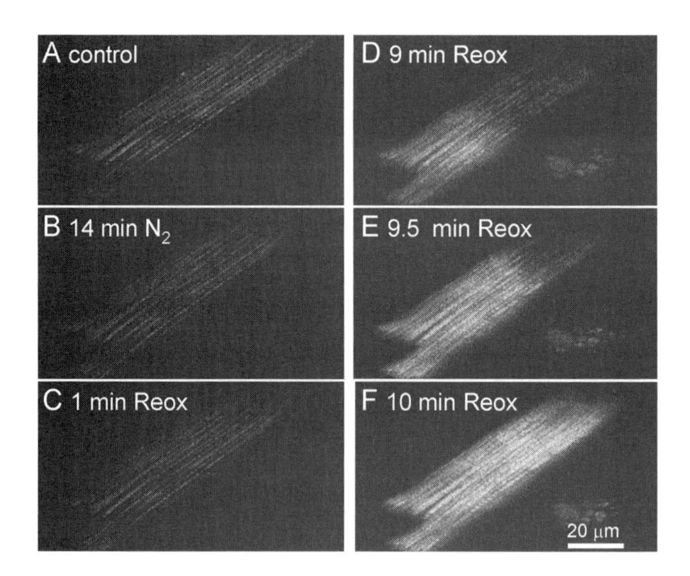

FIGURE 4. $\Delta\psi_m$ depolarization waves during anoxia/reoxygenation in isolated rabbit ventricular myocytes. Confocal images of the fluorescence $\Delta\psi_m$ indicator tetramethylrhodamine methyl ester (TMRM) under aerobic conditions (**A**), after 14 min of anoxia (**B**), after 1 min (**C**), and after 9 to 10 min of reoxygenation (**D–F**). The striped mitochondrial pattern of TMRM fluorescence was maintained during hypoxia. Between 1 and 9 min of reoxygenation, several reversible $\Delta\psi_m$ depolarization waves were observed (not shown). Between 9 and 10 min after reoxygenation, an irreversible $\Delta\psi_m$ depolarization wave (indicated by diffuse bright fluorescence as TMRM leaves the matrix and is unquenched) propagates slowly through the cell, associated ultimately with hypercontracture.

To determine whether the ROS-induced ROS waves were associated with MPT, mitochondria *in situ* were dually loaded with both TMRM and calcein-acetoxymethyl (AM) ester (using Co to quench cytosolic calcein signal).[33] Once trapped in the matrix, calcein (molecular weight [MW] 623) cannot escape unless PTP opening occurs. Under control conditions, the typical striped pattern of mitochondrial loading with both TMRM and calcein was seen (FIG. 5). During the reversible $\Delta\psi_m$ depolarization waves, calcein fluorescence brightened, due to its unquenching when TMRM was released, but remained localized to the mitochondria. After the final irreversible $\Delta\psi_m$ wave, however, calcein fluorescence also diffusely distributed itself throughout the cell. These findings are consistent with the idea that regenerative ROS-induced ROS waves progressively reduce the ability of mitochondria to resist PTP opening, ultimately promoting a final irreversible wave associated with MPT. One possibility that we are currently evaluating is that the ROS-induced ROS waves lead to progres-

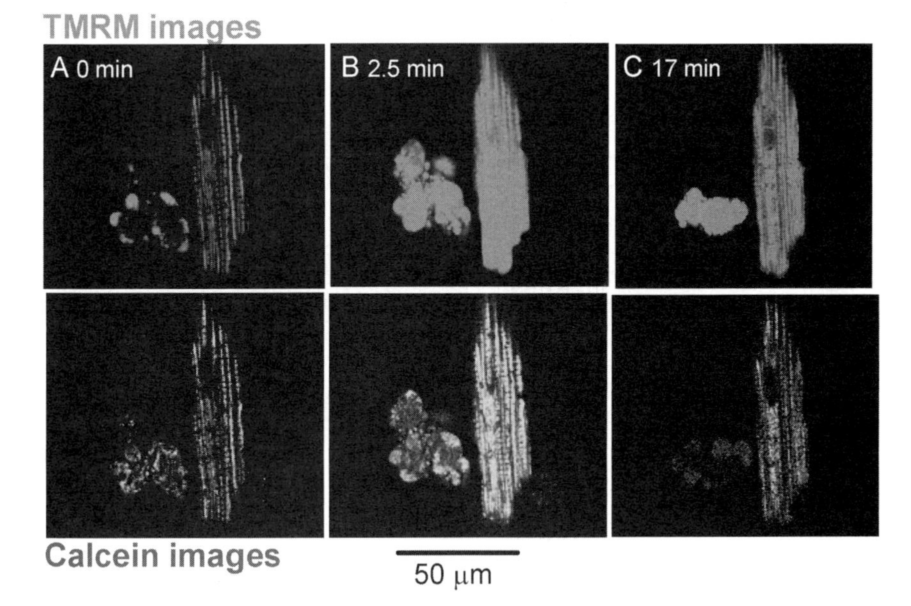

FIGURE 5. Mitochondrial permeability transition (MPT) after $\Delta\psi_m$ depolarization waves, induced by high [Ca^{2+}] and laser-generated reactive oxygen species (ROS) in isolated rabbit ventricular myocytes. Confocal images of a rabbit ventricular myocyte dually loaded with TMRM (*top panels*) to measure $\Delta\psi_m$ and calcein- acetoxymethyl (AM) to monitor inner mitochondrial membrane (IMM) permeability (*lower panels*). (**A**) The control images (*left*) show the typical striped mitochondrial distribution of both TMRM (*upper*) and calcein (*lower*). (**B**) After exposure to high Ca^{2+} and laser light, several reversible $\Delta\psi_m$ depolarization waves occurred, followed by a final irreversible $\Delta\psi_m$ depolarization wave. At this point, the TMRM image showed diffuse bright fluorescence representing unquenching of TMRM after its release into the cytoplasm. The calcein image was also brighter due to its unquenching by TMRM, but still retained a mitochondrial distribution. (**C**) By 17 min, however, the calcein image (*lower right*) was dimmer and diffuse, indicating release of calcein into the cytoplasm and quenching by cytoplasmic TMRM. Calcein (molecular weight 623) release into the cytoplasm is indicative of MPT [in color in Annals Online].

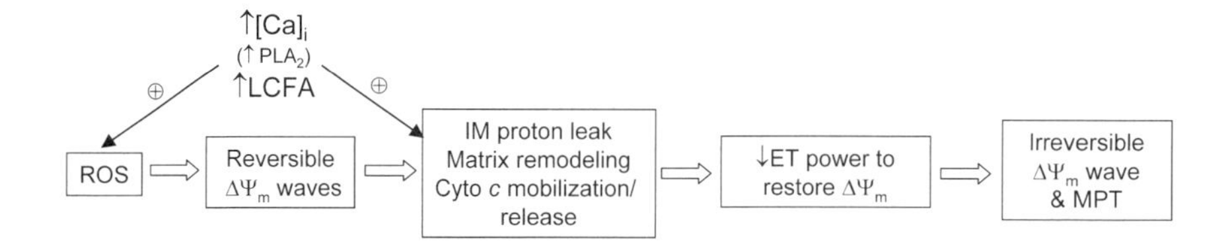

FIGURE 6. Proposed schema of events relating reactive oxygen species–induced $\Delta\psi_m$ depolarization waves to a final irreversible wave associated with mitochondrial permeability transition (MPT). ET, electronic transport; IM, inner mitochondrial membrane; LCFA, long-chain fatty acid; ROS, reactive oxygen species. See text for details.

sive cytochrome *c* mobilization and release, leaving mitochondria with progressively weakened electron transport power and increased susceptibility to PTP opening after each successive wave (FIG. 6).

SUMMARY

Cardiac ischemia/reperfusion results in a variable mixture of injury due to both apoptosis and necrosis that depends on the duration and severity of ischemia, and it can be abrogated by activation of protective pathways during IPC and exposure to various pharmacologic interventions. The targets of ischemic preconditioning are likely multiple but may converge on prevention of MPT after ischemia/reperfusion, thereby preserving mitochondrial function. We postulate that ischemia/reperfusion promotes MPT in two stages: (1) the MPT *priming* phase during ischemia, followed by (2) the MPT *triggering* phase during reperfusion. Drugs that activate mitoK_{ATP} channels exert protective effects on both phases. The mitochondrial network also represents an excitable media that can propagate regenerative waves ($\Delta\psi_m$, Ca,[30,34] ROS[31]) that spread across the entire cell.

ACKNOWLEDGMENTS

This work is supported by the National Institutes of Health (P50 HL-52319, RO1 HL071870-01, P01 HL080111-01), the Laubisch Fund, and Kawata Endowments. The authors thank Tan Duong for technical assistance.

REFERENCES

1. Deaths from coronary heart disease. www.who.int/cardiovascular_diseases/en/cvd_atlas_14_deathHD.pdf. Accessed January 5, 2005.
2. MURRY, C.E., R.B. JENNINGS & K.A. REIMER. 1986. Preconditioning with ischemia: a delay of lethal cell injury in ischemic myocardium. Circulation **74:** 1124–1136.
3. KLONER, R.A. & R.B. JENNINGS. 2001. Consequences of brief ischemia: stunning, preconditioning, and their clinical implications: part 2. Circulation **104:** 3158–3167.
4. VONDRISKA, T.M., J.B. KLEIN & P. PING. 2001. Use of functional proteomics to investigate PKC epsilon-mediated cardioprotection: the signaling module hypothesis. Am. J. Physiol. Heart Circ. Physiol. **280:** H1434–H1441.

5. JUHASZOVA, M., D.B. ZOROV, S.H. KIM, *et al.* 2004. Glycogen synthase kinase-3beta mediates convergence of protection signaling to inhibit the mitochondrial permeability transition pore. J. Clin. Invest. **113:** 1535–1549.
6. ZHAO, Z.Q., J.S. CORVERA, M.E. HALKOS, *et al.* 2003. Inhibition of myocardial injury by ischemic postconditioning during reperfusion: comparison with ischemic preconditioning. Am. J. Physiol. Heart Circ. Physiol. **285:** H579–H588.
7. SATO, T. & E. MARBAN. 2000. The role of mitochondrial K(ATP) channels in cardioprotection. Basic Res. Cardiol. **95:** 285–289.
8. PATTERSON, S.D., C.S. SPAHR, E. DAUGAS, *et al.* 2000. Mass spectrometric identification of proteins released from mitochondria undergoing permeability transition. Cell Death Differ. **7:** 137–144.
9. WEISS, J.N., P. KORGE, H.M. HONDA & P. PING. 2003. Role of the mitochondrial permeability transition in myocardial disease. Circ. Res. **93:** 292–301.
10. NEWMEYER, D.D. & S. FERGUSON-MILLER. 2003. Mitochondria: releasing power for life and unleashing the machineries of death. Cell **112:** 481–490.
11. CROMPTON, M., A. COSTI & L. HAYAT. 1987. Evidence for the presence of a reversible Ca2+-dependent pore activated by oxidative stress in heart mitochondria. Biochem. J. **245:** 915–918.
12. GRIFFITHS, E.J. & A.P. HALESTRAP. 1995. Mitochondrial non-specific pores remain closed during cardiac ischaemia, but open upon reperfusion. Biochem. J. **307:** 93–98.
13. Korge, P., H.M. Honda & J.N. Weiss. 2003. Effects of fatty acids in isolated mitochondria: implications for ischemic injury and cardioprotection. Am. J. Physiol. Heart Circ. Physiol. **285:** H259–H269.
14. SCORRANO, L., M. ASHIYA, K. BUTTLE, *et al.* 2002. A distinct pathway remodels mitochondrial cristae and mobilizes cytochrome c during apoptosis. Dev. Cell **2:** 55–67.
15. OTT, M., J.D. ROBERTSON, V. GOGVADZE, *et al.* 2002. Cytochrome c release from mitochondria proceeds by a two-step process. Proc. Natl. Acad. Sci. U.S.A. **99:** 1259–1263.
16. KUWANA, T., M.R. MACKEY, G. PERKINS, *et al.* 2002. Bid, Bax, and lipids cooperate to form supramolecular openings in the outer mitochondrial membrane. Cell **111:** 331–342.
17. PETIT, P.X., M. GOUBERN, P. DIOLEZ, *et al.* 1998. Disruption of the outer mitochondrial membrane as a result of large amplitude swelling: the impact of irreversible permeability transition. FEBS Lett. **426:** 111–116.
18. VAN DER HEIDEN, M.G., N.S. CHANDEL, P. T. SCHUMACKER & C. B. THOMPSON. 1999. Bcl-xL prevents cell death following growth factor withdrawal by facilitating mitochondrial ATP/ADP exchange. Mol. Cell **3:** 159–167.
19. KORGE, P., H.M. HONDA & J.N. WEISS. 2002. Protection of cardiac mitochondria by diazoxide and protein kinase C: implications for ischemic preconditioning. Proc. Natl. Acad. Sci. U.S.A. **99:** 3312–3317.
20. MURRY, C.E., V.J. RICHARD, K.A. REIMER & R.B. JENNINGS. 1990. Ischemic preconditioning slows energy metabolism and delays ultrastructural damage during a sustained ischemic episode. Circ. Res. **66:** 913–931.
21. ZHANG, D.X., Y.F. CHEN, W.B. CAMPBELL, *et al.* 2001. Characteristics and superoxide-induced activation of reconstituted myocardial mitochondrial ATP-sensitive potassium channels. Circ. Res. **89:** 1177–1183.
22. ARDEHALI, H., Z. CHEN, Y. KO, *et al.* 2004. Multiprotein complex containing succinate dehydrogenase confers mitochondrial ATP-sensitive K+ channel activity. Proc. Natl. Acad. Sci. U.S.A. **101:** 11880–11885.
23. HOLMUHAMEDOV, E.L., L. WANG & A. TERZIC. 1999. ATP-sensitive K+ channel openers prevent Ca2+ overload in rat cardiac mitochondria. J. Physiol. **519:** 347–360.
24. LIU, Y., T. SATO, J. SEHARASEYON, *et al.* 1999. Mitochondrial ATP-dependent potassium channels. Viable candidate effectors of ischemic preconditioning. Ann. N.Y. Acad. Sci. **874:** 27–37.
25. DOS SANTOS, P., A.J. KOWALTOWSKI, M.N. LACLAU, *et al.* 2002. Mechanisms by which opening the mitochondrial ATP-sensitive K(+) channel protects the ischemic heart. Am. J. Physiol. Heart Circ. Physiol. **283:** H284–H295.
26. HAUSENLOY, D.J. & D.M. YELLON. 2004. The mitochondrial permeability transition pore in myocardial preconditioning. Cardiovasc. J. S. Afr. **15**(Suppl 1): S5.

27. KORSHUNOV, S.S., O.V. KORKINA, E.K. RUUGE, *et al.* 1998. Fatty acids as natural uncouplers preventing generation of O2.- and H2O2 by mitochondria in the resting state. FEBS Lett. **435:** 215–218.
28. KORSHUNOV, S.S., V.P. SKULACHEV & A.A. STARKOV. 1997. High protonic potential actuates a mechanism of production of reactive oxygen species in mitochondria. FEBS Lett. **416:** 15–18.
29. KORGE, P., HONDA, H. M.,WEISS, J. N. 2005. K+ dependent regulation of matrix volume improves mitochondrial function under conditions mimicking ischemia-reperfusion. Am J Physiol Heart Circ Physiol. In press.
30. ICHAS, F., L.S. JOUAVILLE & J.P. MAZAT. 1997. Mitochondria are excitable organelles capable of generating and conveying electrical and calcium signals. Cell **89:** 1145–1153.
31. ZOROV, D.B., C.R. FILBURN, L.O. KLOTZ, *et al.* 2000. Reactive oxygen species (ROS)-induced ROS release: a new phenomenon accompanying induction of the mitochondrial permeability transition in cardiac myocytes. J. Exp. Med. **192:** 1001–1014.
32. AON, M.A., S. CORTASSA,, E. MARBAN & B. O'ROURKE. 2003. Synchronized whole cell oscillations in mitochondrial metabolism triggered by a local release of reactive oxygen species in cardiac myocytes. J. Biol. Chem. **278:** 44735–44744.
33. PETRONILLI, V., D. PENZO, L. SCORRANO, *et al.* 2001. The mitochondrial permeability transition, release of cytochrome c and cell death. Correlation with the duration of pore openings in situ. J. Biol. Chem. **276:** 12030–12034.
34. PACHER, P. & G. HAJNOCZKY. 2001. Propagation of the apoptotic signal by mitochondrial waves. EMBO J. **20:** 4107–4121.

Regulation of Cardiac Energetics

Role of Redox State and Cellular Compartmentation during Ischemia

MARCO E. CABRERA,[a,b,c,d] LUFANG ZHOU,[b,d] WILLIAM C. STANLEY,[c,d] AND GERALD M. SAIDEL[b,d]

[a] *Department of Pediatrics, Case Western Reserve University, Cleveland, Ohio 44106-6011, USA*

[b] *Department of Biomedical Engineering, Case Western Reserve University, Cleveland, Ohio 44106-6011, USA*

[c] *Department of Physiology and Biophysics, Case Western Reserve University, Cleveland, Ohio 44106-6011, USA*

[d] *Center for Modeling Integrated Metabolic Systems, Case Western Reserve University, Cleveland, Ohio 44106-6011, USA*

ABSTRACT: The heart is capable of altering its metabolic rate during exercise or ischemia. Under most state transitions, the heart maintains the concentration of adenosine triphosphate (ATP) at relatively constant values, in spite of large fluctuations in metabolic rate or in the delivery of fuels and oxygen. However, the mechanisms responsible for the regulation of cardiac energetics under conditions of increased demand or reduced supply are still under debate. To improve quantitative understanding of the regulation of glycolysis and oxidative phosphorylation under physiological and pathological conditions, it is essential to assess the dynamics of cytosolic and mitochondrial nicotinamide adenine dinucleotide (NAD^+) and its reduced form (NADH) during stress (e.g., ischemia, exercise). However, at present there are no reliable methods to measure the dynamics of redox state *in vivo* in these subcellular compartments. In the present study, computer simulations with a mathematical model of myocardial energy metabolism are used to investigate the role of cytosolic and mitochondrial redox states in regulating cardiac energetics during reduced myocardial blood flow.

KEYWORDS: computer simulations; mathematical model; metabolic regulation

INTRODUCTION

The heart requires a considerable amount of energy per unit time to sustain ionic homeostasis, protein synthesis, and muscular contraction.[1] The heart has also the

Address for correspondence: A. Prof. Marco E. Cabrera, Ph.D., Departments of Pediatrics, Biomedical Engineering, and Physiology and Biophysics, Case Western Reserve University, 11100 Euclid Avenue, RBC-389, Cleveland, OH 44106-6011, USA. Voice: 216-844-5085; fax: 216-844-5478.
marco.cabrera@cwru.edu

**Ann. N.Y. Acad. Sci. 1047: 259–270 (2005). © 2005 New York Academy of Sciences.
doi: 10.1196/annals.1341.023**

capacity to respond quickly to changes in energy demand. It can either increase its metabolic rate several fold during exercise or decrease it gradually during ischemia.[2] Under most state transitions, the heart maintains the adenosine triphosphate (ATP) concentration at an essentially constant level—in spite of large fluctuations of metabolic rate—by matching its rates of ATP utilization and production in a coordinated manner.[3,4] During increased demand, ATP synthesis is powered by the proton-motive force across the inner mitochondrial membrane generated by the oxidation of reducing equivalents derived from the oxidation of glucose, fatty acids, lactate, and ketones.[5] During myocardial ischemia, fuel metabolism is altered dramatically and so is the regulation of the pathways of ATP synthesis. However, the mechanisms responsible for the regulation of cardiac energetics under conditions of increased demand or reduced supply are still under debate.

In the short term, the regulation of cardiac energetics can be studied by investigating the effects of a stress (e.g., ischemia, exercise) on the biochemical species participating in the basic chemical equation for oxidative phosphorylation.[6] Key species participating in energy transfer are nicotinamide adenine dinucleotide (NAD^+) and its reduced form (NADH). Accordingly, at a given rate of ATP hydrolysis (demand), oxygen delivery to the myocardium, the rate of NADH supply from substrate oxidation, the rate of translocation of adenosine diphosphate (ADP) and Pi to the mitochondria, and the rate of ATP formation and translocation to the cytosol, all play a role in regulating cardiac energetics.[7] The dynamics of the chemical species involved in oxidative phosphorylation (NADH, NAD^+, ADP, ATP, Pi, O_2) during an acute stress can thus provide insights into mechanisms of myocardial metabolic regulation. However, biochemical and anatomical analysis of myocardium shows major concentration differences between cytosolic and mitochondrial NADH, NAD^+, ADP, and ATP.[8]

Specifically, the redox states ($NADH/NAD^+$) of the cytosol and mitochondria do not necessarily have the same values nor vary in parallel.[9,10] In addition, the rates of glycolysis and lactate formation/utilization, as well as the rates of pyruvate and fatty acid oxidation and oxidative phosphorylation are modulated by cytosolic and mitochondrial redox changes, respectively.[11,12] Thus, in order to understand the regulation of glycolysis and oxidative phosphorylation under physiological and pathological conditions, it is essential to assess the dynamics of cytosolic and mitochondrial NADH and NAD^+ during a metabolic stress. However, at present there are no reliable methods to measure the dynamics of redox state *in vivo*, with sufficient time resolution, in these subcellular compartments.[13–16] An alternative to experimentation to investigate the roles of factors affecting ATP homeostasis is to use mathematical models of cardiac energy metabolism and computer simulations.[17–20] In the present study, computer simulations with a mathematical model of myocardial energy metabolism are used to investigate the role of cytosolic and mitochondrial redox states in regulating cardiac energetics.

METHODS

Our approach in studying the regulation of cardiac energetics consists of combining computer simulations with a mathematical model of cardiac energy metabolism and *in vivo* experimental measurements in large mammals (e.g., swine) for model validation. To simulate myocardial metabolism during ischemia, the model represents the heart as a perfused tissue with a characteristic myocyte having distinct cytosolic and mitochondrial compartments. In the perfused tissue, convective mass transport occurs in the blood capillaries, while passive transport occurs between plasma and interstitial fluid. Since the characteristic time of capillary interstitial transport is much shorter than that of convection through the capillaries, local chemical equilibrium between plasma and interstitial fluid can be assumed. Therefore, we considered the perfused tissue to consist of a blood ("blood" = plasma + interstitial fluid), a cytosolic, and a mitochondrial compartment. Transport between any two adjacent compartments occurs via passive diffusion or carrier-mediated transport.[21–23] In general, the concentration dynamics of chemical species in blood, cytosol, and mitochondria are described mathematically by mass balances including convection, passive and carrier-mediated diffusion, and distinct chemical reactions involved in energy metabolism. The concentration dynamics for species j in the blood compartment, $C_{bj}(t)$, is represented by

$$V_b \frac{dC_{bj}}{dt} = MBF \cdot (C_{aj} - C_{vj}) - J_{b-c,j} , \tag{1}$$

where V_b is the volume of the blood compartment, MBF is myocardial blood flow, $C_{aj}(t)$ and $C_{vj}(t)$ are the arterial and venous concentration of species j, respectively, and $J_{b-c,j}$ is the mass transport rate from blood to cytosol.

In the cytosol, the concentration of species j, C_{cj}, may change due to reaction processes and mass transfer across the cellular and mitochondrial membranes:

$$V_{cj} \frac{dC_{cj}}{dt} = R_{cj} + J_{b-c,j} - J_{c-m,j} , \tag{2}$$

where V_{cj} is the effective volume of species j in the cytosol, R_{cj} is the net reaction rate of species j, and $J_{c-m,j}$ is the mass transport rate from cytosol to mitochondria. In the mitochondria, the dynamic mass balance equation for species j is

$$V_{mj} \frac{dC_{mj}}{dt} = R_{mj} + J_{c-m,j} , \tag{3}$$

where V_{mj} is the mitochondrial effective volume of species j, C_{mj} its concentration, and R_{mj} its net reaction rate. Detailed description of model equations, biochemical species, and metabolic processes included in this validated model of cardiac energy metabolism can be found in a recent article.[24]

To quantitatively investigate the effect of MBF reductions on energy metabolism and the potential role of redox state in mediating these responses, MBF was reduced from 1.0 $ml \cdot g^{-1} \cdot min^{-1}$ (control) in a step-like manner. The 60 s transition from normal to reduced blood flow was implemented using an exponential function. Accordingly, four sets of simulations were performed, in which MBF was reduced from its initial value to 0.8, 0.6, 0.4, and 0.2 $ml \cdot g^{-1} \cdot min^{-1}$ to simulate various degrees of ischemia. In this study, simulations corresponding to mild (0.6 $ml \cdot g^{-1} \cdot min^{-1}$), moderate (0.4 $ml \cdot g^{-1} \cdot min^{-1}$), and severe (0.2 $ml \cdot g^{-1} \cdot min^{-1}$) ischemia are presented and analyzed. Because our overall goal was to investigate the role of redox state on the regulation of the pathways of ATP synthesis, we simulated the time profiles of (a) myocardial oxygen consumption, (b) mitochondrial and cytosolic $NADH/NAD^+$, (c) the rates of glycogen breakdown, (d) tissue lactate concentrations and lactate-to-pyruvate ratios, and (e) lactate uptake during 65 min of mild, moderate, and severe ischemia corresponding to a 40%, 60%, and 80% reduction in MBF from its normal value of 1.0 $ml \cdot g^{-1} \cdot min^{-1}$.

SIMULATION RESULTS

Myocardial Oxygen Consumption

Flow reductions smaller than 0.2 $ml \cdot g^{-1} \cdot min^{-1}$ (<20%) from normal MBF did not alter myocardial oxygen consumption (MVO_2) significantly (<4%). Larger reductions in MBF resulted in a rapid ($\tau = 54$ s) step-like decrease in MVO_2 (FIG. 1A). The magnitude of the steady-state drop in MVO_2 (ΔMVO_2) with respect to the reduction in MBF (ΔMBF) increased with decreasing MBF (FIG. 2A). Specifically, $\Delta MVO_2/\Delta MBF$ (μmol/ml) was 2.5 at 0.6 $ml \cdot g^{-1} \cdot min^{-1}$, 6.0 at 0.4 $ml \cdot g^{-1} \cdot min^{-1}$, and 12.0 at 0.2 $ml \cdot g^{-1} \cdot min^{-1}$.

Glycogen Breakdown

The net rate of glycogen breakdown displayed a biphasic behavior. First, it increased (phase I) very rapidly to peak values that where linearly related to the degree of MBF reduction and then decreased (phase II) exponentially toward zero. The time constant characterizing the decrease in the net rate of glycogen breakdown in phase II and the magnitude of the decrease were inversely and directly related to the magnitude of MBF reduction, respectively (FIG. 1B, FIG. 2B).

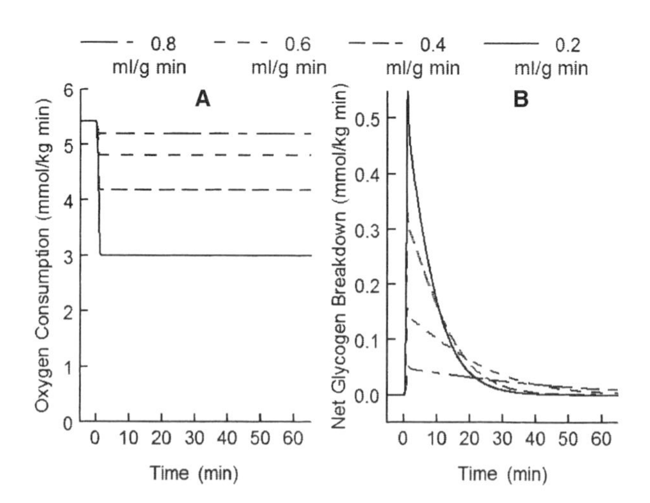

FIGURE 1. Dynamics of myocardial rates of (**A**) oxygen consumption and (**B**) net glycogen breakdown from computer simulations of reductions (0.8, 0.6, 0.4, and 0.2 $ml \cdot g^{-1} \cdot min^{-1}$) in myocardial blood flow.

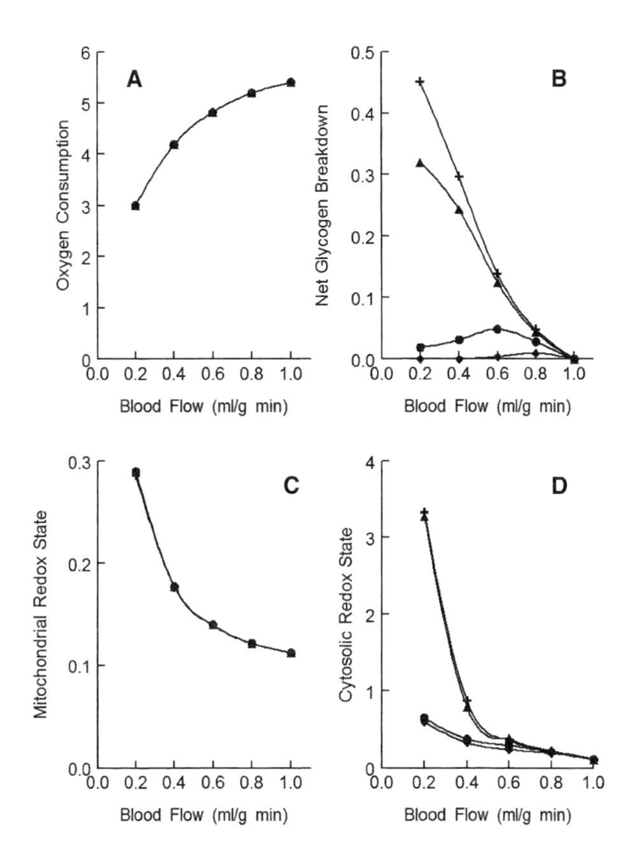

FIGURE 2. Computer-simulated values for (**A**) myocardial oxygen consumption, $mmol \cdot kg^{-1} \cdot min^{-1}$; (**B**) net glycogen breakdown, $mmol \cdot kg^{-1} \cdot min^{-1}$; (**C**) mitochondrial redox state; and (**D**) cytosolic redox state; after 7 min (+), 10 min (*closed triangles*), 30 min (*closed circles*), and 65 min (*closed diamonds*) of the onset of flow reduction plotted as a function of myocardial blood flow (MBF, $ml \cdot g^{-1} \cdot min^{-1}$).

Mitochondrial and Cytosolic Redox State

Mitochondrial redox state ($NADH_m/NAD^+_m$) increased almost as rapidly (τ = 70 s) as MVO_2 decreased and closely mirrored its response (FIG. 3A). Cytosolic redox state ($NADH_c/NAD^+_c$) on the other hand, only followed a pattern similar to that of mitochondrial redox state during mild ischemia, although with a slightly biphasic behavior (FIG. 3B). However, with moderate and severe ischemia, cytosolic redox state clearly displayed a biphasic behavior increasing several fold (phase I) over control values (initial steady state) and then decreasing toward its new steady state (phase II), which was reached at around 30 min (final steady state). However, the magnitude of the increase in redox state in phase I was greatly dependent on the degree of MBF reduction (FIG. 2). The larger the magnitude of MBF reduction from the control value, the larger and sharper the increase in cytosolic redox state in phase I (FIG. 3B). While reducing MBF from 0.6 to 0.4 $ml \cdot g^{-1} \cdot min^{-1}$ resulted in a three-

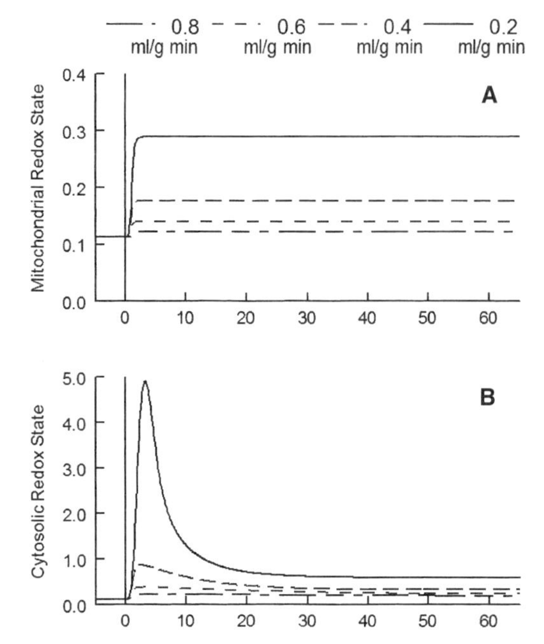

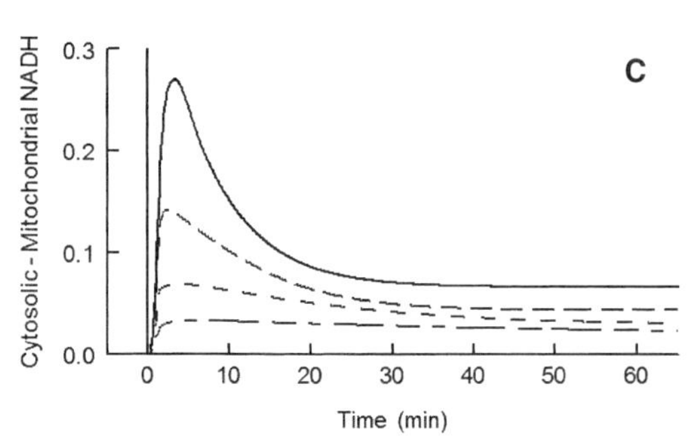

FIGURE 3. Time courses of (**A**) mitochondrial redox state, $[NADH]_m/[NAD^+]_m$; (**B**) cytosolic redox state, $[NADH]_c/[NAD^+]_c$; and (**C**) cytosolic minus mitochondrial [NADH], $mmol \cdot kg^{-1}$ wet weight; from computer simulations of reductions (0.8, 0.6, 0.4, and 0.2 $ml \cdot g^{-1} \cdot min^{-1}$) in myocardial blood flow.

fold increase in redox state from 0.3 to 0.9, an MBF reduction from 0.4 to 0.2 $ml \cdot g^{-1} \cdot min^{-1}$ resulted in over a five-fold increase in cytosolic redox state from 0.9 to 4.9. Similarly, the larger the magnitude of the MBF reduction, the faster the decrease in cytosolic redox state in phase II (FIG. 3B). Thus, the main differences in the responses to ischemia between mitochondrial and cytosolic redox state were: (a) the monophasic versus the biphasic nature of the response and (b) the time taken to reach a steady state (~3 min vs. ~30 min), respectively.

Mitochondrial NADH concentration increased with MBF reduction from 0.045 mM (control) to 0.055 mM (mild), 0.067 mM (moderate), and 0.099 mM (severe). Cytosolic NADH concentration also increased with MBF reduction from 0.045 mM (control) to 0.086 mM (mild), 0.11 mM (moderate), and 0.17 mM (severe). In general, the increase in NADH concentration was larger in the cytosol (FIG. 3C). Since the NAD concentrations in these compartments mirrored the corresponding changes in NADH, the differences in redox state between these two compartments increased (mild: 0.24 vs. 0.14; moderate: 0.33 vs. 0.17; and severe: 0.60 vs. 0.29) with increased blood flow reduction, respectively.

Lactate Metabolism

Myocardial lactate concentration and lactate-to-pyruvate ratio displayed similar behavior, both increasing several fold over the first 5 min of MBF reduction (phase I) and then decreasing in a nearly exponential fashion toward a steady state (phase II), which was reached in 35 to 45 min (FIG. 4). Lactate-to-pyruvate ratio (L/P) displayed a larger increase in phase I and reached a steady state faster than myocardial lactate concentration in phase II. While lactate concentration increases linearly with the degree of blood blow reduction, the lactate-to-pyruvate ratio increases nonlinearly, showing higher L/P values with more severe ischemia (FIG. 4B). The behavior of the lactate-to-pyruvate ratio responses to MBF reduction closely resembles that of the cytosolic redox state responses, but with larger time constants both during phase I and phase II.

Finally, the rate of lactate uptake displayed both a biphasic response and a switch from net lactate uptake to net lactate release and vice versa (FIG. 4C). The changes

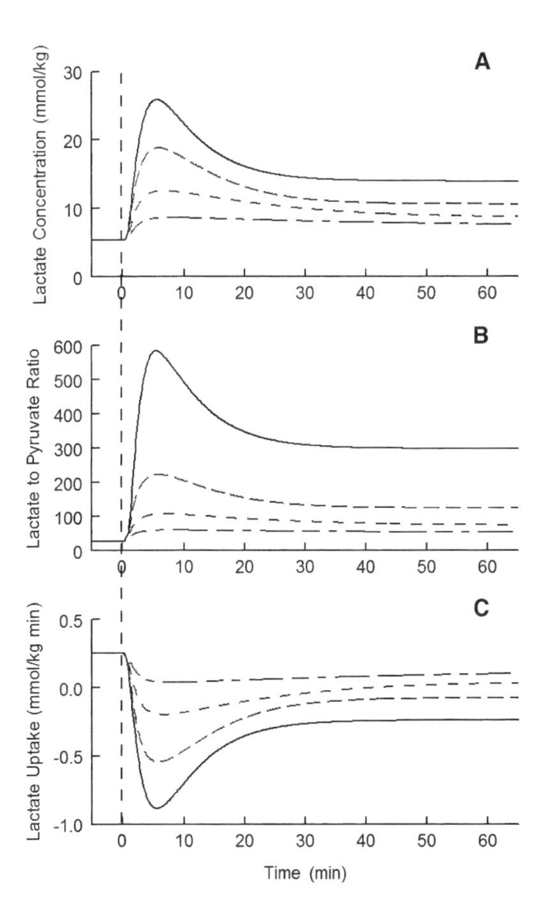

FIGURE 4. Dynamic responses of (**A**) myocardial lactate concentration, (**B**) lactate to pyruvate ratio, and (**C**) net rate of lactate uptake or release, to step changes in myocardial blood flow from the normal control value of 1.0 ml $\cdot$ $g^{-1} \cdot min^{-1}$ to reduced values 0.8, 0.6, 0.4, and 0.2 $ml \cdot g^{-1} \cdot min^{-1}$, representing different degrees of ischemia severity.

in myocardial lactate uptake at the onset of MBF reduction appear to be correlated to the changes in the rates of glycogen breakdown, glycolysis, and lactate production in phase I. The rate of lactate production increased at a faster rate than the rate of lactate release and was larger than the rate of lactate utilization at MBF values lower than 0.7 $ml \cdot g^{-1} \cdot min^{-1}$. This resulted in sustained rates of lactate production and release with moderate and severe ischemia leading to sustained high values of myocardial lactate concentration at these levels of MBF reduction.

DISCUSSION

Compartmentalized Model of Cardiac Energy Metabolism

In the present study, computer simulations with a validated mathematical model of myocardial energy metabolism[25,26] were used to investigate the role of cytosolic and mitochondrial redox states in regulating cardiac energetics during reduced MBF. In this model, the reaction rates of metabolic processes involving energy transfer metabolites (e.g., pyruvate decarboxylation) are coupled to the inter-conversion rates of the metabolites involved (e.g., NAD^+ and NADH).[27] Using this approach, biochemical reactions and metabolic pathways involving energy transfer metabolites are linked to each other and coordinated to match ATP production rates to ATP utilization rates. Our previous model of cardiac energy metabolism assumed a homogeneous intracellular compartment in the cardiac cell. As a consequence, reactions explicitly involving energy transfer metabolites are modulated to the same extent, regardless of whether they occur in the cytosol or in the mitochondria. To date we do not know whether cytosolic and mitochondrial redox states vary in parallel with the same pattern during myocardial ischemia.[28,29] Even though anatomical and biochemical analysis of myocardium has provided evidence demonstrating major concentration differences between cytosolic and mitochondrial NADH, NAD^+, ADP, and ATP, methodological issues have limited the measurement of dynamic changes of these concentrations *in vivo*.[30] Given current experimental limitations, we extended our previous model of cardiac energy metabolism by including cytosolic and mitochondrial compartments and the associated transport processes between them. This enabled us to investigate the role of cytosolic and mitochondrial redox states in regulating cardiac energetics during reduced MBF. In the present study, computer simulations of selected metabolic responses to mild, moderate, and severe ischemia were characterized and compared to elucidate mechanisms regulating the two main pathways of ATP production.

Role of Mitochondrial Redox State in Regulating Cardiac Energetics

The main contribution of this study is the demonstration, via computer simulations, that cytosolic and mitochondrial redox states display different dynamics in response to ischemia, thereby playing distinct roles in regulating cardiac energetics. Specifically, simulations showed that while the mitochondrial redox state responses to step reductions in MBF were monophasic, cytosolic redox state responses displayed a biphasic behavior.

The effect of reduced MBF was an immediate decrease in myocardial oxygen consumption—in spite of reduced energy demand—and a significant increase in net rate of glycogen breakdown. The MVO_2 and $NADH_m/NAD^+_m$ dynamic responses to MBF reduction were very similar and highly correlated, suggesting that the decrease in MVO_2 leads to a lower rate of mitochondrial NADH oxidation than formation and transport into the mitochondria, resulting in NADH accumulation. This supports the hypothesis that changes in mitochondrial NADH/NAD are secondary to changes in myocardial oxygen consumption.[31] This hypothesis has also been supported by other studies reporting extensive NADH oxidation right after energy demand has been increased.[32–34]

Experimental studies on perfused hearts have shown that an increase in mitochondrial NADH can stimulate the rate of mitochondrial respiration, suggesting that enhancing the rate of substrate level reactions can increase the rate of NADH production, leading to increased NADH levels and enhanced rate of ATP synthesis. However, under physiological conditions ATP synthesis and utilization are kept tightly matched and the levels of NADH and other phosphate metabolites do not change significantly;[35] moreover, in living cells under physiological conditions, the flux is largely controlled by the demand.[36,37] Consequently, the assumption that the behavior and control of metabolic pathways studied in isolation will be the same when studied in a cellular or whole organ context under physiological conditions may not be valid.

The mitochondrial phosphorylation state (ADP_m/ATP_m) also displayed a similar response to that of mitochondrial redox. The ischemia-induced changes in mitochondrial redox and phosphorylation states acted as modulator of the MVO_2 steady-state response by increasing this rate and preventing it from dropping even further. This oxygen dependence of MVO_2 and metabolism has been demonstrated previously with theoretical and experimental studies.[38–40] Once the mitochondrial redox and phosphorylation states reach steady state, both cellular states are set to their new values and modulation of the dynamics of any reactions coupled to these states is mainly controlled by their corresponding substrates. This was the case in the present study with respect to the rate of pyruvate and fatty acid oxidation in the mitochondria. However, none of the cytosolic metabolic processes coupled to redox state displayed a dynamic response that resembled that of mitochondrial redox state, suggesting that, under the conditions investigated, the transport of reducing equivalents between mitochondria and cytosol was much slower than the reaction processes altering them in these compartments.

Role of Cytosolic Redox State in Regulating Cardiac Energetics

The dynamic response of cytosolic redox state was quite different from that of mitochondrial redox state. The mitochondrial and cytosolic concentrations of NADH and NAD had the same values only before blood flow was reduced. Even after reaching steady state, the concentration values differed significantly (FIG. 3). This suggests that the processes setting the redox states values in the cytosol and mitochondria are activated differently during ischemia. The rates of glycogenolysis and glycolysis seem to play a major role in setting the cytosolic redox more than mitochondrial oxygen consumption. Moreover, it seems that—during ischemia—the role of redox state is localized, thus modulating only processes that occur in the same cellular compartment. Specifically, the rate of lactate production appears to be largely controlled by the rates of glycolysis and glycogenolysis and modulated by the cytosolic and not the mitochondrial redox. Supporting this view is the fact that in isolated hearts, the degree of mitochondrial reduction is affected more by the availability of pyruvate or lactate than by the cytosolic redox state. Since the cytosolic and mitochondrial redox states are not necessarily linked at all times,[41] there may be conditions in which the rate of lactate production increases, regardless of whether mitochondrial redox increases or decreases concomitantly.

SUMMARY

Analysis of the dynamic responses to reduced MBF generated with our computer model of cardiac energy metabolism indicates that cytosolic and mitochondrial redox states play key roles in the regulation of cardiac energetics. Specifically, characterization of the dynamics of NADH and NAD^+ changes in these subcellular compartments indicates that cytosolic and mitochondrial redox states do not vary concomitantly and seem to be uncoupled from each other. During ischemia, myocardial oxygen consumption appears to be the main determinant of mitochondrial redox state, while glycogen breakdown is the main determinant of cytosolic redox state. Moreover, mitochondrial redox primarily modulates the rates of pyruvate and fatty acid oxidation, while cytosolic redox modulates the rate of lactate formation and utilization.

ACKNOWLEDGMENT

This work was supported by grant GM-66309 from the National Institute of General Medical Sciences for the establishment of the Center for Modeling Integrated Metabolic Systems at Case Western Reserve University.

REFERENCES

1. Brown, G.C. 1992. Control of respiration and ATP synthesis in mammalian mitochondria and cells. Biochem. J. **284:** 1–13.
2. Guth, B.D., R. Schulz & G. Heusch. 1993. Time course and mechanisms of contractile dysfunction during acute myocardial ischemia. Circulation **87:** IV35–IV42.
3. Balaban, R.S., H.L. Kantor, L.A. Katz & R.W. Briggs. 1986. Relation between work and phosphate metabolite in the *in vivo* paced mammalian heart. Science **232:** 1121–1123.
4. Heineman, F.W. & R.S. Balaban. 1990. Control of mitochondrial respiration in the heart *in vivo*. Annu. Rev. Physiol. **52:** 523–542.
5. Stanley, W.C. 2001. Cardiac energetics during ischaemia and the rationale for metabolic interventions. Coron. Artery Dis. **12**(Suppl. 1): S3–S7.
6. Chance, B., J.S. Leigh, Jr., J. Kent, *et al.* 1986. Multiple controls of oxidative metabolism in living tissues as studied by phosphorus magnetic resonance. Proc. Natl. Acad. Sci. U.S.A. **83:** 9458–9462.
7. Balaban, R.S. & F.W. Heineman. 1989. Control of mitochondrial respiration in the heart *in vivo*. Mol. Cell. Biochem. **89:** 191–197.
8. Asimakis, G.K. & L.A. Sordahl. 1981. Intramitochondrial adenine nucleotides and energy-linked functions of heart mitochondria. Am. J. Physiol. **241:** H672–H678.
9. Nuutinen, E.M. 1984. Subcellular origin of the surface fluorescence of reduced nicotinamide nucleotides in the isolated perfused rat heart. Basic Res. Cardiol. **79:** 49–58.
10. Williamson, D.H., P. Lund & H.A. Krebs. 1967. The redox state of free nicotinamide-adenine dinucleotide in the cytoplasm and mitochondria of rat liver. Biochem. J. **103:** 514–527.

11. KATZ, A., A. EDLUND & K. SAHLIN. 1987. NADH content and lactate production in the perfused rabbit heart. Acta Physiol. Scand. **130:** 193–200.
12. KATZ, L.A., J.A. SWAIN, M.A. PORTMAN & R.S. BALABAN. 1989. Relation between phosphate metabolites and oxygen consumption of heart *in vivo*. Am. J. Physiol. **256:** H265–H274.
13. KEDEM, J., A. MAYEVSKY, J. SONN & B.A. ACAD. 1981. An experimental approach for evaluation of the O2 balance in local myocardial regions *in vivo*. Q. J. Exp. Physiol. **66:** 501–514.
14. GRAHAM, T.E. & B. SALTIN. 1989. Estimation of the mitochondrial redox state in human skeletal muscle during exercise. J. Appl. Physiol. **66:** 561–566.
15. KATZ, A., A. EDLUND & K. SAHLIN. 1987. NADH content and lactate production in the perfused rabbit heart. Acta Physiol. Scand. **130:** 193–200.
16. BALABAN, R.S. & L.J. MANDEL. 1988. Metabolic substrate utilization by rabbit proximal tubule. An NADH fluorescence study. Am. J. Physiol. **254:** F407–F416.
17. CH'EN, F.F., R.D. VAUGHAN-JONES, K. CLARKE & D. NOBLE. 1998. Modelling myocardial ischaemia and reperfusion. Prog. Biophys. Mol. Biol. **69:** 515–538.
18. CORTASSA, S., M.A. AON, E. MARBAN, *et al.* 2003. An integrated model of cardiac mitochondrial energy metabolism and calcium dynamics. Biophys. J. **84:** 2734–2755.
19. JAFRI, M.S., S.J. DUDYCHA & B. O'ROURKE. 2001. Cardiac energy metabolism: models of cellular respiration. Annu. Rev. Biomed. Eng. **3:** 57–81.
20. SALEM, J.E., G.M. SAIDEL, W.C. STANLEY & M.E. CABRERA. 2002. Mechanistic model of myocardial energy metabolism under normal and ischemic conditions. Ann. Biomed. Eng. **30:** 202–216.
21. ARNOLD, M.K. 1992. Physiology of the Heart. Raven Press. New York.
22. HORTON, H.R., L.A. MORAN, R.S. OCHS, *et al.* 2002. Principles of Biochemistry. Prentice Hall. Upper Saddle River, NJ.
23. LANOUE, K.F. & A.C. SCHOOLWERTH. 1979. Metabolite transport in mitochondria. Annu. Rev. Biochem. **48:** 871–922.
24. ZHOU, L., J.E. SALEM, G.M. SAIDEL, *et al.* 2005. Mechanistic model of cardiac energy metabolism predicts localization of glycolysis to cytosolic subdomain during ischemia. Am. J. Physiol. Heart Circ. Physiol. **288:** H2400–2411.
25. SALEM, J.E., M.E. CABRERA, M.P. CHANDLER, *et al.* 2004. Step and ramp induction of myocardial ischemia: comparison of *in vivo* and *in silico* results. J. Physiol. Pharmacol. **55:** 519–536.
26. SALEM, J.E., W.C. STANLEY & M.E. CABRERA. 2004. Computational studies of the effects of myocardial blood flow reductions on cardiac metabolism. Biomed. Eng. Online **3:** 15.
27. SALEM, J.E., G.M. SAIDEL, W.C. STANLEY & M.E. CABRERA. 2002. Mechanistic model of myocardial energy metabolism under normal and ischemic conditions. Ann. Biomed. Eng. **30:** 202–216.
28. NUUTINEN, E.M. 1984. Subcellular origin of the surface fluorescence of reduced nicotinamide nucleotides in the isolated perfused rat heart. Basic Res. Cardiol. **79:** 49–58.
29. WILLIAMSON, D.H., P. LUND & H.A. KREBS. 1967. The redox state of free nicotinamide-adenine dinuc leotide in the cytoplasm and mitochondria of rat liver. Biochem. J. **103:** 514–527.
30. ASIMAKIS, G.K. & L.A. SORDAHL. 1981. Intramitochondrial adenine nucleotides and energy-linked functions of heart mitochondria. Am. J. Physiol. **241:** H672–H678.
31. ASHRUF, J.F., J.M. COREMANS, H.A. BRUINING & C. INCE. 1995. Increase of cardiac work is associated with decrease of mitochondrial NADH. Am. J. Physiol. **269:** H856–H862.
32. BRANDES, R. & D.M. BERS. 1996. Increased work in cardiac trabeculae causes decreased mitochondrial NADH fluorescence followed by slow recovery. Biophys. J. **71:** 1024–1035.
33. BRANDES, R. & D.M. BERS. 1997. Intracellular Ca2+ increases the mitochondrial NADH concentration during elevated work in intact cardiac muscle. Circ. Res. **80:** 82–87.
34. NUUTINEN, E.M. 1984. Subcellular origin of the surface fluorescence of reduced nicotinamide nucleotides in the isolated perfused rat heart. Basic Res. Cardiol. **79:** 49–58.

35. BALABAN, R.S., H.L. KANTOR, L.A. KATZ & R.W. BRIGGS. 1986. Relation between work and phosphate metabolite in the *in vivo* paced mammalian heart. Science **232:** 1121–1123.
36. KROUKAMP,O., J.M. ROHWER, J.H. HOFMEYR & J.L. SNOEP. 2002. Experimental supply-demand analysis of anaerobic yeast energy metabolism. Mol. Biol. Rep. **29:** 203–209.
37. HOFMEYR, J.H. & A. CORNISH-BOWDEN. 2000. Regulating the cellular economy of supply and demand. FEBS Lett. 23792, 1–5.
38. WILSON, D.F., M. ERECINSKA, C. DROWN & I.A. SILVER. 1977. Effect of oxygen tension on cellular energetics. Am. J. Physiol. **233:** C135–C140.
39. WILSON, D.F., C.S. OWEN & M. ERECINSKA. 1979. Quantitative dependence of mitochondrial oxidative phosphorylation on oxygen concentration: a mathematical model. Arch. Biochem. Biophys. **195:** 494–504.
40. WILSON, D.F., M. ERECINSKA, C. DROWN & I.A. SILVER. 1979. The oxygen dependence of cellular energy metabolism. Arch. Biochem. Biophys. **195:** 485–493.
41. SCHOLZ, T.D., M.R. LAUGHLIN, R.S. BALABAN, *et al.* 1995. Effect of substrate on mitochondrial NADH, cytosolic redox state, and phosphorylated compounds in isolated hearts. Am. J. Physiol. **268:** H82–H91.

Spatial Regulation of Intracellular pH in the Ventricular Myocyte

PAWEL SWIETACH AND RICHARD D. VAUGHAN-JONES

Burdon Sanderson Cardiac Science Centre, University Laboratory of Physiology, Oxford OX1 3PT, UK

ABSTRACT: In the cardiac myocyte, an adequate intracellular proton mobility is necessary for coupling sarcolemmal proton transport to bulk cytoplasm during intracellular pH regulation. It is also important for dissipating intracellular pH nonuniformity in response to local acid/base disturbances. Because cardiac myocytes have a high buffering capacity, intracellular H^+ mobility is low. Spatial H_i^+ movements occur via a mobile buffer shuttle that most likely involves intracellular dipeptides. In the present work with isolated ventricular myocytes, it is demonstrated that stimulating a large acid efflux on sarcolemmal Na^+–H^+ exchange, results in spatial pH_i gradients of up to 0.1 U. These may have important implications for pH-sensitive processes within the cell, such as contraction. By using computational modeling, it is shown that the gradients can be attributed to the low H_i^+ mobility. Computational modeling is also used to assess the importance of H_i^+ mobility in mediating local recovery of pH following an acidosis in a small region of the cell. Results indicate that local mechanisms for H_i^+ movement will be important in determining the global regulation of intracellular pH.

KEYWORDS: diffusion; modeling; myocyte; pH regulation

INTRODUCTION

Intracellular pH is a key modulator of cardiac function, affecting contraction, cellular excitability, and electrical rhythm.[1–3] This is because protons interact with a wide variety of intracellular proteins, such as enzymes, ion channels, and ion transporters, affecting their three-dimensional folding and hence their physiological activity. Protons also compete with signaling ligands, notably calcium ions, for attachment to specific binding sites on selected proteins. For example, protons can competitively exclude Ca^{2+} binding to troponin C, thereby reducing the force of contraction during intracellular acidosis.[4] In view of the high reactivity of protons, it comes as no surprise that a sophisticated system of H^+-equivalent ion transport proteins is located at the sarcolemma, for the express purpose of regulating and eventually stabilizing intracellular pH.[5] Steady-state pH is typically around 7.1.

The sarcolemmal transporters regulate pH_i by moving excess intracellular acid or base out of the cell in the form of H^+, OH^-, or HCO_3^- ions. Specific transporters include the Na^+–H^+ exchanger (NHE), Na^+–HCO_3^- cotransport (NBC), Cl^-–HCO_3^-

Address for correspondence: Prof. Richard Vaughan-Jones, Ph.D., Burdon Sanderson Cardiac Science Centre, University Laboratory of Physiology, Oxford OX1 3PT, UK. Voice/fax: 01865 272451. richard.vaughan-jones@physiol.ox.ac.uk

Ann. N.Y. Acad. Sci. 1047: 271–282 (2005). © 2005 New York Academy of Sciences. doi: 10.1196/annals.1341.024

exchange, and Cl^-–OH^- exchange.[5] These are able to regulate pH_i because pH in the subsarcolemmal, cytoplasmic microdomain, contiguous with the transporter, is coupled to that of the bulk phase within the cell. This coupling relies on the effective diffusion of intracellular protons. It is surprising, therefore, to learn that significant free H^+ ion diffusion does not occur within the cell.[6–8] This is because almost all intracellular acid or base is strongly buffered, with many of the buffer sites residing on large molecular weight proteins.[9] Almost all spatial movement of protons is achieved via reversible H^+ binding to lower molecular weight mobile buffers. As a result, the effective mobility of intracellular protons is low. This predicts that diffusive coupling of cytoplasmic pH to the sarcolemmal transporters is weak and may result in the development of pH gradients within the cell, particularly when acid/base flux through the transporters is high. Although pH_i gradients generated by membrane transporters have been demonstrated in epithelial cells[10] and neurons,[11] such a phenomenon has not been shown in cardiac cells.

In the present work, we use confocal pH_i-imaging techniques to demonstrate that stimulation of sarcolemmal NHE induces significant pH_i nonuniformity in rat isolated ventricular myocytes. We use a diffusion model to explore the cause of this nonuniformity, and we analyze the relative importance of cytoplasmic diffusion and sarcolemmal transport in the local regulation of pH.

METHODS

Cell Superfusion and Cell Imaging

Experiments were carried out in a 500-µl Plexiglas chamber sealed from beneath by a coverslip (number 1, VWR International, Leicester, UK). To help cells adhere, the coverslip was coated before each experiment with 100 µl of poly-L-lysine (Sigma, Poole, UK). During experiments, the chamber was constantly supplied with superfusate (see Solutions) heated to 37°C by an electrical resistance heater. A surface suction tube connected to a water pump removed superfusate from the far end of the chamber, thereby maintaining a low level of solution (necessary for making fast solution changes). Cells were imaged using an inverted confocal microscope (Leica DM IRBE, Heidelberg, Germany) equipped with a ×40 plano-apochromat objective lens. To measure intracellular pH in the confocal plane, cells were loaded with a novel pH fluorescent dye, carboxy-SNARF-4F (Molecular Probes, Eugene, OR, USA). Like carboxy-SNARF-1, the dye is ratiometric (dual emission at 580 nm and 640 nm; excitation at 514 nm), it has a wide dynamic range and loads well into cells (as the esterified acetoxymethyl-form). Carboxy-SNARF-4F is useful as it has a low pK value (6.86 ± 0.22, $n = 11$), mandating its use in experiments involving severe acidosis. In effect, recordings of pH_i near 6.0 made with carboxy-SNARF-4F are considerably less noisy, as the fluorophore is far from being saturated by protons, unlike those made with carboxy-SNARF-1.

Experimental Protocol

Ventricular myocytes were isolated from ~300 g Wistar rats (killed by cervical dislocation in accordance with UK Home Office guidelines) using a combination of mechanical and enzymatic (0.24 mg/ml Blendzymes III, Roche, Lewes, UK) dispersion. Cells were stored in standard DMEM culture medium at room temperature for up to 10 h. After washing away poly-L-lysine, 200 μl of cells in culture medium were applied to the superfusion chamber together with 1.5 μl of a 1.7 mM solution of carboxy-SNARF-4F in dimethylsulfoxide (DMSO). Myocytes were allowed to load with dye in darkness and at room temperature for 10 min. The cells were then imaged confocally for pH_i using the 514-nm laser line of an Argon laser. To stimulate a large NHE-mediated acid efflux at the sarcolemma, myocytes were exposed to 30 mM ammonium chloride and then returned to Na^+-free normal Tyrode. The duration of exposure to ammonium was varied between 6 and 10 min, depending on the desired level of acidosis. The absence of extracellular Na^+ inhibits NHE. Activity can be restored by the re-addition of sodium to the superfusate. The confocal system was set up to acquire the transmission image of the cell and two fluorescence images (at 580 nm and 640 nm) every 1.05 s at a resolution of 256 by 256 pixels. Each two consecutive images were averaged to reduce noise. Recording would normally start 2 min after the ammonium prepulse and 2 min before the re-addition of Na^+ into the superfusion chamber.

Data Analysis

After experiments, the two stacks of fluorescence images were ratioed (580 nm/640 nm) in a manner similar to that used with carboxy-SNARF-1.[9] Using a macro written for Scion Image (Scion Corp, Frederick, MD, USA), the fluorescence ratio was measured at the edges of a cell (region of interest [ROI]1 and ROI3) and at the midpoint, coinciding with the center of the cell (ROI2). The ROIs were typically 5 by 5 μm. The macro was designed to find the cell edges as a measure to avoid movement artefacts. The time courses of mean ROI ratio were then converted to pH using the calibration for carboxy-SNARF-4F (obtained using the nigericin method[12] with 36 cells superfused with high K^+ solutions at pH 5.5 to 8.5). The macro then calculated the difference between the pH in the center and edges of the cell (i.e., $\Delta pH = [pH_{ROI1} + pH_{ROI3}]/2 - pH_{ROI2}$). To reduce noise, the difference time course was fed through a 5-point rolling-average filter.

Solutions:

(a) HEPES-buffered normal Tyrode contained (in mM): NaCl (135), KCl (4.5), $CaCl_2$ (2), $MgCl_2$ (1), HEPES (20), glucose (11), pH adjusted to 7.4 with 4 M NaOH at 37°C.
(b) Sodium-free Tyrode contained (in mM): *N*-methyl-D-glucamine (140), KCl (4.5), $CaCl_2$ (2), $MgCl_2$ (1), HEPES (20), glucose (11), pH adjusted to 7.4 with 1 M HCl at 37°C.
(c) Ammonium-containing Tyrode contained (in mM): NaCl (105), NH_4Cl (30), KCl (4.5), $CaCl_2$ (2), $MgCl_2$ (1), HEPES (20), glucose (11), pH adjusted to 7.4 with 4 M NaOH at 37°C.

Diffusion Algorithm

To simulate diffusion phenomena in myocytes, we assume cylindrical symmetry about the central axis, a typical rat-myocyte length of 100 μm, and a radius of 12 μm. By taking advantage of axial symmetry, the three dimensional problem simplifies to a two dimensional equation,

$$\frac{\partial u}{\partial t} = \frac{1}{r} \cdot \frac{\partial}{\partial r}\left(r \cdot D \cdot \frac{\partial u}{\partial r}\right) + \frac{\partial}{\partial x}\left(D \cdot \frac{\partial u}{\partial x}\right),$$

where u is the concentration of protons and D is the apparent proton diffusion coefficient ($D_H{}^{app}$, with assumed isotropy), r is the radial space variable and x is the axial variable. $D_H{}^{app}$ was programmed to vary with pH_i, in proportion with the mobile-to-total buffering capacity ratio.[9] The constant of proportionality (equivalent to the mobile buffer diffusion coefficient, D_{mob}) was set to 22.6×10^{-7} cm^2/s in order to fit the data on $D_H{}^{app}$, estimated (using pipette acid loading) to be 9.76×10^{-7} cm^2/s in rat myocytes.[9] The linear pH_i–$D_H{}^{app}$ relationship in the pH_i range 6.5–8 was extrapolated to obtain an estimate at very acidic levels of pH_i:

$$D_H{}^{app} = [\,9.25\,pH_i) - 55.43] \times 10^{-7}\ cm^2/s\ .$$

The initial condition can be set to uniformity (e.g., uniform acid load) or nonuniformity (e.g., regionalized acidosis). The boundary conditions are set to reproduce the pH_i-dependence of rat myocyte NHE (see Ref. 13) and include a leak of +0.83 mM/min accounting for background acid loading and CHE activity:[5]

$$\frac{dpH}{dt} = \frac{34.188 \cdot \dfrac{[H^+]^{1.933}}{[H^+]^{1.933} + (10^{-6.38})^{1.933}} - 0.83}{\beta(pH_i)}\ .$$

The intrinsic buffering capacity, β_i, was derived from previous experimental data[9] for rat myocytes (measured under CO_2/bicarbonate-free conditions). Boundary flux was set to zero at a pH_i of 7.21, defined as the cell's resting pH_i. It is possible to include the above boundary condition throughout the surface area, or confine it to a particular part of the cell. In the latter case, the remainder of the cell membrane would feature a zero-flux (reflection) boundary condition. There is evidence, from skeletal muscle myocytes[14] and preliminary evidence from our laboratory for rat cardiomyocytes, that NHE is not functional in the depths of the T-tubules (Pawel Swietach, Lucy Simmonds, and Richard Vaughan-Jones, unpublished data). This simplifies the implementation of our boundary condition, as it is not necessary to account for the T-system. The diffusion equation was solved using the finite difference method using an explicit scheme[8] for a time step of 0.001 s and space step dx = dr = 1 μm.

RESULTS

Acid Extrusion Generates Heterogeneous pH_i Microdomains

FIGURE 1 illustrates the influence of NHE on pH_i in an isolated ventricular myocyte. An intracellular acid load was induced by prepulsing a cell with 30 mM NH_4Cl. This occurred before the start of the traces illustrated in FIGURE 1A. These show whole-cell pH_i time courses averaged from several experiments (n = 4, 10, 17). Recovery of pH_i from the acid load was prompted by re-adding Na_o^+ in order to reactivate NHE activity. FIGURE 1A shows effects of re-adding Na_o^+ at three different

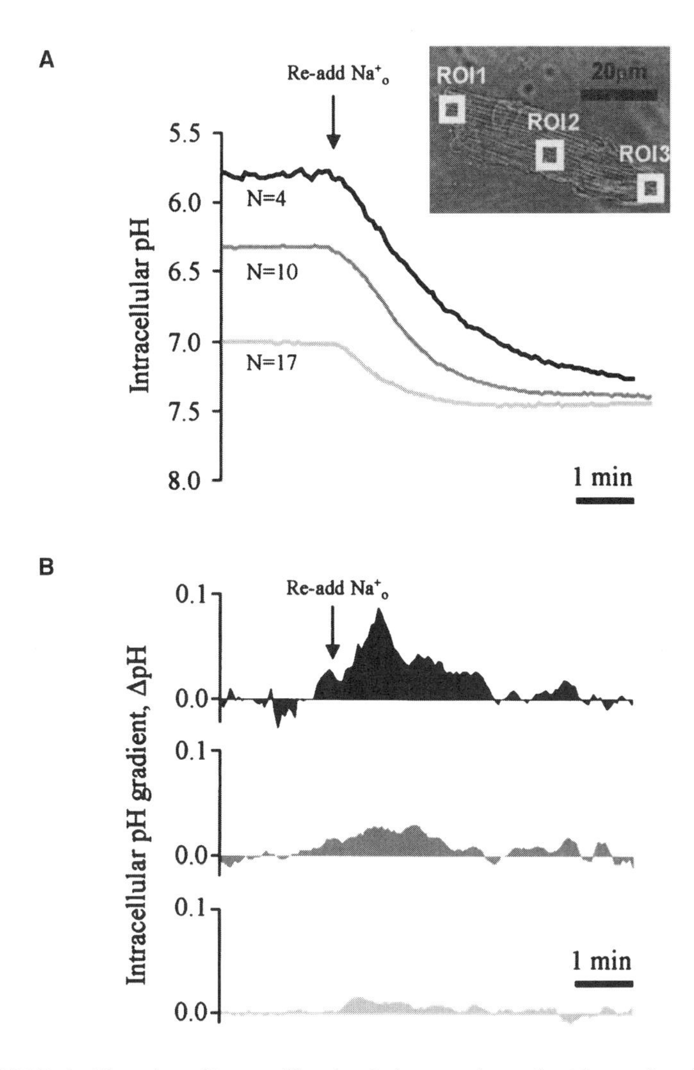

FIGURE 1. Observing pH_i nonuniformity during sarcolemmal acid extrusion. (**A**) Cell-averaged pH during recovery from an acid load. Rat myocytes were first acid loaded using a 30 mM ammonium prepulse, and their intracellular pH was clamped at a constant level by superfusion with a Na^+-free medium. Results were binned according to the size of the acid load ($n = 4$, 10, 17). On re-addition of Na^+, Na^+–H^+ exchanger was dysinhibited and produced sarcolemmal extrusion at a rate related to the acid load. (***Inset***) A typical rat myocyte showing the position of the regions of interest (5 × 5 μm). (**B**) The longitudinal pH_i nonuniformity measured as the difference between the time course of pH_i recovery at the polar regions of the cell and the cell center. Data were binned according to the starting pH_i ($n = 4, 10, 17$).

prepulsed values of pH_i (the starting level of acidosis was related to the duration of the prepulse). Once recovery had been initiated, it was faster at the ends of the cell compared with the middle. This difference was revealed by comparing pH_i in polar and central regions of interest (ROI1 & 3, and ROI2, respectively, see inset to FIG.

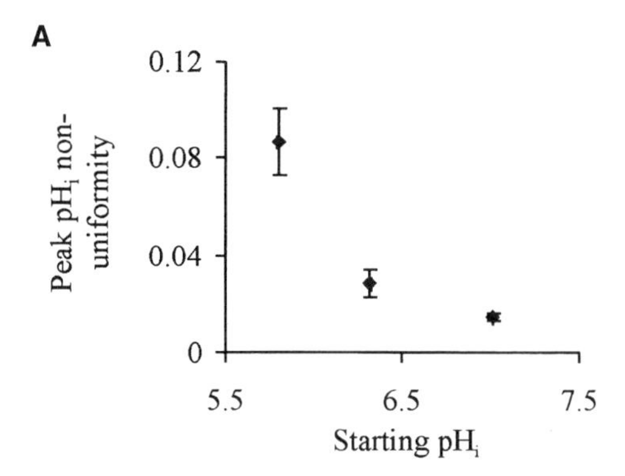

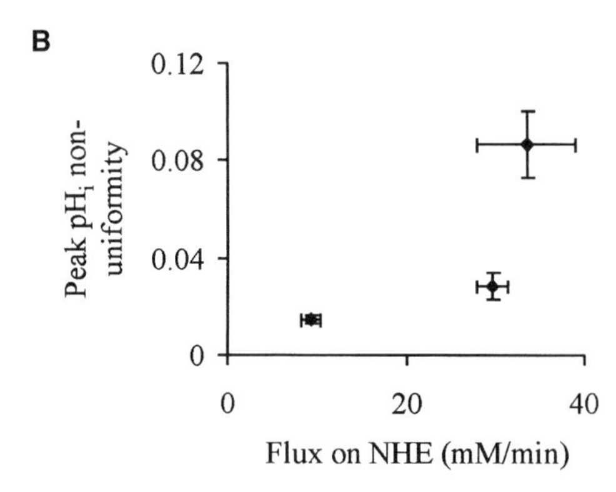

FIGURE 2. The size of the peak pH_i nonuniformity (n = 4, 10, 17) depends on (**A**) the starting pH_i and (**B**) the initial rate of flux carried by Na^+–H^+ exchanger (NHE). All data points were significantly different from zero ($P < 0.05$). A combination of low pH_i (hence, low proton mobility) and high sarcolemmal proton transport rate produce measurable nonuniformity of up to 0.1 pH U.

1A). The regional difference (i.e., from end to middle) has been plotted in FIGURE 1B (averaged traces, n = 4, 10, 17). This shows the transient development of a longitudinal pH_i gradient during the recovery period. The gradient was as large as 0.1 pH U and lasted for a few min. It was symmetrically oriented with respect to the center of the cell (i.e., a similar gradient was detected for ROI1–ROI2 as for ROI3–ROI2 with an intermediate regional nonuniformity between the ROIs). This indicates that, as far as length is concerned, the gradient was biphasic along the cell, with the lowest pH always occurring in the central regions. As shown in FIGURE 2A, the lower the initial starting pH_i, the larger was the peak pH_i gradient during recovery.

Reducing pH_i increasingly stimulates NHE activity.[5,13] It was notable, therefore, that the size of the longitudinal pH_i gradient correlated positively with the initial magnitude of acid extrusion on NHE, estimated as $\beta_i \times dpH_i/dt$, where β_i is intrinsic intracellular buffering power (FIG. 2B). Further analysis showed that a significant radial pH_i gradient was not detectable during the recovery period (not shown, n = 4, 10, 17). Estimates here suggested that gradients from the core of the cell to a peripheral region were less than 0.03 pH U (and not statistically different from zero).

Modeling pH_i Responses to a Sarcolemmal Acid Efflux

A key question is whether a low intracellular H^+ mobility can account for the observed gradients of pH_i. We explored this possibility by modeling spatial proton flux in a myocyte subjected to the same stimuli. Intracellular proton mobility was assumed to fall with pH_i, as described in METHODS. The cell was treated as a cylinder, with NHE activity expressed uniformly at the surface. The results, shown in FIGURE 3B, indicate that a longitudinal gradient was predicted, similar to that measured experimentally (FIG. 3B; this result should be compared with FIG. 1B). The pH_i nonuniformity was quantified as the difference in the simulated pH_i recovery time courses in a square (5 by 5 μm) ROI positioned at the end and in middle of the cell length. The modeled nonuniformity, however, peaked at an earlier time than in the experiments, possibly because of the finite time needed in the latter to restore completely the extracellular Na^+ concentration via the superfusion system. As shown in FIGURE 3C, significant longitudinal pH_i nonuniformity that matches the experimental data could be reproduced only when NHE was expressed uniformly (FIG. 3Ci) in the model sarcolemma. The predicted longitudinal gradient was biphasic (FIG. 3C), as observed experimentally, with higher pH_i values at the polar regions, and the nonuniformity was maintained for a few min, disappearing once pH_i, and, hence, sarcolemmal acid extrusion via NHE had returned to control levels (FIG. 3A). Significant longitudinal pH_i nonuniformity was predicted for a wide range of cell lengths, ranging from 0.07 to 0.09 pH U for typical myocyte lengths between 50 to 200 μm respectively.

Excluding NHE from the ends of the model cell produced much smaller longitudinal pH_i gradients (FIG. 3Cii) and only a small radial gradient (<0.03 U). In general, radial gradients diminished greatly for cells of smaller radii. The pH_i gradients observed experimentally are therefore consistent with a low intracellular proton mobility, equivalent to an apparent diffusion constant, D_H^{app} of about 5×10^{-7} cm^2/s at a pH_i of 6.5. They confirm that sarcolemmal acid efflux can lead to significant pH_i nonuniformity.

Modeling pH_i Responses to a Localized Intracellular Acidosis

The formation of heterogeneous pH_i microdomains within a myocyte indicates that a model of cardiac pH_i regulation must include both the spatial cytoplasmic movement of protons via buffers and the sarcolemmal flux of protons via transporters. In FIGURE 4, we have used our computational model to explore the relative importance of these two processes in regulating the pH of a region of cytoplasm. The pH in a small spherical region of interest (radius 6 μm) within the cylindrical model cell was set instantaneously to pH 6.5 while that of the surrounding cytoplasm remained at 7.21 (FIG. 4A). This simulated an acute localized acidosis. Following this insult, the model predicted a recovery of pH in the ROI (pH_{ROI}), comprising an initial fast and then a slow time course (FIG. 4B; see fast and slow time base). The pH averaged in the rest of the cell ($pH_{surround}$) showed a biphasic change, first falling rapidly and then rising more slowly at a rate similar to that in the ROI. During the local recovery of $pH_{ROI,}$ protons move into surrounding areas via a diffusive shuttling on buffers. This diffusive loss of protons has been plotted in FIG. 4C. Once the acid load has diffused to the sarcolemma, it is transported out of the cell via NHE. The loss of cellular protons via NHE has also been plotted on the same axes in FIG. 4C. Comparison of the two plots indicates that the bulk of the fast recovery of pH_{ROI} shown in panel B is driven by local H^+ shuttling into adjacent areas with little or no sarcolemmal activity, while the slow component is limited by the rate of sarcolemmal extrusion. The times taken for 50% acid loss from the ROI (τ_{ROI}) and the entire

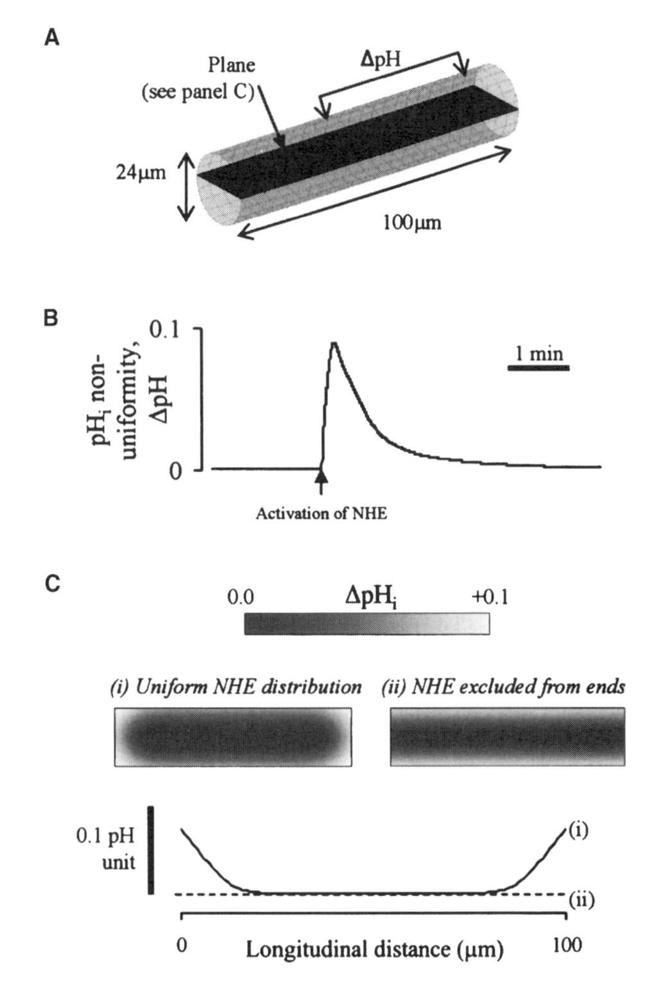

FIGURE 3. Modeling pH$_i$ nonuniformity during recovery from a uniform acid load. (**A**) The model, showing dimensions, geometry, and the position of the slice used to illustrate the pH$_i$ nonuniformity in **C**. (**B**) The model was run to simulate pH recovery from a uniform acid load of 6.0. The difference between the recovery in edge- and middle-located regions of interest (5 × 5 μm) was measured and plotted as a function of time, showing that longitudinal nonuniformity of up to 0.09 pH U could be reproduced, in line with experimental findings. (**C**) The pH distribution in a slice of the cylinder (shown in **A**) plotted at the point of greatest pH nonuniformity. In case (i), Na^+–H^+ exchanger (NHE) is expressed uniformly across the entire surface area of the cell (as per the results presented in **B**). The longitudinal gradient, shown in the *bottom panel*, curve (i), is biphasic. In case (ii), NHE is excluded from the ends of the cell. Under these conditions, the longitudinal gradient is not resolvable, shown in the *bottom panel*, dashed line (ii).

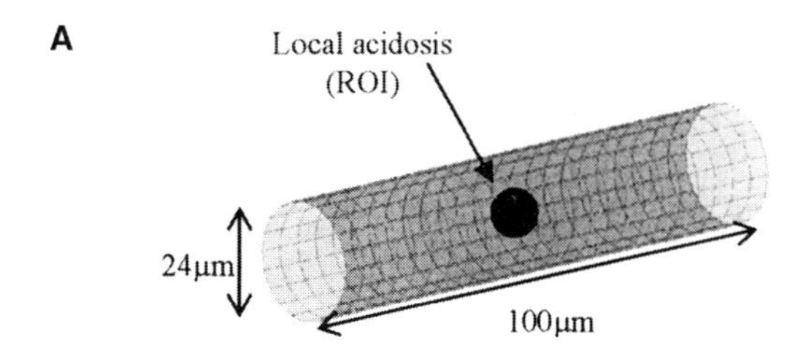

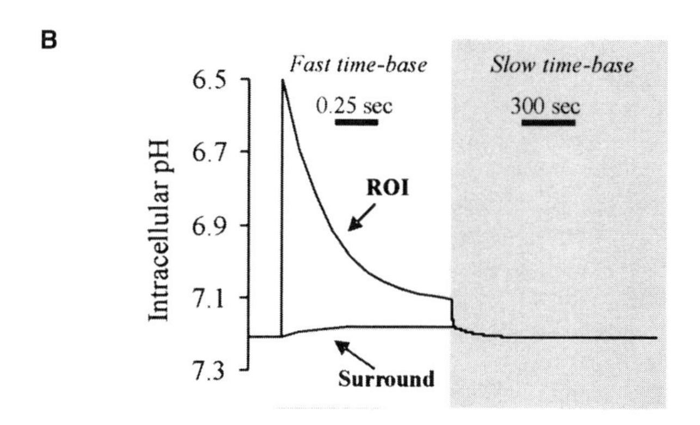

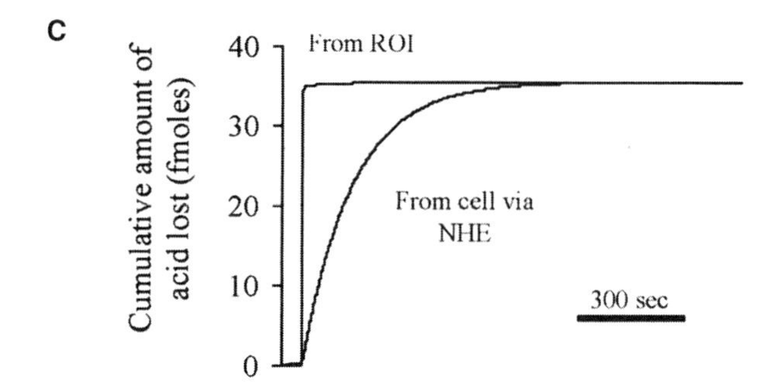

FIGURE 4. Modeling recovery of pH in a localized region of interest (ROI). (**A**) A small spherical (radius 6 μm) acid load of pH 6.5 was centered in the middle of the cell and allowed to dissipate through the cytoplasm and eventually across the sarcolemma. (**B**) The recovery of pH_i in the ROI and its surrounding cytoplasm consisted of a fast diffusive phase (rise of pH_{ROI} and fall of $pH_{surround}$), evident under a fast time base, followed by a much slower phase due to Na^+–H^+ exchanger (NHE)-mediated sarcolemmal extrusion (pH_{ROI} and $pH_{surround}$ rising in phase), evident under a 1,000-fold slower time base. (**C**) The cumulative amount of acid lost from the ROI and from the cell follows a saturating time course, described with a time constant for 50% completion of 0.13 s and 94.7 s, respectively.

cell (τ_{cell}) are 0.13 s and 94.7 s, respectively. The former constant is strongly dependent on the size of the ROI—rising to 0.45 s for an ROI radius of 12 μm (i.e., cell radius). Results of the model indicate the importance of spatial H^+ movement in mediating rapid recoveries of pH_i following a local disturbance.

DISCUSSION

Acid Extrusion Can Generate Heterogeneous pH_i Microdomains

This is the first demonstration of a pH_i gradient induced by a membrane transporter in a cardiac cell. The size of the gradient depends on the magnitude of acid efflux and can be as large as 0.1 pH U. This is a significant pH heterogeneity, when one recalls that a uniform fall of pH_i by 0.1 U from resting pH_i can induce a 25% fall in the force of contraction.[1] The measured nonuniformity would therefore be expected to induce significant spatial nonuniformity of contraction in a myocyte.

Why does the pH_i gradient occur? The answer appears, in part, to be because cytoplasmic H^+ mobility is low, more than two orders of magnitude lower than in pure water.[9] Therefore, stimulation of H^+ efflux across the sarcolemma depletes protons in the subsarcolemmal space faster than they can be replaced by a diffusive flow of H^+ from the bulk cytoplasm. This leads to a spatial pH_i gradient. A gradient was readily predicted by our diffusion model, in which intracellular proton mobility is limited by the concentration, pK, and mobility of intracellular buffers. The possible identity of intrinsic mobile buffers has been discussed previously,[7] the most likely candidates being analogues of the dipeptides, carnosine, anserine, and homocarnosine. These are low molecular weight molecules (~200 Da), containing a histidine residue, which are expressed at relatively high concentration inside cardiac muscle (at a collective concentration of about 18 mM).

Radial distances in a myocyte are sufficiently small (6 to 12 μm) that radial pH_i gradients caused by sarcolemmal acid efflux are predicted to be small (up to 0.03 pH U), consistent with the experimental data, which cannot readily resolve such gradients. It was, therefore, surprising that a biphasic longitudinal gradient was detected. The model reproduces this gradient provided NHE activity is included at the ends as well as the sides of the myocyte. Assuming a uniform membrane density of NHE transporters, an end region of cytoplasm has close access to more transporters than a middle region along the cell length. Hence, acid is more readily extruded from polar regions. The diffusion distance from middle to end (about 50 μm) is considerably greater than the radial distance, so that longitudinal diffusion times for mobile buffers (and hence for protons) are no longer sufficiently brief to prevent the generation of a measurable pH_i gradient. It is therefore likely to be the geometry of the myocyte, combined with a low intracellular proton mobility, that gives rise to the pH_i nonuniformity recorded during acid extrusion. The importance of cell geometry in the generation of pH_i nonuniformity has been discussed previously.[8]

A key question is why NHE should be expressed at the ends of a myocyte, given that the cell usually closely abuts adjacent cells at the intercalated disc. A high NHE-1 density at ventricular junctional regions has been reported previously, using fluorescent antibody markers.[15] We do not know if these NHEs are active *in vivo*. They do not appear to mediate intercellular acid transmission, as high-affinity NHE-1 inhibitors, like cariporide, do not affect cell-to-cell acid movement, most of which occurs through connexin channels.[16] At present, therefore, we are unable to settle the question of whether polar acid extrusion occurs from myocytes when they are coupled within the myocardium.

A further question is whether pH_i gradients during acid efflux would be seen in the presence of carbonic buffer (CO_2/HCO_3^-) as this is a physiological buffer that has been shown to enhance H_i^+ mobility and hence collapse pH_i gradients.[9,17] The present experiments were performed using HEPES-buffered superfusates devoid of CO_2/HCO_3. There are reasons to suspect that pH_i gradients would still occur. At low pH_i, the intracellular bicarbonate concentration is also low (assuming a metabolic acidosis at constant PCO_2, $[HCO_3^-]_i$ would be about 0.9 mM at a pH_i of 6.0, compared with about 14 mM at a pH_i of 7.2). Bicarbonate is a key component of the carbonic buffer shuttle that enhances intracellular H^+ mobility. At low pH_i, the carbonic shuttle may, therefore, start to fail, owing to a lack of sufficient intracellular bicarbonate anions. Under these circumstances, proton mobility would devolve to that defined by the higher concentration of intrinsic buffers, paving the way for a pH_i gradient.

The Spatiotemporal Characteristics of Local pH_i Regulation

In the present work, we have computed the consequences of coupling sarcolemmal acid extrusion to spatial movements of acid within the cell. Two examples have been analyzed. In the first (FIG. 3), the stimulation of a large sarcolemmal acid extrusion led to a shuttling of intracellular acid that created spatial pH_i nonuniformity for 1 to 2 min during pH_i recovery. In the second (FIG. 4), an acute, highly localized intracellular acidosis led, via proton shuttling, to a slow extrusion of acid at the sarcolemma. As a result, pH at the source of the acute disturbance recovered, with initial recovery being generated by a rapid dissipation of acid into the bulk volume of the cell (driven by large spatial gradients of pH_i). A subsequent, slower tail of local pH recovery was limited by the slower kinetics of sarcolemmal NHE (during which time cytoplasmic pH_i had become more uniform). Thus, in the first example, the process of pH_i regulation was initiated at the sarcolemma, whereas in the second it was initiated deep within the cell (this latter example would be analogous to, say, a localized metabolic disturbance in the cytoplasm). In both cases, transient, heterogeneous pH_i microdomains were in evidence. Results of the modeling, therefore, confirm that a weak diffusive coupling between cytoplasm and sarcolemma, caused by low proton mobility, can predispose a cell to significant spatiotemporal nonuniformity of pH_i.

Although the above two examples are deliberately extreme, they demonstrate that pH_i nonuniformity is likely to occur at various times in the life of a cardiac cell, with uncalibrated consequences for its function. The modeling also recognizes that pH_i regulation relies on interactions between cytoplasmic diffusion and membrane transport. In some cases, recovery from acidosis in a given region of the cell will occur via the spatial diffusion of acid, whereas in other cases it will be limited by sarcolemmal transport. This emphasizes that an understanding of intracellular pH regulation, like our understanding of intracellular Ca^{2+} regulation, requires a knowledge of control mechanisms operating at a local level within the cell.

ACKNOWLEDGMENTS

This work was supported by a Programme Grant from the British Heart Foundation (R.D.V.J.), a Welcome Prize Studentship, and an Overseas Research Studentship (P.S.). We thank Ms. Nurindura Banger for her excellent technical assistance.

REFERENCES

1. BOUNTRA C. & R.D. VAUGHAN-JONES. 1989. Effect of intracellular and extracellular pH on contraction in isolated, mammalian cardiac muscle. J. Physiol. **418:** 163–187.
2. CHOI, H.S., A.W. TRAFFORD, C.H. ORCHARD & D.A. EISNER. 2000. The effect of acidosis on systolic Ca^{2+} and sarcoplasmic reticulum calcium content in isolated rat ventricular myocytes. J. Physiol. **529:** 661–668.
3. ORCHARD, C.H. & H.E. CINGOLANI. 1994. Acidosis and arrhythmias in cardiac muscle. Cardiovasc. Res. **28:** 1312–1319.
4. FABIATIO, A. & F. FABIATO. 1978. Effects of pH on the myofilaments and sacroplasmic reticulum of skinned cells from cardiac and skeletal muscles. J. Physiol. **276:** 233–255.
5. LEEM, C.H., D. LAGADIC-GOSSMANN & R.D. VAUGHAN-JONES. 1999. Characterization of intracellular pH regulation in the guinea-pig ventricular myocyte. J. Physiol. **517:** 159–180.
6. JUNGE, W. & S. MCLAUGHLIN. 1987. The role of fixed and mobile buffers in the kinetics of proton movement. Biochim. Biophys. Acta **890:** 1–5.
7. VAUGHAN-JONES, R.D., B.E. PEERCY, J.P. KEENER & K.W. SPITZER. 2002. Intrinsic H^+ ion mobility in the rabbit ventricular myocyte. J. Physiol. **541:** 139–158.
8. SWIETACH, P., M. ZANIBONI, A.K. STEWART, *et al.* 2003. Modelling intracellular H^+ ion diffusion. Prog. Biophys. Mol. Biol. **83:** 69–100.
9. ZANIBONI, M., P. SWIETACH, A. ROSSINI, *et al.* 2003. Intracellular proton mobility and buffering power in cardiac ventricular myocytes from rat, rabbit, and guinea pig. Am. J. Physiol. **285:** H1236–H1246.
10. STEWART, A.K, C.A.R. BOYD & R.D. VAUGHAN-JONES. 1999. A novel role for carbonic anhydrase: pH gradient dissipation in mouse small intestinal enterocytes. J. Physiol. **516:** 209–217.
11. SCHWIENING, C.J. & D. WILLOUGHBY. 2002. Depolarisation induced pH microdomains and their relationship to calcium transients in isolated snail neurons. J. Physiol. **538:** 371–382.
12. THOMAS, J.A., R.N. BUCHSBAUM, A. ZIMNIAK & E. RACKER. 1979. Intracellular pH measurements in Ehrlich ascites tumor cells utilising spectroscopic probes generated in situ. Biochemistry **18:** 2210–2218.
13. YAMAMOTO, T., P. SWIETACH, A. ROSSINI, *et al.* 2004. Functional diversity of electrogenic Na^+-HCO_3^- cotransport in ventricular myocytes from rat, rabbit and guinea pig. J. Physiol. **562:** 455–475.
14. PUTNAM, R.W. 1996. Intracellular pH regulation in detubulated frog skeletal muscle fibres. Am. J. Physiol. **271:** C1358–C1366.
15. PETRECCA, K., R. ATANASIU, S. GRINSTEIN, *et al.* 1999. Subcellular localization of the Na^+/H^+ exchanger NHE1 in rat myocardium. Am. J. Physiol. **276:** H709–H717.
16. ZANIBONI, M., A. ROSSINI, P. SWIETACH, *et al.* 2003b. Proton permeation through the myocardial gap junction. Circ. Res. **93:** 726–735.
17. SPITZER, K.W., R.L. SKOLNICK, B.E. PEERCY, *et al.* 2002. Facilitation of intracellular H^+ ion mobility by CO_2/HCO_3^- in rabbit ventricular myocytes is regulated by carbonic anhydrase. J. Physiol. **541:** 159–167.

Cardiac Systems Biology

ANDREW D. MCCULLOCH[a] AND GIOVANNI PATERNOSTRO[b]

[a] *Department of Bioengineering, University of California San Diego, La Jolla, California 92093, USA*

[b] *The Burnham Institute, La Jolla, California 92037, USA*

ABSTRACT: As more detailed molecular information accumulates on the biology of the heart and other complex systems in health and disease, the need for new integrative analyses and tools is growing. Systems biology and bioengineering seek to use high-throughput technologies and integrative computational analysis to construct networks of the interactions between molecular components in the system, to develop systems models of their functionally integrated biological properties, and to incorporate these systems models into structurally integrated multi-scale models for predicting clinical phenotypes. This review gives examples of recent applications using these approaches to elucidate the electromechanical function of the heart in aging and disease.

KEYWORDS: genomic phenotyping; heart; multi-scale modeling; systems bioengineering

INTRODUCTION

With the success of new technologies for genome sequencing, expression profiling, proteomics, structural biology, and *in vivo* gene targeting, the reductionist emphasis of late twentieth century biology is now being complemented by a new emphasis on integrative biology in the twenty-first century and driven by advances in bioinformatics, systems science, and multi-scale bioengineering modeling.

It is helpful to begin with some working definitions: *Bioinformatics* provides the tools to organize and synthesize information from extensive data on biological components into working knowledge and hypotheses. *Systems biology* combines the molecular components of biological systems and subsystems into their interaction networks and formulates functionally integrated systems models for analyzing the input–output relationships of the networks and understanding how their emergent properties perform required biological functions. *Multi-scale bioengineering* combines systems models with data on biological structure across physical scales and uses physico-chemical principles together with engineering methods to develop structurally and functionally integrative predictive models of the dynamic physiology of living systems in health and disease. The common theme is *integration*. Therefore, bioinformatics facilitates *data integration*; systems biology uses this

Address for correspondence: Prof. Andrew McCulloch, Ph.D., Department of Bioengineering, University of California San Diego, 9500 Gioman Drive, La Jolla, CA 92093-0412, USA. Voice: 858-534-2547; fax: 858-534-5722.

amcculloch@ucsd.edu

Ann. N.Y. Acad. Sci. 1047: 283–295 (2005). © 2005 New York Academy of Sciences.
doi: 10.1196/annals.1341.025

information to develop models that are *functionally integrated* across the molecular components of biological networks and across interacting physiological subsystems; multi-scale bioengineering develops models that are *structurally integrated* across scales of biological organization from molecule to organism or population.

The need for integrative analysis in cardiac physiology and pathophysiology is readily appreciated. Common heart diseases are multi-factorial, multi-genic, and linked to other systemic disorders such as diabetes, hypertension, or thyroid disease. Cardiac structure and function are heterogeneous, and most pathologies are associated with regionally inhomogeneous alterations in function. Even basic physiological functions such as cardiac pacemaker activity involve the coordinated interaction of many gene products in the cell, and indeed the interactions between heterogeneous cells and extracellular matrix. Cardiac pathologies with known molecular etiologies are often seen to depend critically on anatomic substrates for their expression *in vivo*; and of course there are many interacting subsystems at many scales involved in physiological processes. For example, substrate and oxygen delivery influence myocyte metabolism, which strongly affects myocardial mechanoenergetics, in turn altering ventricular stresses, which modulate coronary blood flow thus affecting substrate and oxygen delivery.

We summarize here some novel examples of data integration strategies for network discovery, functionally integrated systems models, and structurally integrated multi-scale modeling that have been used recently to investigate cardiac genotype-phenotype relations in acquired and genetic diseases and aging.

HIGH-THROUGHPUT CARDIAC PHENOTYPING AND DATA INTEGRATION FOR GENE DISCOVERY AND NETWORK RECONSTRUCTION

While we have lots of genomic and proteomic data, discovering their functions in a complex system like the heart is an exhaustive process often based on genetic manipulation and gene targeting in model organisms like the mouse. These approaches have been enormously informative, but a single transgenic mouse takes months to develop and study; a knockout mouse, over a year. The phenotypic consequences of genetic defects manifest gradually with age. Mice can live to three years of age, making studies very long and costly. Occasionally, two genetically engineered strains have been crossed to investigate pair-wise gene interactions. Back-crossing into different backgrounds has demonstrated the profound effects that genetic polymorphisms can have on phenotype, but does not immediately point to the interacting genes. Systematically profiling phenotypic variations in consomic mouse[1] or rat[2] strains—in which single chromosomes have been swapped between inbred strains—is one systematic approach to this problem, but this still requires the painstaking process of breeding and studying numerous congenic mice to isolate polymorphic alleles on the chromosome of interest.

In simpler systems, such as *Escherichea coli*, *Caenorhabditis elegans*, and *Saccharomyces cerevesiae*, that are amenable to large scale mutagenesis and rapid phenotyping, a strategy known as *genomic phenotyping* has been developed. When phenotypic characteristics can be assayed for genome scale mutant libraries, a large number of candidate alleles for further investigation can be found. Combining this new phenotypic information with genomic or proteomic data can then be used to help discover novel networks and pathways. For example, in *S. cerevisiae*, Begley and co-workers[3] used genome-wide phenotyping to identify hundreds of alleles important for cellular recovery after exposure to mutagens. By integrating the phenotypic data on cell recovery in over 1,600 gene deletion mutants with a database of the yeast interactome, they identified several multi-protein networks important for damage recovery. Some networks were associated with DNA metabolism and repair, but most were associated with unexpected functions such as cytoskeletal remodeling and lipid metabolism. In this way, systematic phenotypic assays can be linked to underlying molecular pathways.

Might this data integration approach be feasible for studying cardiac biology? There is reason to believe it might be. The fruit fly (*Drosophila melanogaster*) was the first organism with a functioning circulation to have its genome sequenced.[4]

The heart of the fly consists of a tubular structure that contracts spontaneously throughout the insect's lifespan and has the main function of circulating the hemolymph, which transports energy substrates from the abdomen to the thorax and head.[5]

Several human disease models have been developed in *Drosophila*, particularly for neurological diseases.[6–8] *Drosophila* is also commonly employed as a model organism for studying the genetics of aging, partly because it represents a genetically tractable organism with a short life span.[9] Genetic screens have allowed the identification of single genes that control life span in flies.[10,11]

We made the first attempt to exploit *Drosophila melanogaster* for investigations of adult cardiac dysfunction.[12,13] Lakatta commented that this work constitutes the "first big step" toward establishing *Drosophila* as a valuable new model in the study of cardiac aging.[12] Studies of larval and pupal fly hearts, using different techniques, had been reported previously.[14] We developed methods for studying cardiac function *in vivo* in adult flies. Using two different cardiovascular stress methods (elevated ambient temperature and external electrical pacing), we found that maximal heart rate is significantly and reproducibly reduced with aging in *Drosophila*, analogous to observations in elderly humans.[15] The temperature stress is, in insects, a physiological stress on cardiac function. Insects are poikilotherms—their body temperature changes according to the environmental temperature. Therefore, a change in ambient temperature would affect the metabolic rate and require an increase in cardiac output. Enzymatic activity does indeed increase in insects with temperature over the range we studied.[16] An increase in oxygen consumption has also been shown in *Drosophila* with increasing ambient temperature.[17]

The main advantages of *Drosophila melanogaster* as a model organism are as follows:

- Large unbiased genetic screens can be performed.
- Genetic interactions can be identified by crossing the mutant of interest with collections of other mutants and identifying modifier genes (suppressors or enhancers).
- The short life span facilitates the study of the effects of aging.
- Many mutants and chromosomal markers are available.
- Genetic techniques are very powerful and have been perfected since the beginning of the last century.

- The genome has been sequenced, greatly speeding efforts to identify the genes responsible for mutant phenotypes.

The main disadvantage is the evolutionary distance from man. However, many important findings for human medicine and biology have originated from studies in *Drosophila* or other invertebrates. Examples are the identification of genes regulating embryonal development in *Drosophila*,[18] the initial identification of several components of the apoptotic machinery,[19] and elucidation of gene pathways involved in neurogenesis.[20] In all these examples, rapid progress in the understanding of complex problems was made possible by initial investigations in *Drosophila* and subsequent extensions of the findings to mammals and humans. Several surveys have shown remarkable conservation of human genes in the fly genome, including cardiac disease-relevant genes.[21,22]

The cardiac functional measurement obtained from the *Drosophila* model that is best suited for future studies in mice and humans is the decline of maximum heart rate with age. This is a very important index of cardiac function, as it is one of the main causes of the decrease in the capacity for physical work in older people.[23] The other determinant of cardiac output, the stroke volume (the amount of blood pumped with each heart beat), does not change with age in humans.[23] This age-related change is also present in rodents,[24,25] and it is easily measured in rodents and humans using non-invasive methods.[15,24,25]

Scientifically rigorous human mortality trials for anti-aging interventions would require a very large population of subjects and to our knowledge have never been successfully performed. The problem is that the statistical power of survival analysis depends on the number of deaths within the study period. A physiological parameter such as maximal heart rate, however, could be measured in each individual, and therefore provide a more realistic end point for human studies.

Using a systems biology approach, after a complete list of the parts of the system is obtained, repeated cycles of modeling and systematic experiments take place. The three essential components of systems biology are

(1) High-throughput technology to acquire biological data;
(2) Biological tools for genome level perturbation studies, which are available in a few model organisms: yeast, *Drosophila*, *C. elegans*, and mouse; and
(3) Quantitative models.

As an initial quantitative model, we recently proposed a computational model of aging that can explain several of the properties of aging genes, including their conservation across evolutionary distant species.[26] The model is supported by an analysis of aging genes and interaction databases.[26]

The broad spectrum of biological dysfunction that occurs during aging suggests that gene network level properties are especially important. We also showed that local network connectivity (the number of links per node) is significantly higher in aging genes,[26] as also reported by Promislow.[27]

Many aging genes have been found from unbiased screens in model organism. Genetic interventions promoting longevity are usually quantitative, the number of genes involved is large, and the effect of the genetic background is strong. It is not clear how optimal interventions can be devised on such a complex genetics. In the case of aging research, therefore, the need for a quantitative model to guide and help interpret experiments is especially important.

The computational model presented in Ferrarini *et al.*[26] suggests a general conclusion: restoring some nodes of a damaged network can have a large effect on the function of the network, uniquely because of their topological properties. In other words, higher connectivity might not just be a property of aging genes but actually the reason why they can affect the generalized dysfunction of aging.

This functional network model (based on network dynamics) can be further developed to allow the study of real biological networks and the integration of biological data. We are also developing high-throughput technology for the measurement of cardiac aging in *Drosophila*. A systems biology approach can therefore be used for the study of cardiac aging in *Drosophila*, iterating modeling and data collection.

FUNCTIONALLY INTEGRATED SYSTEMS MODELS OF SIGNAL TRANSDUCTION IN CARDIAC MYOCYTES

Cardiac Myocyte Models

The first cardiac myocyte models, published in 1960,[28] used the formalism developed by Hodgkin and Huxley[29] to model the contributions of sodium and potassium currents to the action potential (AP) and to investigate the mechanistic basis of the plateau of the cardiac AP. As new experimental data on myocyte electrophysiology were obtained, the models were refined and extended, and their results in turn informed new experiments. The latest generation of cardiac myocyte ionic models include upward of twenty ionic fluxes and forty or more ordinary differential equations. They also include separate compartments representing the sarcoplasmic reticulum (SR), the narrow subsarcolemmal dyadic space between the sarcolemmal dihydropyridine receptor and the ryanodine receptor on the SR, and the bulk myoplasm. But these compartments are lumped, and these "common pool" models have no spatial structure. They are functionally integrated systems models.

In addition to integrating across the cellular components responsible for ion fluxes and AP generation, systems models have also been developed that integrate the functional components of cellular subsystems responsible for calcium handling, myofilament interactions, and energy metabolism. This then creates the opportunity to develop systems models that not only integrate the components within a single subsystem but that also integrate the functions of two or more interacting subsystems. Excitation–contraction (E-C) coupling is one obvious such interaction, and mechanistic systems models of cardiac E-C coupling have been published since the early 1980s.[30] Since then, more detailed models have been developed of myofilament activation by calcium and the effects of feedback from crossbridge binding.[31] Bluhm *et al.*[32] coupled the Luo-Rudy ionic model to a model of myofilament activation to predict the time-course of isometric tension development following an increase in sarcomere length.

This progress naturally suggests the integration of electrophysiological models with models of energy metabolism in the cardiac myocyte. The first comprehensive example was published by Ch'en and colleagues.[33] The analysis gave valuable insights into the arrhythmogenic mechanisms of ischemia, especially during the highly vulnerable reperfusion period. More recently, a comprehensive thermokinetic model by Cortassa and co-workers[34] analyzed control of cardiac mitochondrial bioenergetics by combining equations for the tricarboxylic acid (TCA) cycle, oxidative phosphorylation, and mitochondrial calcium handling. The model reproduced observations on mitochondrial bioenergetics, calcium dynamics, and respiratory control and demonstrated how calcium feedback provides a mechanism for matching mitochondrial energy production with the metabolic demand of the myocyte as workload changes. Michailova and McCulloch[35] extended the model of the ventricular myocyte by Winslow *et al.*[36] by incorporating equations for calcium and magnesium buffering and transport by adenosine triphosphate (ATP) and adenosine diphosphate (ADP) and equations for MgATP regulation of the sodium-potassium pump, and the sarcolemmal and sarcoplasmic calcium pumps. Under normal conditions, the model showed that calcium binding by low-affinity ATP and diffusion of CaATP might affect the amplitude and time-course of intracellular calcium signals. Some of these predictions were subsequently supported by experimental observations.[37] More recently, this model has also been used to study the effects of magnesium on ventricular excitation-contraction coupling.[38]

Systems Models of Signal Transduction Pathways

Computational models can help to provide a quantitative understanding of how signaling networks control cellular physiology. We distinguish between models of cell signaling at two levels of detail: large-scale top-down models and bottom-up mechanistic systems models. Each approach has advantages and disadvantages, while some hybrid approaches such as Bayesian network modeling offer advantages from each modeling strategy.[39] Top-down modeling, often based on noisy proteomic or protein interaction data such as yeast two-hybrid studies, shows promise for understanding the qualitative properties of signaling networks.[40,41] Top-down models may also be useful for discovering novel pathways.[42]

Mechanistic systems models typically begin by formulating biochemical reactions with kinetic rate laws as differential equations using available experimental data on describing reaction kinetics and parameters. They should be able to make quantitative, experimentally testable predictions.

In silico analysis of network perturbations allows us to formulate new hypotheses regarding signaling pathways. Hoffmann *et al.*[43] used this approach to predict functional differences between IκB isoforms *in silico*, and then used the model to design an experiment to test this prediction. Bhalla and Iyengar[44] showed that feedback and crosstalk between neuronal signaling pathways could lead to emergent properties such as bistability, which may lead to long-term potentiation. When mechanisms are unknown and several hypotheses are plausible, systems models representing alternative hypotheses may be compared quantitatively. This approach was used to identify transitions between mode 1 and mode 2 gating of the L-type calcium channel during β-adrenergic stimulation.[45]

However, mechanistic systems models are invariably data limited, especially for less well-characterized signaling networks. Optimizing a reduced model with sparse data is possible but often difficult. Inconsistency of experimental data from different sources may prevent any single model from predicting all data. While high-throughput genomic and proteomic approaches promise vast amounts of data, the data will be structural rather than functional. Mechanistic models of new signaling networks will require uniform high-throughput data and extensive validation against independent measurements.

Systems Models of Myocyte Signaling Pathways

While not yet as comprehensive as databases of metabolic networks, public data resources on cell signaling, such as the Alliance for Cellular Signaling, are emerging.[46] (For another example, see http://www.signaling-gateway.org/.) This is opening the way for more detailed systems models of signal transduction pathways. By integrating models of signaling networks with models of cardiac myocytes, it is now possible to investigate neurohormonal regulation of E-C coupling. We developed a new model of β_1-adrenergic regulation of cardiac E-C coupling by combining a systems model of ionic currents and calcium handling in rat ventricular myocytes, with a novel mechanistic model of cAMP-mediated cell signaling via β_1-adrenergic receptor that included the L-type calcium channel, phospholamban, and inhibitor-1 as phosphorylation targets of protein kinase A (PKA).[47] Later, we extended this model to include troponin-I and the ryanodine receptor (RyR) as additional PKA targets, though the analysis suggested that the controversial role of RyR phosphorylation during adrenergic stimulation may be fairly insignificant under normal conditions in the intact cell, owing to the effects of calcium autoregulation by the sarcoplasmic reticulum.[48] This new class of models may be especially important for elucidating the pathogenesis of congestive heart failure, where dysregulated calcium handling in the myocyte is accompanied by downregulation of β-adrenergic signaling.

Cardiac cellular models are beginning to contribute to understanding phenotypic consequences of gene mutations linked to inherited arrhythmias,[49,50] notably long Q-T syndrome (LQTS), a family of mutations causing prolonged AP duration by affecting genes that encode the subunits of the fast and slow activating delayed rectifier potassium channels (I_{Kr} and I_{Ks}) and the fast sodium channel (I_{Na}). Of these, the most prevalent is LQT1, affecting I_{Ks}, for which several specific defects in KCNQ1 have been identified in man. A complication in extending this work in reconstituted systems such as transgenic mice is that I_{Kr} and I_{Ks} are not present in mouse, a species which depends mainly on I_{to} and I_{Ksus} for repolarization.[51] Therefore, computational analyses become more valuable and have helped elucidate arrhythmogenic mechanisms in LQT3,[52] LQT2, and LQT6[50] by incorporating biophysical properties of mutant channels (in heterologous systems) into ventricular cell models. Interestingly, in a study of LQT1 patients, 82% of lethal cardiac events occurred during exercise or emotional stress, suggesting a connection with sympathetic stimulation.[53]

Recently, a mutation identified in a Finnish population (KCNQ1-G589D) was shown to disrupt binding of a PKA anchoring protein (AKAP9) known as Yotaio, to a leucine zipper motif near the C-terminus of the pore-forming subunit of the slow activating delayed rectifier potassium channel, I_{Ks}.[54] Yotaio was shown in this study to target PKA and phosphatase-1 to the channel. This mutation therefore selectively disables β-adrenergic regulation of I_{Ks}, which is the main mechanism of APD shortening with increasing heart rate during sympathetic stimulation. To investigate how this defect may lead to ventricular arrhythmia during sympathetic stimulation, we reconstituted it in an *in silico* model of beta-adrenergic signaling and E-C coupling in the rabbit ventricular myocyte.[55] While the KCNQ1-G589D mutation alone did not prolong the resting AP when coupled with β-adrenergic stimulation in the whole cell, the mutation induced AP prolongation instead of shortening and after-depolarizations associated with reactivation of the L-type calcium channel. These changes are recognized cellular mechanisms for arrhythmogenesis.

STRUCTURALLY INTEGRATED MULTI-SCALE MODELS OF WHOLE HEART PHENOTYPE

Multicellular Models of Myocardial Electromechanics

Multicellular models come in three main varieties: cellular automata, resistively coupled networks, and continuum models. Cellular automata and resistively coupled networks use physical properties or rules that approximate them to combine individual cells modeled as systems of ordinary differential equations as described earlier. These models are computationally tractable and have been particularly informative in elucidating the effects of electrical loading by neighboring cells on the propagation of the AP from cell to cell via current flux through gap junctions. Continuum models of electrical impulse propagation typically use bi-domain or monodomain theory, in which the intracellular and extracellular conductivities of the tissue are lumped into diffusion tensors, which are anisotropic and may differ between intracellular and extracellular domains. Continuum methods are also widely used for modeling regional biomechanics in the heart and other tissues.[56] Within the continuum framework, however, a constitutive model of the contributions of cellular and extracellular constituents is required to relate the passive and contractile stresses in the tissue to the state of deformation (strain). While many constitutive models are phenomenological relations, curve-fitted to multiaxial experimental measurements of tissue mechanical responses,[57] it is possible to apply micromechanical principles and quantitative histological measurements to derive microstructural constitutive relations.[58]

The development of anatomically detailed models of ventricular geometry and muscle fiber architecture in dog,[59] rabbit,[60] and pig,[61] and atrial geometry and fiber bundle architecture in man[62] has enabled investigators to develop structurally integrated continuum models of cardiac electrical impulse propagation[62,63] and wall mechanics[61] by using finite volume, finite difference, or finite element methods. As cellular models become more biophysically detailed and functionally integrated, the opportunity is arising for structurally integrated models that are also functionally integrated, such as continuum models of coupled ventricular electromechanics.[64–68]

Returning to LQT1, it is apparent that the cellular defect alone is not sufficient to explain the incidence of malignant tachyarryhthmias. A growing body of evidence, especially from isolated perfused myocardial "wedge" preparations, suggests that transmural heterogeneity of cell type can lead to increased dispersion of repolarization and "phase 2 reentry" in the setting of drugs that prolong QT interval.[69,70] In the absence of animal models, studies of drug-induced QT prolongation have also been the main model systems for understanding the mechanisms of malignant arrhythmias in the genetic LQTSs associated with loss of I_{Kr} and I_{Ks} function. Therefore, we incorporated the cellular model of the KCNQ1-G589D mutation into a three-dimensional continuum model of AP propagation in a rabbit ventricular wedge with transmurally varying muscle fiber orientations.[55] While the KCNQ1-G589D mutation alone did not prolong the QT interval when coupled with β-adrenergic stimulation in the whole cell, the mutation induced QT prolongation and after-depolarizations, which are potential cellular mechanisms for arrhythmogenesis. These alterations amplified transmural repolarization heterogeneities in a three-dimensional rabbit ventricular wedge model and led to other T-wave abnormalities on simulated elec-

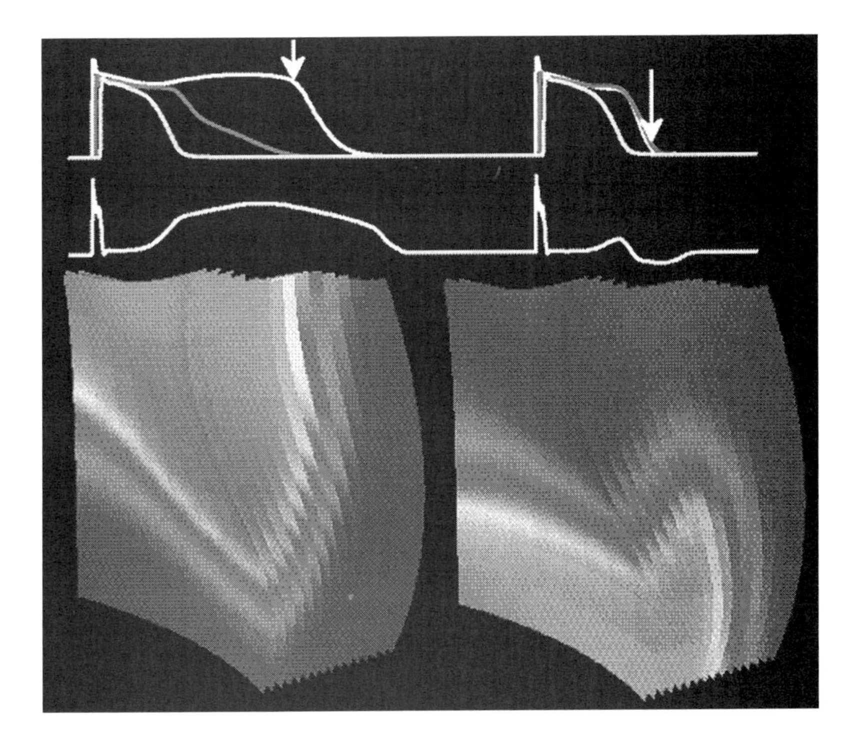

FIGURE 1. Proarrhythmic repolarization abnormalities predicted by a computational model of KCNQ1-G589D mutation in a ventricular wedge with β-adrenergic stimulation.[55] Action potentials (*upper tracings*) from endocardium (*earliest upstroke, white*), midmyocardium (*gray, red, in Annals online*), and epicardium (*latest upstroke, yellow*) and corresponding electrocardiogram (*lower tracing*) for the wedge model. Snapshots of repolarization at the times indicated by arrows reveal an early afterdepolarization confined to the endocardium (*left side of each wedge, hot colors*), creating a very broad T-wave (*left*). This causes altered three-dimensional repolarization sequence and T-wave inversion due to abbreviated endocardial action potential duration on the subsequent beat (*right*) [in color in Annals Online].

trocardiograms (FIG. 1). This model was is one of the first examples in which a single gene defect has been reconstituted into a ventricular tissue model computationally before it was done experimentally.

CONCLUSION

Bioinformatics, systems biology, and multi-scale bioengineering approaches are demonstrating potential to contribute to an integrative understanding of normal cardiac biology and the changes that occur during aging and genetic and acquired diseases. As computational power and high-throughput technologies continually improve, we will eventually be able to reconstruct and model more comprehensive and fully characterized cellular networks and use these systems models in multi-scale predictions of organ system phenotype in human heart diseases.

ACKNOWLEDGMENTS

We acknowledge the contributions of our colleagues and students Jake Feala, Luca Ferrarini, Anushka Michailova, Jeffrey Saucerman, Sarah Healy, and Taras Usyk. This research was supported by the National Institutes of Health through the National Biomedical Computation Resource (P41-RR08605), the National Science Foundation (BES-0086482), and a grant from the Defense Advanced Research Projects Agency (executed by the U.S. Army Medical Research and Materiel Command/ TATRC Cooperative Agreement, Contract # W81XWH-04-2-0012).

REFERENCES

1. Singer, J.B., A.E. Hill, L.C. Burrage, *et al.* 2004. Genetic dissection of complex traits with chromosome substitution strains of mice. Science **304:** 445–448.
2. Cowley, A.W., Jr., M. Liang, R.J. Roman, *et al.* 2004. Consomic rat model systems for physiological genomics. Acta Physiol. Scand. **181:** 585–592.
3. Begley, T.J., A.S. Rosenbach, T. Ideker & L.D. Samson. 2002. Damage recovery pathways in *Saccharomyces cerevisiae* revealed by genomic phenotyping and interactome mapping. Mol. Cancer Res. **1:** 103–112.
4. Adams, M.D., S.E. Celniker, R.A. Holt, *et al.* 2000. The genome sequence of Drosophila melanogaster. Science **287:** 2185–2195.
5. Rizki, T.M. 1978. The circulatory system and associated cells and tissues. *In* The Genetics and Biology of Drosophila, Vol. 2B. M. Ashburner & T.R.F. Wright, Eds.: 397–452. Academic Press. London.
6. Feany, M.B. & W.W. Bender. 2000. A Drosophila model of Parkinson's disease. Nature **404:** 394–398.
7. Kazemi-Esfarjani, P. & S. Benzer. 2000. Genetic suppression of polyglutamine toxicity in Drosophila. Science **287:** 1837–1840.
8. Min, K.T. & S. Benzer. 1999. Preventing neurodegeneration in the Drosophila mutant bubblegum. Science **284:** 1985–1988.
9. Rose, M.R. 1991. Evolutionary Biology of Aging. Oxford University Press. New York.
10. Lin, Y.J., L. Seroude & S. Benzer. 1998. Extended life-span and stress resistance in the Drosophila mutant methuselah. Science **282:** 943–946.
11. Rogina, B., R.A. Reenan, S.P. Nilsen & S.L. Helfand. 2000. Extended life-span conferred by cotransporter gene mutations in drosophila. Science **290:** 2137–2140.
12. Lakatta, E.G. 2001. Heart aging: a fly in the ointment? Circ. Res. **88:** 984–986.
13. Paternostro, G., C. Vignola, D.U. Bartsch, *et al.* 2001. Age-associated cardiac dysfunction in Drosophila melanogaster. Circ. Res. **88:** 1053–1058.
14. White, L.A., J.M. Ringo & H.B. Dowse. 1992. Effects of deuterium oxide and temperature on heart rate in Drosophila melanogaster. J. Comp. Physiol. [B]. **162:** 278–283.
15. Fleg, J.L., F. O'Connor, G. Gerstenblith, *et al.* 1995. Impact of age on the cardiovascular response to dynamic upright exercise in healthy men and women. J. Appl. Physiol. **78:** 890–900.
16. Evans, W.A.L. & D.W. Payne. 1964. Carbohydrases of the alimenetary tract of the desert locust, *Schistocerca gregaria*. J. Insect Physiol. **10:** 657–674.
17. Miquel J., P.R. Lundgren, K.G. Bensch & H. Atlan. 1976. Effects of temperature on the life span, vitality and fine structure of Drosophila melanogaster. Mech. Ageing Dev. **5:** 347–370.
18. Nusslein-Volhard, C. & E. Wieschaus. 1980. Mutations affecting segment number and polarity in Drosophila. Nature **287:** 795–801.
19. White, K., M.E. Grether, J.M. Abrams, *et al.* 1994. Genetic control of programmed cell death in Drosophila. Science **264:** 677–683.
20. Artavanis-Tsakonas, S., M.A. Muskavitch & B. Yedvobnick. 1983. Molecular cloning of Notch, a locus affecting neurogenesis in Drosophila melanogaster. Proc. Natl. Acad. Sci. U.S.A. **80:** 1977–1981.

21. FORTINI, M.E., M.P. SKUPSKI, M.S. BOGUSKI & I.K. HARIHARAN. 2000. A survey of human disease gene counterparts in the drosophila genome. J. Cell Biol. **150:** F23–F30.
22. REITER, L.T., L. POTOCKI, S. CHIEN, *et al.* 2001. A systematic analysis of human disease-associated gene sequences in Drosophila melanogaster. Genome Res. **11:** 1114–1125.
23. LAKATTA, E.G. 2002. Introduction: chronic heart failure in older persons. Heart Fail. Rev. **7:** 5–8.
24. CORRE, K.A., H. CHO & R.J. BARNARD. 1976. Maximum exercise heart rate reduction with maturation in the rat. J. Appl. Physiol. **40:** 741–744.
25. BARNARD, R.J., H. DUNCAN & A.T. THORSTENSSON. 1974. Heart rate responses of young and old rats to various levels of exercise. J. Appl. Physiol. **36:** 472–474.
26. FERRARINI, L., L. BERTELLI, J. FEALA, *et al.* 2005 A more efficient search strategy for aging genes based on connectivity. Bioinformatics **21:** 338–348
27. PROMISLOW, D.E. 2004. Protein networks, pleiotropy and the evolution of senescence. Proc. R. Soc. Lond. B. Biol. Sci. **271:** 1225–1234.
28. NOBLE, D. 1960. Cardiac action and pacemaker potentials based on the Hodgkin-Huxley equations. Nature **188:** 495–497.
29. HODGKIN, A.L. & A.F. HUXLEY. 1952. A quantitative description of membrane current and its application to conduction and excitation in nerve. J. Physiol. **117:** 500–544.
30. WONG, A.Y. 1981. A model of excitation-contraction coupling of mammalian cardiac muscle. J. Theor. Biol. **90:** 37–61.
31. RICE, J.J. & P.P. DE TOMBE. 2004. Approaches to modeling crossbridges and calcium-dependent activation in cardiac muscle. Prog. Biophys. Mol. Biol. **85:** 179–195.
32. BLUHM, W.F., W.Y. LEW, A. GARFINKEL & A.D. MCCULLOCH. 1998. Mechanisms of length-history dependent tension in an ionic model of the cardiac myocyte. Am. J. Physiol. **274:** H1032–H1040.
33. CH'EN, F., K. CLARKE, R. VAUGHAN-JONES & D. NOBLE. 1997. Modeling of internal pH, ion concentration, and bioenergetic changes during myocardial ischemia. *In* Analytical and Quantitative Cardiology. S. Sideman & R. Beyar, Eds.: 281–290. Plenum. New York.
34. CORTASSA, S., M.A. AON, E. MARBAN, *et al.* 2003. An integrated model of cardiac mitochondrial energy metabolism and calcium dynamics. Biophys. J. **84:** 2734–2755.
35. MICHAILOVA, A. & A. MCCULLOCH. 2001. Model study of ATP and ADP buffering, transport of Ca(2+) and Mg(2+), and regulation of ion pumps in ventricular myocyte. Biophys. J. **81:** 614–629.
36. WINSLOW, R.L., J. RICE, S. JAFRI, *et al.* 1999. Mechanisms of altered excitation-contraction coupling in canine tachycardia-induced heart failure, II: model studies. Circ. Res. **84:** 571–586.
37. YANG, Z. & D.S. STEELE. 2001. Effects of cytosolic ATP on Ca(2+) sparks and SR Ca(2+) content in permeabilized cardiac myocytes. Circ. Res. **89:** 526–533.
38. MICHAILOVA, A., M.E. BELIK & A. MCCULLOCH. 2004. Effects of magnesium on cardiac excitation-contraction coupling. J. Am. Coll. Nutr. **23:** 514S–517S
39. SACHS, K., D. GIFFORD, T. JAAKKOLA, *et al.* 2002. Bayesian network approach to cell signaling pathway modeling. Sci. STKE. **2002:** PE38.
40. IDEKER, T. & D. LAUFFENBURGER. 2003. Building with a scaffold: emerging strategies for high- to low-level cellular modeling. Trends Biotechnol. **21:** 255–262.
41. WUCHTY, S., Z.N. OLTVAI & A.L. BARABASI. 2003. Evolutionary conservation of motif constituents in the yeast protein interaction network. Nat. Genet. **35:** 176–179.
42. IDEKER, T., O. OZIER, B. SCHWIKOWSKI & A.F. SIEGEL. 2002. Discovering regulatory and signalling circuits in molecular interaction networks. Bioinformatics. **18**(Suppl. 1): S233–S240.
43. HOFFMANN, A., A. LEVCHENKO, M.L. SCOTT & D. BALTIMORE. 2002. The IkappaB-NF-kappaB signaling module: temporal control and selective gene activation. Science. **298:** 1241–1245.
44. BHALLA, U.S. & R. IYENGAR. 1999. Emergent properties of networks of biological signaling pathways. Science **283:** 381–387.
45. HERZIG, S., P. PATIL, J. NEUMANN, *et al.* 1993. Mechanisms of beta-adrenergic stimula-

tion of cardiac Ca2+ channels revealed by discrete-time Markov analysis of slow gating. Biophys. J. **65:** 1599–1612.
46. Papin, J. & S. Subramaniam. 2004. Bioinformatics and cellular signaling. Curr. Opin. Biotechnol. **15:** 78–81.
47. Saucerman, J.J., L.L. Brunton, A.P. Michailova & A.D. McCulloch. 2003. Modeling beta-adrenergic control of cardiac myocyte contractility *in silico*. J. Biol. Chem. **278:** 47997–48003.
48. Saucerman, J.J. & A.D. McCulloch. 2004. Mechanistic systems models of cell signaling networks: a case study of myocyte adrenergic regulation. Prog. Biophys. Mol. Biol. **85:** 261–278.
49. Rudy, Y. 2004. From genetics to cellular function using computational biology. Ann. N.Y. Acad. Sci. **1015:** 261–270.
50. Mazhari, R., J.L. Greenstein, R.L. Winslow, *et al.* 2001. Molecular interactions between two long-QT syndrome gene products, HERG and KCNE2, rationalized by *in vitro* and *in silico* analysis. Circ. Res. **89:** 33–38.
51. Fiset, C., R.B. Clark, T.S. Larsen & W.R. Giles. 1997. A rapidly activating sustained K+ current modulates repolarization and excitation-contraction coupling in adult mouse ventricle. J. Physiol. (Lond.) **504:** 557–563.
52. Clancy, C.E. & Y. Rudy. 1999. Linking a genetic defect to its cellular phenotype in a cardiac arrhythmia. Nature **400:** 566–569.
53. Schwartz, P.J., S.G. Priori, C. Spazzolini, *et al.* 2001. Genotype-phenotype correlation in the long-QT syndrome: gene-specific triggers for life-threatening arrhythmias. Circulation **103:** 89–95.
54. Marx, S.O., J. Kurokawa, S. Reiken, *et al.* 2002. Requirement of a macromolecular signaling complex for beta adrenergic receptor modulation of the KCNQ1-KCNE1 potassium channel. Science **295:** 496–499.
55. Saucerman, J.J., S.N. Healy, M.E. Belik, *et al.* 2004. Proarrhythmic consequences of a KCNQ1 AKAP-binding domain mutation. Computational models of whole cells and heterogeneous tissue. Circ. Res. **95:** 1216–1224.
56. Usyk, T.P. & A.D. McCulloch. 2003. Computational methods for soft tissue biomechanics. *In* Biomechanics of Soft Tissue in Cardiovascular Systems, Vol. 441. G.A. Holzapfel & R.W. Ogden, Eds.: 273–342. Springer. Vienna.
57. Lin, D.H.S. & F.C.P. Yin. 1998. A multiaxial constitutive law for mammalian left ventricular myocardium in steady-state barium contracture or tetanus. J. Biomech. Eng. **120:** 504–517.
58. MacKenna, D.A., S.M. Vaplon & A.D. McCulloch. 1997. Microstructural model of perimysial collagen fibers for resting myocardial mechanics during ventricular filling. Am. J. Physiology. **273:** H1576–H1586.
59. Nielsen, P.M.F., I.J. Le Grice, B.H. Smaill & P.J. Hunter. 1991. Mathematical model of geometry and fibrous structure of the heart. Am. J. Physiol. **260:** H1365–H1378.
60. Vetter, F.J. & A.D. McCulloch. 1998. Three-dimensional analysis of regional cardiac function: a model of rabbit ventricular anatomy. Prog. Biophys. Mol. Biol. **69:** 157–183.
61. Stevens, C. & P.J. Hunter. 2003. Sarcomere length changes in a 3D mathematical model of the pig ventricles. Prog. Biophys. Mol. Biol. **82:** 229–241.
62. Harrild, D. & C. Henriquez. 2000. A computer model of normal conduction in the human atria. Circ. Res. **87:** E25–E36.
63. Xie, F., Z. Qu, J. Yang, *et al.* 2004. A simulation study of the effects of cardiac anatomy in ventricular fibrillation. J. Clin. Invest. **113:** 686–693.
64. Usyk, T.P., I.J. LeGrice & A.D. McCulloch. 2002. Computational model of three-dimensional cardiac electromechanics. Comput. Vis. Sci. **4:** 249–257.
65. Usyk, T.P. & A.D. McCulloch. 2003. Electromechanical model of cardiac resynchronization in the dilated failing heart with left bundle branch block. J. Electrocardiol. **36:** 57–61.
66. Usyk, T.P. & A.D. McCulloch. 2003. Relationship between regional shortening and asynchronous electrical activation in a three-dimensional model of ventricular electromechanics. J. Cardiovasc. Electrophysiol. **14:** S196–S202.

67. KERCKHOFFS, R.C., P.H. BOVENDEERD, J.C. KOTTE, *et al.* 2003. Homogeneity of cardiac contraction despite physiological asynchrony of depolarization: a model study. Ann. Biomed. Eng. **31:** 536–547.
68. KERCKHOFFS, R.C., O.P. FARIS, P.H. BOVENDEERD, *et al.* 2003. Timing of depolarization and contraction in the paced canine left ventricle: model and experiment. J. Cardiovasc. Electrophysiol. **14:** S188–S195.
69. ANTZELEVITCH, C. 2004. Cellular basis and mechanism underlying normal and abnormal myocardial repolarization and arrhythmogenesis. Ann. Med. **36**(Suppl. 1): 5–14.
70. FENICHEL, R.R., M. MALIK, C. ANTZELEVITCH, *et al.* 2004. Drug-induced torsades de pointes and implications for drug development. J. Cardiovasc. Electrophysiol. **15:** 475–495.

Measuring and Mapping Cardiac Fiber and Laminar Architecture Using Diffusion Tensor MR Imaging

PATRICK HELM,[a] MIRZA FAISAL BEG,[b] MICHAEL I. MILLER,[b] AND RAIMOND L. WINSLOW[a]

[a] *The Center for Cardiovascular Bioinformatics & Modeling, The Johns Hopkins University School of Medicine and Whiting School of Engineering, Baltimore, Maryland 21218, USA*

[b] *The Center for Imaging Sciences, The Johns Hopkins University School of Medicine and Whiting School of Engineering, Baltimore, Maryland 21218, USA*

ABSTRACT: The ventricular myocardium is known to exhibit a complex spatial organization, with fiber orientation varying as a function of transmural location. It is now well established that diffusion tensor magnetic resonance imaging (DTMRI) may be used to measure this fiber orientation at high spatial resolution. Cardiac fibers are also known to be organized in sheets with surface orientation varying throughout the ventricles. This article reviews results on use of DTMRI for measuring ventricular fiber orientation, as well as presents new results providing strong evidence that the tertiary eigenvector of the diffusion tensor is aligned locally with the cardiac sheet surface normal. Considered together, these data indicate that DTMRI may be used to reconstruct both ventricular fiber and sheet organization. This article also presents the large deformation diffeomorphic metric mapping (LDDMM) algorithm and shows that this algorithm may be used to bring ensembles of imaged and reconstructed hearts into correspondence (e.g., registration) so that variability of ventricular geometry, fiber, and sheet orientation may be quantified. Ventricular geometry and fiber structure is known to be remodeled in a range of disease processes; however, descriptions of this remodeling have remained subjective and qualitative. We anticipate that use of DTMRI for reconstruction of ventricular anatomy coupled with application of the LDDMM method for image volume registration will enable the detection and quantification of changes in cardiac anatomy that are characteristic of specific disease processes in the heart. Finally, we show that epicardial electrical mapping and DTMRI imaging may be performed in the same hearts. The anatomic data may then be used to simulate electrical conduction in a computational model of the very same heart that was mapped electrically. This facilitates direct comparison and testing of model versus experimental results and opens the door to quantitative measurement, modeling, and analysis of the ways in which remodeling of ventricular microanatomy influences electrical conduction in the heart.

KEYWORDS: computational modeling; fiber orientation; MRI; sheet orientation; ventricular anatomy

Address for correspondence: Prof. Raimond L. Winslow, Ph.D., Center for Cardiovascular Bioinformatics & Modeling, Johns Hopkins University School of Medicine, Baltimore, MD 21218, USA. Voice: 410-516-5417; fax: 410-516-5294.
rwinslow@bme.jhu.edu

Ann. N.Y. Acad. Sci. 1047: 296–307 (2005). © 2005 New York Academy of Sciences.
doi: 10.1196/annals.1341.026

INTRODUCTION

The fiber structure of the cardiac ventricles plays a critical role in determining properties of electrical conduction in the heart. Conduction is anisotropic, with current spread being most rapid in the direction of the fiber long axis.[1,2] Conduction is also influenced by properties of tissue geometry such as expansion and contraction,[3–5] as well as by the spatial rate of change of fiber orientation.[6] Remodeling of ventricular geometry and fiber organization is a prominent feature of several cardiac pathologies. These alterations may figure importantly in arrhythmogenesis.[7] A detailed knowledge of ventricular fiber structure, how it may be remodeled in cardiac pathology, and the effects of this remodeling on ventricular conduction is, therefore, of fundamental importance to the understanding of cardiac electromechanics in health and disease.

Ventricular fiber orientation was first studied by means of fiber dissections[8–10] and histologic measurements in transmural plugs of ventricular tissue.[11–15] The principle conclusions of this work were that cardiac fibers are arranged as counterwound helices encircling the ventricular cavities and that fiber orientation is a function of transmural location, with fiber direction being oriented predominantly in the base–apex direction on the epicardial and endocardial surfaces and rotating to a circumferential direction in the midwall. These initial fiber dissection studies were restricted to a small number of myocardial measurement sites. Nielsen *et al.*[16] overcame this limitation by using a custom-built histologic apparatus to measure inclination angle at up to ~14,000 points throughout the myocardium, with a resolution of ~500, 2,500, and 5,000 μm in the radial, longitudinal, and circumferential directions, respectively. This work led to the first complete reconstruction of ventricular geometry and fiber orientation—an achievement that has had significant impact on the cardiac mechanics and electrophysiology communities.

Recently, diffusion tensor magnetic resonance imaging (DTMRI), a technique for measuring the self-diffusion of protons in fibrous tissue, has emerged as a powerful new tool for the rapid measurement of cardiac geometry and fiber structure at high spatial resolution. In this article, we will summarize the current state of DTMRI for use in reconstruction of ventricular anatomy. We first review results regarding the application of DTMRI to measure fiber orientation in the cardiac ventricles. These results demonstrate that the primary eigenvector of the diffusion tensor measured at each image voxel is aligned with the direction of the cardiac fiber long axis and that both ventricular geometry and fiber orientation may be measured rapidly and at high spatial resolution using DTMRI. Second, we present new data providing strong support for the hypothesis that the tertiary eigenvector of the diffusion tensor measured at each image voxel is aligned with the surface normal to the cardiac sheets.[17] Third, we describe a quantitative mathematical approach known as large deformation diffeomorphic metric mapping (LDDMM), by which ensembles of imaged and reconstructed hearts may be registered so that statistical analysis of variation of ventricular geometry and shape may be performed. Finally, we demonstrate how DTMRI data may be used in modeling of electrical conduction in the cardiac ventricles.

METHODS

DTMRI Imaging Methods

DTMRI methods have been described previously.[18] Anesthesia was induced with sodium thiopental and maintained after intubation with isofurane. A midline thoracotomy was performed, animals were heparinized and cardiac arrest was induced with a bolus of potassium chloride. Following explantation, a Langendorff preparation was used to perfuse the heart with saline solution until coronary vessels were clear. Vinyl polysiloxane was injected into each of the four chambers via the tricuspid and mitral valves in order to maintain diastolic conformation during imaging. The hearts were then perfused with and stored in a 7% by volume isotonic solution of formaldehyde. Care was taken to fix hearts in the same cardiac phase so that their fiber orientations would be similar. Small variations in cardiac phase during fixation are not problematic since fiber angles do not change significantly between diastole and systole.[11] Each heart was placed in an acrylic container filled with the perfluoropolyether Fomblin (the low dielectric effect and minimal MR signal of Fomblin increases contrast and eliminates unwanted susceptibility artifacts near the boundaries of the heart). The long axis of each heart was aligned with the z axis of the scanner, and images were acquired with a 4-element phased array coil on a 1.5 T GE CV/I MRI Scanner (GE Medical Systems, Wausheka, WI). The system had a 40 mT/m maximum gradient amplitude and a 150 T/m/s slew rate. Hearts were placed in the center of the coil and a 3-D fast spin-echo sequence was used to acquire diffusion images.[18] Total scan time was approximately 60 h per heart.

Reconstruction of Ventricular Geometry and Fiber Organization

After imaging, the three principle eigenvalues and eigenvectors were computed for each voxel from the diffusion tensors.[17,19] Voxels in the 3-D DTMRI images representing compact myocardium were identified using a parametric active contouring method implemented as part of *HeartWorks*.[20] Ventricular geometry and fiber orientation were modeled using 3-D finite elements as described by Nielsen *et al.*[16] using a graphical user interface implemented in Matlab (Mathworks, Natick, MA). Diffusion tensor eigenvectors corresponding to voxels within the finite element boundaries were used to study the fiber structure of the ventricles. These eigenvectors were transformed into local material coordinates of the model as described by LeGrice *et al.*[21] Fiber inclination angle θ (the angle formed by the circumferential epicardial tangent vector t_1 and the projection of the primary eigenvector onto the surface tangent plane defined by vectors t_1 and the axial tangent vector t_2) was used as a measure of fiber orientation and was computed as shown in FIGURE 1. The circumferential tangent vector t_1 was constrained to the xy plane. Intersection angle ϕ was used as a measure of cardiac sheet orientation and was defined relative to the tertiary eigenvector where $\frac{\pi}{2}+\phi$ was the angle formed by the surface normal and the projection of the tertiary eigenvector onto the plane defined by the surface normal n and t_2. A total of 11 normal and 7 tachycardia pacing-induced failing canine hearts were reconstructed using these methods. Data was loaded into the Cardiac Anatomic Database System (CADS)[22] for visualization and analysis.

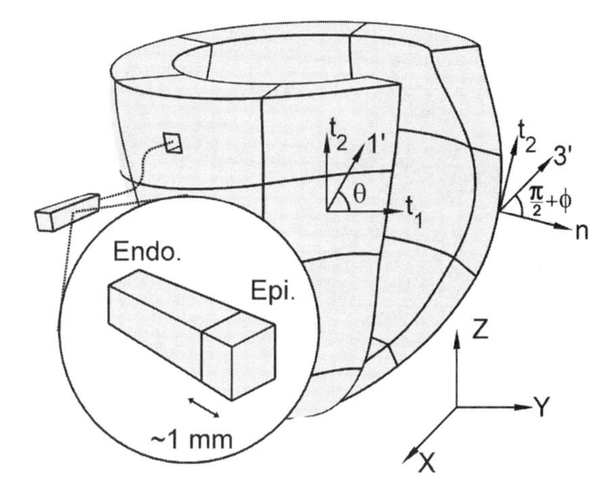

FIGURE 1. The coordinate system used to define the inclination angle, θ, and the intersection angle, ϕ.

Anatomical Registration and Structural Analysis of Imaged Hearts

We have recently developed a mathematical method known as the LDDMM algorithm for accurate registration of ensembles of imaged and reconstructed heart anatomies.[18] In this method, an anatomy is defined as collection of landmarks, curves, surfaces, subvolumes and tensors. A "template" anatomy is selected from a database of imaged hearts, and large deformation mappings are generated and used to deform this template onto "target" anatomies (e.g., other imaged hearts in the database). The transformations are constrained to be 1:1 and differentiable with differentiable inverse so that every point in the template has a corresponding point in the target, connected sets of template image points remain connected in the target, surfaces are mapped as surfaces, and volumes are mapped as volumes, thereby maintaining the global relationships between these structures. Regional anatomic variability is then studied as a property of the local variability of these transformations. The advantages of the LDDMM method over small-deformation mapping schemes are that the computed transformations are invertible and, therefore, the choice of an anatomic template is arbitrary and LDDMM computes a geodesic that is a shortest-length path in the space of transformations matching the given imagery, thereby quantifying deformation via a scalar metric distance.

RESULTS

Substantial data now confirms the hypothesis that the primary eigenvector of the diffusion tensor is aligned locally with the long axis of cardiac fibers.[17,20,23,24] As an example, FIGURE 2A shows fiber inclination angle (ordinate) as a function of transmural depth (abscissa). Open symbols show histological measurements from transmural tissue plugs, and filled circles show DTMRI estimates obtained at the same locations of these transmural tissue plugs. Histological and DTMRI-based measurements are in close agreement. We have also shown that DTMRI can be used to image and reconstruct fiber orientation throughout the cardiac ventricles.[18,20] FIGURE 2B [in color in Annals Online] shows an example of reconstruction of both ven-

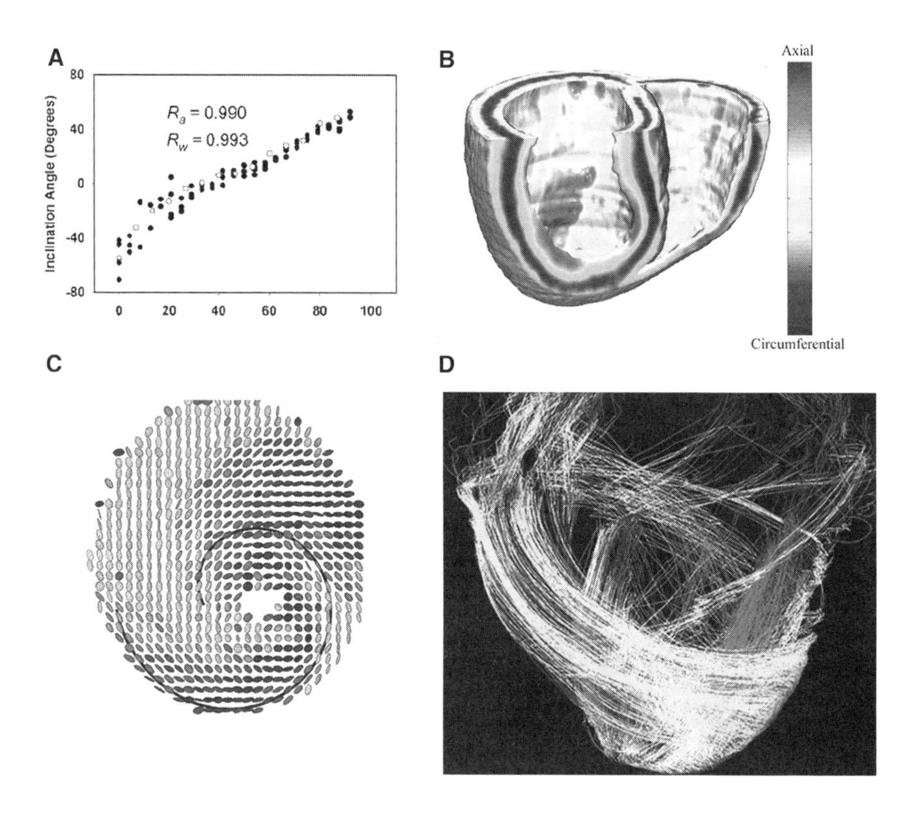

FIGURE 2. (**A**) Comparison of histological and diffusion tensor magnetic resonance imaging (DTMRI)-based estimates of fiber inclination angle at corresponding transmural location. (**B**) Reconstruction of fiber inclination angle throughout the canine cardiac ventricles. (**C**) Map of fiber inclination angle at the apex of a canine heart. (**D**) Fiber tracing performed in a canine heart using DTMRI data [in color in Annals Online].

tricular geometry and fiber orientation in a normal canine heart. Fiber inclination angle varies from the circumferential direction to the base–apex direction. Imaging resolution is ~300 μm in plane and ~800 μm out of plane. Imaging at this resolution enables measurement of fiber orientation at roughly two orders of magnitude more points than is possible with histological reconstruction methods.[16] FIGURE 2C shows that fiber orientation may be reconstructed accurately even in regions of the ventricles where fiber curvature is large (e.g., the apex of the heart). Finally, FIGURE 2D shows that fiber-tracing methods may be used to map fiber trajectories within the heart in a matter reminiscent of the original fiber dissections of MacCallum,[8] Mall,[9] and Fox and Hutchins.[10]

Cardiac fibers are known to be organized in sheets approximately four cells in thickness that radiate outward from the endocardial to the epicardial surface.[21,25] Local surface orientation of these sheets varies spatially. In previous work, we compared the intersection angle obtained by projection of the tertiary eigenvector of the diffusion tensor onto a plane defined by the epicardial and axial tangent vectors (see

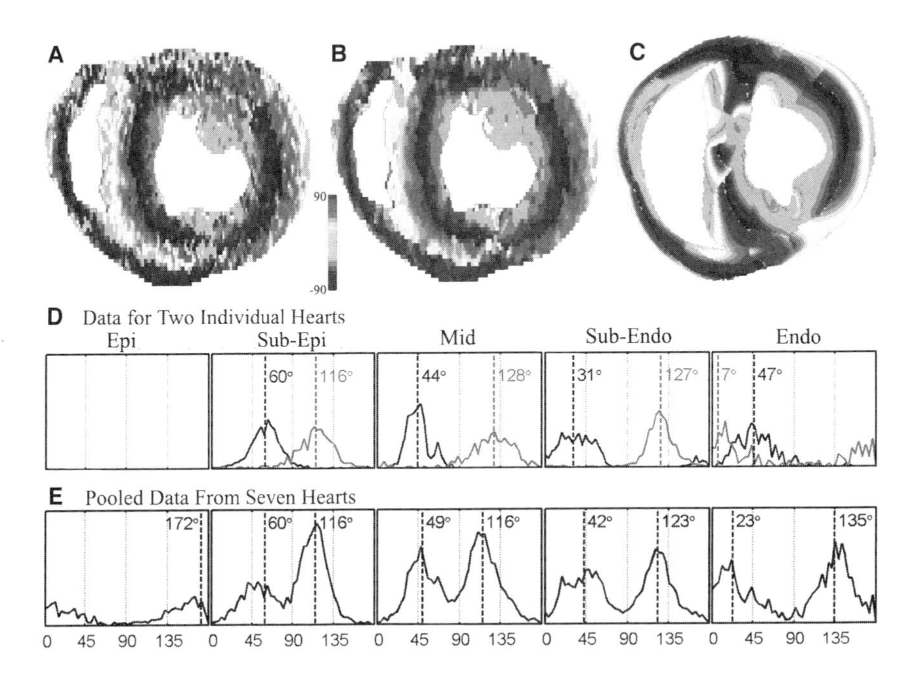

FIGURE 3. (**A**) Intersection angle measured using diffusion tensor magnetic resonance imaging in a short-axis section of the rabbit ventricle, (**B**) smoothed using a median filter. (**C**) Intersection angle measured by LeGrice *et al.*[21] in a short-axis section of the canine heart. (**D**) Histograms of intersection angle measured in the indicated regions of canine myocardium for two different hearts (gray vs. black lines). (**E**) Pooled data, as in **D**, from seven normal canine hearts [in color in Annals Online].

FIG. 1) with the intersection angle generated by projection of surface normal to the cardiac sheets measured histologically by Nielsen *et al.*[16] These data are shown in FIGURE 3A–C. FIGURE 3A shows DTMRI-based estimates of this inclination angle in a single short-axis section of the rabbit heart, and FIGURE 3B shows the data of FIGURE 3A smoothed using a median filter.[17] FIGURE 3C shows interpolated intersection angles of the surface normal to the ventricular sheets determined by LeGrice *et al.* in canine heart.[21,25] The short-axis section of FIGURE 3C was selected to be positioned at roughly the same percentage distance from base to apex as was the section of FIGURE 3A and B. The qualitative correlation between intersection angles of the tertiary eigenvector and the sheet normals is striking, despite being not only from different hearts, but also from different species.

More recent theoretical analyses coupled with histological reconstructions have predicted and verified the existence of two populations of sheets differing in orientation by approximately 90° in the canine[26] and ovine[27] ventricles. We, therefore, sought to determine if DTMRI may be used to reconstruct these two differing populations of sheet angles (Helm PA., H.J. Tseng, L. Younes, E. McVeigh, R. Winslow, unpublished data). To do this, we measured intersection angles for the tertiary eigenvector of the diffusion tensor at ventricular locations similar to those where previous histological measurements of dual sheet structure have been reported.[26,27] In these prior experimental studies, left ventricular transmural blocks of tissue were isolated and sectioned in ~1-mm increments parallel to the epicardial tangent plane, and measurements of intersection angles were averaged within these regions. Using a similar approach, we isolated 24 transmural regions by selecting 4 transmural regions each from within the anterior base, lateral base, posterior base, anterior apex, lateral apex, and posterior apex of the left ventricle. These 24 regions were divided into ~1-mm cube subregions each containing, on average, 291 tensor measurements. Based on normalized transmural depth, we assigned each cube as being in the epicardium (0% to 20%), subepicardium (20% to 40%), midwall (40% to 60%), subendocardium (60% to 80%), or endocardium (80% to 100%).

FIGURE 3D–E shows analyses of the distribution of intersection angles computed from the tertiary eigenvector in the four transmural regions located in the apical-anterior left ventricular free wall for two hearts. We observed that within each region of each heart, distribution of intersection angle exhibited a single dominant peak (FIG. 3D). The dominant intersection angle differed at various anatomic locations. This finding is similar to observations made using histological reconstruction methods in which each report indicates the presence of a dominant muscle layer.[26,27] Dominant populations were located on average at approximately 45° and 117°. Histological measurements have indicated similar values for sheet angles. Specifically, Dokos *et al.* identified a dominant midwall muscle layer at ~45° and another at ~110°,[28] and Ashikaga *et al.* observed angles at 36° and another at ~70° to 90°.[29] Finally, when the distribution angles for all hearts were pooled, a bimodal distribution of angles per location emerged (FIG. 3E), with the two dominant angles being equally likely. This is similar to the results of Arts *et al.* in which pooled histological measurements were obtained from six canine hearts. Peaks of pooled intersection angles were observed at 45° and 135° per location with similar probabilities.[26] These data provide strong evidence that the tertiary eigenvector of the diffusion tensor corresponds to the surface normal of the laminar sheets.

The data presented above demonstrate how DTMRI may be used to reconstruct the detailed anatomic structure of the cardiac ventricles. Using this approach, we have imaged and reconstructed a population of 11 normal and 7 failing canine hearts (failure was induced using the tachycardia pacing procedure). Our goal is to first assess within-class variability of these anatomies (normal vs. normal, failing vs. failing) and then, using estimates of within-class variability, extend the analyses to identify between-class differences of both geometry, fiber, and sheet orientation between normal versus failing hearts.

FIGURE 4 shows results of applying the LDDMM algorithm to register a template to a target heart. The left panel of FIGURE 4A shows the structure of a normal canine heart selected as the template. The right panel shows a view of a failing canine heart selected as the "target." The panel labeled "deformation metric" is a representation of the displacements needed to bring every point in the template into 1:1 correspondence with a point in the target heart. The panel labeled "deformed template" shows the results of deforming the template onto the target heart. Comparison of the two right panels shows that the mapping computed using the LDDMM algorithm leads to accurate registration of target and template hearts.

Registration of template and target hearts enables comparison of anatomic structure at corresponding points in both hearts. As an example, FIGURE 4B shows fiber

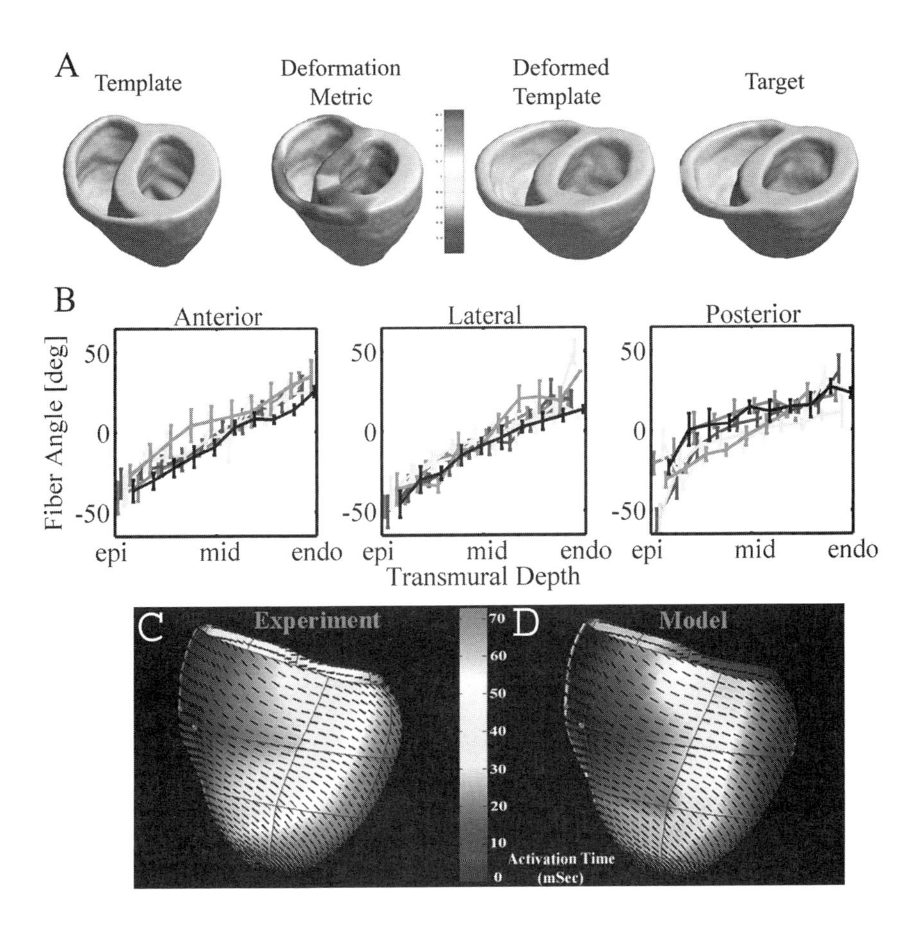

FIGURE 4. (**A**) Results of computing and applying LDDMM, taking the template onto the target heart. (**B**) Plots of inclination angle (ordinate) as a function of percent transmural depth (abscissa) for the six normal target canine hearts following registration to the template. (**C**) Plot of experimentally measured activation time. (**D**) Plot of simulated activation time [in color in Annals Online].

inclination of 6 normal canine hearts taken at corresponding transmural locations of the left ventricle. Locations were selected by choosing a transmural set of voxels in the anterior, lateral, and posterior left ventricular walls of the normal template heart and then using the LDDMM mapping to locate corresponding sets of voxels in each of the six normal target hearts. The ordinate plots average fiber inclination angle at each transmural location (abscissa) for each of the target hearts. Error bars indicate the standard deviation of inclination angle calculated per heart at each transmural location. This variation reflects both anatomical variability as well as measurement noise. The mean standard deviation calculated over all transmural locations and hearts is 6.0, 5.9, and 5.7 degrees for anterior, lateral, and posterior left ventricular locations, respectively. The standard deviation of mean inclination angle calculated over all target hearts and transmural locations is 3.3, 4.7, and 7.2 degrees for anterior, lateral, and posterior left ventricular locations, respectively. These data, therefore, offer an example of how the LDDMM algorithm may be used to register hearts in order to assess anatomic variability at corresponding locations. In particular, these data indicate that variations of fiber inclination angle at corresponding points over an ensemble of hearts is surprisingly small.

The DTMRI data provide important information that may be used to constrain computational models of electrical conduction in the cardiac ventricles. Electrical conduction may be modeled by solving the monodomain equation over the myocardial domain H,

$$\frac{\partial v(\underline{x},t)}{\partial t} = \frac{1}{C_m}\left[-I_{ion}(\underline{x},t) - I_{app}(\underline{x},t) + \frac{1}{\beta}\left(\frac{\kappa}{\kappa+1}\right)\nabla \cdot (M_i(\underline{x})\nabla v(\underline{x},t))\right] \text{ on H}, \quad (1)$$

where $I_{ion}(\underline{x},t)$ is membrane ionic current as defined in the canine myocyte model of Winslow *et al.*[30] The conductivity tensors at each myocardial point $\underline{x}$ are then defined as

$$M_i(\underline{x}) = P(\underline{x})G_i(\underline{x})P^T(\underline{x}), \quad (2)$$

where $G_i(\underline{x})$ is a diagonal matrix with elements $\sigma_{1,i}$, $\sigma_{2,i}$ and $\sigma_{3,i}$, ($\sigma_{1,i}$ is longitudinal, and $\sigma_{2,i}$ and $\sigma_{3,i}$ are transverse intracellular conductivities), and $P(\underline{x})$ is the transformation matrix from local to scanner coordinates at each point $\underline{x}$.[31,32] When working from DTMRI data, the columns of $P(\underline{x})$ are set equal to the eigenvectors of the diffusion tensor estimated at point $\underline{x}$.[31,32] Coupling conductances are set as in previous models[33,34] and refined to yield measured epicardial conduction velocities. The reaction–diffusion monodomain equation (Eqs. 1 and 2) is then solved using methods described previously.[35]

FIGURE 4 shows the results of applying these methods to the analysis of conduction in a normal canine heart. FIGURE 4C shows activation time measured experimentally using a 256-electrode sock array in response to a right ventricular stimulus pulse applied at the epicardial location marked as spheres. Following electrical mapping, the heart was excised and imaged using DTMRI, then a finite-element model was fit to the resulting geometry and fiber orientation data sets, and EQUATIONS 1 and 2 were solved—in response to the same stimulus as studied experimentally—on the myocardial domain. FIGURE 4D shows results of simulating conduction using a computational model of the very same heart that was mapped electrically in FIGURE 4C. Results are seen to be in qualitative agreement.

SUMMARY

It is now clear that the primary eigenvector of the diffusion tensor is aligned locally with the cardiac fiber long axis, and, therefore, DTMRI may be used for rapid measurement of fiber orientation throughout the canine ventricles at high spatial resolution. In addition, the data of FIGURE 3 showing that intersection angles measured using DTMRI agree closely with those measured using histologic reconstruction provide strong evidence supporting the hypothesis that the tertiary eigenvector of the diffusion tensor is aligned locally with the cardiac sheet surface normal. Considered together, these observations suggest that DTMRI may be used to reconstruct both ventricular fiber and sheet organization. We also show that the LDDMM algorithm may be used to register imaged and reconstructed hearts so that variability of ventricular geometry, fiber, and sheet orientation may be quantified over ensembles of imaged hearts. Ventricular geometry and fiber structure is known to be remodeled in a range of disease processes,[36–38] however, descriptions of this remodeling have remained subjective and qualitative. Future research will combine use of DTMRI with the LDDMM method to detect statistically significant changes of cardiac anatomy that are characteristic of specific disease processes in the heart. Finally, we show that epicardial electrical mapping and DTMRI imaging may be performed in the same hearts. The anatomic data may then be used to simulate electrical conduction in a computational model of the very same heart that was mapped electrically. This facilitates direct comparison and testing of model versus experimental results and opens the door to quantitative measurement, modeling, and analysis of the ways in which remodeling of ventricular microanatomy influences electrical conduction in the heart.

ACKNOWLEDGMENTS

This work was supported by NIH Grants HL-70894 and HL-52307, and by the Whitaker Foundation, the Falk Medical Trust, and IBM Corporation. All DTMRI datasets employed in this study and the Cardiac Anatomic Database System are downloadable from http://www.ccbm.jhu.edu.

REFERENCES

1. Franzone, P.C., L. Guerri, M. Pennacchio & B. Taccardi. 1998. Spread of excitation in 3-D models of the anisotropic cardiac tissue. II. Effects of fiber architecture and ventricular geometry. Math. Biosci. **147:** 131–171.
2. Kanai, A. & G. Salama. 1995. Optical mapping reveals that repolarization spreads anisotropically and is guided by fiber orientation in guinea pig hearts. Circ. Res. **77:** 784–802.
3. Rohr, S. & B.M. Salzberg. 1994. Characterization of impulse propagation at the microscopic level across geometrically defined expansions of excitable tissue: multiple site optical recording of transmembrane voltage (MSORTV) in patterned growth heart cell cultures. J. Gen. Physiol. **104:** 287–309.
4. Rohr, S., A.G. Kleber & J.P. Kucera. 1999. Optical recording of impulse propagation in designer cultures. Cardiac tissue architectures inducing ultra-slow conduction. Trends Cardiovasc. Med. **9:** 173–179.
5. Kucera, J.P., A.G. Kleber & S. Rohr. 1998. Slow conduction in cardiac tissue, II: effects of branching tissue geometry. Circ. Res. **83:** 795–805.
6. Taccardi, B., E. Macchi, R.L. Lux, *et al.* 1994. Effect of myocardial fiber direction on epicardial potentials. Circulation **90:** 3076–3090.
7. Chen, P.S., Y.M. Cha, B.B. Peters & L.S. Chen. 1993. Effects of myocardial fiber orientation on the electrical induction of ventricular fibrillation. Am. J. Physiol. **264:** H1760–H1773.
8. MacCallum, J.B. 1900. On the muscular architecture and growth of the ventricles of the heart. Johns Hopkins Hosp. Rep. **9:** 307–335.
9. Mall, F. 1911. On the muscular architecture of the ventricles of the human heart. Am. J. Anat. **11:** 211–266.

10. Fox, C.C. & G.M. Hutchins. 1972. The architecture of the human ventricular myocardium. Johns Hopkins Med. J. **25:** 289–299.
11. Streeter, D., H.M. Spotnitz, D.P. Patel, *et al.* 1969. Fiber orientation in the canine left ventricle during diastole and systole. Circ. Res. **24:** 339–347.
12. Streeter, D.D. & W.T. Hanna. 1973. Engineering mechanics for successive states in canine left ventricular myocardium. II. Fiber angle and sarcomere length. Circ. Res. **33:** 656–664.
13. Ross, A.A. & D.D. Streeter, Jr. 1979. Myocardial fiber disarray [letter]. Circulation **60:** 1425–1426.
14. Streeter, D. 1979. Chapter 4. Gross morphology and fiber geometry of the heart. *In* Handbook of Physiology, The Cardiovascular System I. R. Berne, Ed.: 61–112. American Physiological Society. Bethesda, MD.
15. Streeter, D.D. & C. Ramon. 1983. Muscle pathway geometry in the heart wall. J Biomechanical Engineering **105:** 367–373.
16. Nielsen, P.M., I.J. LeGrice, B.H. Smaill & P.J. Hunter. 1991. Mathematical model of geometry and fibrous structure of the heart. Am. J. Physiol. **260:** H1365–H1378.
17. Scollan, D., A. Holmes, R.L. Winslow & J. Forder. 1998. Histological validation of myocardial microstructure obtained from diffusion tensor magnetic resonance imaging. Am. J. Physiol. **275:** H2308–H2318.
18. Beg, F.M., P.A. Helm, E. McVeigh, *et al.* 2004. Computational cardiac anatomy using magnetic resonance imaging. Mag. Res. Med. **52:** 1167–1174.
19. Basser, P., J. Mattiello & D. LeBihan. 1994. Estimation of the effective self-diffusion tensor from the NMR spin echo. J. Magn. Reson. B. **103:** 247–254.
20. Scollan, D., A. Holmes, F. Zhang & R. Winslow. 2000. Reconstruction of cardiac ventricular geometry and fiber orientation using GRASS and diffusion-tensor magnetic resonance imaging. Ann. Biomed. Eng. **28:** 934–944.
21. Legrice, I.J., P.J. Hunter & B.H. Smaill. 1997. Laminar structure of the heart: a mathematical model. Am. J. Physiol. **272:** H2466–H2476.
22. Winslow, R., P.A. Helm, W. Baumgartner, *et al.* 2002. Imaging-based integrative models of the heart: closing the loop between experiment and simulation. *In* Novartis Foundation Symposium 247 on In Silico Simulation of Biological Processes, London, 2002. 129–140.
23. Hsu, E.W., A.L. Muzikant, S.A. Matulevicius, *et al.* 1998. Magnetic resonance myocardial fiber-orientation mapping with direct histological correlation. Am. J. Physiol. **274:** H1627–H1634.
24. Holmes, A., D. Scollan & R. Winslow. 1999. Direct histological validation of diffusion tensor MRI in formalin-fixed myocardium. Mag. Res. Med. **44:** 157–160.
25. LeGrice, I.J., B.H. Smaill, L.Z. Chai, *et al.* 1995. Laminar structure of the heart: ventricular myocyte arrangement and connective tissue architecture in the dog. Am. J. Physiol. **269:** H571–H582.
26. Arts, T., K.D. Costa, J.W. Covell & A.D. McCulloch. 2001. Relating myocardial laminar architecture to shear strain and muscle fiber orientation. Am. J. Physiol. Heart Circ. Physiol. **280:** H2222–H2229.
27. Harrington, K.B., F. Rodriguez, A. Cheng, *et al.* 2005. Direct measurement of transmural laminar architecture in the anterolateral wall of the ovine left ventricle: new implications for wall thickening mechanics. Am. J. Physiol. Heart Circ. Physiol. **288:** H1324–H1330.
28. Dokos, S., B.H. Smaill, A.A. Young & I.J. LeGrice. 2002. Shear properties of passive ventricular myocardium. Am. J. Physiol. Heart Circ. Physiol. **283:** H2650–H2659.
29. Ashikaga, H., J.C. Criscione, J.H. Omens, *et al.* 2004. Transmural left ventricular mechanics underlying torsional recoil during relaxation. Am. J. Physiol. Heart Circ. Physiol. **286:** H640–H647.
30. Winslow, R.L., J. Rice, S. Jafri, *et al.* 1999. Mechanisms of altered excitation-contraction coupling in canine tachycardia-induced heart failure. II. Model Studies Circ. Res. **84:** 571–586.
31. Winslow, R.L., D.F. Scollan, A. Holmes, *et al.* 2000. Electrophysiological modeling of cardiac ventricular function: from cell to organ. Annu. Rev. Biomed. Eng. **2:** 119–155.

32. WINSLOW, R., D.F. SCOLLAN, J. GREENSTEIN, *et al.* 2001. Mapping, modeling and visual exploration of structure-function relationships in the heart. IBM Systems J. **40:** 342–359.
33. HENRIQUEZ, C.S. 1993. Simulating the electrical behavior of cardiac tissue using the bidomain model. Crit. Rev. Biomed. Eng. **21:** 1–77.
34. HENRIQUEZ, C.S., A.L. MUZIKANT & C.K. SMOAK. 1996. Anisotropy, fiber curvature, and bath loading effects on activation in thin and thick cardiac tissue preparations: simulations in a three-dimensional bidomain model. J. Cardiovasc. Electrophysiol. **7:** 424–444.
35. YUNG, C. 2000. Application of a stiff, operator-splitting scheme to the computational modeling of electrical properties in cardiac ventricles. MS thesis, Johns Hopkins University, Baltimore, MD.
36. MARON, B.J., N. SATO, W.C. ROBERTS, *et al.* 1979. Quantitative analysis of cardiac muscle cell disorganization in the ventricular septum. Comparison of fetuses and infants with and without congenital heart disease and patients with hypertrophic cardiomyopathy. Circulation **60:** 685–696.
37. MARON, B.J., T.J. ANAN & W.C. ROBERTS. 1981. Quantitative analysis of the distribution of cardiac muscle cell disorganization in the left ventricular wall of patients with hypertrophic cardiomyopathy. Circulation **63:** 882–894.
38. ROBERTS, W.C., R.J. SIEGEL & B.M. MCMANUS. 1987. Idiopathic dilated cardiomyopathy: analysis of 152 necropsy patients. Am. J. Cardiol. **60:** 1340–1355.

Electrotonic Cell–Cell Interactions in Cardiac Tissue

Effects on Action Potential Propagation and Repolarization

YORAM RUDY

Cardiac Bioelectricity and Arrhythmia Center, Washington University in St. Louis, St. Louis, Missouri 63130-4899, USA

ABSTRACT: This article characterizes through computer simulations, the role of cell–cell interactions and of tissue structure in determining properties of action potential propagation and repolarization in cardiac tissue. The results demonstrate strong interactions between membrane excitatory processes and electrical loading by the passive tissue structure. These interactions modulate the ionic mechanism of conduction and influence conduction velocity and robustness. Cell–cell interactions also have a strong effect on the dispersion of repolarization in cardiac tissue.

KEYWORDS: cardiac action potential; cardiac arrhythmias; gap junctions; tissue structure

INTRODUCTION

Propagation of excitation in cardiac tissue is determined by local source-sink relations. Successful propagation requires that the amount of charge supplied by the source must be equal to or exceed the charge required to excite the membrane at the sink location. At the cellular level, this condition is determined by the state of membrane excitability and the electrical coupling between cells through gap junctions. At the more macroscopic scale of the multicellular tissue, this condition is influenced strongly by structural properties of the cardiac myocardium, such as branching of fibers or heterogeneities of gap-junctional coupling over tissue segments. In addition to affecting action potential conduction, intercellular coupling affects the action potential duration and its spatial variation (dispersion) across the tissue.

In this review article, mathematical models are used to examine the role of intercellular electrical coupling and of tissue structural elements in the propagation and repolarization of the action potential in cardiac tissue. The emphasis is on characterizing the safety (robustness) of conduction and its underlying ionic mechanism for different levels of gap-junction coupling and during the crossing of structural dis-

Address for correspondence: Prof. Yoram Rudy, Ph.D., Washington University in St. Louis, Cardiac Bioelectricity Center, 290 Whitaker Hall, Campus Box 1097, One Brookings Drive, St. Louis, MO 63130-4899, USA. Voice: 314-935-8160; fax: 314-935-8168.
rudy@wustl.edu

Ann. N.Y. Acad. Sci. 1047: 308–313 (2005).
doi: 10.1196/annals.1341.027

continuities (e.g., branching points). In addition, the effect of gap-junction coupling on the dispersion of action potential duration is evaluated. This article is a conference report that summarizes previously published results.[1–4]

METHODS

Simulations were conducted in a multicellular one-dimensional model that incorporates membrane ionic fluxes through ion channels, pumps, and exchanges, and the intercellular flow of current through gap junctions.[1] The cell's electrical activity is represented by the dynamic Luo-Rudy (LRd) ventricular cell model.[5] In addition to membrane ionic fluxes, this model accounts for dynamic changes in concentrations of Na^+, K^+, and Ca^{2+} (Ca^{2+} handling by the sarcoplasmic reticulum and calcium buffers are included in the model). The model cells are of realistic dimensions (100 microns long) and are connected by gap functions with 80-angstrom-long resistive channels (connexons). The gap-junction conductance can be modulated to simulate varying degrees of intercellular coupling.

Structural inhomogeneities are introduced starting from cell 81 in a fiber containing 160 cells.[2] They include an increase in gap-junction conductance or tissue expansion due to branching. To characterize the asymmetry introduced by the structural inhomogeneities, in certain simulations the fiber is stimulated separately from either one of its two ends (cell 1 or, alternatively, cell 160). Dispersion of repolarization is studied in a fiber containing an endocardial region (cells 1–80), mid-myocardial (M cell) region (cells 81–110), and an epicardial region (cells 111–190).[3] Heterogeneity of repolarization between these regions results from heterogeneity in the density of I_{Ks} (the slow delayed rectifier K^+ current), with M cells having the lowest density and slowest repolarization.

RESULTS

Microscopic Discontinuities: Cell–Cell Communication[1]

Propagation across areas of a healed infarct[6] and propagation transverse to fiber orientation[7] are examples of conduction in a substrate with reduced intercellular coupling. Such conduction is discontinuous, with long conduction delays across the gap junctions and an almost simultaneous activation of the entire cell. Thus, conduction velocity is dominated by the gap junctions, rather than by the state of membrane excitability.

To characterize the robustness of conduction for various degrees of intercellular coupling, we define and compute a safety factor (SF) index as the ratio of charge generated to charge consumed by a cell during its excitation cycle. SF > 1 indicates successful conduction, with the fraction of SF above 1 indicating the margin of safety. When SF = 1, conduction is critical, and when SF < 1, propagation fails.

Our simulations show that both conduction velocity and SF vary with the degree of intercellular coupling at gap junctions. As coupling is reduced, velocity decreases as a result of longer delays in crossing the gap junctions. However, the SF *increases* to a maximum of almost 3 before it decreases sharply toward 1, below which conduction fails. SF is maximum at a gap junction conductance of 0.023 μs, a value that is about 100 times smaller than the normal physiological value of 2.5 μs. The high SF at greatly reduced coupling when conduction is abnormally slow seems paradoxical. It results from reduced electrical loading on the depolarizing cell. Due to the decreased coupling, current leakage out of the cell to the downstream fiber is minimized. This results in a long conduction delay and, at the same time, charge is conserved to depolarize the cell undergoing excitation. This process supports very safe conduction that is also very slow. The result is the ability to sustain extremely slow conduction velocities, below 1 cm/s. Such slow velocities have been measured experimentally under conditions of reduced gap-junction coupling[8] and have been observed in healed infarcts, making possible the formation of microreentry loops in such substrate. It should be added here that conduction can also be slowed by reducing membrane excitability (e.g., during acute ischemia). However, the minimum velocity before conduction failure is 17 cm/s, only one-third of control velocity (54 cm/s) and not very slow.[9,10]

The ionic basis of slow conduction during reduced coupling differs from that of normal conduction, which depends solely on the fast sodium current. At highly reduced coupling, the depolarizing charge generated by the L-type calcium current can equal or even exceed that generated by the sodium current. This strong dependence on the calcium current results from the long intercellular conduction delays across the gap junctions. Due to the long delay, the downstream cell depolarizes when the upstream source-cell is deep into the plateau phase of its action potential. At this phase, the calcium current is the source of depolarizing charge (the sodium current is mostly inactivated within less than 1 ms) that supports conduction in the fiber. This dependence on the calcium current in the presence of long conduction delays has been confirmed experimentally.[11–14]

Structural Heterogeneities[2]

Heterogeneity of myocardial structure is accentuated by pathology and remodeling. Tissue expansion through branching of fibers or a region of increased gap-junction coupling both constitute an increased electrical load and have similar effects on action potential propagation. Simulation results show that propagation into a region of increased loading involves long conduction delays across the transition zone and a sharp local decrease of SF. Thus, such transition sites are the Achilles' heel of propagation. Along the homogeneous regions of the fiber, away from the transition zone, propagation is supported solely by the sodium current. For cells in the transition region, the L-type calcium current becomes a major contributor of depolarizing charge to support conduction (a consequence of the long conduction delay, the long duration of the calcium current, and the fast inactivation of the sodium current that makes its contribution negligible beyond 1 ms). The strong dependence of conduction through the transition zone on the calcium current is demonstrated by the fact that 30% reduction of this current is sufficient to cause conduction failure in this region. Interestingly, the magnitude of the calcium current is sharply increased in cells just proximal to the transition site, where it is needed for sustaining conduction. This increase is a consequence of increased driving force for the current caused by reduced plateau potentials in these cells. The lower plateau is caused by the large load on the cells during the slow depolarization of cells distal to the transition site. Because of the long transmission delay, distal cells still depolarize and consume charge from the proximal cells when these cells are in their plateau phase, "pulling down" the plateau potential. It should be emphasized that these properties apply only to propagations *into* regions of increased electrical loading. In the reverse direction, the action potential encounters a decrease in electrical load, SF is high and propagation is robust and depends solely on the sodium current in this direction. Such asymmetries provide a substrate for the development of unidirectional block and reentrant arrhythmias.

Dispersion of Repolarization[3]

Electrophysiological heterogeneity due to nonuniform expression of ion channels is a property of ventricular myocardium. In particular, a subpopulation of cells (midmyocardial M cells) has been described to display a longer action potential duration and a steeper dependence of this duration, or rate, than the other (epicardial or endocardial) ventricular cell types.[15] Whereas these properties are well established in isolated cells, in the intact heart, the cells are coupled through gap junctions and are subject to electrical loading that affects their behavior.

Simulation results show a difference of 90 ms between the action potential durations of isolated M cells and epicardial cells. This difference is greatly reduced to only 18 ms when the cells are well coupled (gap-junction conductance of 2.5 μs; conduction velocity of 56 cm/s, typical for propagation along fibers in normal myocardium). However, with reduced coupling, the difference increases sharply. At gap-junction conductance of 0.25 μs (velocity = 18 cm/s, typical for propagation transverse to fibers) the difference between M cell and epicardial cell action potential duration is 44 ms. When cells are very poorly coupled (conductance of 0.025 μs; velocity = 3.5 cm/s, observed under pathological conditions such as infarction) the difference increases to 71 ms, approaching the value for isolated cells. Thus, dispersion of repolarization, an asymmetrical property of the myocardium that can lead to unidirectional block and reentry, depends strongly on the degree of intercellular coupling through gap junctions.

DISCUSSION

In this article, we review results from mathematical modeling studies that provide insights into the properties of action potential propagation and repolarization in the multicellular myocardium. The emphasis is on the effects of structural factors, including cellular interactions on a microscopic scale and structural inhomogeneities on a more macroscopic scale. Both factors are shown to cause local conduction delays that make propagation discontinuous. Importantly, when long conduction delays are present, the membrane switches from sodium-supported conduction to mostly calcium-supported conduction. This phenomenon illustrates the intimate interaction between the passive network properties (i.e., gap-junction coupling, heterogeneities of tissue structure) and the excitatory membrane currents during conduction. It shows that altered network properties can cause major changes in the action of membrane currents. In this respect, this constitutes a feedback mechanism from the sink (the passive network) to the source of excitation (the ionic currents that generate conduction). Another example of a (positive) feedback response is the augmentation of the calcium current just proximal to a structural transition zone, where conduction relies on this current. In this case, the structural inhomogeneity dictates reliance of conduction on the calcium current, and the membrane responds by augmenting this current.

The principles that are established by the simulations have clinical implications. As shown, reduced intercellular coupling supports very slow conduction and induces large dispersion of repolarization. These two properties support the occurrence of unidirectional block and reentry and may promote arrhythmogenesis in the infracted and aging myocardium, where intercellular coupling is reduced and fibrosis is present. Similarly, macroscopic tissue heterogeneities are accentuated in the structurally diseased myocardium, also due to fibrosis and scar tissue. The electrical asymmetries and long conduction delays introduced by such heterogeneities also provide a substrate for unidirectional block and reentry. Finally, the demonstrated involvement of the L-type calcium current in conduction in such substrate identifies this current as a possible target for antiarrhythmic drug therapy.

SUMMARY

This review article explores, through computer simulations, the effects of electrical interactions between cells and with structural inhomogeneities on action potential propagation and repolarization in cardiac tissue. The following principles are established:

(1) Reduced cell–cell coupling through gap junctions provides a substrate for slow but robust conduction of the cardiac action potential.
(2) Conduction in such substrates is discontinuous and involves long conduction delays across the intercellular gap junctions.
(3) Conduction in such substrates relies on the L-type calcium current.
(4) On a more macroscopic scale, conduction is discontinuous with long delays across heterogeneities of the tissue structure (e.g., branching of fibers, tissue expansion, increased cell–cell coupling).
(5) In such regions, the safety factor for conduction is reduced and conduction relies on the L-type calcium current, which is augmented locally.
(6) The structural inhomogeneities introduce asymmetries in electrical loading that promote the occurrence of unidirectional block and consequently increase the propensity for reentrant arrhythmias.
(7) Dispersion of repolarization shows strong dependence on the degree of cell–cell coupling through gap junctions.

ACKNOWLEDGMENTS

The work summarized here and the preparation of this article were supported by Grants R01-HL49054 and R37-HL33343 from the NIH–National Heart, Lung, and Blood Institute. Thanks to Jennifer Godwin for preparing the manuscript for submission.

REFERENCES

1. Shaw, R.M. & Y. Rudy. 1997. Ionic mechanisms of propagation in cardiac tissue. Roles of the sodium and L-type calcium currents during reduced excitability and decreased gap junction coupling. Circ. Res. **81:** 727–741.
2. Wang, Y. & Y. Rudy. 2000. Action potential in inhomogeneous cardiac tissue: safety factor considerations and ionic mechanism. Am. J. Physiol. Heart Circ. Physiol. **278:** H1019–H1029.
3. Viswanathan, P.C., R.M. Shaw & Y. Rudy. 1999. Effects of I_{Kr} and I_{Ks} heterogeneity on action potential duration and its rate dependence: a simulation study. Circulation **99:** 2466–2474.
4. Kléber, A.G. & Y. Rudy. 2004. Basic mechanisms of cardiac impulse propagation and associated arrhythmias. Physiol. Rev. **84:** 431–488.
5. Luo, C.H. & Y. Rudy. 1994. A dynamic model of the cardiac ventricular action potential. I. Simulations of ionic currents and concentration changes. Circ. Res. **74:** 1071–1096.
6. Dillon, S.M., M.A. Allessie, P.C. Ursell, *et al.* 1988. Influences of anisotropic tissue structure on reentrant circuits in the epicardial border zone of subacute canine infarcts. Circ. Res. **63:** 182–206.
7. Spach, M.S., P.C. Dolber & J.F. Heidlage. 1988. Influence of the passive anisotropic properties on directional differences in propagation following modification of the sodium conductance in human atrial muscle. A model of reentry based on anisotropic discontinuous propagation. Circ. Res. **62:** 811–832.
8. Rohr, S., J.P. Kucera & A.G. Kléber. 1997. Slow conduction in cardiac tissue. I. Effects of a reduction of excitability versus a reduction of electrical coupling on microconduction. Circ. Res. **83:** 781–794.
9. Shaw, R.M. & Y. Rudy. 1997. Electrophysiologic effects of acute myocardial ischemia. A mechanistic investigation of action potential conduction and conduction failure. Circ. Res. **80:** 124–138.
10. Kagiyama, Y., J.L. Hill & L.S. Gettes. 1982. Interaction of acidosis and increased extracellular potassium on action potential characteristics and conduction in guinea pig ventricular muscle. Circ. Res. **51:** 614–623.
11. Joyner, R.W., R. Kumar, R. Wilders, *et al.* 1996. Modulating L-type calcium current affects discontinuous cardiac action potential conduction. Biophys. J. **71:** 237–245.
12. Kumar, R. & R.W. Joyner. 1995. Calcium currents of ventricular cell pairs during action potential conduction. Am. J. Physiol. Heart Circ. Physiol. **268:** H2476–H2486.
13. Sugiura, H. & R.W. Joyner. 1992. Action potential conduction between guinea pig ventricular cells can be modulated by calcium current. Am. J. Physiol. Heart Circ. Physiol. **263:** H1591–H1604.
14. Rohr, S. & J. Kucera. 1997. Involvement of the calcium inward current in cardiac impulse propagation: induction of unidirectional conduction block by nifedipine and reversal by Bay K 8644. Biophys. J. **72:** 754–766.
15. Sicouri, S. & C.A. Antzelevitch. 1991. A subpopulation of cells with unique electrophysiological properties in the deep subepicardium of the canine ventricle. The M cell. Circ. Res. **68:** 1729–1741.

Modulation of Transmural Repolarization

CHARLES ANTZELEVITCH

Masonic Medical Research Laboratory, Utica, New York 13501, USA

ABSTRACT: Ventricular myocardium in larger mammals has been shown to be comprised of three distinct cell types: epicardial, M, and endocardial. Epicardial and M cell action potentials differ from endocardial cells with respect to the morphology of phase 1. These cells possess a prominent I_{to}-mediated notch responsible for the "spike and dome" morphology of the epicardial and M cell response. M cells are distinguished from the other cell types in that they display a smaller I_{Ks}, but a larger late I_{Na} and $I_{Na\text{-}Ca}$. These ionic distinctions underlie the longer action potential duration (APD) and steeper APD-rate relationship of the M cell. Difference in the time course of repolarization of phase 1 and phase 3 are responsible for the inscription of the electrocardiographic J wave and T wave, respectively. These repolarization gradients are sensitively modulated by electrotonic communication among the three cells types, $[K^+]_o$, and the presence of drugs that either reduce or augment net repolarizing current. A reduction in net repolarizing current generally leads to a preferential prolongation of the M cell action potential, responsible for a prolongation of the QT interval and an increase in transmural dispersion of repolarization (TDR), which underlies the development of torsade de pointes arrhythmias. An increase in net repolarizing current can lead to a preferential abbreviation of the action potential of epicardium in the right ventricle (RV), and endocardium in the left ventricle (LV). These actions also lead to a TDR that manifests as the Brugada syndrome in RV and the short QT syndrome in LV.

KEYWORDS: Brugada syndrome; cardiac heterogeneity; electrocardiogram; long QT syndrome; M cell; short QT syndrome

INTRODUCTION

Intrinsic Electrical Heterogeneities in Ventricular Myocardium

Our understanding of the electrophysiology and pharmacology of the ventricular myocardium has advanced in recent years with the delineation of three distinct myocardial cell types: epicardial, M, and endocardial. Differences in the electrophysiologic characteristics and pharmacologic profiles of these three myocardial cell types have been described in the dog, guinea pig, rabbit, and human ventricles.[1]

The three cell types differ with respect to currents that contribute to the early repolarization phase, phase 1. The action potentials of epicardial and M cells display a prominent transient outward current (I_{to})-mediated phase 1 that is absent in endocardial cells (see Ref. 1 for references). The early repolarization phase gives the

Address for correspondence: Dr. Charles Antzelevitch, Ph.D., Masonic Medical Research Laboratory, 2150 Bleecker Street, Utica, NY 13501-1787, USA. Voice: 315-735-2217; fax: 315-735-5648. ca@mmrl.edu

Ann. N.Y. Acad. Sci. 1047: 314–323 (2005). © 2005 New York Academy of Sciences.
doi: 10.1196/annals.1341.028

epicardial action potential a notched appearance. In the canine heart, I_{to} and the action potential notch are much larger in right versus left ventricular epicardium[2] and M[3] cells.

The principal feature of the M cell is the ability of its action potential to prolong more than that of epicardium or endocardium with slowing of rate. In the early 1990s, the M cells became the focus of intense investigation after their identification and characterization in the deep structures of the canine ventricle.[4–6] M cell distribution in the ventricular wall has been investigated in greatest detail in the canine left ventricle. M cells with the longest action potential duration (APD) are typically found in the deep subepicardium to midmyocardium in the lateral wall, deep subendocardium to midmyocardium in the anterior wall, and throughout the wall in the region of the outflow tracts. M cells are also present in the deep layers of papillary muscles, trabeculae, and interventricular septum.[7] Tissue slices isolated from the M region display an APD at 90% repolarization (APD_{90}) that is more than 100 ms longer than tissues isolated from the epicardium or endocardium at basic cycle lengths of 2000 ms or greater. In the intact ventricular wall, this disparity in APD_{90} is less pronounced due to electrotonic coupling of cells. The transmural increase in APD is relatively gradual, except between the epicardium and subepicardium where there is often a sharp increase in APD. This has been shown to be due to an increase in tissue resistivity in this region,[8] which may be related to the sharp transition in cell orientation in this region as well as to reduced expression of connexin 43,[9,10] which is principally responsible for intracellular communication in the ventricular myocardium. Thus, both the degree of electrotonic coupling and intrinsic action potential durations play a key role in the expression of electrical heterogeneity in the ventricular myocardium.

The prolonged APD of M cells has been shown to be due to a number of ionic distinctions, including a smaller I_{Ks} and a larger late I_{Na}[11,12] and sodium–calcium exchange current ($I_{Na\text{-}Ca}$)[13] compared to epicardial and endocardial cells. The net result is a decrease in repolarizing current during phases 2 and 3 of the M cell action potential. These ionic distinctions sensitize the M cells to a variety of pharmacological agents. Agents that block I_{Kr}, I_{Ks}, or increase I_{Ca} or late I_{Na} produce much greater prolongation in M cell APD than on the APD of epicardial or endocardial cells.

Electrical Heterogeneity Contributes to Inscription of J Wave and T Wave of ECG

Transmural differences in the time course of repolarization of the three predominant myocardial cell types have been shown to be largely responsible for the inscription of the J wave and T wave of the ECG. The transmural gradient resulting from the presence of an I_{to}-mediated notch in epicardium but not endocardium gives rise to the J wave, or Osborne wave.[14] Voltage gradients developing as a result of the different time course of repolarization of phases 2 and 3 in the three cell types give rise to opposing voltage gradients on either side of the M region, which are in large part responsible for the inscription of the T wave.[15] When the T wave is upright, the epicardial response is the earliest to repolarize and the M cell action potential is the latest. Full repolarization of the epicardial action potential coincides with the peak of the T wave and repolarization of the M cells is coincident with the end of the T wave. It follows that the duration of the M cell action potential determines the QT interval, whereas the duration of the epicardial action potential determines the QT_{peak} interval. Another interesting finding stemming from these studies is that the T_{peak}–T_{end} interval may provide an index of transmural dispersion of repolarization (TDR).[5,15]

The available data suggest that T_{peak}–T_{end} measurements should be limited to precordial leads since these leads may more accurately reflect TDR. Recent studies have also provide guidelines for the estimation of TDR in the case of more complex T waves, including negative, biphasic, and triphasic T waves.[16] In these cases, the interval from the nadir of the first component of the T wave to the end of the T wave provides an accurate electrocardiographic approximation of TDR.

The clinical applicability of these concepts remains to be fully validated. An important step toward validation of the T_{peak}–T_{end} interval as an index of transmural dispersion was provided in a report by Lubinski *et al.*,[17] which showed an increase of this interval in patients with congenital long QT syndrome (LQTS). Recent studies suggest that the T_{peak}–T_{end} interval may be a useful index of transmural dispersion and thus may be prognostic of arrhythmic risk under a variety of conditions.[18–23] Takenaka *et al.*[22] recently demonstrated exercise-induced accentuation of the T_{peak}–T_{end} interval in LQT1 patients, but not LQT2. These observation coupled with those of Schwartz *et al.*[24] demonstrating an association between exercise and risk for torsade de pointes (TdP) in LQT1 but not LQT2 patients, again point to the potential value of T_{peak}–T_{end} in forecasting risk for the development of TdP.

Direct evidence in support of T_{peak}–T_{end} as a valuable index to predict TdP in patients with LQTS was provided by Yamaguchi and co-workers.[25] These authors concluded that T_{peak}–T_{end} is more valuable than QTc and QT dispersion as a predictor of TdP in a patient with acquired LQTS. Shimizu *et al.* demonstrated that T_{peak}–T_{end}, but not QTc, predicted sudden cardiac death in patients with hypertrophic cardiomypathy.[21] Most recently, Watanabe *et al.* demonstrated that prolonged T_{peak}–T_{end} is associated with inducibility as well as spontaneous development of ventricular tachycardia (VT) in high-risk patients with organic heart disease.[26] Although additional work is needed to assess the value of these noninvasive indices of electrical heterogeneity and their prognostic value in the assignment of arrhythmic risk, evidence is mounting in support of the hypothesis that TDR rather QT prolongation underlies the substrate responsible for the development of TdP.[27–31]

AMPLIFICATION OF TRANSMURAL HETEROGENEITY AS THE BASIS FOR VT/VF

Brugada Syndrome

The Brugada syndrome is characterized by (1) an accentuated ST segment elevation or J wave appearing principally in the right precordial leads (V1–V3), often followed by a negative T wave; (2) very closely coupled extrasystoles; and (3) rapid polymorphic VT (PVT), which at times may be indistinguishable from ventricular fibrillation (VF).[32] The ECG sign of the Brugada syndrome is dynamic and often concealed, but can be unmasked by potent sodium channel blockers such as ajmaline, flecainide, procainamide, disopyramide, propafenone, and pilsicainide.[33–35]

The arrhythmogenic substrate responsible for the development of extrasystoles and PVT in the Brugada syndrome is believed to be secondary to amplification of heterogeneities intrinsic to the early phases (phase 1–mediated notch) of the action potential of cells residing in different layers of the right ventricular wall of the heart. Rebalancing of the currents active at the end of phase 1 is thought to underlie the accentuation of the action potential notch in right ventricular epicardium, which is responsible for the augmented J wave and ST segment elevation associated with the Brugada syndrome (see Ref. 36 for references). The presence of a transient outward current (I_{to})-mediated spike and dome morphology, or notch, in ventricular epicardium, but not endocardium, creates a transmural voltage gradient responsible for the inscription of the electrocardiographic J wave.[14,37] The ST segment is normally isoelectric due to the absence of transmural voltage gradients at the level of the action potential plateau. Accentuation of the right ventricular action potential notch under pathophysiologic conditions leads to exaggeration of transmural voltage gradients and thus to accentuation of the J wave or to J point elevation. If the epicardial action potential continues to repolarize before that of endocardium, the T wave remains positive, giving rise to a saddleback configuration of the ST segment elevation. Further accentuation of the notch is accompanied by a prolongation of the epicardial action potential causing it to repolarize after endocardium, thus leading to inversion of the T wave. The down-sloping ST segment elevation, or accentuated J wave, observed in experimental wedge models often appears as an R′, suggesting that the appearance of a right bundle branch block morphology in Brugada patients may be due in large part to early repolarization of right ventricular epicardium, rather than major delays in impulse conduction in the right bundle.[38] Despite the appearance of a typical Brugada sign, accentuation of the right ventricular epicardial action potential notch alone does not give rise to an arrhythmogenic substrate. The arrhythmogenic substrate may develop with a further shift in the balance of current leading to loss of the action potential dome at some epicardial sites but not others. As a consequence, a marked TDR develops, creating a vulnerable window that, when captured by a premature extrasystole, can trigger a reentrant arrhythmia. Because loss of the action potential dome in epicardium is generally heterogeneous, epicardial dispersion of repolarization develops as well. Conduction of the action potential dome from sites where it is maintained to sites where it is lost causes local reexcitation via phase 2 reentry, leading to the development of a closely-coupled extrasystole capable of capturing the vulnerable window across the ventricular wall, thus triggering a circus movement reentry in the form of VT/VF.[39,40] Support for these hypotheses derives from experiments involving the arterially perfused right ventricular wedge preparation[39] and from recent studies in which monophasic action potential electrodes were positioned on the epicardial and endocardial surfaces of the right ventricular outflow tract in patients with the Brugada syndrome.[41,42]

Long QT Syndrome

The long QT syndrome (LQTS) is characterized by the appearance of long QT intervals in the ECG, an atypical PVT known as TdP, and a high risk for sudden cardiac death.[43–45] Congenital LQTS is subdivided into seven genotypes distinguished by mutations in at least six different ion genes and a structural anchoring protein located on chromosomes 3, 4, 7, 11, 17, and 21.[46–51] Acquired LQTS refers to a syndrome similar to the congenital form but caused by exposure to drugs that prolong the duration of the ventricular action potential,[52] or QT prolongation secondary to bradycardia or an electrolyte imbalance. In recent years this syndrome has been extended to encompass the reduced repolarization reserve attending remodeling of the ventricular myocardium that accompanies dilated and hypertrophic cardiomyopathies.[53–57]

Amplification of spatial dispersion of repolarization within the ventricular myocardium is thought to generate the principal arrhythmogenic substrate. The accentuation of spatial dispersion is typically secondary to an increase of transmural and transseptal dispersion of repolarization and the development of early afterdepolarization-induced triggered activity (TA) underlie the substrate and trigger for the development of TdP arrhythmias observed under LQTS conditions.[1,58] Models of the LQT1, LQT2, and LQT3 forms of the LQTS syndrome have been developed using the canine arterially perfused left ventricular wedge preparation.[59] These models have shown that in these three forms of LQTS, preferential prolongation of the M cell APD leads to an increase in the QT interval as well as an increase in TDR, the latter providing the substrate for the development of spontaneous as well as stimulation-induced TdP.[60–62]

The response to sympathetic activation displays a very different time course in the case of LQT1 and LQT2, both in experimental models and in the clinic.[58,63] In LQT1, isoproterenol produces an increase in TDR that is most prominent during the first 2 min, but which persists, to a lesser extent, during steady-state. TdP incidence is enhanced during the initial period as well as during steady-state. In LQT2, isoproterenol produces only a transient increase in TDR that persists for less than 2 min. TdP incidence is, therefore, enhanced only for a brief period of time. These differences in time course may explain the important differences in autonomic activity and other gene-specific triggers that contribute to events in patients with different LQTS genotypes.[64,65]

Short QT Syndrome

The short QT syndrome (SQTS) was proposed as a new clinical entity by Gussak *et al.* in 2000.[66] SQTS is an inherited syndrome characterized by a QTc $\leq$ 300 ms and high incidence of VT/VF in infants, children, and young adults.[67] The familial nature of this sudden death syndrome was confirmed by Gaita *et al.* in 2003.[68] The first genetic defect responsible for the short QT syndrome, reported by Brugada *et al.* in 2004, involved two different missense mutations resulting in the same amino acid substitution in HERG (N588K), which caused a gain of function in the rapidly activating delayed rectifier channel, I_{Kr}.[69] A second gene was recently reported by Bellocq *et al.*[70] A missense mutation in KCNQ1 (KvLQT1) caused a gain of function in I_{Ks}.

A distinctive electrocardiographic feature of the short QT syndrome is the appearance of tall peaked symmetrical T waves. The augmented T_{peak}–T_{end} interval associated with this electrocardiographic feature of the syndrome suggests that TDR is increased. Recent preliminary data from a wedge model of the short QT syndrome has provided evidence in support of the hypothesis that an increase in outward repolarizing current can increase TDR and thus create the substrate for reentry. The potassium channel opener pinacidil causes a heterogeneous abbreviation of APD among the different cell types spanning the ventricular wall, thus creating the substrate for the genesis of VT under conditions associated with short QT intervals. PVT could be readily induced with programmed electrical stimulation. The increase in TDR was further accentuated by isoproterenol, leading to easier induction and more persistent VT/VF. The latter is likely due to the reduction in the wavelength of the reentrant circuit, which reduces the pathlength required for maintenance of reentry.[71]

Catecholaminergic Polymorphic VT

Catecholaminergic, or familial, polymorphic ventricular tachycardia (CPVT) is a rare, autosomal dominant inherited disorder, predominantly affecting children or adolescents with structurally normal hearts. It is characterized by bidirectional ventricular tachycardia (BVT), PVT, and a high risk of sudden cardiac death (30% to 50% by the age of 20 to 30 years).[72,73] Recent molecular genetic studies have identified mutations in genes encoding for the cardiac ryanodine receptor 2 or calsequestrin 2 in patients with this phenotype.[74–77] Several lines of evidence point to delayed afterdepolarization (DAD)-induced TA as the mechanism underlying monomorphic VT (MVT) or BVT in these patients. The cellular mechanisms underlying the various ECG phenotypes, and the transition of MVT to PVT or VF, were recently elucidated with the help of the wedge preparation.[78] The wedge was exposed to low-dose caffeine to mimic the defective calcium homeostasis encountered under conditions that predispose to CPVT. The combination of isoproterenol and caffeine led to the development of DAD-induced TA arising from epicardium, endocardium, or the M region. Migration of the source of ectopic activity was responsible for the transition from MVT to slow PVT. Alternation of epicardial and endocardial source of ectopic activity gave rise to a BVT. Epicardial VT was associated with an increased T_{peak}–T_{end} interval and TDR due to reversal of the normal transmural activation sequence, thus creating the substrate for reentry, which permitted the induction of a more rapid PVT with programmed electrical stimulation. Propranolol or verapamil suppressed arrhythmic activity.[78]

CONCLUSION

Amplification of transmural heterogeneities of repolarization in ventricular myocardium can predispose to the development of potentially lethal reentrant arrhythmias. The long QT, short QT, Brugada, and catecholaminergic VT syndromes are examples of pathologies that have very different phenotypes and etiologies, but share a common final pathway in causing sudden death. These same mechanisms are likely responsible for life-threatening arrhythmias in a variety of other cardiomyopathies ranging from heart failure and hypertrophy, which may involve mechanisms very similar to those operative in LQTS, to ischemia and infarction, which may involve mechanisms more closely resembling those responsible for the Brugada syndrome.

ACKNOWLEDGMENTS

This work was supported by grants from the National Institutes of Health (HL-47678), American Heart Association–New York state affiliate, and the Masons of New York state and Florida.

REFERENCES

1. ANTZELEVITCH, C. & R. DUMAINE. 2002. Electrical heterogeneity in the heart: physiological, pharmacological and clinical implications. *In* Handbook of Electrophysiology: The Heart. E. Page, H. Fozzard & R.J. Solaro, Eds.: 654–692. Oxford University Press, New York.
2. DI DIEGO, J.M., Z.Q. SUN & C. ANTZELEVITCH. 1996. I_{to} and action potential notch are smaller in left vs. right canine ventricular epicardium. Am J Physiol. **271:** H548–H561.
3. VOLDERS, P.G., K.R. SIPIDO, E. CARMELIET, *et al.* 1999. Repolarizing K+ currents ITO1 and IKs are larger in right than left canine ventricular midmyocardium. Circulation **99:** 206–210.
4. SICOURI, S. & C. ANTZELEVITCH. 1991. A subpopulation of cells with unique electrophysiological properties in the deep subepicardium of the canine ventricle. The M cell. Circ. Res. **68:** 1729–1741.
5. ANTZELEVITCH, C. W. SHIMIZU, G.X. YAN, *et al.* 1999. The M cell: its contribution to the ECG and to normal and abnormal electrical function of the heart. J. Cardiovasc. Electrophysiol. **10:** 1124–1152.
6. ANYUKHOVSKY, E.P., E.A. SOSUNOV, R.Z. GAINULLIN, *et al.* 1999. The controversial M cell. J. Cardiovasc. Electrophysiol. **10:** 244–260.
7. SICOURI, S., J. FISH & C. ANTZELEVITCH. 1994. Distribution of M cells in the canine ventricle. J. Cardiovasc. Electrophysiol. **5:** 824–837.
8. YAN, G.X., W. SHIMIZU & C. ANTZELEVITCH. 1998. Characteristics and distribution of M cells in arterially-perfused canine left ventricular wedge preparations. Circulation **98:** 1921–1927.
9. POELZING, S., F.G. AKAR, E. BARON, *et al.* 2004. Heterogeneous connexin43 expression produces electrophysiological heterogeneities across ventricular wall. Am. J. Physiol. Heart Circ. Physiol. **286:** H2001–H2009.
10. YAMADA, K.A., E.M. KANTER, K.G. GREEN, *et al.* 2004. Transmural distribution of connexins in rodent hearts. J. Cardiovasc. Electrophysiol. **15:** 710–715.
11. ZYGMUNT, A.C., G.T. EDDLESTONE, G.P. THOMAS, *et al.* 2001. Larger late sodium conductance in M cells contributes to electrical heterogeneity in canine ventricle. Am. J. Physiol. **281:** H689–H697.
12. LIU, D.W. & C. ANTZELEVITCH. 1995. Characteristics of the delayed rectifier current (IKr and IKs) in canine ventricular epicardial, midmyocardial, and endocardial myocytes. Circ. Res. **76:** 351–365.
13. ZYGMUNT, A.C., R.J. GOODROW & C. ANTZELEVITCH. 2000. $I_{Na\text{-}Ca}$ contributes to electrical heterogeneity within the canine ventricle. Am. J. Physiol. **278:** H1671–H1678.
14. YAN, G.X. & C. ANTZELEVITCH. 1996. Cellular basis for the electrocardiographic J wave. Circulation **93:** 372–379.
15. YAN, G.X. & C. ANTZELEVITCH. 1998. Cellular basis for the normal T wave and the electrocardiographic manifestations of the long QT syndrome. Circulation **98:** 1928–1936.
16. EMORI, T. & C. ANTZELEVITCH. 2001. Cellular basis for complex T waves and arrhythmic activity following combined I(Kr) and I(Ks) block. J. Cardiovasc. Electrophysiol. **12:** 1369–1378.
17. LUBINSKI, A., E. LEWICKA-NOWAK, M. KEMPA, *et al.* 1998. New insight into repolarization abnormalities in patients with congenital long QT syndrome: the increased transmural dispersion of repolarization. Pacing Clin. Electrophysiol. **21:** 172–175.
18. WOLK, R., S. STEC & P. KULAKOWSKI. 2001. Extrasystolic beats affect transmural electrical dispersion during programmed electrical stimulation. Eur. J. Clin. Invest. **31:** 563–569.
19. TANABE, Y., M. INAGAKI, T. KURITA, *et al.* 2001. Sympathetic stimulation produces a greater increase in both transmural and spatial dispersion of repolarization in LQT1 than LQT2 forms of congenital long QT syndrome. J. Am. Coll. Cardiol. **37:** 911–919.
20. FREDERIKS, J., C.A. SWENNE, J.A. KORS, *et al.* 2001. Within-subject electrocardiographic differences at equal heart rates: role of the autonomic nervous system. Pflugers Arch. **441:** 717–724.
21. SHIMIZU, M., H. INO, K. OKEIE, *et al.* 2002. T-peak to T-end interval may be a better

predictor of high-risk patients with hypertrophic cardiomyopathy associated with a cardiac troponin I mutation than QT dispersion. Clin. Cardiol. **25:** 335–339.
22. TAKENAKA, K., T. AI, W. SHIMIZU, *et al.* 2003. Exercise stress test amplifies genotype-phenotype correlation in the LQT1 and LQT2 forms of the long-QT syndrome. Circulation **107:** 838–844.
23. WATANABE, N., Y. KOBAYASHI, K. TANNO, *et al.* 2004. Transmural dispersion of repolarization and ventricular tachyarrhythmias. J. Electrocardiol. **37:** 191–200.
24. SCHWARTZ, P.J., S.G. PRIORI, C. SPAZZOLINI, *et al.* 2001. Genotype-phenotype correlation in the long-QT syndrome: gene-specific triggers for life-threatening arrhythmias. Circulation **103:** 89–95.
25. YAMAGUCHI, M., M. SHIMIZU, H. Ino, *et al.* 2003. T wave peak-to-end interval and QT dispersion in acquired long QT syndrome: a new index for arrhythmogenicity. Clin. Sci. (Lond) **105:** 671–676.
26. WATANABE, N., Y. KOBAYASHI, K. TANNO K, *et al.* 2004. Transmural dispersion of repolarization and ventricular tachyarrhythmias. J. Electrocardiol. **37:** 191–200.
27. ANTZELEVITCH, C., L. BELARDINELLI, A.C. ZYGMUNT, *et al.* 2004. Electrophysiologic effects of ranolazine: a novel anti-anginal agent with antiarrhythmic properties. Circulation **110:** 904–910.
28. DI DIEGO, J.M., L. BELARDINELLI & C. ANTZELEVITCH. 2003. Cisapride-induced transmural dispersion of repolarization and torsade de pointes in the canine left ventricular wedge preparation during epicardial stimulation. Circulation **108:** 1027–1033.
29. ANTZELEVITCH, C. 2004. Drug-induced channelopathies. *In* Cardiac Electrophysiology: From Cell to Bedside, 4th ed. Douglas P. Zipes & Jose Jalife, Eds.: 151–157. W. B. Saunders, New York.
30. BELARDINELLI, L., C. ANTZELEVITCH & M.A. VOS. 2003. Assessing predictors of drug-induced torsade de pointes. Trends Pharmacol. Sci. **24:** 619–625.
31. FENICHEL, R.R., M. MALIK, C. ANTZELEVITCH, *et al.* 2004. Drug-induced torsade de pointes and implications for drug development. J. Cardiovasc. Electrophysiol. **15:** 1–21.
32. BRUGADA, P. & J. BRUGADA. 1992. Right bundle branch block, persistent ST segment elevation and sudden cardiac death: a distinct clinical and electrocardiographic syndrome: a multicenter report. J. Am. Coll. Cardiol. **20:** 1391–1396.
33. BRUGADA, R., J. BRUGADA, C. ANTZELEVITCH, *et al.* 2000. Sodium channel blockers identify risk for sudden death in patients with ST-segment elevation and right bundle branch block but structurally normal hearts. Circulation **101:** 510–515.
34. SHIMIZU, W., C. ANTZELEVITCH, K. SUYAMA, *et al.* 2000. Effect of sodium channel blockers on ST segment, QRS duration, and corrected QT interval in patients with Brugada syndrome. J. Cardiovasc. Electrophysiol. **11:** 1320–1329.
35. PRIORI, S.G., C. NAPOLITANO, M. GASPARINI, *et al.* 2000. Clinical and genetic heterogeneity of right bundle branch block and ST-segment elevation syndrome: a prospective evaluation of 52 families. Circulation **102:** 2509–2515.
36. ANTZELEVITCH, C. 2001. The Brugada syndrome: ionic basis and arrhythmia mechanisms. J. Cardiovasc. Electrophysiol. **12:** 268–272.
37. FISH, J.M. & C. ANTZELEVITCH. 2004. Role of sodium and calcium channel block in unmasking the Brugada syndrome. Heart Rhythm **1:** 210–217.
38. GUSSAK, I., C. ANTZELEVITCH, P. BJERREGAARD, *et al.* 1999. The Brugada syndrome: clinical, electrophysiologic and genetic aspects. J. Am. Coll. Cardiol. **33:** 5–15.
39. YAN, G.X. & C. ANTZELEVITCH. 1999. Cellular basis for the Brugada syndrome and other mechanisms of arrhythmogenesis associated with ST-segment elevation. Circulation **100:** 1660–1666.
40. LUKAS, A. & C. ANTZELEVITCH. 1996. Phase 2 reentry as a mechanism of initiation of circus movement reentry in canine epicardium exposed to simulated ischemia. Cardiovasc. Res. **32:** 593–603.
41. ANTZELEVITCH, C., P. BRUGADA, J. BRUGADA, *et al.* 2002. Brugada syndrome: a decade of progress. Circ. Res. **91:** 1114–1119.
42. KURITA, T., W. SHIMIZU, M. INAGAKI, *et al.* 2002. The electrophysiologic mechanism of ST-segment elevation in Brugada syndrome. J. Am. Coll. Cardiol. **40:** 330–334.
43. SCHWARTZ, P.J. 1985. The idiopathic long QT syndrome: progress and questions. Am. Heart J. **109:** 399–411.

44. Moss, A.J., P.J. Schwartz, R.S. Crampton, *et al.* 1991. The long QT syndrome: prospective longitudinal study of 328 families. Circulation **84:** 1136–1144.
45. Zipes, D.P. 1991. The long QT interval syndrome. A Rosetta stone for sympathetic related ventricular tachyarrhythmias. Circulation **84:** 1414–1419.
46. Wang, Q., J. Shen, I. Splawski, *et al.* 1995. *SCN5A* mutations associated with an inherited cardiac arrhythmia, long QT syndrome. Cell **80:** 805–811.
47. Mohler, P.J., J.J. Schott, A.O. Gramolini, *et al.* 2003. Ankyrin-B mutation causes type 4 long-QT cardiac arrhythmia and sudden cardiac death. Nature **421:** 634–639.
48. Plaster, N.M., R. Tawil, M. Tristani-Firouzi, *et al.* 2001. Mutations in Kir2.1 cause the developmental and episodic electrical phenotypes of Andersen's syndrome. Cell **105:** 511–519.
49. Curran, M.E., I. Splawski, K.W. Timothy, *et al.* 1995. A molecular basis for cardiac arrhythmia: *HERG* mutations cause long QT syndrome. Cell **80:** 795–803.
50. Wang, Q., M.E. Curran, I. Splawski, *et al.* 1996. Positional cloning of a novel potassium channel gene: *KVLQT1* mutations cause cardiac arrhythmias. Nat. Genet. **12:** 17–23.
51. Splawski, I., M. Tristani-Firouzi, M.H. Lehmann, *et al.* 1997. Mutations in the hminK gene cause long QT syndrome and suppress I_{Ks} function. Nat. Genet. **17:** 338–340.
52. Bednar, M.M., E.P Harrigan, R.J. Anziano, *et al.* 2001. The QT interval. Prog. Cardiovasc. Dis. **43:** 1–45.
53. Tomaselli, G.F. & E. Marban. 1999. Electrophysiological remodeling in hypertrophy and heart failure. Cardiovasc. Res. **42:** 270–283.
54. Sipido, K.R., P.G. Volders, S.H. De Groot, *et al.* 2000. Enhanced Ca(2+) release and Na/Ca exchange activity in hypertrophied canine ventricular myocytes: potential link between contractile adaptation and arrhythmogenesis. Circulation **102:** 2137–2144.
55. Volders, P.G., K.R. Sipido, M.A. Vos, *et al.* 1999. Downregulation of delayed rectifier K(+) currents in dogs with chronic complete atrioventricular block and acquired torsades de pointes. Circulation **100:** 2455–2461.
56. Undrovinas, A.I., V.A. Maltsev & H.N. Sabbah. 1999. Repolarization abnormalities in cardiomyocytes of dogs with chronic heart failure: role of sustained inward current. Cell. Mol. Life Sci. **55:** 494–505.
57. Maltsev, V.A., H.N. Sabbah, R.S. Higgins, *et al.* 1998. Novel, ultraslow inactivating sodium current in human ventricular cardiomyocytes. Circulation **98:** 2545–2552.
58. Antzelevitch, C. & W. Shimizu. 2002. Cellular mechanisms underlying the Long QT syndrome. Curr. Opin. Cardiol. **17:** 43–51.
59. Shimizu, W. & C. Antzelevitch. 2000. Effects of a K(+) channel opener to reduce transmural dispersion of repolarization and prevent torsade de pointes in LQT1, LQT2, and LQT3 models of the long-QT syndrome. Circulation **102:** 706–712.
60. Shimizu, W. & C. Antzelevitch. 1998. Cellular basis for the ECG features of the LQT1 form of the long QT syndrome: effects of b-adrenergic agonists and antagonists and sodium channel blockers on transmural dispersion of repolarization and torsade de pointes. Circulation **98:** 2314–2322.
61. Shimizu, W. & C. Antzelevitch. 1997. Sodium channel block with mexiletine is effective in reducing dispersion of repolarization and preventing torsade de pointes in LQT2 and LQT3 models of the long-QT syndrome. Circulation **96:** 2038–2047.
62. Shimizu, W. & C. Antzelevitch. 2000. Differential effects of beta-adrenergic agonists and antagonists in LQT1, LQT2 and LQT3 models of the long QT syndrome. J. Am. Coll. Cardiol. **35:** 778–786.
63. Noda, T., H. Takaki, T. Kurita, *et al.* 2002. Gene-specific response of dynamic ventricular repolarization to sympathetic stimulation in LQT1, LQT2 and forms of congenital long QT syndrome. Eur. Heart J. **23:** 975–983.
64. Schwartz, P.J., S.G. Priori, C. Spazzolini, *et al.* 2001. Genotype-phenotype correlation in the long-QT syndrome: gene-specific triggers for life-threatening arrhythmias. Circulation **103:** 89–95.
65. Ali, R.H., W. Zareba, A. Moss, *et al.* 2000. Clinical and genetic variables associated with acute arousal and nonarousal-related cardiac events among subjects with long QT syndrome. Am. J. Cardiol. **85:** 457–461.

66. GUSSAK, I., P. BRUGADA, J. BRUGADA, *et al.* 2000. Idiopathic short QT interval: a new clinical syndrome? Cardiology **94:** 99–102.
67. GUSSAK, I., P. BRUGADA, J. BRUGADA, *et al.* 2002. ECG phenomenon of idiopathic and paradoxical short QT intervals. Cardiac. Electrophysiol. Rev. **6:** 49–53.
68. GAITA, F., C. GIUSTETTO, F. BIANCHI, *et al.* 2003. Short QT Syndrome: a familial cause of sudden death. Circulation **108:** 965–970.
69. BRUGADA, R., K. HONG, R. DUMAINE, *et al.* 2004. Sudden death associated with short-QT syndrome linked to mutations in HERG. Circulation **109:** 30–35.
70. BELLOCQ, C., A. VAN GINNEKEN, C.R. BEZZINA, *et al.* 2004. Mutation in the KCNQ1 gene leading to the short QT-interval syndrome. Circulation **109:** 2394–2397.
71. EXTRAMIANA, F. & C. ANTZELEVITCH. 2004. Transmural electrophysiological heterogeneties underlying arrhythmogenesis associated with short QT intervals. Circulation **110:** 3661–3666.
72. LEENHARDT, A., V. LUCET, I. DENJOY, *et al.* 1995. Catecholaminergic polymorphic ventricular tachycardia in children: a 7-year follow-up of 21 patients. Circulation **91:** 1512–1519.
73. SWAN, H., K. PIIPPO, M. VIITASALO, *et al.* 1999. Arrhythmic disorder mapped to chromosome 1q42-q43 causes malignant polymorphic ventricular tachycardia in structurally normal hearts. J. Am. Coll. Cardiol. **34:** 2035–2042.
74. PRIORI, S.G., C. NAPOLITANO, M. MEMMI, *et al.* 2002. Clinical and molecular characterization of patients with catecholaminergic polymorphic ventricular tachycardia. Circulation **106:** 69–74.
75. PRIORI, S.G., C. NAPOLITANO, N. TISO, *et al.* 2001. Mutations in the cardiac ryanodine receptor gene (hRyR2) underlie catecholaminergic polymorphic ventricular tachycardia. Circulation **103:** 196–200.
76. LAITINEN, P.J., K.M. BROWN, K. PIIPPO, *et al.* 2001. Mutations of the cardiac ryanodine receptor (RyR2) gene in familial polymorphic ventricular tachycardia. Circulation **103:** 485–490.
77. POSTMA, A.V., I. DENJOY, T.M. HOORNTJE, *et al.* 2002. Absence of calsequestrin 2 causes severe forms of catecholaminergic polymorphic ventricular tachycardia. Circ. Res. **91:** e21–e26.
78. NAM, G.B., A. BURASHNIKOV & C. ANTZELEVITCH. 2004. Cellular mechanisms underlying the development of catecholaminergic ventricular tachycardia. Circulation. In press.

Species- and Preparation-Dependence of Stretch Effects on Sino-Atrial Node Pacemaking

PATRICIA J. COOPER AND PETER KOHL

Laboratory of Physiology, University of Oxford, Oxford OX1 3PT, UK

ABSTRACT: Acute dilation of the right atrium (e.g., via increased venous return) raises spontaneous beating rate (BR) of the heart in many species. Neural mechanisms contribute to this behavior *in vivo*, but a positive chronotropic response to stretch can also be observed in isolated right atrial tissue preparations and even at the level of single sino-atrial node (SAN) cells. The underlying mechanism has previously been reported to be compatible with stretch-activation of cation nonselective ion channels (SAC). This review reports species peculiarities in the chronotropic response of isolated SAN tissue strips to stretch: in contrast to guinea pig, murine SAN preparations respond to distension with a reduction in spontaneous BR. This differential response need not necessarily involve disparate (sub-)cellular mechanisms, as SAC activation would occur against the background of very different SAN electrophysiology in the two species. On the basis of single SAN cell action potential recordings, this review illustrates how this may give rise to potentially opposing effects on spontaneous BR. Interestingly, streptomycin (a useful SAC blocker in isolated cells) has no effect on stretch-induced chronotropy *in situ*, and this is interpreted as an indication of protection of SAC, in native tissue, from interaction with the drug.

KEYWORDS: guinea pig; *in situ*; mouse; streptomycin; stretch activated channels; volume activated channels

INTRODUCTION

Changes in venous return to the heart do not only affect the volume available for rapid ventricular filling and subsequent atrial contraction, but also diastolic atrial dimensions. Increased right atrial filling distends the atrial wall, including the sino-atrial node (SAN), which impinges on pacemaker function. Francis Bainbridge showed in 1915 that injection of fluids into the jugular vein of anaesthetised dogs elevates venous return and increases cardiac beating rate (BR).[1] This was attributed to a vagal response and is referred to as the *Bainbridge Reflex*. Several investigators repeated Bainbridge's experiments, and the majority indeed observed tachycardia upon increased venous return (although some found little change, or even bradycardic responses; for reviews, see Refs. 2 and 3).

Address for correspondence: Dr. Peter Kohl, M.D., Ph.D., Laboratory of Physiology, University of Oxford, Oxford OX1 3PT, UK. Voice: 44 1865 272500; fax: 44 1865 272554.
peter.kohl@physiol.ox.ac.uk

Ann. N.Y. Acad. Sci. 1047: 324–335 (2005). © 2005 New York Academy of Sciences.
doi: 10.1196/annals.1341.029

Subsequent studies in isolated hearts[4] and SAN tissue[5] revealed that—even in denervated preparations—stretch of the SAN pacemaker may give rise to a positive chronotropic response. More recent work, involving acute and chronic pharmacological blockade of intrinsic cardiac autonomic pathways, illustrated that this intrinsic component of the positive chronotropic response to stretch is not mediated via the activity of cardiac neurones, either.[6] Furthermore, experiments in isolated, spontaneously active SAN pacemaker cells confirmed that axial stretch may reliably and reversibly increase their spontaneous BR via a mechanism that involves stretch activation of an ion current with the properties of cation non-selective stretch activated channels (SAC).[7] This data confirms that Bainbridge's observation is encoded, in part at least, at the level of SAN pacemaker cells.

The stretch-induced increase in BR observed in single cells (+5% of initial spontaneous BR[7]) is significantly smaller than that seen in multicellular preparations (+15% to +40%[4,5]) and whole animal studies (+32% in dog;[1] +23.5% in man[8]). Clearly, a multitude of factors—such as the level of investigation, biological preparation, amplitude of stretch, experimental techniques and tools—affect SAN mechanosensitivity and subsequent electrophysiological behavior.

There are, however, similarities in the electrophysiological responses reported from SAN preparations at various levels of functional integration. Thus, the electrophysiological signature of the SAN's positive chronotropic response to stretch is (1) a reduction in maximum diastolic potential (less negative), (2) a reduction in maximum systolic potential (less positive), and (3) an increase in BR.

Also, at each level of investigation, the response to stretch is affected by the characteristics of mechanical stimulation (the larger the rate of rise and/or amplitude of stretch, the more pronounced the BR response) and by background BR (the lower the initial BR, the larger the potential effect of stretch within a given species). The latter has been investigated in some depth by Coleridge and Linden,[3] who studied the influence of anaesthetics employed in whole animal experiments (dog) to replicate the observations of Bainbridge.[1] They found that the type of anaesthetic influenced background BR of the heart, and that bolus fluid injections could increase BR at low initial BR (50–100 bpm) or, conversely, decrease BR at high initial BR (150–200 bpm). This may help to explain some of the variability of findings reported by researchers who attempted to reproduce Bainbridge's observations *in vivo*.

Here, we investigate whether in addition to (or instead of) variations in background BR, species differences in action potential (AP) characteristics may affect SAN responses to mechanical stimulation. This is assessed using SAN tissue isolated from species with very different cardiac electrophysiological background characteristics: guinea pig and mouse.

METHODS

Right atrial tissue containing the SAN was dissected from Langendorff-perfused hearts of either guinea pig (n = 46) or mouse (n = 25), after 5 min of coronary perfusion with physiological saline solution. Standard solution contained (in mM): NaCl 118, KCl 4.8, $MgSO_4$ 1.2, KH_2PO_4 1.2, $NaHCO_3$ 25, Glucose 11, $CaCl_2$ 1.8; it was bubbled with 95% O_2/5% CO_2, maintained at 37°C, and checked for correct osmolarity (297–305 mOsm, A0300; Knauer, Berlin, Germany).

The SAN region was identified as the area on the posterior wall of the right atrium between the superior and inferior *venae cavae*, *Crista terminalis* (*CT*) and atrial septum. One to three SAN tissue strips were cut perpendicular to the *CT*, near the leading pacemaker site. Silk suture loops (7/0, Ethicon Johnson and Johnson; St-Stevens Woluwe, Belgium) were tied to either end of the strips (*CT* and atrial septum), and attached to miniature hooks on a vertical force transducer (FORT10; WPI, Stevenage, UK) and a micrometer. Manual adjustment of the micrometer allowed setting of resting diastolic tension (0.1 g) and permitted length changes for experimental intervention (usually between 25%–60% of strip length, except for FIGURE 1 where length was increased by up to 123% of strip length for reference purposes). The tension developed by the preparation was recorded online using Acknowledge 3.7 software and Biopac MP150 data acquisition equipment (Biopac Systems Inc., Goleta, CA). BR was determined in real-time as the inverse of the peak-to-peak interval of active force recordings.

Sino-atrial node tissue preparations were placed in a water-jacketed organ bath containing 50 mL of physiological saline solution (at 37°C, unless otherwise stated) and allowed to equilibrate for 30 min for stabilization of BR. Pharmacological agents, when used, were added directly to the superfusate and allowed to equilibrate for at least 10 min (in some cases this was subsequent to coronary perfusion with the same drug for 5 min prior to SAN dissection). Complete exchange of the organ bath solution occurred every 15 min.

All stretch protocols were performed on the background of a stable BR. Based on previous data in the literature on rabbit SAN stretch,[9] passive tension was adjusted (via the micrometer screw) to the following levels relative to control diastolic tension: +0.5 g, +1.0 g, +1.5 g, and +2.0 g. Tissue strips were held at the new length for about 30 s (during which passive tension would, after an initial tissue-viscosity related peak, decline to a quasi-steady state), before length was returned to the original (pre-stretch) level. A recovery period of at least 5 min followed each intervention.

Within this range of passive diastolic tension increases, the BR response rose linearly with the amplitude of stretch (see FIG. 1B). It is important to note that these tension steps, modelled after a prior report,[9] cause very significant strain of the SAN tissue strips (e.g., +28 to +64% at +0.5 g, and +85 to +123% at +2.0 g). This extent of SAN tissue lengthening clearly exceeds expected (patho-)physiologically relevant changes *in situ* (a rise in right atrial pressure by 10 mmHg, observed in heavy exercise,[10] has been related to a 30% change in SAN strip length).[5]

Subsequent investigations, therefore, employed the smallest tension level at which a rate response could be reliably observed (+0.5 g, unless otherwise stated), to avoid non-physiological strain and possible injury-related artefacts.

All drugs (except GsMTx-4, courtesy of Dr. Fred Sachs) used in this study were sourced from Sigma-Aldrich Ltd. (Poole, UK) and dissolved appropriately to prepare stock solutions on the day of the experiment. Final concentrations used were as follows: 1 mM 9-anthracene carboxylate (9-AC); 40 to 500 μM streptomycin sulphate; 0.1 to 2 μM acetylcholine.

Background BR was determined by averaging the rate of contractions during 15 s prior to application of stretch. During stretch, both peak BR (as in Arai *et al.*[9]) and sustained changes (averaged over 30 s) were assessed. Recovery BR at original tissue length was monitored for 5 min after return to control length. Analysis of peak

BR responses to stretch did not provide qualitatively different results to the sustained responses reported in the following text (although both the overall amplitude and the variability of the response were larger).

Where data is presented as a group's mean, the error is expressed as the standard error of the mean (mean ± SEM). Statistical analysis was performed in Prism (v4.0; GraphPad, San Diego, CA), using Student's t-test or ANOVA, where appropriate, and significance of null-hypothesis rejection was accepted at $p < 0.05$.

RESULTS

Effect of Stretch

In order to describe the effect of stretch characteristics on SAN pacemaking rate in guinea pig tissue strips, mechanical stimulation was initially applied at a number of tension levels. Usually, BR peaked with maximum passive tension and then declined during the subsequent reduction in tissue stress (attributed to gradual lengthening of tissue viscous elements; see example in FIG. 1A). Upon release of stretch, BR recovery was not necessarily instantaneous and could require a period of up to 15 s.

FIGURE 1B illustrates that the stretch-induced increase in guinea pig SAN BR is linearly related, within the tested range of mechanical stimuli, to the extent of passively applied tension, and reaches 5.0% to 14.4% of initial BR values ($p = 0.006$). This is similar (although slightly less pronounced) to the increase in peak BR observed by Kodama *et al.*[11] and Arai *et al.*,[9] who used rabbit SAN tissue strips. Dif-

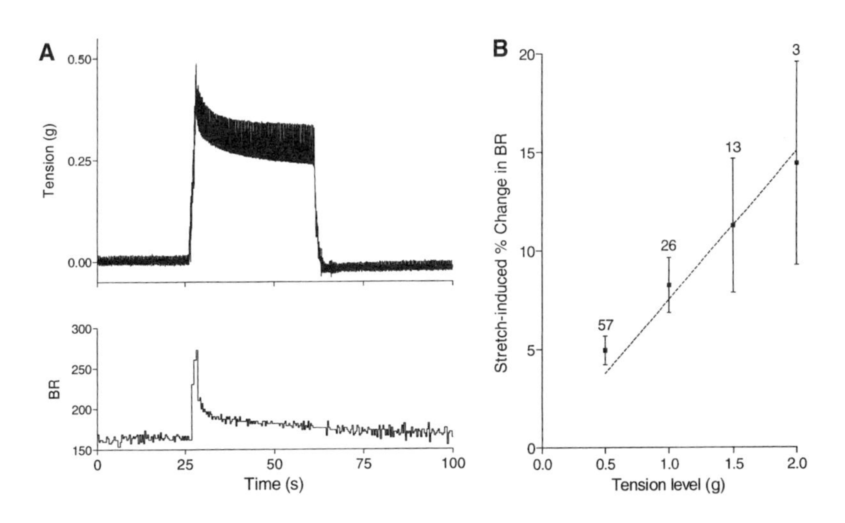

FIGURE 1. Effect of stretch on guinea pig sino-atrial node tissue. (**A**) *Upper panel* shows stretch-induced change in tissue tension (by <+0.5 g), and the *lower panel* shows the concomitant increase in beating rate (BR). (**B**) Mean (± SEM) percentage increase in BR recorded at different tension levels ($p = 0.006$, ANOVA, across all groups; numbers above columns refer to n values).

ferences may be species dependent or related to minor differences in experimental procedures or data analysis (e.g., peak vs. maintained response).

In comparison, axial stretch of single rabbit SAN cells (by up to 8% of cell length, using carbon fibers) causes an increase in spontaneous BR of approximately 5%,[7] which is quite compatible with the tissue response observed in this investigation at the lower levels of mechanical stimulation.

Mechanisms

It has been suggested that mechano-dependent ion channels may underlie the stretch-induced increase in heart rate.[12] There are at least two possible candidates for this response: cation non-selective SAC[7] and cell-volume activated chloride channels (VAC).[9]

Arai and colleagues[9,11] found that application of various VAC blockers caused a reduction in the stretch-induced increase in BR of rabbit isolated SAN tissue (which was significant only at high levels of tissue distension). To reassess this, we applied mechanical stimuli at the lower, more physiological range reported in the literature (+0.5 g).[9] Even this intervention causes strains that potentially exceed the levels normally observed *in situ*.[5] Application of 1 mM 9-AC (a blocker of VAC) has *no effect* on the stretch-induced increase in BR (FIG. 2), supporting the notion that VAC do not underlie the SAN response to (near-) physiological stress or strain.

As in the study by Arai *et al.*,[9] pharmacological block of VAC reduces background BR (i.e., in the absence of mechanical stimulation). This is different from observations made in single isolated SAN pacemaker cells, where a 9-AC induced reduction in spontaneous BR was only observed subsequent to hyposmotic cell swelling.[13] It is likely that tissue dissection and/or experimental conditions (i.e., colloid-free perfusion) will cause cell swelling in multicellular preparations (despite checking osmotic pressure of solutions). In any case, the observed reduction in BR in the presence of 9-AC would, if anything, favor the identification of a stretch-induced increase in BR. Importantly, it confirms that 9-AC is likely to have reached

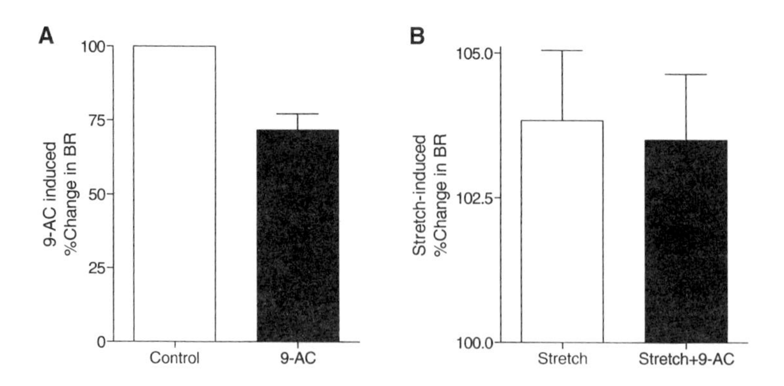

FIGURE 2. Effect of 9-AC (1 mM) on guinea pig sino-atrial node. (**A**) 9-AC reduced initial heart rate to 71.7 ± 5.5% (n = 3, p = 0.035). (**B**) 9-AC had no significant effect on the stretch-induced increase in beating rate (n = 3, p = 0.865).

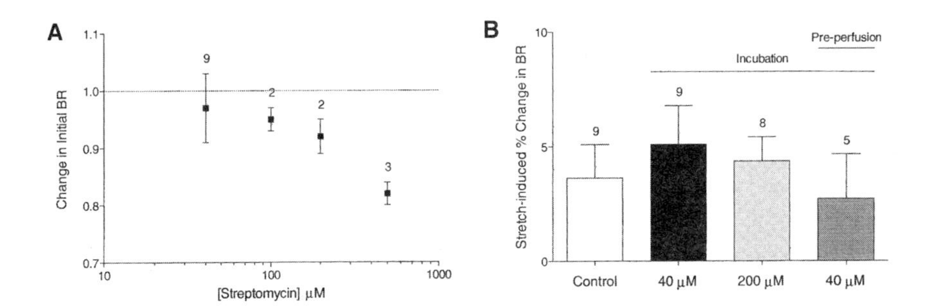

FIGURE 3. Effect of streptomycin on guinea pig sino-atrial node (SAN) tissue. (**A**) Negative chronotropic effect of high streptomycin concentrations (100–500 μM, $p = 0.006$) on spontaneous beating rate (BR) in guinea pig SAN tissue strips. (**B**) Streptomycin (40 and 200 μM) did not affect the stretch-induced increase in spontaneous BR ($p = 0.76$), regardless of whether the drug was applied to the superfusate only (Incubation), or additionally by 5 min coronary perfusion prior to SAN tissue extraction (Pre-perfusion). Numbers above columns indicate n values for each group.

VAC inside the SAN tissue strip, thereby making it less probable that the lack of a 9-AC effect on stretch-induced chronotropy is a false-negative finding.

To investigate the alternative possibility that activation of SAC might underlie the stretch-induced increase in BR, we applied streptomycin at concentrations from 40 to 500 μM.[14] At the lowest concentration (40 μM), streptomycin blocks SAC relatively selectively in isolated cardiac cells. At higher concentrations the drug is known to affect Ca^{2+} transients (200 μM or more) and to inhibit the L-type Ca^{2+} current (by 50% at 2 mM).[15] In keeping with the latter, high concentrations (100–500 μM) of streptomycin reliably and reversibly reduced spontaneous BR ($p = 0.006$; FIG. 3A) and contractility ($p = 0.02$; not shown) of guinea pig SAN tissue strips. This observation lends credibility to the suggestion that streptomycin reaches pacemaker cells inside the isolated SAN tissue strips. The streptomycin-induced reduction in background BR was observed regardless of whether the drug was applied by coronary perfusion prior to tissue dissection, or via the superfusate only (where the effect occurred within 5 min).

Interestingly, streptomycin did *not* affect the ability of SAN tissue to respond to distension by an increase in spontaneous BR, neither at low concentrations that are known to block SAC in isolated cells (40 μM), nor at levels that are high enough to affect background BR in the present multi-cellular preparation (200 μM; see FIG. 3B). This was, again, observed regardless of the drug application protocol, including coronary pre-perfusion and incubation for durations of up to 50 min.

Species Differences

The overall response to stretch, observed in the guinea pig preparations, is similar to previous observations in other slow-beating species, such as the rabbit. To assess the effect of species differences, we compared stretch effects on BR in guinea pig to those in mouse hearts, which show a much higher average BR (240 bpm and 450 bpm, respectively). FIGURE 4 shows a summary of stretch responses in guinea pig and mouse SAN tissue strips. Some of the experiments were performed at a reduced tem-

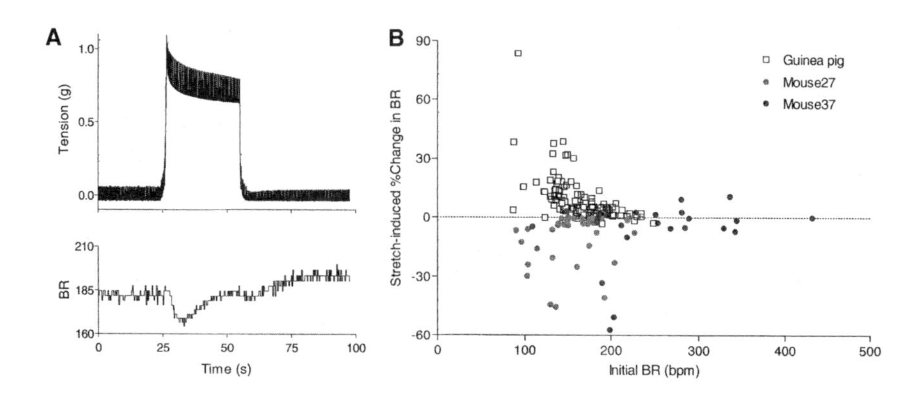

FIGURE 4. Species differences in sino-atrial node (SAN) beating rate (BR) response to stretch. (**A**) *Upper panel* shows change in passive tension level (here to <+1.0 g) with a stretch intervention, and the *lower panel* shows the concomitant BR reduction in a mouse SAN strip, recorded at 27°C. (**B**) Summary of stretch responses in guinea pig at 37°C and mouse SAN tissue strips recorded at either 37°C (Mouse37) or 27°C (Mouse27), plotted as a function of initial BR.

perature (27°C instead of 37°C) or after application of acetylcholine (0.1–2 μM) in order to extend the range of background BR and their overlap between the two species.

In line with earlier suggestions, reduced background BR appears to enhance the stretch-induced effects on chronotropy if studied *within* a species. Thus, in guinea pig the average stretch-induced increase in BR was +13.8% from a control BR of 58 ± 5 bpm (at 27°C), yet only +9.8% from a control BR of 150 ± 12 bpm (at 37°C, n = 4, p = 0.0004, paired observations; not shown).

More striking, however, is the observation that the BR-dependence of stretch effects on BR cannot be extrapolated between species. Thus, murine SAN tissue strips overwhelmingly responded to stretch with a *reduction* in BR—a negative chronotropic response (FIG. 4)—which is also more pronounced at lower background BR. The stretch-induced negative chronotropy was observed over a wide range of background BR levels, which overlapped completely with the BR range of guinea pig (FIG. 4B) or rabbit[9] (two species that show a positive chronotropic response to stretch).

These responses (either positive or negative chronotropy) are independent of neuronal influences, as experiments conducted in the presence of 1 μM atropine and 1 μM propranolol did not alter the response (guinea pig: n = 6, p = 0.67; mouse: n = 12, p = 0.77; not shown).

DISCUSSION

In many species employed in classical physiology studies (including man[8]), the heart responds to stretch of the SAN region with an increase in spontaneous BR. This has been observed in whole animals, isolated hearts, SAN tissue strips, and single SAN cells. However, the mechanisms underlying this response have not yet been comprehensively identified.

In this study, we used tissue strips from the right atrium, including the SAN region, from guinea pig and mouse. In order to keep the strain of SAN preparations closer to physiological limits, we employed changes in tissue tension at the low end of those used in previous experimental investigations. Additionally, stretch interventions were applied only up to three times per strip (during control, drug application, and washout, for analysis of paired observations). This is different from previous work, where multiple stretch interventions across large tension ranges (from +0.5 g, over +1.0 g, and +1.5 g, to +2.0 g) were repeated 8 to 12 times per strip.[9]

We found in this study that repeated mechanical stimulation—in particular after high levels of stretch—required larger length changes in order to elicit matching tension levels, suggesting irreversible (or at least long-lasting) changes in visco-elastic properties of the tissue strip. This is not surprising if one considers that a +2.0 g stress causes tissue strain by 73% to 123%. Such distension is very likely to cause cell and/or tissue damage. This may potentially be associated with stretch-induced swelling, which may have contributed to some of the controversy on the effects of VAC block on the chronotropic response of the heart to stretch, which has been observed only at non-physiologically high levels of tissue distension.[9]

We focused our analysis on the average BR response over a 30 s period of stretch. The sustained response to stretch, both in the presence and absence of pharmacological agents, shows the same *qualitative* behavior as the peak BR, albeit at a smaller amplitude. Hence, we are reporting conservative values for stretch-induced responses, compared to other published experimental work that focussed solely on peak changes in BR.

Based on the previous identification, in rabbit single SAN pacemaker cells, of VAC (activated by cell inflation via a pipette in ruptured patch mode),[16] it had been suggested that VAC might underlie the positive chronotropic response of the heart to right atrial distension. Mechanical activation of VAC has also been demonstrated using centrifugal magnetic bead stretching of β1-integrins in ventricular myocytes,[17] and underlying channel properties have been reported in atrial cells.[18]

In contrast, Lei and Kohl showed that hyposmotic swelling of *spontaneously* beating rabbit SAN pacemaker cells *slows* their spontaneous BR.[13] This suggests that cell swelling may be an inappropriate intervention to mimic effects caused by changes in diastolic wall stress or strain. In addition, Sasaki *et al.*[19] found no activation of VAC in ventricular myocytes by direct cell membrane manipulation; only in the presence of cell volume changes was VAC activation observed. As there is no evidence to suggest that SAN cell volume changes during the *normal* cardiac cycle of contraction and relaxation, it is probable that other mechanisms must be involved. These mechanisms could be related to mechanical activation of SAC.

In order to identify a potential contribution by SAC, we applied streptomycin. This antibiotic has been shown to inhibit stretch-mediated effects in isolated ventricular myocytes at concentrations of 40 μM.[14,20] At higher concentrations (2 mM), voltage-dependent ion channels (e.g., the L-type Ca^{2+} channel and the delayed rectifier K^+ channel[15]) are affected in isolated cardiac cells. In whole heart preparations, however, varied results have been obtained. Salmon *et al.* found that 200 μM (but not 50 μM) streptomycin causes a significant reduction in wall-stress induced arrhythmias in working rat heart.[21] In contrast, Sung *et al.* saw no effect of 200 μM streptomycin on acute load-induced changes in conduction velocity and AP duration in rabbit whole heart preparations.[22]

In our superfused SAN tissue preparations, high concentrations of streptomycin (100–500 μM) caused a reduction in control BR (and contractility), suggesting an inhibitory effect on the L-type Ca^{2+} channels. However, streptomycin had no effect on the stretch-induced positive chronotropic response, neither at 40 μM (where it blocks single cell SAC), nor at 200 μM (where it reduces BR in our multi-cellular preparations, compatible with the previously reported blocking action on L-type Ca^{2+} channels).

This observation may be interpreted in two principally different ways: either SAC play no role in stretch-induced changes in SAN electrophysiology, or streptomycin does not succeed in blocking SAC *in situ*. Given that (1) the axial stretch induced changes in single SAN pacemaker cell electrophysiology are *fully consistent* with SAC activation, (2) implementation of matching SAC currents in mathematical models of SAN cell and tissue activity is sufficient to *fully reproduce* experimentally observed stretch effects, and (3) the signature of single cell stretch effects (reduced maximum diastolic potential, reduced AP amplitude, increased BR) is *fully compatible* with isolated SAN tissue electrophysiology responses,[5,7] we believe that the first of the previously listed interpretations (lack of SAC effect in SAN tissue) is unlikely to be correct. Alternatively, SAC may be protected, *in situ*, from the blocking action of streptomycin. Such an interpretation would be in keeping with the observation that vestibular inhibition in patients taking aminoglycoside antibiotics (such as streptomycin) is usually observed only after long-term exposure to the drug.[23]

To assess the principal plausibility of this suggestion, we conducted a pilot experiment using the novel peptide blocker of SAC, GsMTx-4.[24] This drug can be used in multi-cellular preparations[25] and has been reported to be highly selective for cationic SAC.[24] In our guinea pig SAN preparation, 400 nM GsMTx-4 caused full and reversible inhibition of the stretch-induced increase in BR. Using a +1.0 g mechanical stimulus, the stretch-induced change in BR was: +11.1% during control conditions (initial BR 140 bpm), −2.0% after 5 min exposure to GsMTx-4 (initial BR 119 bpm), and +11.9% after 6 min wash-out (initial BR 121 bpm).

It should be noted, therefore, that streptomycin—while being a useful drug to dissect cation non-selective SAC effects in single cells—is likely to give false-negative results at higher levels of functional integration, including native tissue preparations, isolated heart, and whole animal studies.

In our experiments, the magnitude of stretch-induced BR changes was correlated to background BR (and mechanical stimulus amplitude) within each species. The direction of change, however, was opposite, with guinea pig SAN tissue showing a positive chronotropic response to stretch, and murine preparations overwhelmingly a negative chronotropic response (see FIG. 4B; note that only 1% of 322 individual stretch applications caused negative chronotropy in guinea pig SAN tissue, even in tissue maintained at 27°C [not shown]).

This difference in stretch-induced changes in BR may be a consequence of the very different underlying electrophysiology of species with moderate (guinea pig, rabbit) and very high BR (mouse). Thus, murine pacemaking involves a significant contribution of the fast Na^+ current[26] and has much faster repolarization dynamics (no plateau), giving its repolarization a "convex," rather than "concave," appearance.

To illustrate this point, FIGURE 5 shows a superimposition of pacemaker AP, recorded from rabbit and mouse single SAN cells. Overlaying this with an indication of the SAC reversal potential, one can appreciate that rabbit SAN pacemaking is

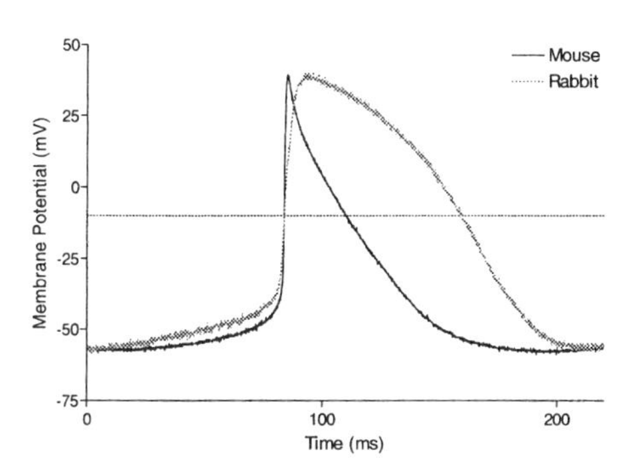

FIGURE 5. Superimposition of isolated sino-atrial node (SAN) cell action potentials from rabbit (*gray*) and mouse (*black*). The dashed line at −10 mV indicates the expected reversal potential for cation non-selective stretch-activated currents (as identified in rabbit SAN cells[7]).

dominated by periods of time during which the SAN AP moves *toward* the stretch-activated reversal potential. The corresponding potential change would be *accelerated* by activation of SAC, thereby shortening cycle length. In contrast, in murine cells the duration of time during which the SAN AP moves *away* from the SAC reversal potential is more pronounced and often exceeds 50% of the cardiac cycle.[27] Thus, SAC activation would have the potential of *slowing* murine pacemaking.

The same mechanisms—SAC activation by mechanical stimulation—may therefore be expected to have differential effects on overall BR in different species and could underlie both the positive chonotropic response to stretch in hearts with low to moderate BR *and* the negative chronotropy seen in species with very high BR.

Finally, in order to resolve any potential contribution of differences in cardiac intrinsic nerve system activity, we performed experiments in the presence of atropine and propranolol (1 μM each), which had no effect on the initial BR. Stretch-induced responses in both guinea pig and mouse SAN tissue strips were not affected by this autonomic block (as previously reported for rat SAN tissue).[6] This underscores the suggestion that intracardiac neurons are unlikely to have a significant effect on mechanical modulation of BR in isolated tissue.

CONCLUSION

The chronotropic response of the heart to stretch can be explained on the basis of SAC activation, while VAC do not appear to play a significant role under physiological conditions. Likewise, the response does not require a contribution of the intrinsic cardiac autonomous nervous system.

Streptomycin, a suitable blocker of SAC in isolated cells, appears ineffective for experimental identification of acute SAC effects *in situ*.

Background BR affects stretch responses, with lower BR generally promoting more pronounced (relative and absolute) stretch-induced changes in chronotropy, whether positive or negative. There is not an absolute relation between background BR and stretch-induced changes, and extrapolation across species boundaries may be misleading.

The principally different effects of stretch on SAN BR in guinea pig (acceleration) and mouse (deceleration) do not necessarily need to involve disparate underlying mechanisms, but may be a consequence of species-dependent differences in SAN electrical activity. Further studies, involving axial distension of isolated SAN cells from species with moderate and fast BR, are required to verify this hypothesis.

ACKNOWLEDGMENTS

We thank Drs. Fred Sachs (for kindly providing us with a sample of GsMTx-4), Alan Garny (for help with modeling studies), and Ming Lei (who contributed to mouse SAN cell electrophysiology work). Peter Kohl is a Royal Society Research Fellow. This work was supported by the British Heart Foundation and the UK Medical Research Council.

REFERENCES

1. BAINBRIDGE, F.A. 1915. The influence of venous filling upon the rate of the heart. J. Physiol. **50:** 65–84.
2. HAKUMÄKI, M.O. 1987. Seventy years of the Bainbridge reflex. Acta. Physiol. Scand. **130:** 177–185.
3. COLERIDGE, J.C.G. & R.J. LINDEN. 1955. The effect of intravenous infusions upon the heart rate of the anaesthetized dog. J. Physiol. **128:** 310–319.
4. BLINKS, J.R. 1956. Positive chronotropic effect of increasing right atrial pressure in the isolated mammalian heart. Am. J. Physiol. **186:** 299–303.
5. DECK, K.A. 1964. Dehnungseffekte am spontanschlagenden, isolierten Sinusknoten. Pflugers Arch. **280:** 120–130.
6. WILSON, S.J. & C.P. BOLTER. 2002. Do cardiac neurons play a role in the intrinsic control of heart rate in the rat? Exp. Physiol. **87:** 675–682.
7. COOPER, P.J., M. LEI, L.X. CHENG, *et al.* 2000. Axial stretch increases spontaneous pacemaker activity in rabbit isolated sinoatrial node cells. J. Appl. Physiol. **89:** 2099–2104.
8. DONALD, D.E. & J.T. SHEPHERD. 1978. Reflexes from the heart and lungs: physiological curiosities or important regulatory mechanisms. Cardiovasc. Res. **12:** 449–469.
9. ARAI, A., I. KODAMA & J. TOYAMA. 1996. Roles of Cl^- channels and Ca^{2+} mobilization in stretch-induced increase of SA node pacemaker activity. Am. J. Physiol. **270:** H1726–H1735.
10. BOLTER, C.P. 1996. Effect of changes in transmural pressure on contraction frequency of the isolated right atrium of the rabbit. Acta. Physiol. Scand. **156:** 45–50.
11. KODAMA, I., A. ARAI & J. TOYAMA. 1995. Mechanical regulation of pacemaking activity in the sinoatrial node: roles of stretch-sensitive channels and calcium dynamics. Heart Vessels **9**(Suppl. 1): 212–213.
12. KOHL, P., P. HUNTER & D. NOBLE. 1999. Stretch-induced changes in heart rate and rhythm: clinical observations, experiments and mathematical models. Prog. Biophys. Mol. Biol. **71:** 91–138.
13. LEI, M. & P. KOHL. 1998. Swelling-induced decrease in spontaneous pacemaker activity of rabbit isolated sino-atrial node cells. Acta. Physiol. Scand. **164:** 1–12.
14. GANNIER, F., E. WHITE, A. LACAMPAGNE, *et al.* 1994. Streptomycin reverses a large stretch induced increase in $[Ca^{2+}]_i$ in isolated guinea pig ventricular myocytes. Cardiovasc. Res. **28:** 1193–1198.
15. BELUS, A. & E. WHITE. 2002. Effects of streptomycin sulphate on I_{CaL}, I_{Kr} and I_{Ks} in guinea-pig ventricular myocytes. Eur. J. Pharmacol. **445:** 171–178.
16. HAGIWARA, N., H. MASUDA, M. SHODA, *et al.* 1992. Stretch-activated anion currents of rabbit cardiac myocytes. J. Physiol. **456:** 285–302.
17. BROWE, D.M. & C.M. BAUMGARTEN. 2003. Stretch of β1 integrin activates an outwardly rectifying chloride current via FAK and Src in rabbit ventricular myocytes. J. Gen. Physiol. **122:** 689–702.

18. SATO, R. & S. KUOMI. 1998. Characterization of the stretch-activated chloride channel in isolated human atrial myocytes. J. Membr. Biol. **163:** 67–76.
19. SASAKI, N., T. MITSUIYE & A. NOMA. 1992. Effects of mechanical stretch on membrane currents of single ventricular myocytes of guinea-pig heart. Jpn. J. Physiol. **42:** 957–970.
20. BELUS, A. & E. WHITE. 2003. Streptomycin and intracellular calcium modulate the response of single guinea-pig ventricular myocytes to axial stretch. J. Physiol. **546:** 501–509.
21. SALMON, A.H.J., J.L. MAYS, G.R. DALTON, *et al*. 1997. Effect of streptomycin on wall-stress-induced arrhythmias in the working rat heart. Cardiovasc. Res. **34:** 493–503.
22. SUNG, D., R.W. MILLS, J. SCHETTLER, *et al*. 2003. Ventricular filling slows epicardial conduction and increases action potential duration in an optical mapping study of the isolated rabbit heart. J. Cardiovasc. Electrophysiol. **14:** 739–749.
23. RYBAK, L.P. 1986. Drug ototoxicity. Ann. Rev. Pharmacol. Toxicol. **26:** 79–99.
24. SUCHYNA, T.M., J.H. JOHNSON, K. HAMER, *et al*. 2000. Identification of a peptide toxin from *Grammostola spatulata* spider venom that blocks cation-selective stretch-activated channels. J. Gen. Physiol. **115:** 583–598.
25. BODE, F., F. SACHS & M.R. FRANZ. 2001. Tarantula peptide inhibits atrial fibrillation. Nature **409:** 35–36.
26. MAIER, S.K., R.E. WESTENBROEK, T.T. YAMANUSHI, *et al*. 2003. An unexpected requirement for brain-type sodium channels for control of heart rate in the mouse sinoatrial node. Proc. Natl. Acad. Sci. U.S.A. **100:** 3507–3512.
27. COOPER, P.J. & P. KOHL. 2005. Mechanical modulation of sino-atrial node pacemaking. *In* Cardiac Mechano-Electric Feedback and Arrhythmias: From Pipette to Patient. P. Kohl, M.R. Franz & F. Sachs, Eds.: 72–82. Elsevier. Philadelphia.

Dependence of Electrical Coupling on Mechanical Coupling in Cardiac Myocytes

Insights Gained from Cardiomyopathies Caused by Defects in Cell–Cell Connections

JEFFREY E. SAFFITZ

Department of Pathology and Center for Cardiovascular Research, Washington University School of Medicine, St. Louis, Missouri 63110, USA

ABSTRACT: Cardiac myocytes are electrically coupled by large gap junctions to ensure safe conduction. Membrane regions containing gap junction channels are rigid and susceptible to fragmentation in response to shear stress. Thus, gap junctions in cardiac myocytes are located in close proximity to points of cell–cell adhesion within the intercalated disk. These adhesion junctions mechanically stabilize the sarcolemmas of adjacent cells to allow formation and maintenance of large arrays of intercellular channels. It has been proposed that the extent to which cardiac myocytes are coupled mechanically at cell–cell adhesion junctions is an important determinant of the extent to which cells can become electrically coupled at gap junctions. This hypothesis has been tested by analyzing gap junctions in human cardiomyopathies caused by mutations in plakoglobin and desmoplakin, intracellular proteins that link adhesion molecules at cell–cell junctions to the cardiac myocyte cytoskeleton. Marked remodeling of cardiac gap junctions, despite the presence of normal intracellular levels of connexin43, was observed. This suggests that defects in cell–cell adhesion, or the presence of discontinuities between adhesion junctions and the cytoskeleton, destabilize gap junctions and diminish electrical coupling. This could contribute to the high incidence of ventricular arrhythmias and sudden death known to occur in these cardiomyopathies.

KEYWORDS: Carvajal syndrome; gap junctions; intercalated disks; mechanical junctions; Naxos disease

INTRODUCTION

Cardiac muscle is composed of individual cells, each invested with an insulating lipid bilayer. Electrical activation of the myocardium, therefore, requires cell–cell transfer of current at gap junctions.[1] These specialized regions of the sarcolemma contain arrays of densely packed intercellular channels that directly connect the cytoplasmic compartments of neighboring cells and permit cell–cell passage of ions

Address for correspondence: Prof. Jeffrey E. Saffitz, M.D., Ph.D., Department of Pathology, Box 8118, Washington University School of Medicine, 660 South Euclid Avenue, St. Louis, MO 63110, USA. Voice: 314-362-7728; fax: 314-362-4096.
saffitz@pathology.wustl.edu

Ann. N.Y. Acad. Sci. 1047: 336–344 (2005). © 2005 New York Academy of Sciences.
doi: 10.1196/annals.1341.030

and small molecules.[1] Because transfer of depolarizing current can occur only at gap junctions, it follows that the number and spatial distribution of gap junction channels are important determinants of the velocity of impulse propagation and the three-dimensional temporal/spatial pattern of electrical activation of the heart.[1]

Mechanisms regulating expression of gap junction channel proteins and the distribution of intercellular electrical junctions in the normal and diseased heart are not understood in detail. It is likely, however, that multiple mechanisms participate in determining the extent to which cardiac myocytes are electrically and metabolically coupled to their neighbors. These include transcriptional and posttranslational mechanisms regulating connexin protein synthesis and assembly into channels, channel conductance states and open probability times, and removal of connexins from gap junctions and their degradation. Other factors may also exert a powerful effect on gap junctional coupling in cardiac myocytes. We have proposed that the extent to which cardiac myocytes are coupled mechanically at cell–cell adhesion junctions is an important determinant of the extent to which they can become electrically coupled at gap junctions.[2] Here, we present current evidence in support of this hypotheses based on studies of human cardiomyopathies caused by mutations in proteins responsible for linking adhesion molecules at the intercalated disk to the intracellular cytoskeletal network in cardiac myocytes.

THE DEPENDENCE OF ELECTRICAL COUPLING AND MECHANICAL COUPLING: A WORKING HYPOTHESIS

Working ventricular myocytes are extensively interconnected by numerous gap junctions that are among the largest in any mammalian tissues.[3] Presumably, these large arrays of low-resistance channels are required to ensure safe conduction. However, a gap junctional plaque approaches the theoretical maximum amount of protein that can be packed within the lipid bilayer. As a result, membrane regions containing these channel arrays are rigid and highly susceptible to fragmentation in response to shear stress, which could be significant in large junctions situated between contracting ventricular myocytes. The heart has solved this problem by surrounding gap junctions with cell–cell adhesion junctions that stabilize the sarcolemmas of interconnected cells and create membrane regions that favor formation and maintenance of large arrays of gap junction channels.

A considerable body of evidence supports the general concept that the extent to which neighboring cells are coupled by gap junctions depends on the structure and function of adhesion junctions. Intercellular adhesion seems to play a critical role in the relationship between intercellular communication and cell growth. For example, inhibition of gap junctional communication is a marker of tumor progression and abolition of contact inhibition between adjacent cells.[4,5] Highly malignant tumors such as "signet ring" cell carcinomas of the GI tract and small cell carcinomas of the lung are typically composed of "dyscohesive" cells which are communication-deficient. In contrast, less malignant, better differentiated tumors show more normal tissue architecture and exhibit more intact gap junction communication.[4,5]

A similar dependency of gap junction formation on mechanical junctions exists in cardiac myocytes.[6] For example, when freshly disaggregated adult rat ventricular myocytes are placed in culture and allowed to reestablish contact, the initial event in reformation of the intercalated disk is development of fibrillar connections in the intercellular space and subsequent appearance of subplasmalemmal plaque-like structure. Immunohistochemistry of these nascent adherens junctions reveals the presence of N-cadherin, α-catenin, and plakoglobin, but neither connexin43 (Cx43) immunoreactive signal nor ultrastructurally discernible gap junctions are present at this stage.[6] It is not until definitive fascia adherens junctions become well formed that Cx43 signal localizes to intercellular junctions and gap junctions arise, and the newly formed gap junctions are invariably flanked by adhesive junctions.[6] These observations suggest that formation of adhesion junctions is a prerequisite of gap junction formation within the intercalated disk.

Fascia Adherens Junctions and Desmosomes

Fascia adherens junctions and desmosomes, located mainly in the large intercalated disks that connect myocytes in end-to-end fashion, are the two types of mechanical junctions responsible for intercellular adhesion in cardiac myocytes. Both fascia adherens junctions and desmosomes consist of intercellular adhesion molecules, which span the sarcolemma and bind in the extracellular space to connect adjacent cells, and intracellular proteins, which link the cytoplasmic domains of the adhesion molecules to components of the cytoskeleton. Normal intercellular adhesion depends on the normal expression, distribution, and function of all three components of these macromolecular complexes—adhesion molecules, linker proteins, and cytoskeletal proteins. Defects in any single component causing discontinuities in the network of cell–cell adhesion junctions and the cytoskeleton in cardiac myocytes may produce pathophysiological effects. In general, the human heart diseases caused by defects in cell–cell junction proteins are arrhythmogenic right ventricular or dilated cardiomyopathies associated with a high incidence of ventricular arrhythmias and sudden cardiac death.[7]

The most numerous and conspicuous cell–cell mechanical junctions are the fascia adherens junctions that are located at the ends of sarcomeres.[8] The transmembrane adhesion molecules of fascia adherens junctions are the N-cadherins. These proteins are attached to actin in sarcomeres by intracellular linker proteins, including plakoglobin (γ-catenin), β-catenin, and α-actinin. The other class of mechanical junctions, desmosomes, are smaller more discrete structures than fascia adherens junctions.[8] The transmembrane adhesion molecules in desmosomes are the desmosomal cadherins: desmocollins and desmogleins. These cadherins are connected to the intermediate filament component of the cytoskeleton, which, in cardiac myocytes, is composed of desmin. The major linker protein connecting desmosomal cadherins to the desmin cytoskeleton is desmoplakin. Plakoglobin and plakophilin also contribute to this linkage.

REMODELING OF GAP JUNCTIONS IN HUMAN CARDIOMYOPATHIES CAUSED BY DEFECTIVE LINKAGE BETWEEN ADHESION JUNCTIONS AND THE CYTOSKELETON

Carvajal Syndrome and Naxos Disease

Carvajal syndrome and Naxos disease are rare familial cardiocutaneous syndromes consisting of the clinical triad of woolly hair, palmoplantar keratoderma, and heart disease.[9,10] These monogenic cardiomyopathies are characterized by profound structural alterations in the heart and a high incidence of ventricular arrhythmias and sudden death, often occurring in children or young adults. Analysis of diseased human cardiac tissues provides an opportunity to test the hypothesis that defects in cell–cell adhesion or discontinuities in the linkage between cell–cell mechanical junctions and the intracellular cytoskeletal network lead to remodeling of gap junctions.

Carvajal syndrome is a familial syndrome characterized by recessive inheritance of woolly hair, epidermolytic palmoplantar keratoderma, and dilated cardiomyopathy.[9] It is caused by a mutation in the gene encoding desmoplakin, the most abundant of the desmosomal proteins.[11] Desmoplakin is a member of the plakin family of cytolinkers, large modular proteins that link cytoskeletal filaments to each other and to membrane-associated adhesion junctions.[12] The disease-causing mutation in Carvajal syndrome is a deletion of nucleotide 7901 in exon 24.[11] This produces a premature stop codon located 18 amino acids from the point mutation that truncates the C-terminal domain,[11] the region known to interact with desmin.[13]

We have recently described the structural and molecular pathology of the cardiomyopathy in Carvajal syndrome.[14] The heart in Carvajal syndrome exhibits ventricular hypertrophy, ventricular dilatation, and discrete focal ventricular aneurysms. There is no apparent fibrofatty replacement of right ventricular myocardium, but the right ventricle shows focal areas of wall thinning and aneurysmal dilatation in a distribution similar to that seen in arrhythmogenic right ventricular cardiomyopathy. Intercalated disks connecting ventricular myocytes show markedly decreased amounts of specific immunoreactive signal for desmoplakin (the mutant protein in Carvajal syndrome), plakoglobin (another linker protein), and the gap junction protein, Cx43 (FIG. 1). The intermediate filament protein, desmin, has a normal intracellular distribution but fails to localize at intercalated disks (FIG. 2). These observations suggest that altered protein–protein interactions caused by deletion of the C-terminal domain of desmoplakin results in failure of plakoglobin to localize to the intercalated disk and disrupts normal connections between desmosome and the desmin cytoskeleton.

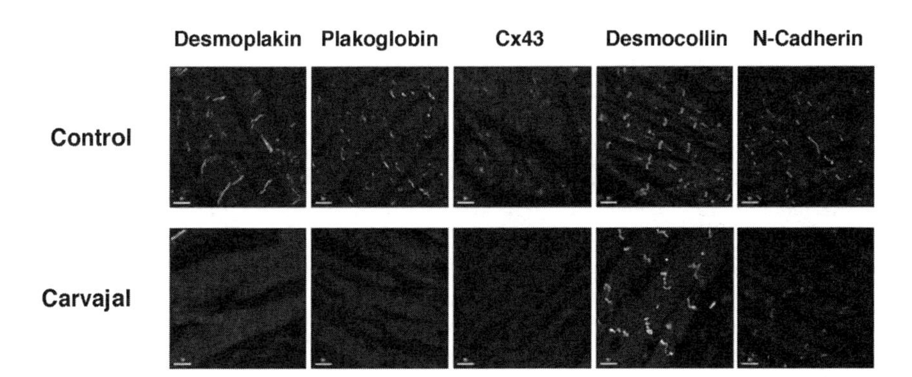

FIGURE 1. Confocal immunofluorescence images showing the distribution of desmoplakin, plakoglobin, connexin43 (Cx43), desmocollin, and N-cadherin at intercalated disks in left ventricular myocardium from Carvajal syndrome and a normal control heart [in color in Annals Online]. Bars = 20 μm.

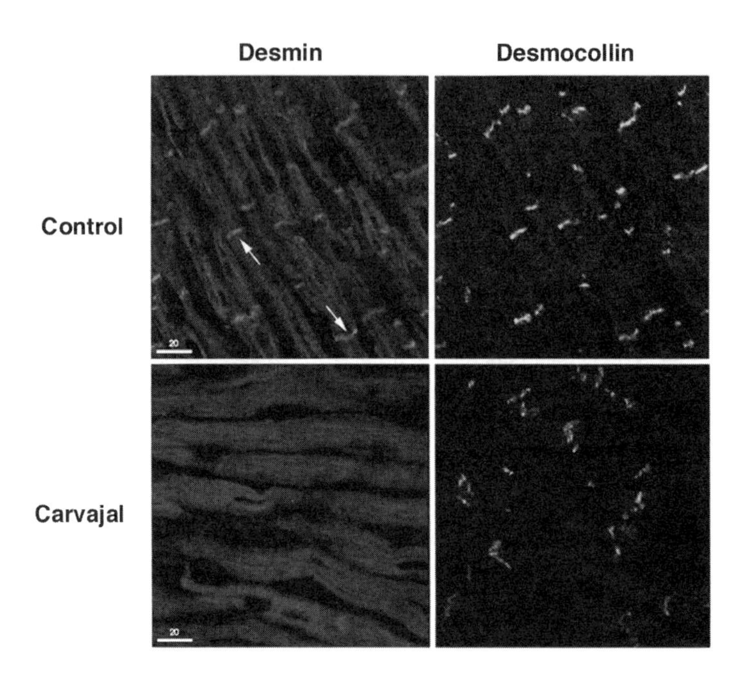

FIGURE 2. Confocal immunofluorescence images of double-stained left ventricular sections showing the distribution of desmin and desmocollin in Carvajal syndrome and in a normal control heart. Desmin is distributed in a sarcomeric pattern in both Carvajal syndrome and control myocardium. Desmin signal is also concentrated at intercalated disks in control tissue (*arrows*) in a pattern identical to that of desmocollin. In contrast, desmin signal is absent at intercalated disks in Carvajal syndrome [in color in Annals Online].

Diminished Cx43 immunoreactivity at intercalated disks further suggests that defective linkage between mechanical junctions and the cytoskeleton affects formation or maintenance of electrical junctions, which, in turn, could contribute to conduction abnormalities and promote arrhythmogenesis.

We have also recently characterized expression of cell–cell junction proteins in cardiac tissues from patients with Naxos disease.[15] This recessive familial cardiocutaneous syndrome consists of the clinical triad of woolly hair, palmoplantar keratoderma, and arrhythmogenic right ventricular cardiomyopathy.[10] It is associated with a particularly high incidence of sudden cardiac death. Approximately 60% of affected children present with syncope and/or sustained ventricular tachycardia and the annual risk of arrhythmic death is 2.3%.[16]

The mutation in Naxos disease is a deletion of nucleotypes 2157 and 2158 in the gene encoding plakoglobin.[17] This mutation causes a frame shift resulting in premature termination of translation and truncation of the C-terminal domain by 56 residues. Also known as γ-catenin, plakoglobin is a component of both desmosomes and fascia adherens junctions in cardiac myocytes where it functions as an intracellular linker protein responsible for connecting adhesion molecules in cell–cell junctions to actin and desmin in the cytoskeleton.[8]

We analyzed expression of Cx43 and other intercellular junction proteins in cardiac tissue from four patients with Naxos disease.[15] We also performed immunoblotting analysis and electron microscopy in one patient who died in childhood before overt arrhythmogenic right ventricular cardiomyopathy had developed. Cx43 expression at intercellular junctions was reduced significantly in both the right and left ventricles in all patients with Naxos disease (FIG. 3).

Electron microscopy revealed smaller and fewer gap junctions interconnecting ventricular myocytes. Immunoblotting showed that truncated plakoglobin was expressed in the myocardium but failed to localize normally at intercellular junctions (FIG. 4). Although Cx43 immunoreactive signal at gap junctions was dramatically reduced, immunoblotting showed little or no reduction in the tissue content of Cx43 (FIG. 4). However, the highly phosphorylated P2-isoform of Cx43 was absent in Naxos disease myocardium. This isoform is selectively located in gap junctions, whereas nonphosphorylated Cx43 resides largely in nonjunctional, including intracellular, pools.[18] These results indicate that remodeling of myocyte gap junctions occurs in early Naxos disease, presumably because of abnormal linkage between mechanical junctions and the cytoskeleton. The coupling defect and the subsequent development of pathologic changes in the myocardium could create a highly arrhythmogenic substrate and enhance the risk of sudden death in Naxos disease.

Previously, diminished expression of Cx43 signal at intercalated disks and remodeling of gap junctions has only been observed in structurally abnormal hearts in patients with advanced heart disease.[19] This has invariably been associated with a decrease in the total content of Cx43 in the tissue,[19] presumably reflecting changes in the rates of Cx43 mRNA and/or protein synthesis, or in the balance between Cx43 production and degradation. Demonstration of a marked decrease in Cx43 signal at intercalated disks, despite apparently normal intracellular levels of Cx43, suggests that remodeling of gap junctions in Naxos disease is related primarily to an inability to assemble and/or maintain normally large gap junction channel arrays.

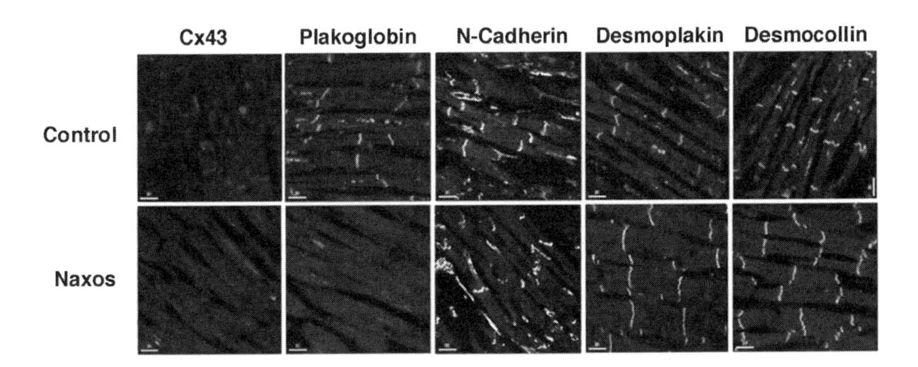

FIGURE 3. Confocal immunofluorescence images showing the distribution of connexin43 (Cx43), plakoglobin, N-cadherin, desmoplakin, and desmocollin at intercalated disks in left ventricular tissue from a normal control heart and from Naxos disease [in color in Annals Online].

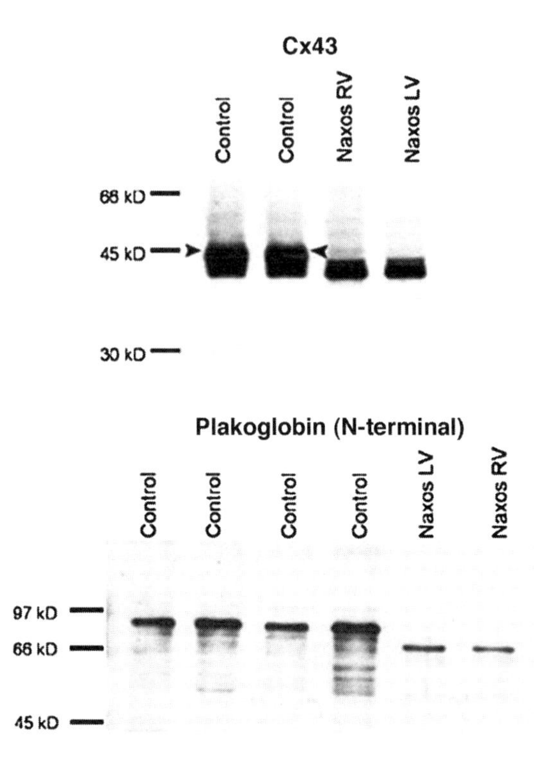

FIGURE 4. Immunoblots of connexin43(Cx43) and plakoglobin in left and right ventricular myocardium from a patient with Naxos disease and controls. Highly phosphorylated Cx43 (*arrowheads*) is absent in Naxos disease. Truncated plakoglobin, identified with an antibody against the N-terminus, is expressed in both left and right ventricles in Naxos disease [in color in Annals Online].

FUTURE DIRECTIONS

There are several fascinating and potentially informative but, as yet, largely unexplored mouse lines with genetically altered expression of cell–cell junction and cytoskeletal proteins implicated in human cardiomyopathies. In fact, germline knockouts of genes encoding plakoglobin, desmoplakin, desmin, and vinculin were created in mice years before it was known that these proteins played a role in human heart disease. Plakoglobin-null mice exhibit lethal embryonic phenotypes that include ventricular rupture and hemopericardium.[20] Germline knockout of desmoplakin results in early embryonic lethality associated with profound defects in gastrulation and mesenchymal cell migration.[21]

Future studies may be directed at development of cardiac-specific conditional knockouts of these genes. Other approaches may include development of transgenic mouse models expressing the human mutations. Our hypothesis predicts that genetic defects in expression of plakoglobin and desmoplakin will cause remodeling of gap junctions at intercalated disks, conduction abnormalities, and enhanced risk of arrhythmias.

ACKNOWLEDGMENT

This work was supported by a grant from the March of Dimes.

REFERENCES

1. SAFFITZ, J.E., D.L. LERNER & K.A. YAMADA. 2004. Gap junction distribution and regulation in the heart. *In* Cardiac Electrophysiology: From Cell-to-Bedside. D.P. Zipes & J. Jalife, Eds.: 181–191. 4th ed. Saunders. Philadelphia.
2. SAFFITZ, J.E. 2003. Dependence of electrical coupling on mechanical coupling in cardiac myocytes. *In* Advances in Cardiovascular Medicine. G. Thiene & A.C. Dessina, Eds.: 15–28. Universitá degli Studi di Padova.
3. HOYT, R.H., M.L. COHEN & J.E. SAFFITZ. 1989. The distribution and three-dimensional structure of intercellular junctions in canine myocardium. Circ. Res. **64:** 565–574.
4. YAMASAKI, H., M. MESNIL, Y. OMORI, *et al.* 1995. Intercellular communication and carcinogenesis. Mutat. Res. **333:** 181–188.
5. TROSKO, J.E. & R.J. RUCH. 1998. Cell-cell communication in carcinogenesis. Front. Biosci. **15:** D208–D236.
6. KOSTIN, S., S. HEIN, E.P. BAUER, *et al.* 1999. Spatiotemporal development and distribution of intercellular junctions in adult rat cardiomyocytes in culture. Circ. Res. **85:** 154–167.
7. FATKIN, D. & R.M. GRAHAM. 2002. Molecular mechanisms of inherited cardiomyopathies. Physiol. Rev. **82:** 945–980.
8. ZHURINSKY, J., M. SHTUTMAN & A. BEN-ZE'EV. 2000. Plakoglobin and β-catenin: protein interactions, regulation and biological roles. J. Cell Sci. **113:** 3127–3139.
9. CARVAJAL-HUERTA, L. 1998. Epidermolytic palmoplantar keratoderma with woolly hair and dilated cardiomyopathy. J. Am. Acad. Dermatol. **39:** 418–421.
10. PROTONOTARIOS, N., A. TSATSOPOULOU, P. PATSOURAKOS, *et al.* 1986. Cardiac abnormalities in familial palmoplantar keratosis. Br. Heart J. **56:** 321–326.
11. NORGETT, E.E., S.J. HATSELL, L. CARVAJAL-HUERTA, *et al.* 2000. Recessive mutation in desmoplakin disrupts desmoplakin-intermediate filament interactions and causes dilated cardiomyopathy, woolly hair and keratoderma. Hum. Mol. Genet. **9:** 2761–2766.
12. LEUNG, C.L., R.K.H. LIEM, D.D.D. PARRY, *et al.* 2001. The plakin family. J. Cell Sci. **114:** 3409–3410.
13. KOWALCZYK, A.P., E.A. BORNSLAEGER, S.M. NORVELL, *et al.* 1999. Desmosomes: intercellular adhesive junctions specialized for attachment of intermediate filaments. Int. Rev. Cyt. **185:** 237–302.
14. KAPLAN, S.R., J.J. GARD, L. CARVAJAL-HUERTA, *et al.* 2004. Structural and molecular pathology of the heart in Carvajal syndrome. Cardiovasc. Pathol. **13:** 26–32.
15. KAPLAN, S.R., J.J. GARD, N. PROTONOTARIOS, *et al.* 2004. Remodeling of gap junctions in arrhythmogenic right ventricular cardiomyopathy due to a deletion in plakoglobin (Naxos disease). Heart Rhythm **1:** 3–11.
16. PROTONOTARIOS, N., A. TSATSOPOULOU, A. ANASTASAKIS, *et al.* 2001. Genotype-phenotype assessment in autosomal recessive arrhythmogenic right ventricular cardiomyopathy (Naxos disease) caused by a deletion in plakoglobin. J. Am. Coll. Cardiol. **38:** 1477–1484.
17. MCKOY, G., N. PROTONOTARIOS, A. CROSBY, *et al.* 2000. Identification of a deletion of plakoglobin in arrhythmogenic right ventricular cardiomyopathy with palmoplantar keratoderma and woolly hair (Naxos disease). Lancet **355:** 2119–2124.
18. MUSIL, L.S. & D.A. GOODENOUGH. 1991. Biochemical analysis of connexin43 intracellular transport, phosphorylation, and assembly into gap junctional plaques. J. Cell Biol. **115:** 1357–1374.

19. PETERS, N.S., C.R. GREEN, P.A. POOLE-WILSON, *et al.* 1993. Reduced content of connex43 gap junctions in ventricular myocardium from hypertrophied and ischemic human hearts. Circulation **88:** 864–875.
20. RUIZ, P., V. BRINKMANN, B. LEDERMANN, *et al.* 1996. Targeted mutation of plakoglobin in mice reveals essential functions of desmosomes in the embryonic heart. J. Cell Biol. **135:** 215–225.
21. GALLICANO, G.I., P. KOUKLIS, C. BAUER, *et al.* 1998. Desmoplakin is required early in development for assembly of desmosomes and cytoskeletal linkage. J. Cell Biol. **143:** 2009–2022.

Spatial Nonuniformity of Contraction Causes Arrhythmogenic Ca^{2+} Waves in Rat Cardiac Muscle

HENK E.D.J. TER KEURS,[a] YUJI WAKAYAMA,[b] MASAHITO MIURA,[b] BRUNO D. STUYVERS,[a] PENELOPE A. BOYDEN,[c] AND AMIR LANDESBERG[d]

[a] *Departments of Medicine, Physiology, and Biophysics, Health Sciences Centre, University of Calgary, Calgary, Alberta T2N 4N1, Canada*

[b] *First Department of Internal Medicine, Tohoku University School of Medicine, Sendai, Japan*

[c] *Department of Pharmacology, Center for Molecular Therapeutics, Columbia University, New York, New York, USA*

[d] *Faculty of Biomedical Engineering, Technion, Israel Institute of Technology, Haifa, Israel*

ABSTRACT: Landesberg and Sideman's four state model of the cardiac cross-bridge (XB) hypothesizes a feedback of force development to Ca^{2+} binding by troponin C (TnC). We have further modeled this behavior and observed that the force (F)–Ca^{2+} relationship as well as the F–sarcomere length (SL) relationship and the time course of F and Ca^{2+} transients in cardiac muscle can be reproduced faithfully by a single effect of F on deformation of the TnC-Ca complex and, thereby, on the dissociation rate of Ca^{2+}. Furthermore, this feedback predicts that rapid decline of F in the activated sarcomere causes release of Ca^{2+} from TnC-Ca^{2+}, which is sufficient to initiate arrhythmogenic Ca^{2+} release from the sarcoplasmic reticulum (SR). This work investigated the initiation of Ca^{2+} waves underlying triggered propagated contractions (TPCs) in rat cardiac trabeculae under conditions that simulate functional nonuniformity caused by mechanical or ischemic local damage of the myocardium. A mechanical discontinuity along the trabeculae was created by exposing the preparation to a small constant flow jet of solution that reduces excitation–contraction coupling in myocytes within that segment. Force was measured, and SL as well as $[Ca^{2+}]_i$ were measured regionally. When the jet contained caffeine, 2,3-butanedione monoxime or low-$[Ca^{2+}]$, muscle-twitch F decreased and the sarcomeres in the exposed segment were stretched by shortening the normal regions outside the jet. During relaxation, the sarcomeres in the exposed segment shortened rapidly. Short trains of stimulation at 2.5 Hz reproducibly caused Ca^{2+} waves to rise from the borders exposed to the jet. Ca^{2+} waves started during F relaxation of the last stimulated twitch and propagated into segments both inside and outside of the jet. Arrhythmias, in the form of nondriven rhythmic activity, were triggered when the amplitude of the Ca^{2+} wave increased by raising

Address for correspondence: Prof. Henk E.D.J. ter Keurs, M.D., Ph.D., Dept. of Medicine, CardioVascular Sciences, Physiology and Biophysics, University of Calgary Health Science Centre, 3330 Hospital Drive N.W., Calgary, Alberta T2N 4N1, Canada. Voice: 403-220-4521; fax: 403-270-0313.
terkeurs@ucalgary.ca

Ann. N.Y. Acad. Sci. 1047: 345–365 (2005). © 2005 New York Academy of Sciences.
doi: 10.1196/annals.1341.031

$[Ca^{2+}]_o$. The arrhythmias disappeared when the muscle uniformity was restored by turning the jet off. These results show that nonuniform contraction can cause Ca^{2+} waves underlying TPCs, and suggest that Ca^{2+} dissociated from myofilaments plays an important role in the initiation of arrhythmogenic Ca^{2+} waves.

KEYWORDS: **arrhythmias; Ca^{2+} waves; nonuniformity; rat trabeculae; troponin C**

INTRODUCTION

Cardiac disease leads invariably to nonuniformity of myocardium. The role of electrical nonuniformity of the myocardium in reentry arrhythmias is well established.[1] Less is known about the role played by nonuniform myocardial stress and strain distributions and by nonuniform excitation–contraction (E-C) coupling in mechanisms underlying extrasystoles that initiate arrhythmias. It is well known that tens of µmol/l of Ca^{2+} shuttle during the cardiac cycle between the sarcoplasmic reticulum (SR) and the cytosol where troponin C (TnC) is the dominant ligand. Hence, it is conceivable that nonuniformity of myocardium may lead to extrasystoles by several mechanisms, including abnormal SR-Ca^{2+} transport following damage and/or due to the mechanical events occurring in nonuniform myocardium that dissociate Ca^{2+} from TnC.

It has been shown that "spontaneous" SR-Ca^{2+} release causes both transient inward currents and arrhythmogenic delayed afterdepolarizations (DADs) as well as aftercontractions.[2,3] A sufficiently large SR-Ca^{2+} load in cells at the rim of a damaged region could create an unstable state where spontaneous SR-Ca^{2+} release may become so large that the resulting transient inward current depolarizes the cells enough to trigger a new action potential, which perpetuates itself as a triggered arrhythmia.[4] Alternatively, events that result from the tug-of-war between normal myocardium and weak cells in the ischemic zone could trigger the Ca^{2+} release and lead to arrhythmias. This tug-of-war may play a role in Ca^{2+} release, triggered in damaged regions of isolated rat ventricular and human atrial trabeculae, resulting in Ca^{2+} release that appears to be initiated after stretch of the damaged region during the regular twitch and propagates into neighboring myocardium by the combination of Ca^{2+} diffusion and Ca^{2+} induced SR-Ca^{2+} release (CICR). The elevation of $[Ca^{2+}]_i$, which accompanies this process (FIG. 1) causes a propagating contraction (triggered propagated contractions [TPC]; FIG. 1)[5,6] and may depolarize the cell beyond the threshold for action potential generation.[7]

It is clear that damage of a cardiac cell causes loss of integrity of the cell membrane and allows Ca^{2+} entry into damaged cells and their neighbors, which in its turn will induce SR-Ca^{2+} overload and cause spontaneous microscopic Ca^{2+} release.[6] The spontaneous Ca^{2+} release increases resting force (F) and decreases F of the next twitch,[8,9] which sets the scene for a tug-of-war where weakened myocytes are stretched by normal myocytes to which they are linked. The observation that the TPCs always start shortly after the rapid shortening of the damaged areas during the relaxation phase suggests that it is, in fact, the shortening and F decrease during relaxation that initiates a TPC. In order to understand this phenomenon, we have proposed the existence of reverse excitation–contraction coupling (RECC)[10] based on the classical concept of E-C coupling (FIG. 2). The observation[11–15] that rapid short-

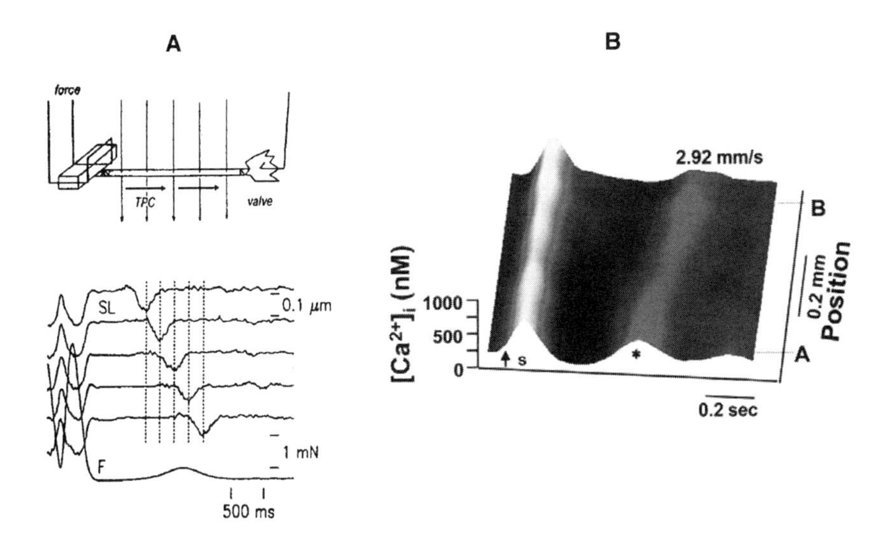

FIGURE 1. (**A**) Sarcomere length (SL) recordings at five different sites (each 300 μm apart) along a 2.94-mm-long trabecula during a triggered propagated contractions (TPC) with a propagation velocity of 1.4 mm/s. The interval between peak sarcomere shortening due to the TPC (*vertical dashed lines*) was constant from site to site, indicating that propagation velocity remained constant along the preparation. $[Ca^{2+}]_o$ 1.0 mM, temperature 21°C. Initial SL varied less than 0.05 μm between the sites of measurement. F = force. (Modified from Daniels.[69]) (**B**) The Ca^{2+} transients as a function of distance along the preparation at different times. The propagating nature of the Ca^{2+} wave is evident from the figures [in color in Annals Online]. (Modified from Miura *et al.*[23])

ening and F decline of a contracting muscle causes a surge of Ca^{2+} release from the myofilaments provides a candidate mechanism for initiation of TPCs[5] (FIG. 2, *middle*). Ca^{2+} that dissociates from the contractile filaments due to the quick release of the damaged areas during relaxation could initiate a wave of Ca^{2+} release, if, at that time, the SR has recovered sufficiently[16] to allow CICR to amplify the initial Ca^{2+} surge in the damaged region and/or the border zone (FIG. 2, *middle*). The propagating Ca^{2+} transient, in turn, will activate arrhythmogenic Ca^{2+} dependent depolarizing currents (FIG. 2, *right*).[3,17] The observation that neither initiation of TPCs nor their propagation is affected by Gd^{3+} ions suggests that stretch activated channels play little or no role in the initiation or propagation of TPCs.[18]

A detailed study of the role of strong and weak muscle regions of damaged muscle in the initiation of arrhythmogenic Ca^{2+} waves is hampered by the difficulty of controlling the extent and severity of damage, and, as such, neither sarcomere length (SL) nor $[Ca^{2+}]^i$ can be measured reliably. Here, we show early experimental results using a novel model of controlled nonuniformity in rat trabeculae. Using this approach, we show that controlled initiation of propagating Ca^{2+} waves underlying TPCs can trigger nondriven repetitive regular spontaneous contractions in cardiac muscle, that is, arrhythmogenicity of nonuniform muscle. Then, we show that a cardiac cross-bridge (XB) model,[19,20] based on the effect of the feedback of XB-force (F) to the binding of Ca^{2+} to TnC suggests that initiation of arrhythmogenic Ca^{2+} waves can be explained by

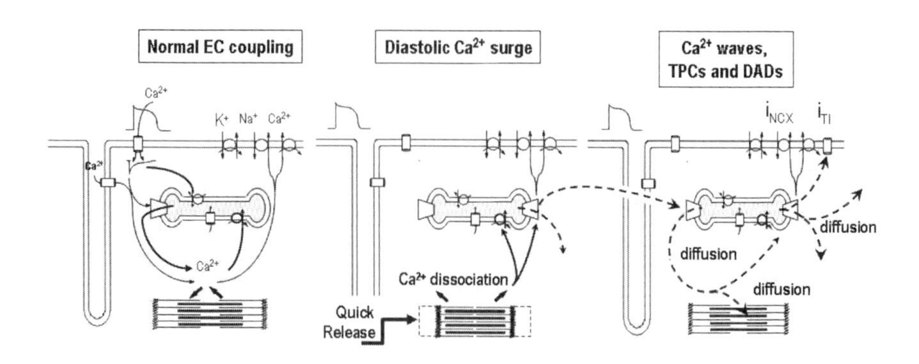

FIGURE 2. Model of reverse excitation–contraction coupling (RECC). ***(Left)*** The events during the normal twitch. During the action potential a transient Ca^{2+} influx enters the cells followed by a maintained component of the slow inward current *(dashed line)*. Ca^{2+} entry does not lead directly to force development as the Ca^{2+} that enters is rapidly bound to binding sites on the sarcoplasmic reticulum (SR). The rapid influx of Ca^{2+} via the T-tubuli induces release of Ca^{2+} from the SR, by triggering opening of Ca^{2+} channels in the terminal cisternae, thus activating the contractile filaments to contract. Rapid relaxation follows because the cytosolic Ca^{2+} is sequestered rapidly by the SR and partly extruded through the cell membrane by the Na^+/Ca^{2+} exchanger and by the low-capacity high-affinity Ca^{2+} pump. This process loads the SR. Na^+/Ca^{2+} exchange is electrogenic so that Ca^{2+} extrusion through the exchanger leads to a depolarizing current. ***(Middle)*** RECC in nonuniform muscle during triggering of the triggered propagated contractions (TPC). Nonuniform muscle contains weak segments that are stretched by strong segments during the twitch. During rapid relaxation, the relatively weak segment rapidly shortens owing to relaxation of stronger myocardium. This quick release of the weak sarcomeres leads to dissociation of Ca^{2+} from the contractile filaments during the relaxation phase. The SR is enough recovered to respond to this Ca^{2+} surge by Ca^{2+} induced SR-Ca^{2+} release (CICR). The resultant elevation of $[Ca^{2+}]_i$ causes diffusion of Ca^{2+} to adjacent sarcomeres. ***(Right)*** Arrival of diffusing Ca^{2+} after release in the damaged region leads to CICR by the SR in the adjacent sarcomeres. Ca^{2+} diffuses again the next sarcomere, while causing a local contraction as well as a delayed afterdepolarization (DAD) due to electrogenic Na^+/Ca^{2+} exchange and activation of Ca^{2+} sensitive nonselective channels in the sarcolemma. Diffusion of Ca^{2+} along its gradient maintains the propagation of the TPC.

the nonuniformity in E-C coupling leading to a Ca^{2+} surge from TnC, as a result of rapid F decline in the weak regions of nonuniform muscle during relaxation of stronger regions. In conclusion, we propose that troponin C is the source of a "funny" Ca^{2+} surge that leads to arrhythmogenic Ca^{2+} waves in mechanically nonuniform cardiac muscle.

METHODS

Measurements of F, SL, and $[Ca^{2+}]_i$ in Rat Trabeculae

Trabeculae were dissected from the right ventricle of Lewis Brown Norway rats and mounted between a motor arm and F transducer in a bath perfused by HEPES solution on an inverted microscope. SL was measured by laser diffraction techniques.[21] Measurement of $[Ca^{2+}]_i$ has been described previously.[22–25] Briefly, fura-2 salt was microinjected iontophoretically into the trabecula.[25] Excitation light of

340, 360, or 380 nm was used and fluorescence was collected using an image intensified charge coupled device (CCD) camera (model 330; General Scanning Inc., Watertown, MA) to assess local $[Ca^{2+}]_i$.[25] Variations in $[Ca^{2+}]_i$ along the trabeculae were calculated from the calibrated ratio of F_{360}/F_{380}.

Reduction of Local Contraction and Induction of Ca^{2+} Waves

To produce nonuniform E-C coupling, a restricted region was exposed to a small "jet" of solution (≈0.06 ml/min) that had been directed perpendicularly to a small muscle segment (300 μm; FIG. 3) using a syringe pump connected to a fine glass pipette (FIG. 3A) (Wakayama *et al.*[24] in review). The jet solution contained standard HEPES solution as well as either (1) caffeine (CF, 5 mmol/l) to deplete SR of Ca^{2+} through opening of the SR-Ca^{2+} release channels;[26–29] (2) 2,3-butanedione monoxime (BDM, 20 mmol/l) to suppress myosin ATPase and, thus, activation of XBs[30–32] while reducing the SR-Ca^{2+} content modestly; or (3) low $[Ca^{2+}]$ (low-$[Ca^{2+}]_{jet}$, 0.2 mmol/l) to reduce Ca^{2+} availability for E-C coupling.[28] The Ca^{2+} concentration in the jet ($[Ca^{2+}]_{jet}$) was usually identical to the bath solution ($[Ca^{2+}]_o$), except for low-$[Ca^{2+}]_{jet}$ solution or unless mentioned otherwise. During exposure to the jet solutions, Ca^{2+} waves underlying TPCs were induced by stimulation of the muscle at 2.5 Hz for 7.5 s repeated every 15 s at $[Ca^{2+}]_o$ of 2 to 3 mmol/l at 24°C.[7] Measurement of $[Ca^{2+}]_i$ commenced within 10 min, as soon as the amplitude of stimulated twitches, TPCs, and underlying Ca^{2+} waves were constant.

Data Analysis

Data were expressed as mean ± SEM. Statistical analysis was performed using ANOVA followed by a post hoc test. Differences were considered significant when $p < 0.05$.

Model Simulations

Experimentally observed Ca^{2+} transients and twitch kinetics were simulated to gain insight into the parameters of the Ca^{2+} release and removal processes that dictate their time course. The model was simplified after the model of Michailova *et al.*[33] and assumed release, ligand binding, uptake, and diffusion of Ca^{2+}. Ca^{2+} channels opening was operator initiated.

We assumed a pulse shaped Ca^{2+} release flux (R) with an exponential rise and fall:

$$R = A \cdot \left(1 - e^{-(t - t_s/\tau_{on})}\right) \cdot e^{-(t - t_s/\tau_{off})} \quad (1)$$

Released Ca^{2+} binds to ATP, Fluo-4, and TnC according to the following reaction of the general form:

$$(Ca^{2+} + \text{Ligand}) \underset{K_-^L}{\overset{K_+^L}{\rightleftharpoons}} (Ca - \text{Ligand})$$

Ligand concentrations and rate constants (K_+^L, K_-^L) were as given by Csernoch *et al.*[34] adjusted to 26°C, assuming a Q_{10} of 2. $[Mg^{2+}]_i$ was assumed constant (1 mM)

TABLE 1. Parameters, variables, and units

T	Time (s)
t_s	Start time (s)
A	Magnitude of Ca^{2+} release (mol/l)
τ_{on}	On-time constant (s)
τ_{off}	Off-time constant (s)
$[Ca]_{Rest}$	Resting $[Ca^{2+}]$ (nmol/l)
$[Ca]_t$	Instantaneous $[Ca^{2+}]$ (nmol/l)
R	Ca^2 release flux (number of Ca^{2+} ions/s)
U	Ca^{2+} uptake flux (number of Ca^{2+} ions/s)
U_{max}	Maximal Ca^{2+} uptake flux (number of Ca^{2+} ions/s)
EC_{U50}	Ca^{2+} at half maximal uptake (nmol/l)
f and g	Rate of attachment and detachment of XBs
$k_+^{ATP}, k_+^{TnC}, k_+^{Fluo4}$	Ca^{2+} on-rate constants for ATP, TnC, and Fluo-4
$k_-^{ATP}, k_-^{TnC}, k_-^{Fluo4}$	Ca^{2+} off-rate constants for ATP, TnC, and Fluo-4
[ATP], [TnC], [Fluo-4], [Myosin]	ATP, TnC, Fluo-4, and myosin concentrations (nmol/l)
[Ca·ATP], [Ca·TnC], [Ca·Fluo-4]	Ca^{2+} ATP, Ca^{2+} TnC, Ca^{2+} Fluo-4 concentrations (nmol/l)

and ATP only binds Ca^{2+} and Mg^{2+}. The flux of Ca^{2+} elimination (U) was assumed to follow Hill kinetics. We used the parameters for Ca^{2+} uptake by the SR measured for cardiac myocytes in the laboratory.[35]

$$U = \frac{U_{max} \cdot [Ca]_N^{Hill}}{EC_{U50}^{Hill} + [Ca]_N^{Hill}} \tag{2}$$

The Hill coefficient (Hill = 2.2) was assumed to be constant. EC_{U50} and U_{max} were fitted to the experimental data. [Ca·TnC] and [Ca·Fluo-4] were calculated from the following:

$$\begin{aligned} \frac{d[Ca]}{dt} &= R - U - k_+^{ATP}([ATP] - [Ca \cdot ATP])[Ca] + k_-^{ATP}[Ca \cdot ATP] \\ &- k_+^{TnC}([TnC] - [Ca \cdot TNC])([Ca] + k_-^{TnC})[Ca \cdot TNC] - k_+^{Fluo4} \\ &([Fluo4] - [Ca \cdot Fluo4])[Ca] + k_-^{Fluo4}[Ca \cdot Fluo4] \end{aligned} \tag{3}$$

Force, Actin Strain, and Ca^{2+} Binding on TnC

We assumed that [Ca·TnC] enables XB formation and F development, calculated from

$$d[XB]/dt = ([TnCa]/[TnC]_{total})f[XB_{free}] - g[XB_{att}]\,, \tag{4}$$

where f and g are the rates of XB attachment and detachment, respectively. We assumed that F generated by XBs is proportional to the unitary force per XB (F_{XB} = 2–6 pN) and is transmitted by actin, assumed to exhibit a stiffness constant = κ, resulting in an exponential strain (ε) in response to F:

$$\varepsilon = e^{(F \cdot \kappa)} \tag{5}$$

It is known that Ca^{2+} binding to TnC is diffusion limited.[33] Hence, we chose to create a feedback of F to Ca^{2+} TnC kinetics by assuming that deformation of actin is accompanied by a structural change of TnC that reduces the rate of dissociation of bound Ca^{2+}.

$$k_{-} = \frac{k'_{-}}{\varepsilon} \tag{6}$$

where k'_{-} is the rate of dissociation of Ca^{2+} from TnC on the unstrained actin.

The cytosolic $[Ca^{2+}]_i$ was assumed to be 70 nM,[36] and the calculations were started with the buffers in equilibrium. For simplification, we have neither incorporated other ligands nor other buffering systems known to exist in cardiac myocytes.[28,37] The calculations were performed with an integration interval of 1 μs. Rise times of $[Ca^{2+}]_i$ transients were simulated by fitting the rate constants of Ca^{2+} channel opening and closing, their open time (Δt), and R to the experimentally observed rising phase of the Ca^{2+} transient. The decline of a Ca^{2+} transient was simulated by fitting U_{max} and EC_{U50} to the observed decline of the transients in the absence of F.[38]

F-pCa Relationships: Time Course of the Twitch and Effects of Quick Releases

For the F-pCa steady-state relationship, we assumed Hill's behavior, while the interaction between F and the K_D of Ca·TnC was strain dependent:

$$F = F_{max}[Ca^{2+}]_i/(K_D + [Ca^{2+}]_i) \text{ and } K_D = k_{-}/k_{+} = k'_{-}/\varepsilon \tag{7}$$

F_{max} at saturating $[Ca^{2+}]_i$ is a function of actomyosin overlap, which depends on the geometry of the sarcomere.[39] The feedback coefficient, ε, depends on the stiffness coefficient, κ, of actin and is identical for the simulations of the individual relationships.

EXPERIMENTAL RESULTS

Experimental Nonuniformity and Sarcomere Mechanics

FIGURE 3 shows the experimental model of nonuniformity in a cardiac trabecula using a jet of fluid that reached only a small segment of the muscle, while the remainder was exposed to the main solution that perfused the muscle bath. The jet reached one short muscle segment (~300 μm). The fluid flow from the pipette using a solution with composition similar to the HEPES solution in the bath had no effect on F or SL by itself. Although, we have studied the effects of BDM (FIG. 3), caffeine, and low-$[Ca^{2+}]$ in the jet applied from aside to the muscle, we will focus on the effects of BDM here.

When a jet containing either BDM (FIG. 3A) or caffeine or low-[Ca^{2+}] (data not shown) was applied to the stimulated trabeculae, sarcomere stretch replaced rapidly the normal active shortening of the sarcomeres in the exposed segment, while peak F, F/F_{max}, decreased by 10% to 35% depending on the contents of the jet solution. The jet solution affected resting SL very little. All effects were rapidly reversible. Sarcomere dynamics along muscles exposed to a jet revealed three distinct regions during the stimulated twitch (FIG. 3B): (1) a region located >200 μm from the jet where sarcomeres exhibited typical shortening; (2) the segment exposed to the jet where sarcomeres were stretched; (3) in a "border zone" (BZ) between these two segments, sarcomeres shortened early during the twitch and then were stretched (FIG. 3B). Sarcomere stretch in the BZ was less than in the jet-exposed segment. The BZ extended 1 to 2 cell lengths beyond the jet-exposed region. Similar changes in regional sarcomere dynamics were observed in caffeine and low-$[Ca^{2+}]_{jet}$ experiments.

Nonuniformity and $[Ca^{2+}]_i$ Transient

In contrast to the similarity of the effects of the various jet solutions on sarcomere dynamics, jets of caffeine, BDM, or low-$[Ca^{2+}]_o$ solution had distinct effects on $[Ca^{2+}]_i$. Robust electrically driven $[Ca^{2+}]_i$-transients occurred in the regions outside the jet independent of the composition of the jet solution (FIG. 3C). As expected, both the caffeine-jet and low-$[Ca^{2+}]_{jet}$ decreased the peak of the stimulated $[Ca^{2+}]_i$ transient; this contrasted the effect of BDM, which decreased the $[Ca^{2+}]_i$ transient only slightly (FIG. 3C). The effect of caffeine to increase diastolic $[Ca^{2+}]_i$ in the jet region was opposite to that of low-$[Ca^{2+}]_{jet}$ and BDM, which both decreased diastolic $[Ca^{2+}]_i$. The $[Ca^{2+}]_i$ changes were smaller in the BZ, consistent with a gradient between regions due to diffusion of the contents of the jet. Ca^{2+} waves started systematically in the BZ after the decline of the last stimulated Ca^{2+} transient (FIG. 3D). These waves propagated into the regions outside and—in the cases of BDM and low-$[Ca^{2+}]_{jet}$—inside the jet exposed region. FIGURE 3C (BDM jet) clearly shows two initiation sites of four Ca^{2+} waves in the BZ and symmetric propagation into regions outside and inside the jet.

Initiation of Ca^{2+} Waves

FIGURE 3 shows the initiation of Ca^{2+} waves in the BZ of a BDM exposed trabeculae. All muscles responded reproducibly to increasing $[Ca^{2+}]_o$: at $[Ca^{2+}]_o$ = 1 mmol/l (FIG. 3C, *top*), only a localized transient in $[Ca^{2+}]_i$ (≈300 nmol/l)—denoted as Ca^{2+} surge (see *arrow*)—occurred along ~100 to 150 μm of the BZs without apparent propagation of Ca^{2+} waves. The Ca^{2+} surge took place ~325 ms after the stimulus, during the relaxation phase of the twitch, when F had declined by 70% to 80% (FIG. 3D). Increasing $[Ca^{2+}]_o$ led to a decrease of the latency and an increase of the Ca^{2+} surge (FIG. 3C, *bottom*) and led to development of bidirectional propagating Ca^{2+} waves. The initial Ca^{+} surge always occurred in the BZs late during relaxation of the last stimulated Ca^{2+} transient. Increasing $[Ca^{2+}]_o$ further also increased the velocity of propagation of the Ca^{2+} waves. Similar observations were made in low-$[Ca^{2+}]_{jet}$ exposed muscles. In caffeine exposed muscles, the Ca^{2+} waves did not propagate into the jet region. This precluded determination of the site of origin of Ca^{2+} waves, but the earliest Ca^{2+} surge was also observed in the BZ. These observations suggest strongly that the Ca^{2+} surge in the BZ was the initiating event of Ca^{2+}

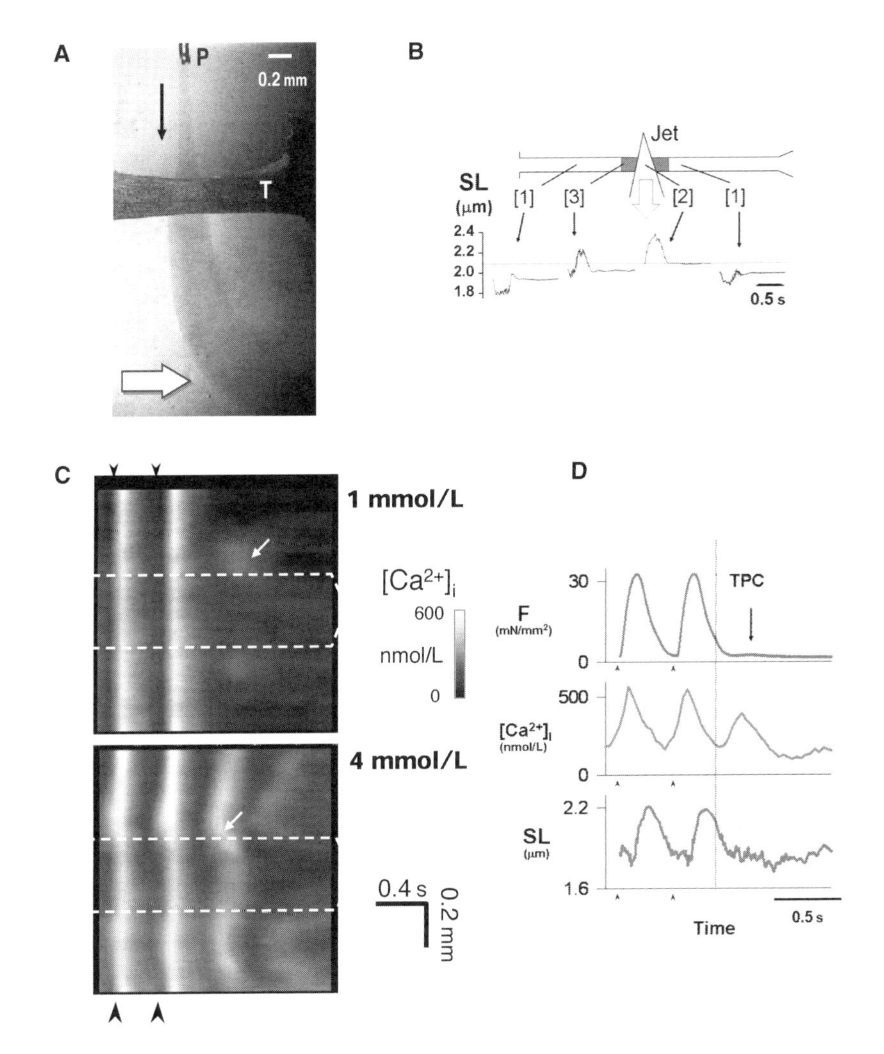

FIGURE 3. Nonuniform cardiac muscle and $[Ca^{2+}]_i$. (**A**) Micrograph of a trabecula in a bath that is perfused with HEPES solution (*white arrow*). A micropipette provides a jet of solution (*dark grey*) oriented perpendicular to the long axis of the muscle (*black arrow*). The composition of the jet is chosen so as to modify excitation–contraction (E-C) coupling in the region of the muscle exposed to the jet. (**B**) Sarcomere length (SL) changes in three regions of the muscle during the twitch, during exposure to a jet containing 2,3-butanedione monoxime (BDM) (20 mmol/l). Segment [1] is not affected by the jet; segment [2] is in the center of the jet; segments [3] form the border zone between region [1] and [2]. Normal contraction in segments [1] is accompanied by stretch in [2], while segment [3] shows initial contraction followed by later stretch during the twitch. (**C**) Initiating events of Ca^{2+} waves induced by local BDM (20 mmol/l) exposure. At low $[Ca^{2+}]$ in the bath (1 mmol/l, *top panel*) only a local Ca^{2+} surge (starting 360 ms) after stimulation is observed. Increasing $[Ca^{2+}]_o$ to 4 mmol/l (*bottom panel*) led to the initiation of bidirectional Ca^{2+} waves, which propagate into the segment inside the jet and into the normal muscle. Both amplitude of the initial and propagating transient as well as propagation velocity increased with increase of $[Ca^{2+}]_o$, while the latency of onset of the Ca^{2+} transient decreased (300 ms). *Arrows* indicate initiation sites of propagating waves. (**D**) Time course of force $[Ca^{2+}]_i$ and SL during the twitch in the border zone [3]. It is clear that the $[Ca^{2+}]_i$ transient of the triggered propagated contraction (TPC) is initiated following the quick release of force and SL in this region [in color in Annals Online].

waves. The delay between peak of the last stimulated Ca^{2+} transient and the start of the propagating Ca^{2+} transient in the BZ decreased inversely with the amplitude of the initial Ca^{2+} surge in all jet exposures.

Propagation of Ca^{2+} Waves

Propagation velocity of the Ca^{2+} waves as in FIGURE 3 outside and inside the jet region, ranged from 0.2 to 2.8 mm/s, (i.e., comparable to Ca^{2+} waves observed in studies of damaged muscle).[22–24] V_{prop} correlated with the $[Ca^{2+}]_i$ increase seen in the BZ during the Ca^{2+} surge. Furthermore, V_{prop} correlated with the amplitude of the waves both inside and outside the jet[22] (FIG. 3C). Propagation into the normal region occurred often with a gradual decline in amplitude. Only the fast waves propagated (1.6 ± 0.3 mm/s) into the normal region outside the jet with a negligible decrease in amplitude. Small waves propagated at lower V_{prop} and with decrement ($V_{prop} = 0.8 \pm 0.1$ mm/s), but completely through the region of observation (≈450 μm). The slowest waves ($V_{prop} = 0.5 \pm 0.1$ mm/s) stopped after 200 to 300 μm. Frequently, waves propagating inside the jet region collided with the wave arriving from the opposite BZ and then terminated (see FIG. 3C).

Nonuniformity and Arrhythmias

Whenever TPCs occur in muscles rendered nonuniform by damage, depolarizations accompany them similar to DADs with a duration that correlated exactly with the time during which the TPCs travel through the trabeculae; the amplitude of the DADs also correlated exactly with the amplitude of the TPCs.[7] This tight correspondence between TPCs and DADs suggests that the DAD is elicited by a Ca^{2+} dependent current as has been proposed by Kass *et al.*[3] Hence, if the $[Ca^{2+}]_i$ transient is large enough, it is expected to lead to action potential formation. Clearly, nonuniformity due to damage initiates a chain of subcellular events leading to arrhythmogenic oscillations of $[Ca^{2+}]_i$ (FIG. 4A).

Experiments with muscles rendered nonuniform by a jet of solution containing an inhibitor of E-C coupling prove that mechanical nonuniformity is a sufficient requirement for arrhythmogenicity. FIGURE 4B shows that nonuniformity of E-C coupling created by the jet induced rapidly reversible nondriven rhythmic activity. The arrhythmia consisted of spontaneous twitches at regular intervals starting after an aftercontraction that followed the last stimulated contraction. The arrhythmia (muscle exposed to BDM) continued until the next stimulus train (7.5 s; FIG. 4B). The intervals between nondriven contractions were usually slightly longer than those of the preceding stimulus train. As shown in FIGURE 4B, these arrhythmias terminated abruptly when the jet was turned off and the uniformity of E-C coupling restored.

Simulation of E-C Coupling with Feedback between F and Ca^{2+} Dissociation from TnC

We have tested whether feedback of the F to TnC Ca^{2+} kinetics[20] reproduces the Ca^{2+} surge that appears to trigger the Ca^{2+} waves. In order to test the general applicability of the model, we have first studied whether the model reproduces the characteristic features of cardiac muscle in the steady state and during the twitch. For the twitch, we have simplified the model compared to existing models to a single release

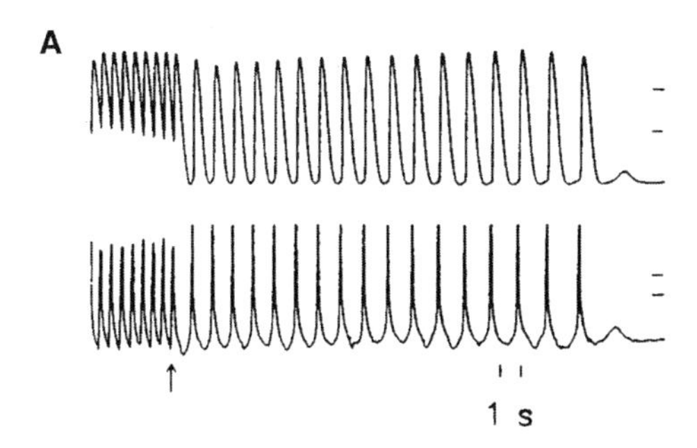

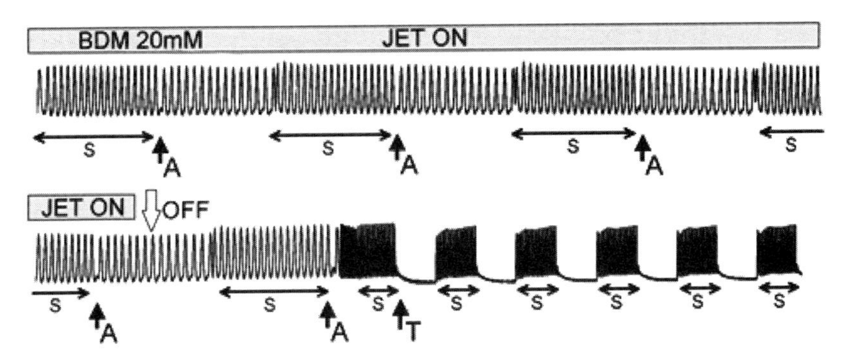

FIGURE 4. Triggered propagated contractions (TPCs) cause arrhythmias. (**A**) Force (*top tracing*) and membrane potential (*bottom tracing*) recordings during a train of conditioning stimuli (ending at the *arrow*) and a subsequent triggered arrhythmia in muscle with a damaged end. Note the initial slow upstroke in both force and membrane potential of triggered twitches, suggestive of an underlying TPC and delayed afterdepolarization (DAD). The triggered arrhythmia terminated spontaneously with an increase in the interval between triggered beats, followed by a TPC and DAD; $[Ca^{2+}]_o$ 2.25 mM. Resting membrane potential –71 mV. Modified from Daniels.[69] (**B**) Nonuniform E-C coupling caused by the jet containing 2,3-butanedione monoxime (BDM) (20 mmol/l) is arrhythmogenic. Recording of force showing that stimulus trains during local exposure to BDM (*gray bars* above the tracings) consistently induced arrhythmias (A). An expanded force tracing showing that spontaneous contractions were both preceded and followed by aftercontractions induced by the stimulus train. OFF (*white arrow*) indicates when the jet was turned off; this rapidly eliminated the contractile nonuniformity and its arrhythmogenic effects. S indicates stimulus trains (2.5 Hz-7.5 s) repeated every 15 s. $[Ca^{2+}]_o = 3.5$ mmol/l; temperature 25.8°C.

pulse, and we have simplified the relaxation phase by assuming a single Ca^{2+} removal process. These assumptions generated the time course of $[Ca^{2+}]_i$ similar to that of unloaded cardiac myocytes.[33] Intracellular Ca^{2+} was assumed to bind to intracellular ligands including TnC and ATP. Ca^{2+} was assumed to bind to a single low-affinity site on TnC with a fixed-on rate constant, but to dissociate with an off-rate that was inversely proportional to the Ca·TnC deformation caused by XB-F that strains the actin filament (κ). The rate of formation (f) of F developing cross-bridges

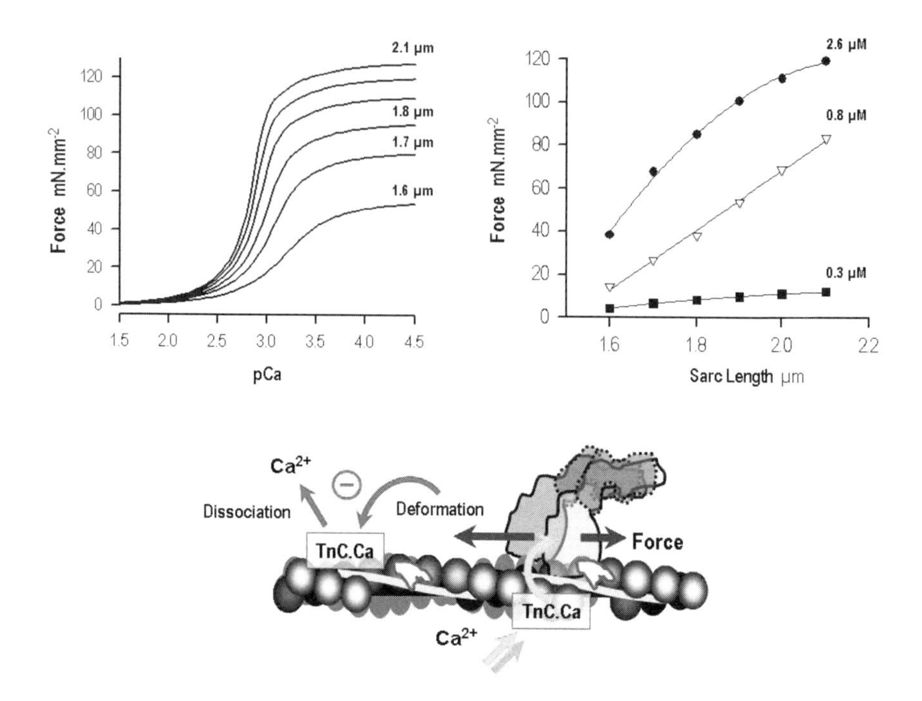

FIGURE 5. ***(Bottom)*** cartoon of the cross-bridge (XB)-actin-troponin C (TnC) interaction. We assume in the model that force (F) development by XBs feeds back on dissociation of Ca^{2+} ions from Ca·TnC by inducing a deformation of C complex, which reduces the off-rate of Ca^{2+} from Ca^{2+}·Tnc. ***(Top left)*** The F-pCa curves that result from this feedback (κ = 0.025). These curves are identical to published F-pCa curves.[41] ***(Top right)*** The F–sarcomere length (SL) curves at three levels of $[Ca^{2+}]_i$ are also near-identical to the published F-SL curves both in skinned trabeculae and in intact cardiac trabeculae [in color in Annals Online].[41]

was fit to the rate of F development of trabeculae after a quick release, but ratio of f and g was taken from the literature.[40]

FIGURE 5 shows that simulation of the feedback of F to Ca^{2+} dissociation from Ca^{2+}·TnC through deformation of the actin filament and TnC on the filament by a single feedback of actin strain κ predicts fundamental properties known for cardiac muscle. The first is a realistic reproduction of the steady-state F-pCa relationship that is known from experiments at controlled SL.[41] FIGURE 5 (*top left panel*) shows that an increase of SL causes an apparent shift of the F-pCa relationship to lower pCa when the muscle is stretched as a result of an increase of the apparent Hill coefficient (or steepness of the curves), although TnC still is assumed to bind only one Ca^{2+} ion. It is not surprising that the feedback predicts realistic F-SL relationships at varied activation levels for $[Ca^{2+}]_i$ (FIG. 5, *top right panel*). Increased F development increases the duration of Ca^{2+} binding to TnC and, hence, the duration of the twitch. Such a progressive prolongation of the twitch has indeed been shown for cardiac muscle twitches at constant SL and is reproduced accurately by the introduction of a tight relationship between the actin strain κ and a decrease of the dissociation rate of Ca

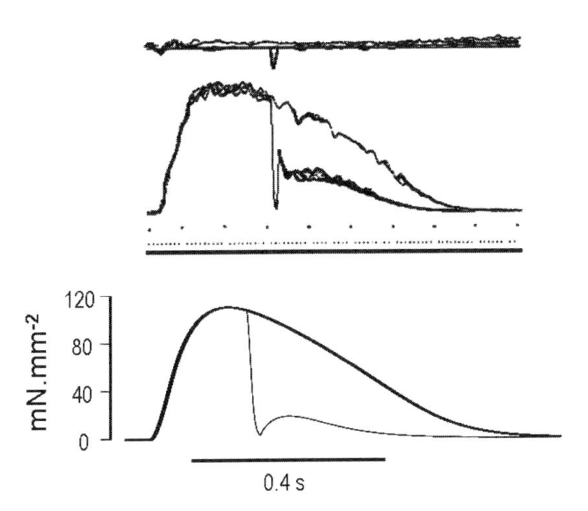

FIGURE 6. (***Bottom***) The simulated time course of force (F) during a twitch *in the presence* of feedback of F to the off-rate of Ca^{2+} from Ca·TnC ($\kappa = 0.025$). Both the time course of F and the effect of a quick release are identical to the effect of an experimental quick release during a twitch (***top***) that starts at constant sarcomere length of 2.15 μm in which a quick release of F is induced by a 30-ms-lasting shortening transient 0.1 μm.[42]

from Ca^{2+}·TnC.[20,42] This observation holds both in model and experiments irrespective of whether the increased F is caused by varied SL, or $[Ca^{2+}]_o$ or by other interventions.[43] FIGURE 6 shows that twitch prolongation is accompanied by a proportional prolongation of the time during which TnC is occupied by Ca^{2+} ions (FIG. 6, *bottom left*). Conversely, any F decrease is expected to cause accelerate dissociation of Ca from Ca^{2+}·TnC and to decrease to subsequent F. This phenomenon is well known as the deactivation response to shortening and is realistically replicated by the model (FIG. 6). The degree of F reduction during shortening deactivation depends on the redistribution of Ca^{2+} over ligands and Ca^{2+} extrusion.

FIGURE 7 also shows that the model reproduces our observation[25] and the observation by Julian's group[44] that the $[Ca^{2+}]_i$ transient during the twitch exhibits a prominent plateau at higher F levels; this observation is remarkable in the model as well (FIG. 7). These authors explained their observation on the basis of cooperativity between F development and affinity of TnC for Ca^{2+} ions, but we are not aware of previous quantitative modeling explaining this observation. This property is completely lost in the absence of the feedback (FIG. 7).

Finally, FIGURE 7 shows the possible importance of this mechanism; the response of F and $[Ca^{2+}]_i$ to a quick reduction of F,[45] in that the model predicts large $[Ca^{2+}]_i$ transients upon a quick release owing to Ca^{2+} dissociation from the Ca^{2+}·TnC. The Ca^{2+} transient is well above the $[Ca^{2+}]_i$ level that is required to induce propagating CICR.[36] We postulate that these $[Ca^{2+}]_i$ transients occur in nonuniform muscle and cause reverse E-C coupling, which is responsible for Ca^{2+} waves TPCs, DADs, and arrhythmias.

DISCUSSION

Experimental Nonuniform E-C Coupling

We have used in this study as a novel model of nonuniform E-C coupling in cardiac muscle to study the initiation of arrhythmogenic Ca^{2+} waves underlying TPCs

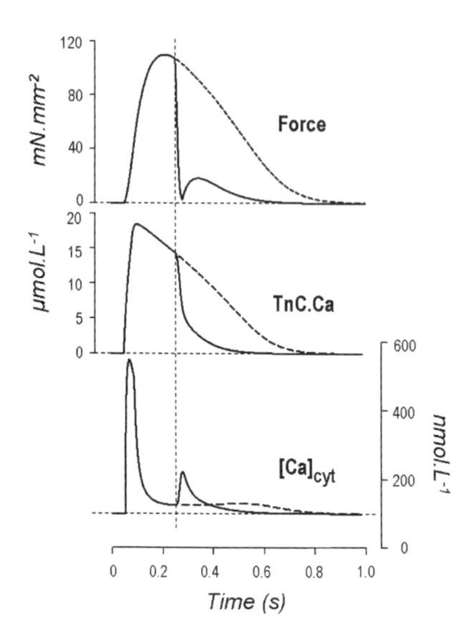

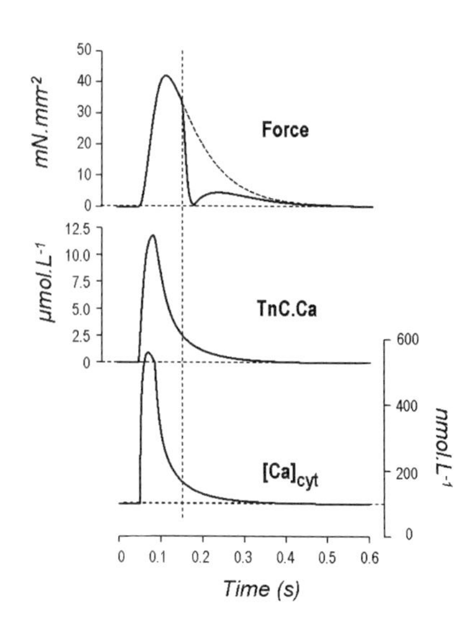

FIGURE 7. (***Left***) The simulated time course of force (F), free $[Ca^{2+}]_I$, and [TnC·Ca] during a twitch *in the presence* of feedback of F to the off-rate of Ca^{2+} from TnC.Ca (κ = 0.025). The time course of F is identical to the time course of F during the published sarcomere isometric contraction.[70] The time course of $[Ca^{2+}]_i$ is identical to the published time course of $[Ca^{2+}]_i$,[63] including the plateau of the $[Ca^{2+}]_i$ transient. The effect of feedback of F on dissociation of Ca^{2+} ions from Ca·TnC leads to prolonged binding of Ca^{2+} to TnC, which exceeds the time course of the $[Ca^{2+}]_i$ transient. The effect of a quick release of the muscle 200 ms after onset of contraction (indicated by the *vertical dashed line* through the tracings) causes a rapid drop of F and a rapid and large decrease of the amount of Ca^{2+} bound to Ca·TnC, which causes a substantial increase in free $[Ca^{2+}]_i$, denoted here as the Ca^{2+} surge. (***Right***) The time course of F, free $[Ca^{2+}]_i$, and [TnC·Ca] during the twitch *in the absence* of feedback of F to the off-rate of Ca^{2+} from Ca·TnC. Peak F is smaller and the [TnC·Ca] now follows the $[Ca^{2+}]_i$ transient. A quick release (at the *dashed line*) still causes a rapid drop of force but fails to elicit a $[Ca^{2+}]_i$ transient during relaxation.

in cardiac muscle. We created nonuniformity in E-C coupling by exposing a small segment of the muscle to caffeine, BDM or low $[Ca^{2+}]_o$. We expected that (1) low-$[Ca^{2+}]_{jet}$ would reduce Ca^{2+} currents and thereby SR-Ca^{2+} content,[28] (2) caffeine would open SR-Ca^{2+} release channels and thereby deplete the SR,[26–28] and (3) BDM would modestly affect Ca^{2+} transport[30–32] and inhibit cross-bridge cycling.[31,32] Consistent with these expectations, the amplitude of stimulated Ca^{2+} transients, which reflect the SR-Ca^{2+} load,[28] decreased dramatically in regions exposed to caffeine and low-$[Ca^{2+}]_{jet}$, but only slightly with BDM (FIG. 3). The effects of BDM are emphasized here because the drug quite apparently reduced F development significantly—at this concentration—in the jet-exposed region and left the Ca^{2+} transients relatively unaffected.

Each of these perturbations reduced muscle F due to creation of a muscle segment that developed less twitch F than the normal cells remote from the jet, as is witnessed by stretch of the weakened sarcomeres in the jet by the fully activated sarcomeres outside the exposed region. These regions were connected mechanically by a border zone, BZ of one to two cells, where the sarcomeres first contracted and, then, were stretched. The diffraction pattern of sarcomeres in the BZ showed a clear single peak during both shortening and lengthening, strongly suggesting that sarcomere contraction in the BZ was also partially suppressed, probably owing to diffusion of the contents of each jet solution.

Nonuniform E-C Coupling and the Ca^{2+} Surge

Ca^{2+} waves and TPCs have been closely related to Ca^{2+} overload in damaged regions and the resultant nonuniformity of muscle contraction.[10] However, in that model it is difficult to investigate the underlying mechanisms because damage is difficult to control. This study shows clearly that Ca^{2+} waves are reversibly initiated in regions without damage and, more specifically, from the border zone, in which contraction is partially suppressed. The common effect of the three protocols was to suppress contraction and reduce sarcomere F; the latter occurred *with* (caffeine and low $[Ca^{2+}]_{jet}$) or *without* change of the Ca^{2+} transient (BDM), and *with* (caffeine) or *without* change of diastolic $[Ca^{2+}]_i$ (BDM). The observation that the effect of F was common to all three interventions, whereas the effect on $[Ca^{2+}]_i$ and on the Ca^{2+} transient was dramatically different between the interventions, makes it reasonable to conclude that Ca^{2+} waves are initiated as a result of nonuniformity of sarcomere F generation and the resultant sequence of stretch and release of sarcomeres in the BZ by contraction of normal cells in the region remote from the jet. By varying $[Ca^{2+}]_o$, we detected a small localized $[Ca^{2+}]_i$ surge in the BZ, which developed into a propagating wave when $[Ca^{2+}]_o$ was increased (FIG. 3C). Once initiated, these Ca^{2+} waves traveled from the region with the localized $[Ca^{2+}]_i$ rise proving that this region constitutes the initiation site for Ca^{2+} waves.

The initiating Ca^{2+} surge took place late during the relaxation phase of the twitch when both F and free Ca^{2+} in the cytosol had decayed by 70% to 80% (FIG. 3D). By this time, the SR-Ca^{2+} channels had partially recovered[46] and were able to support CICR and Ca^{2+} wave generation.[28] However, the delay between the start of the stimulus and the Ca^{2+} surge makes it highly unlikely that Ca^{2+} entry via L-type Ca^{2+} channels causes CICR from the SR involved in the initial Ca^{2+} surge. Furthermore, Ca^{2+} waves never started in jet-exposed regions where sarcomeres were maximally stretched, even if the amplitude of the stimulated Ca^{2+} transients witnessed a robust SR-Ca^{2+} load.[28] Ca^{2+} waves never started simultaneously with the peak of the stretch (FIG. 4C) making it unlikely that a stretch-related mechanism, such as activation of Gd^{3+}-sensitive stretch-activated channels[21,24,25,47,48] is involved in the initial Ca^{2+} surge. Finally, Ca^{2+} waves started from the BZ with reduced SR-Ca^{2+} release (CF and low-$[Ca^{2+}]_{jet}$) and/or reduced $[Ca^{2+}]^i$ (low-$[Ca^{2+}]_{jet}$ and BDM), suggest to us that another mechanism of wave initiation different from "spontaneous SR-Ca^{2+} release" was involved.

Initiation and Propagation of Ca^{2+} Waves

We propose an alternative explanation on the basis of these experiments and the results of model studies. It is likely that quick-release-induced Ca^{2+} dissociation from the myofilaments, demonstrated in uniform cardiac muscle, is applicable to the chain of cells in the nonuniform muscle exposed to the jet. Rapid sarcomere shortening during the F decline occurred both in the jet region and in the BZ, but led only to a Ca^{2+} surge and Ca^{2+} wave initiation in the BZ (FIG. 3C), making it probable that quick-release-induced Ca^{2+} dissociation from TnC, caused by the decline of F in the shortening BZ sarcomeres, led to the local Ca^{2+}surge.[11,13,25,49–51] The region inside the jet, where E-C coupling was all but abolished, probably contained either little TnC-Ca^{2+} (caffeine or low-$[Ca^{2+}]_{jet}$) or only few Ca^{2+} activated F generating cross-bridges (BDM), which would render a quick release of this region unable to generate a Ca^{2+} surge and Ca^{2+} wave. The BZ, on the other hand, could generate a Ca^{2+} surge that is large enough to induce local CICR and thus a Ca^{2+} wave, even if only a fraction of TnC[37,52] were occupied with Ca^{2+}.[10]

A precise quantitative analysis of the relation between quick-release dynamics and onset of the Ca^{2+} surge and/or waves may shed further light on this mechanism of initiation of Ca^{2+} waves. Our present study allows for an estimate of the latency (~40 ms) between rapid F and SL changes and the initial rise of $[Ca^{2+}]_i$ during the Ca^{2+} surge in a small BZ region ~100 μm (FIG. 3D). Cannell and Allen have shown that it requires ≈20 ms for an intrasarcomeric (≈1 μm) Ca^{2+} gradient to dissipate.[53] Hence, the minimum value for the delay between Ca^{2+} dissociation of the filaments and SR–Ca^{2+} release could correspond to Ca^{2+} diffusion from the myofilaments to SR-Ca^{2+} release sites over, at most, a few sarcomere lengths.

Ca^{2+} waves propagated in this model at slightly lower velocities (0.2 to 2.8 mm/s) than those of previous studies of regionally damaged muscles.[23,24,54] In this study, the amplification of the Ca^{2+} signal by SR-Ca^{2+} release required for propagation may have been lower in the absence of Ca^{2+} loading of the muscle by damaged areas.[5,6,10,22,47,52,55–57] A lower cellular Ca^{2+} load would explain the occurrence of smaller Ca^{2+} waves that propagated more slowly and with a gradual decline of their amplitude and then disappeared after a few hundred μm. Ca^{2+} waves did propagate into the region with normal SR-Ca^{2+} release, such as inside the BDM jet (FIG. 3D), and did propagate slowly inside the low-$[Ca^{2+}]$ jet where the SR-Ca^{2+} load is expected to be reduced. Ca^{2+} waves did not propagate through the region exposed to caffeine (data not shown), which is consistent with the effect of caffeine to deplete SR-Ca^{2+} required for wave propagation.[22]

Feedback between Cross-Bridge F and Ca^{2+} Dissociation from TnC

FIGURE 4 shows the fundamental property of XB TnC interaction that underlies the model: feedback of F to Ca^{2+} dissociation from Ca^{2+}·TnC through the strain κ of the actin filament and subsequent deformation of TnC on the filament predicts fundamental properties known for cardiac muscle. In view of the realistic reproduction of the steady-state F-pCa relationship (FIG. 5),[58] it is not surprising that the feedback predicts realistic F-SL relationships at varied activation levels for $[Ca^{2+}]^i$. The prediction by the model that the F-SL exhibits little sensitivity to shortening, as has been observed,[21] needs to be tested in order to ascertain its merit in explaining both the end-systolic pressure volume relationship described by Suga and Sagawa in Ref. 59.

The twitches generated by the model are similar to twitches generated by rat cardiac trabeculae at 25°C. The increase of the duration of Ca^{2+} binding to TnC, in response to F development, explains that progressive prolongation of the twitch has indeed been shown for cardiac muscle twitches at varied SL, which was held constant during the twitch and was reproduced accurately by the introduction of a tight relationship between the actin strain κ and a decrease of the dissociation rate of Ca from Ca^{2+}·TnC. This observation holds both in model and experiments irrespective of whether the increased F is caused by varied SL, or $[Ca^{2+}]_o$, or varied pH.[43] Conversely, any F decrease is expected to cause accelerate dissociation of Ca^{2+} from Ca^{2+}·TnC and to decrease to subsequent F. This phenomenon is well known as the deactivation response to shortening and is realistically replicated by the model (FIG. 6). The degree of F reduction during shortening deactivation depends on the redistribution of Ca^{2+} over ligands and Ca^{2+} extrusion.

The model also reproduced our observation,[25] confirmed by Julian's group,[44] that the $[Ca^{2+}]_i$ transient during the twitch exhibits a prominent plateau at higher F levels;[44] this observation is remarkable in the model as well (FIG. 7, left). These authors explained their observation on the basis of cooperativity between F development and affinity of TnC for Ca^{2+} ions, but we are not aware of previous quantitative modeling explaining this observation. This property is completely lost in the absence of the feedback (FIG. 7). The combination of these predictions is no proof for the postulated F-Ca.TnC feedback, but makes it a useful working model for cardiac muscle and should stimulate further studies of the dynamics of the interaction between TnC and Ca^{2+}.

CONCLUSIONS

A striking finding of this study is that the arrhythmogenic nature of mechanically nonuniform myocardium independent of any damage, is clearly due to the induction of Ca^{2+} waves by the Ca^{2+} surge (FIG. 3). In this study of controlled nonuniformity of muscle contraction, we identify nonuniform contraction in cardiac muscle for the first time as a possible arrhythmogenic mechanism owing to a Ca^{2+} surge that results from Ca^{2+} dissociation from Ca^{2+}·TnC during relaxation. Simulations using a model that assumes that XB-F retards Ca^{2+} dissociation from the Ca·TnC complex, indeed, predict a large $[Ca^{2+}]_i$ transient upon a quick release owing to Ca^{2+} dissociation from the Ca^{2+}·TnC. The Ca^{2+} transient is well above the $[Ca^{2+}]_i$ level that is required to induce propagating CICR.[36] We postulate that these $[Ca^{2+}]_i$ transients occur in nonuniform muscle and cause reverse E-C coupling (FIG. 2), which is responsible for Ca^{2+} waves TPCs, DADs, and arrhythmias. Such Ca^{2+} waves are known to cause transient depolarizations[7,24,60–62] and cause arrhythmias[7] (FIG. 4).

Whether this mechanism contributes to arrhythmias in diseased heart where nonuniform segmental wall motion[29,63] may result from ischemia, nonuniform electrical activation or nonuniform adrenergic activation[63,64] remains to be proven, although the arrangement of the cardiac wall in muscle fascicles, which transmit F longitudinally and, therefore, are subject to comparable constraints as the trabeculae in this study makes this possibility highly likely. Importantly, our model of RECC also predicts that the initial Ca^{2+} surge from TnC will lead more readily to arrhythmias in hearts with abnormal SR-Ca^{2+} storage[65–67] and/or with increased open probability of the SR-Ca^{2+} release channels[68] owing to either mutation of the channel gene[61,63,65] or to posttranslational changes of the channel in heart failure.[66]

ACKNOWLEDGMENTS

This work was supported by grants from the Janssen Research Foundation and the Alberta Heart and Stroke Foundation. Dr. H.E.D.J. ter Keurs is a medical scientist of the Alberta Heritage Foundation for Medical Research (AHFMR).

REFERENCES

1. SPOONER, P.M. & M.R. ROSEN. 2001. Foundations of Cardiac Arrhythmias. Basic Concepts and Clinical Approaches. Marcel Dekker. New York/Basel Switzerland.
2. FERRIER, G.R. 1976. The effects of tension on acetylstrophanthidin-induced transient depolarizations and aftercontractions in canine myocardial and Purkinje tissues. Circ. Res. **38:** 156–162.
3. KASS, R.S., W.J. LEDERER, R.W. TSIEN & R.WEINGART. 1978. Role of calcium ions in transient inward currents and aftercontractions induced by strophanthidin in cardiac purkinje fibres. J. Physiol. **281:** 187–208.
4. CRANEFIELD, P.F. 1977. Action potentials, afterpotentials, and arrhythmias. Circ. Res. **41:** 415–423.
5. DANIELS, M.C.G. & H.E.D.J. TER KEURS. 1990. Spontaneous contractions in rat cardiac trabeculae; trigger mechanism and propagation velocity. J. Gen. Physiol. **95:** 1123–1137.
6. MULDER, B.J.M., P.P.DE TOMBE & H.E.D.J. TER KEURS. 1989. Spontaneous and propagated contractions in rat cardiac trabeculae. J. Gen. Physiol. **93:** 943–961.
7. DANIELS, M.C.G., D. FEDIDA, C. LAMONT & H.E.D.J. TER KEURS. 1991. Role of the sarcolemma in triggered propagated contractions in rat cardiac trabeculae. Circ. Res. **68:** 1408–1421.
8. KORT, A.A. & E.G. LAKATTA. 1984. Calcium-dependent mechanical oscillations occur spontaneously in unstimulated mammalian cardiac tissues. Circ. Res. **54:** 396–404.
9. STERN, M.D., A.A. KORT, G.M. BHATNAGAR & E.G. LAKATTA. 1983. Scattered-light intensity fluctuations in diastolic rat cardiac muscle caused by spontaneous calcium-dependent cellular mechanical oscillations. J. Gen. Physiol. **82:** 119–153.
10. TER KEURS, H.E.D.J., Y.M. ZHANG & M. MIURA. 1998. Damage induced arrhythmias: reversal of excitation-contraction coupling. Cardiovasc. Res. **40:** 444–455.
11. HOUSMANS, P.R., N.K.M. LEE & J.R. BLINKS. 1983. Active shortening retards the decline of the intracellular calcium transient in mammalian heart muscle. Science **221:** 159–161.
12. ALLEN, D.G. & J.C. KENTISH. 1985. The cellular basis of the length-tension relation in cardiac muscle. J. Mol. Cell. Cardiol. **17:** 821–840.
13. ALLEN, D.G. & S. KURIHARA. 1982. The effects of muscle length on intracellular calcium transients in mammalian cardiac muscle. J. Physiol. **327:** 79–94.
14. ALLEN, D.G. & J.C. KENTISH. 1988. Calcium concentration in the myoplasm of skinned ferret ventricular muscle following changes in muscle length. J. Physiol. **407:** 489–503.
15. BACKX, P.H.M., W.D. GAO, M.D. AZAN-BACKX & E. MARBAN. 1995. The relationship between contractile force and intracellular [Ca++] in intact rat cardiac trabeculae. J. Gen. Physiol. **105:** 1–19.
16. BANIJAMALI, H.S., W.D. GAO, B.R. MACINTOSH & H.E.D.J. TER KEURS. 1991. Force-interval relations of twitches and cold contractures in rat cardiac trabeculae: influence of ryanodine. Circ. Res. **69:** 937–948.
17. DANIELS, M.C.G., D. FEDIDA, C. LAMONT & H.E.D.J. TER KEURS. 1991. Role of the sarcolemma in triggered propagated contractions in rat cardiac trabeculae. Circ. Res. **68:** 1408–1421.
18. ZHANG, Y. & H.E.D.J. TER KEURS. 1994. Are stretch activated channels involved in triggered propagated contractions in rat cardiac trabeculae? Biophys. J. **66:** A317.
19. LANDESBERG, A. & S. SIDEMAN. 1994. Coupling calcium binding to troponin-C and cross-bridge cycling in skinned cardiac cells. Am. J. Physiol. **266:** H1260–H1271.

20. LANDESBERG, A. & S. SIDEMAN. 1994. Mechanical regulation in the cardiac muscle by coupling calcium kinetics with crossbridge cycling; a dynamic model. Am. J. Physiol. **267:** H779–H795.
21. TER KEURS, H.E.D.J., W.H. RIJNSBURGER, R. VAN HEUNINGEN & M.J. NAGELSMIT. 1980. Tension development and sarcomere length in rat cardiac trabeculae. Evidence of length-dependent activation. Circ. Res. **46:** 703–714.
22. MIURA, M., P.A. BOYDEN & H.E.D.J. TER KEURS. 1999. Ca^{2+} waves during triggered propagated contractions in intact trabeculae; determinants of the velocity of propagation. Circ. Res. **84:** 1459–1468.
23. MIURA, M., P.A. BOYDEN & H.E.D.J. TER KEURS. 1998. Ca^{2+} waves during triggered propagated contractions in intact trabeculae. Am. J. Physiol. **274:** H266–H276.
24. WAKAYAMA, Y., Y. SUGAI, Y. KAGAYA, *et al.* 2001. Stretch and quick release of cardiac trabeculae accelerates Ca2+ waves and triggered propagated contractions. Am. J. Physiol. Heart Circ. Physiol. **281:** H2133–2142.
25. BACKX, P.H. & H.E.D.J. TER KEURS. 1993. Fluorescent properties of rat cardiac trabeculae microinjected with fura-2 salt. Am. J. Physiol. **264:** H1098–H1110.
26. KONISHI, M., S. KURIHARA & T. SAKAI. 1984. The effects of caffeine on tension development and intracellular calcium transients in rat ventricular muscle J. Physiol. **355:** 605–618.
27. SITSAPESAN, R. & A.J. WILLIAMS. 1990. Mechanisms of caffeine activation of single calcium-release channels of sheep cardiac sarcoplasmic reticulum. J. Physiol. **423:** 425–439.
28. BERS, D.M. 2001. Excitation-contraction coupling and cardiac contractile force. 2nd edition. Kluwer. Dordrecht, the Netherlands.
29. YOUNG, A.A., S. DOKOS, K.A. POWELL, *et al.* 2001. Regional heterogeneity of function in non-ischemic dilated cardiomyopathy. Cardiovasc. Res. **49:** 308–318.
30. SELLIN, L.C. & J.J. MCARDLE. 1994. Multiple effects of 2,3-butanedione monoxime. Pharmacol. Toxicol. **74:** 305–313.
31. HERRMANN, C., J. WRAY, F. TRAVERS & T. BARMAN. 1992. Effect of 2,3-butanedione monoxime on sarcoplasmic reticulum of saponin-treated rat cardiac muscle. Biochemistry **31:** 12227–12232.
32. BACKX, P.H., W.D. GAO, M.D. AZAN-BACKX & E. MARBAN. 1994. Mechanism of force inhibition by 2,3-butanedione monoxime in rat cardiac muscle: roles of [Ca2+]i and cross-bridge kinetics. J. Physiol. **476:** 487–500.
33. MICHAILOVA, A., F. DELPRINCIPE, M. EGGER & E. NIGGLI. 2002. Spatiotemporal features of Ca2+ buffering and diffusion in atrial cardiac myocytes with inhibited sarcoplasmic reticulum. Biophys. J. **83:** 3134–3151.
34. CSERNOCH, L., J. ZHOU, M.D. STERN, *et al.* 2004. The elementary events of Ca2+ release elicited by membrane depolarization in mammalian muscle. J. Physiol. **557:** 43–58.
35. DAVIDOFF, A.W., P.A. BOYDEN, K. SCHWARTZ, *et al.* 2004. Congestive heart failure after myocardial infarction in the rat: cardiac force and spontaneous sarcomere activity. Ann. N.Y. Acad. Sci. **1015:** 84–95.
36. STUYVERS, B., M. MIURA & H.E.D.J. TER KEURS. 1997. Dynamics of viscoelastic properties of rat cardiac sarcomeres during the diastolic interval: involvement of Ca2+. J. Physiol. **502:** 661–667.
37. FABIATO, A. 1983. Calcium-induced release of calcium from the cardiac sarcoplasmic reticulum. Am. J. Physiol. **245:** C1–C14.
38. ISHIDE, N., M. MIURA, M. SAKURAI & T. TAKISHIMA. 1992. Initiation and development of calcium waves in rat myocytes. Am. J. Physiol. **263:** H327–H332.
39. TER KEURS, H.E.D.J., E.H. HOLLANDER & M.H.C. TER KEURS. 2000. The effects of sarcomere length on the force-cytosolic Ca2+ relationship in intact rat trabeculae. *In* Skeletal Muscle Mechanics: From Mechanisms to Function. W. Herzog, Ed.: 53–70. John Wiley and Sons. Chichester.
40. WOLEDGE, R.C., N.A. CURTIN & E. HOMSHER. 1985. Basic facts and ideas. *In* Energetic Aspects of Muscle Contraction. R.C. Woledge, N.A. Curtin & E. Homsher, Eds.: 1–26. Academic Press. New York.
41. KENTISH, J.C., H.E.D.J. TER KEURS, L. RICCIARDI, *et al.* 1986. Comparison between the

sarcomere length-force relations of intact and skinned trabeculae from rat right ventricle. Circ. Res. **58:** 755–768.
42. TER KEURS, H.E.D.J., W.H. RIJNSBURGER & R. VAN HEUNINGEN. 1980. Restoring forces and relaxation of rat cardiac muscle. Eur. Heart J. **1**(Suppl. A): 67–80.
43. BUCX, J.J.J. 1995. Ischemia of the heart: a study of sarcomere dynamics and cellular metabolism. Ph.D. thesis, Erasmus University of Rotterdam, the Netherlands.
44. JIANG, Y., M.F. PATTERSON, D.L. MORGAN & F.J. JULIAN. 1998. Basis for late rise in fura 2 R signal reporting [Ca2+]i during relaxation in intact rat ventricular trabeculae. Am. J. Physiol. **274:** C1273–C1282.
45. RAIMONDI, G.A. & A.C. RAIMONDI. 1977. PEEP and pulmonary vascular resistance. N. Engl. J. Med. **3:** 1011.
46. BANIJAMALI, H., W.D. GAO, B. MACINTOSH & H.E.D.J. TER KEURS. 1998. Force-interval relations of twitches and cold contractures in rat cardiac trabeculae; effect of ryanodine. Circ. Res. **69:** 937–948.
47. ZHANG, Y., M. MIURA & H.E.D.J. TER KEURS. 1996. Triggered propagated contractions in rat cardiac trabeculae; inhibition by octanol and heptanol. Circ. Res. **79:** 1077–1085.
48. ZHANG, Y. & H.E.D.J. TER KEURS. 1996. Effects of gadolinium on twitch force and triggered propagated contractions in rat cardiac trabeculae. Cardiovasc. Res. **32:** 180–188.
49. KURIHARA, S. & K. KOMUKAI. 1995. Tension-dependent changes of the intracellular Ca^{2+} transients in ferret ventricular muscles. J. Physiol. **489:** 617–625.
50. ALLEN, D.G. & J.C. KENTISH. 1988. Calcium concentration in the myoplasm of skinned ferret ventricular muscle following changes in muscle length. J. Physiol. **407:** 489–503.
51. LAB, M.J., D.G. ALLEN & C.H. ORCHARD. 1984. The effects of shortening on myoplasmic calcium concentration and on the action potential in mammalian ventricular muscle. Circ. Res. **55:** 825–829.
52. BACKX, P.H.M., P.P. DE TOMBE, J.H.K. VAN DEEN, *et al.* 1989. A model of propagating calcium-induced calcium release mediated by calcium diffusion. J. Gen. Physiol. **93:** 963–977.
53. CANNELL, M.B. & D.G. ALLEN. 1984. Model of calcium movements during activation in the sarcomere of frog skeletal muscle. Biophys. J. **45:** 913–925.
54. RAISARO, A., A. VILLA, F. RECUSANI, *et al.* 1993. [The remodeling of the heart in heart failure: from thoracic radiography to magnetic resonance]. G. Ital. Cardiol. **23:** 167–175.
55. LAMONT, C., P.W. LUTHER, C.W. BALKE & G.W. WIER. 1998. Intercellular Ca^{2+} waves in rat heart muscle. J. Physiol. **512:** 669–676.
56. WIER, W.G., H.E.D.J. TER KEURS, E. MARBAN, *et al.* 1997. Ca^{2+} 'sparks' and waves in intact ventricular muscle resolved by confocal imaging. Circ. Res. **81:** 462–469.
57. KANEKO, T., H. TANAKA, M. OYAMADA, *et al.* 2000. Three distinct types of Ca^{2+} waves in Langendorff-perfused rat heart revealed by real-time confocal microscopy. Circ. Res. **86:** 1093–1099.
58. KENTISH, J.C., H.E.D.J. TER KEURS, L. RICCIARDI, *et al.* 1986. Comparison between the sarcomere length-force relations of intact and skinned trabeculae from rat right ventricle. Circ. Res. **58:** 755–768.
59. TER KEURS, H.E.D.J. & M.I.M. NOBLE, Eds. 1988. Developments in Cardiovascular Medicine, Vol. 89. Starling's Law of the Heart Revisited. Kluwer Academic Publishers. Dordrecht.
60. MIURA, M., N. ISHIDE, H. ODA, *et al.* 1993. Spatial features of calcium transients during early and delayed afterdepolarizations. Am. J. Physiol. **265:** H439–H444.
61. SCHLOTTHAUER, K. & D.M. BERS. 2000. Sarcoplasmic reticulum Ca(2+) release causes myocyte depolarization. Underlying mechanism and threshold for triggered action potentials. Circ. Res. **87:** 774–780.
62. LAKATTA, E.G. 1992. Functional implications of spontaneous sarcoplasmic reticulum Ca2+ release in the heart. Cardiovasc. Res. **26:** 193–214.
63. SIOGAS, K., S. PAPPAS, G. GRAEKAS, *et al.* 1998. Segmental wall motion abnormalities alter vulnerability to ventricular ectopic beats associated with acute increases in aortic pressure in patients with underlying coronary artery disease. Heart **79:** 268–273.

64. JIANG, Y. & F.J. JULIAN. 1997. Pacing rate, halothane, and BDM affect fura 2 reporting of [Ca2+]i in intact rat trabeculae. Am. J. Physiol. **273:** C2046–C2056.
65. LAITINEN, P.J., K.M. BROWN, K. PIIPPO, *et al.* 2001. Mutations of the cardiac ryanodine receptor (RyR2) gene in familial polymorphic ventricular tachycardia. Circulation **103:** 485–490.
66. MARKS, A.R. 2001. Ryanodine receptors/calcium release channels in heart failure and sudden cardiac death. J. Mol. Cell. Cardiol. **33:** 615–624.
67. POSTMA, A.V., I. DENJOY, T.M. JOORNTJE, *et al.* 2002. Absence of calsequesterin 2 causes severe forms of catecholaminergic polymorphic ventricular tachycardia. Circ. Res. **91:** e21–e26.
68. BOYDEN, P.A. & H.E.D.J. TER KEURS. 2001. Reverse excitation contract coupling: Ca2+ ions as initiators of arrhythmias. J. Cardiovasc. Electrophysiol. **12:** 382–385.
69. DANIELS, M.C.G. 1991. Mechanism of triggered arrhythmias in damaged myocardium. Ph.D. thesis, University of Utrecht, the Netherlands.
70. VAN HEUNINGEN, R., W.H. RIJNSBURGER & H.E.D.J. TER KEURS. 1982. Sarcomere length control in striated muscle. Am. J. Physiol. **242:** H411–H420.

Ryanodine Receptor-Targeted Anti-Arrhythmic Therapy

XANDER H.T. WEHRENS,[a] STEPHAN E. LEHNART,[a] AND ANDREW R. MARKS[a,b]

[a] *Department of Physiology and Cellular Biophysics, Center for Molecular Cardiology, College of Physicians and Surgeons of Columbia University, New York, New York 10032, USA*

[b] *Department of Medicine, College of Physicians and Surgeons of Columbia University, New York, New York 10032, USA*

ABSTRACT: Cardiac arrhythmia is an important cause of death in patients with heart failure (HF) and inherited arrhythmia syndromes, such as catecholaminergic polymorphic ventricular tachycardia (CPVT). Alterations in intracellular calcium handling play a prominent role in the generation of arrhythmias in the failing heart. Diastolic calcium leak from the sarcoplasmic reticulum (SR) via cardiac ryanodine receptors (RyR2) may initiate delayed afterdepolarizations and triggered activity leading to arrhythmias. Similarly, SR Ca^{2+} leak through mutant RyR2 channels may cause triggered activity during exercise in patients with CPVT. Novel therapeutic approaches, based on recent advances in the understanding of the cellular mechanisms underlying arrhythmias in HF and CPVT, are currently being evaluated to specifically correct defective Ca^{2+} release in these lethal syndromes.

KEYWORDS: arrhythmia; calstabin2; CPVT; delayed afterdepolarization; heart failure; JTV519; ryanodine receptor; triggered activity

INTRODUCTION

Fatal ventricular arrhythmias occur mostly in patients suffering from heart failure (HF), but may also strike apparently healthy individuals without known cardiac abnormalities. Such sudden cardiac deaths (SCD) account for about half of all fatalities from HF.[1–3] Moreover, lethal ventricular tachycardia may also occur in the structurally normal hearts of patients with inherited arrhythmogenic syndromes such as catecholaminergic polymorphic ventricular tachycardia (CPVT).[4,5] Novel therapeutic approaches, based on recent advances in the understanding of the cellular mechanisms underlying arrhythmias in HF and CPVT, are currently being evaluated to specifically correct defective Ca^{2+} handling in these lethal syndromes.

Address for correspondence: Prof. Andrew R. Marks, M.D., Department of Physiology and Cellular Biophysics, Columbia University College of Physicians & Surgeons, P&S 11-511, 630 West 168th Street, New York, NY 10032, USA. Voice: 212-305-0272; fax: 212-305-3690.
arm42@columbia.edu

**Ann. N.Y. Acad. Sci. 1047: 366–375 (2005). © 2005 New York Academy of Sciences.
doi: 10.1196/annals.1341.032**

MECHANISMS OF VENTRICULAR ARRHYTHMIAS

From clinical and experimental investigations, it has become clear that abnormal impulse generation that can initiate ventricular arrhythmias generally originates from either reentry or triggered activity.[6] Triggered activity, by definition, occurs following a previous action potential (AP) since it is "triggered" by the previous impulse. Triggered impulses result from membrane depolarizations, termed afterdepolarizations. Two types of afterdepolarizations have been distinguished: early (EADs) and delayed afterdepolarizations (DADs).

EADs are generated during the AP and have been defined as "oscillations at the plateau level of membrane potential or later during phase 3 of repolarization."[6,7] EADs typically occur in the setting of prolonged repolarization due to alterations in K^+ currents (e.g., long QT syndrome), drug-induced QT prolongation, metabolic abnormalities (e.g., hypokalemia), or structural heart disease (e.g., hypertrophy). EADs are believed to be due to reopening of L-type Ca^{2+} channels ($I_{Ca,L}$) or Na^+ channels (I_{Na}) during the AP plateau phase.[8] On the other hand, DADs are defined as "oscillations in membrane potential that occur after complete repolarization of an action potential" (FIG. 1). Triggered activity due to DADs is more likely to occur during fast heart rates, increased sympathetic tone, or intracellular Ca^{2+} overload.[9]

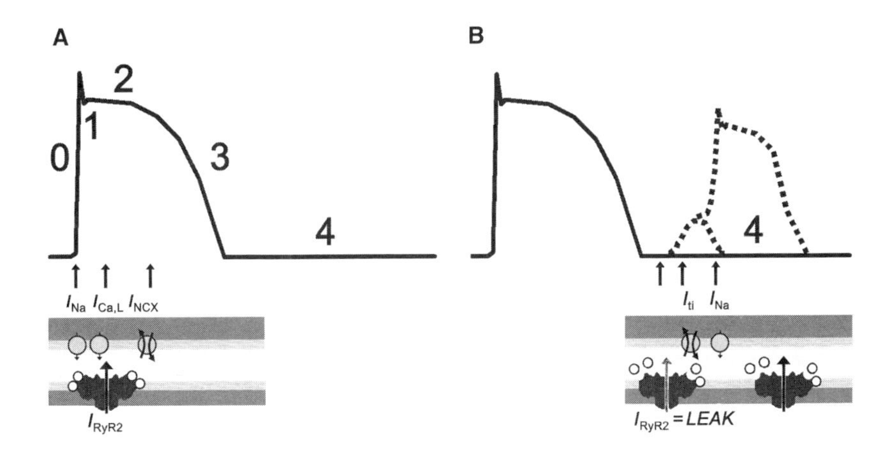

FIGURE 1. Depolarizing ionic currents and intracellular Ca^{2+} release contributing to the cardiac action potential and delayed afterdepolarizations. (**A**) During sinus rhythm, propagation of a depolarizing wavefront activates inward sodium current (I_{Na}) during phase 0 of the cardiac action potential (AP), and L-type Ca^{2+} current ($I_{Ca,L}$) during the plateau phase of the AP. Ca^{2+} influx into the myocyte triggers a several-times-larger intracellular Ca^{2+} release current via ryanodine receptors (RyR2), a process described as excitation–contraction coupling. Ca^{2+} influx from the extracellular space is compensated by depolarizing, forward mode Na^+/Ca^{2+} exchanger (NCX) Ca^{2+} extrusion. (**B**) During heart failure (HF) or catecholaminergic polymorphic ventricular tachycardia (CPVT), intracellular Ca^{2+} leak via defective RyR2 channels due to decreased levels of stabilizing calstabin2 activates depolarizing forward mode NCX current, providing a transient inward current (I_{ti}). If the membrane potential reaches the threshold level for I_{Na} depolarization, an aberrant AP is triggered, which leads to additional RyR2 calcium release and premature contraction [in color in Annals online].

Electrophysiological Alterations in Heart Failure

Altered electrophysiological properties of diseased cardiomyocytes may provide a substrate for contractile dysfunction or fatal arrhythmias in patients with cardiac hypertrophy or HF.[10,11] One of the best-documented changes in humans and animals with HF is prolongation of the action potential duration (APD) at low heart rates.[12–14] The changes in ionic currents underlying the APD prolongation include decreased transient outward and delayed rectifier K^+ currents (for more extensive review, see Refs. 10 and 15). In addition to delayed repolarization, increased dispersion of the APD and refractoriness may further predispose to reentrant arrhythmias.[13,16] Indeed, decreased conduction velocities and heterogeneous repolarization have been demonstrated in patients with idiopathic dilated cardiomyopathy[17,18] and experimental models of nonischemic HF.[15,19]

Although these changes in APD and conduction velocities suggest reentry as a possible mechanism, many studies favor triggered arrhythmias in nonischemic HF as well as ~50% of ischemic HF.[17,20] In a rabbit HF model (combined pressure and volume overload), Pogwizd[21] has shown spontaneously occurring ventricular tachycardias (VTs) initiated and maintained by triggered activity. These rabbits exhibit moderate to severe left ventricular dysfunction similar to nonischemic HF in humans, and display spontaneously occurring VTs in 90% that is further enhanced by β-adrenergic stimulation.[22] Three-dimensional mapping studies in patients with dilated cardiomyopathy at the time of heart transplantation demonstrated that the initiation and maintenance of spontaneous VTs occurred due to focal triggered activity.[23]

Triggered activity in the failing heart likely involves both early and delayed afterdepolarizations. Increased activity of the Na^+/Ca^{2+} exchanger (NCX) may be involved in the genesis of EADs in the hypertrophic and failing heart, due to (1) enhanced Ca^{2+} loading via reverse mode NCX due to AP prolongation, or (2) spontaneous sarcoplasmic reticulum (SR) Ca^{2+} release that initiates a transient inward current, I_{ti} (carried by NCX).[24] On the other hand, it is believed that DADs are likely triggers for nonreentrant arrhythmias in HF.[20] Arrhythmias resulting from a transient inward current can be evoked by spontaneous Ca^{2+} release from the SR[25,26] (see below).

Altered Ca^{2+} Handling in Failing Cardiomyocytes

Alterations in intracellular calcium handling play a prominent role in depressing intracellular Ca^{2+} transients and cardiac function in the failing heart.[27] Bers *et al.*[28] proposed that the three most important abnormalities in intracellular Ca^{2+} handling are reduced sarcoplasmic reticulum ATPase (SERCA2a) function, enhanced forward mode Na^+/Ca^{2+} exchange, and increased SR Ca^{2+} leak via ryanodine receptors (RyRs). These alterations all contribute to a reduction in the SR Ca^{2+} load and, thus, to a reduced intracellular Ca^{2+} transient. The relative contributions of these three mechanisms may vary among different models of HF. In a pacing-induced rabbit HF model, NCX expression and function is enhanced ~2-fold increasing Ca^{2+} extrusion and SERCA2a function depressed by ~25% decreasing SR Ca^{2+} uptake.[22] Diastolic SR Ca^{2+} leak was significantly enhanced in intact HF rabbit myocyte at any given SR Ca^{2+} load.[29] These findings are consistent with enhanced RyR2 open probabilities reported in human and canine HF[30,31] (see below).

Abnormal SR Ca^{2+} Release Causes Delayed Afterdepolarizations

The cellular basis for DADs is Ca^{2+} after-transients resulting from spontaneous SR Ca^{2+} release.[30,32,33] Abnormal Ca^{2+} released during the DAD is removed from the cell by the electrogenic Na^+/Ca^{2+} exchanger[22,34] or possibly by a Ca^{2+} activated Cl^- current,[35] providing the depolarizing transient inward current (I_{ti}) that causes the DAD. DADs have been reported in failing human cardiomyocytes.[36,37] Triggered activity arising from DADs in response to catecholamines has been demonstrated in experimental models of HF.[36,37] Several explanations have been proposed for the paradox that normally a high SR Ca^{2+} content is required for spontaneous Ca^{2+} release, whereas in HF, SR Ca^{2+} stores are usually depleted. Pogwizd *et al.*[37] and Pogwizd and Bers[38] explained this paradox by a downregulated but functional β-adrenergic responsiveness, which in the presence of an acutely increased adrenergic drive will increase the SR Ca^{2+} content to reach the threshold of spontaneous Ca^{2+} release by ryanodine receptors. This is consistent with experimental findings that DADs occur prominently in the presence of norepinephrine,[33,36] and that catecholamine levels in patients with HF are generally elevated.[39] On the other hand, certain alterations in Ca^{2+} handling in failing cardiomyocytes could predispose to the development of DADs, particularly at faster heart rates. For example, decreased I_{K1} current densities and increased forward NCX contribute to a reduced depolarizing threshold required to trigger potentially arrhythmogenic action potentials.[40] Moreover, an increased open probability of cardiac RyRs due to increased sensitivity to Ca^{2+} dependent activation may also contribute to arrhythmogenesis in failing hearts[30] (see below).

Abnormal Ryanodine Receptor Function in Heart Failure

Abnormal structure and function of ryanodine receptors has been reported in failing hearts.[11] Release of Ca^{2+} from the SR via RyR2 initiates contraction of the cardiomyocytes.[41] RyRs are homotetrameric channels located on the SR membrane, each ryanodine receptor monomer comprising a 560-kDa molecule characterized by a large cytosolic N-terminal scaffold and a smaller C-terminal domain containing the transmembrane segments forming the channel pore. Various channel modulators are associated with the N-terminal domain, and enable rapid and localized regulation of RyR2 Ca^{2+} release (see review).[11]

Highly conserved leucine/isoleucine zipper (LIZ) motifs in RyR2 bind to similar LIZs in adaptor proteins for kinases (e.g., protein kinase A [PKA]) and phosphatases (e.g., PP1 and PP2A) that regulate RyR2 function.[42] The channel-stabilizing subunit calstabin2 (also known as FKBP12.6) binds to RyR2 with a stoichiometry of one calstabin2 bound to each RyR2 monomer. Binding of calstabin2 to RyR2 stabilizes the channel in the closed state during the resting phase of the heart (diastole).[26,43] Activation of β-adrenergic receptors by catecholamines results in the activation of cAMP-dependent signal amplification cascades and phosphorylation of target proteins (ion channels, transporters) by PKA. Phosphorylation of serine 2809 on RyR2 (RyR2-Ser^{2809}) by PKA decreases the binding affinity for calstabin2 and causes dissociation of the channel-stabilizing subunit, which increases the sensitivity to Ca^{2+} dependent activation.[26,30,44]

Patients with HF have chronically elevated plasma concentrations of catecholamines, which contribute to the alterations in intracellular Ca^{2+} handling.[39] Chronic activation of the β-adrenergic pathway results in maladaptive changes in the heart, including downregulation and uncoupling of β-adrenoceptors.[45] Downstream effects of increased intracellular kinase activity include PKA-mediated hyperphosphorylation of RyR2.[30] Reduced levels of PP1 and PP2A in the RyR2 macromolecular complex might contribute to the maintenance of long-term hyperphosphorylation of RyR2 by PKA.[30] Hyperphosphorylation of RyR2 by PKA results in the dissociation of calstabin2, which causes a leftward shift in the Ca^{2+} sensitivity of the channel. In addition, RyR2 can open aberrantly during diastole, leading to SR Ca^{2+} "leak," which could result in triggered arrhythmias due to delayed afterdepolarizations.[26] Bers *et al.*[28] have recently shown evidence of diastolic SR Ca^{2+} leak in isolated cardiomyocytes from failing hearts.

Triggered Arrhythmias in Patients with CPVT

Bidirectional ventricular tachycardia is a rare and unusual arrhythmia, consistently associated with CPVT but better known for its association with digoxin toxicity.[46] This type of arrhythmia is characterized by broad complex tachycardia with alternating QRS complex polarity. DADs are believed to be the underlying cause of bidirectional ventricular tachycardia and, hence, arrhythmia generation in CPVT.[47] Similar to DADs in the failing heart, a transient inward current (I_{ti}) is evoked by spontaneous Ca^{2+} release from the SR under conditions that favor accumulation of intracellular Ca.[9] If the amplitude of the I_{ti} exceeds a threshold potential, depolarization will occur (FIG. 1B). Propagation of this abnormal impulse can subsequently cause a triggered arrhythmia.

Several factors increase the amplitude of the DAD and, hence, the chance that the threshold potential for sodium channel depolarization will be reached. These factors include increasing the rate of the triggering action potential (e.g., an increased heart rate) and increasing intracellular Ca^{2+} loading.[48] Increased heart rates may explain the association between exercise and arrhythmias in CPVT; increased Ca^{2+} loading is brought about by cardiac glycoside toxicity and might explain triggered activity in patients with digoxin toxicity. Interestingly, hypercalcemia alone does not cause bidirectional ventricular tachycardia, presumably because increased Ca^{2+} entry into the cell is sufficiently buffered or balanced by Ca^{2+} efflux. However, aberrant diastolic release of SR Ca^{2+} due to CPVT-associated RyR2 mutations may trigger these arrhythmias due to temporo-spacial proximity and functional coupling of RyR2 leak with depolarizing membrane transport mechanisms (FIG. 1).

Mutations in Ryanodine Receptors Cause Arrhythmias in CPVT

Mutations in the cardiac ryanodine receptor (RyR2) have recently been demonstrated in patients with CPVT.[4,5] This genetic arrhythmogenic disorder of the heart is characterized by stress- or exercise-induced ventricular tachycardias that lead to SCD.[49] Affected individuals typically present during adolescence with repetitive stress-induced syncopal events leading to SCD in 30% to 50% by 30 years of age.[50–52] Genetic linkage studies and DNA sequence analysis have identified mutations in the cardiac ryanodine receptor gene (*RyR2*)[4,5] and calsequestrin (CASQ2)[53] in affected individuals with CPVT.

Analysis of the biophysical properties of CPVT-mutant RyR2 channels has provided new insights into the molecular basis of triggers that initiate ventricular arrhythmias[26] Under resting conditions, CPVT-mutant RyR2 channels are indistinguishable from wild-type channels,[26,54] consistent with the clinical observation that CPVT patients do not develop arrhythmias at rest.[51] In contrast, CPVT-linked RyR2 mutant channels display increased single-channel function following PKA phosphorylation.[26,54] Mutant RyR2 channels linked to CPVT also have a decreased affinity for the channel-stabilizing subunit calstabin2, compared to wild-type channels.[26,52] Together, these findings suggest that during exercise, calstabin2-depleted CPVT-mutant channels might trigger SR Ca^{2+} leak, which could initiate delayed afterdepolarizations and ventricular arrhythmias. Indeed, DADs were observed during electrophysiological testing in patients with CPVT.[47] Consistent with this model is the finding that calstabin2-deficient mice consistently develop ventricular arrhythmias

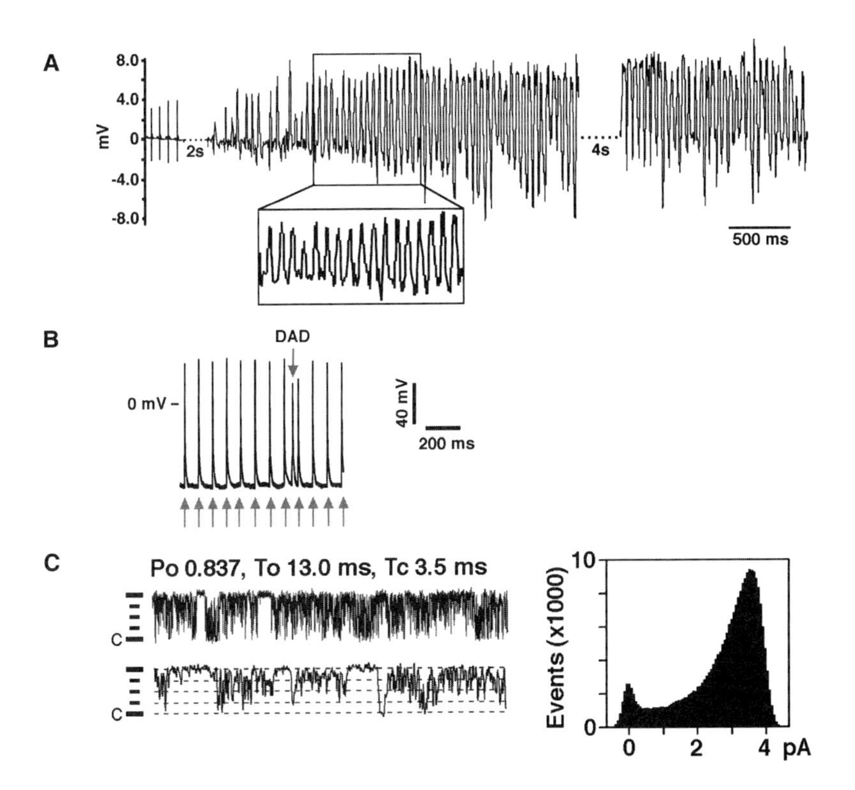

FIGURE 2. Triggered arrhythmias in calstabin$2^{-/-}$ mice. (**A**) In calstabin2-deficient mice, an exercise-stress test triggers ventricular arrhythmias and sudden cardiac death, which does not occur in wild-type mice undergoing the same protocol. (**B**) Isolated cardiomyocytes from calstabin$2^{-/-}$ mice show delayed afterdepolarizations (DADs) when paced at 750 beats per min (similar to heart rates in an exercising mouse). (**C**) Single RyR2 channels isolated from the heart of calstabin$2^{-/-}$ mice after exercise-stress testing show an abnormally high open probability, which provides the basis for DADs and triggered arrhythmias. (From Wehrens *et al.*[26] Reproduced by permission.)

and sudden cardiac death during exercise-stress testing[26] (FIG. 2). Delayed afterdepolarizations were observed in cardiomyocytes isolated from calstabin2$^{-/-}$ mice after exposure to isoproterenol. Thus, this rare genetic syndrome further suggests a direct link between abnormalities in Ca^{2+} handling and triggers for cardiac arrhythmias.

RYANODINE RECEPTOR-BASED THERAPY FOR VENTRICULAR ARRHYTHMIAS

The identification of calstabin2-depleted RyR2 as a source of diastolic SR Ca^{2+} leak in both HF and CPVT has led to the hypothesis that increasing calstabin2 binding to RyR2 could be a potential new molecular target to treat triggered arrhythmias.[26,55] In proof-of-principle experiments, we have recently demonstrated that a genetically altered calstabin2 protein is capable of binding to PKA-phosphorylated CPVT-mutant RyR2.[26] Association of the mutant calstabin2 to RyR2 restored defective single channel gating of CPVT-mutant or PKA-hyperphosphorylated RyR2 channels.[26,52] These findings suggested that increasing calstabin2 binding to phosphorylated RyR2 channels might provide a very specific therapeutic strategy for preventing triggered arrhythmias in HF and CPVT.[55]

The 1,4-benzothiazepine derivative JTV519 also effectively enhances calstabin2 binding to PKA-phosphorylated wild-type and CPVT-mutant RyR2 channels.[52,56,57] Nanomolar concentrations of JTV519 dramatically increase the binding affinity of calstabin2 to PKA-phosphorylated RyR2.[57] JTV519 has been shown to inhibit diastolic Ca^{2+} leakage from the SR in a canine model of pacing-induced HF.[56,58] Moreover, normalized SR Ca^{2+} release via calstabin2-bound RyR2 was associated with improved contractile function in these dogs with HF.[58]

Using a genetic mouse model of catecholamine-induced ventricular arrhythmias, we recently showed that ventricular arrhythmias induced by an exercise-stress protocol in haplo-insufficient calstabin2$^{+/-}$ mice could be effectively inhibited by treatment with JTV519. Catecholamine-induced ventricular arrhythmias evoked during programmed electrical stimulation could also be inhibited in calstabin2$^{+/-}$ mice pretreated with JTV519.[57] Co-immunoprecipitation experiments showed increased binding of calstabin2 to the RyR2 channel complex in calstabin2$^{+/-}$ mice that were protected against ventricular arrhythmias.[57] It is believed that JTV519 induces a conformational change in RyR2 that allows calstabin2 to rebind to the PKA-phosphorylated channel, which inhibits diastolic Ca^{2+} leak from the SR and prevents cardiac arrhythmias.[57] As discussed before, mutant RyR2 channels linked to CPVT typically show a reduced calstabin2 binding affinity and are hyperactive following PKA phosphorylation.[26] In the presence of JTV519, calstabin2 could reassociate with CPVT-mutant RyR2 channels, which normalized single channel gating in planar lipid bilayers.[52] Taken together, these findings indicate that JTV519 and its derivatives may constitute a novel class of drugs that could lead to the development of a gene-defect-specific treatment for arrhythmias in patients with HF or CPVT.[3,59]

CONCLUSION

Altered RyR2 function has an important role in HF and genetic forms of arrhythmias. RyR2 dysfunction is related to intracellular Ca^{2+} leak, which is associated with electrical instability of the plasma membrane that facilitates the development of arrhythmias in HF. HF results in RyR2 PKA hyperphosphorylation and chronic depletion of the stabilizing subunit calstabin2 (FKBP12.6) from the channel complex. Interventions that rescue RyR2 function through inhibition of PKA phosphorylation also normalize calstabin2 binding. Pharmacologic interventions like JTV519 increase calstabin2 binding to RyR2 and thereby inhibit DADs and triggered arrhythmias implying rescue of intracellular Ca^{2+} leak as a new therapeutic strategy in CPVT and HF.

REFERENCES

1. PARMLEY, W.W. 1987. Factors causing arrhythmias in chronic congestive heart failure. Am. Heart J. **114:** 1267–1272.
2. DE MARIA, R., A. Gavazzi, A. Caroli, *et al.* 1992. Ventricular arrhythmias in dilated cardiomyopathy as an independent prognostic hallmark. Italian Multicenter Cardiomyopathy Study (SPIC) Group. Am. J. Cardiol. **69:** 1451–1457.
3. FARR, M.A. & C.T. BASSON. 2004. Sparking the failing heart. N. Engl. J. Med. **351:** 185–187.
4. LAITINEN, P.J., K.M. Brown, K. Piippo, *et al.* 2001. Mutations of the cardiac ryanodine receptor (RyR2) gene in familial polymorphic ventricular tachycardia. Circulation **103:** 485–490.
5. PRIORI, S.G., C. Napolitano, N. Tiso, *et al.* 2001. Mutations in the cardiac ryanodine receptor gene (hRyR2) underlie catecholaminergic polymorphic ventricular tachycardia. Circulation **103:** 196–200.
6. WIT, A.L. & M.R. ROSEN. 1983. Pathophysiologic mechanisms of cardiac arrhythmias. Am. Heart J. **106:** 798–811.
7. VOLDERS, P.G., M.A. Vos, B. Szabo, *et al.* 2000. Progress in the understanding of cardiac early afterdepolarizations and torsades de pointes: time to revise current concepts. Cardiovasc. Res. **46:** 376–392.
8. JANUARY, C.T. & J.M. RIDDLE. 1989. Early afterdepolarizations: mechanism of induction and block. A role for L-type $Ca2^{+}$ current. Circ. Res. **64:** 977–990.
9. SCOOTE, M. & A.J. WILLIAMS. 2002. The cardiac ryanodine receptor (calcium release channel): emerging role in heart failure and arrhythmia pathogenesis. Cardiovasc. Res. **56:** 359–372.
10. TOMASELLI, G.F. & E. MARBAN. 1999. Electrophysiological remodeling in hypertrophy and heart failure. Cardiovasc. Res. **42:** 270–283.
11. WEHRENS, X.H. & A.R. MARKS. 2003. Altered function and regulation of cardiac ryanodine receptors in cardiac disease. Trends Biochem. Sci. **28:** 671–678.
12. BEUCKELMANN, D.J., M. NABAUER & E. ERDMANN. 1993. Alterations of K^+ currents in isolated human ventricular myocytes from patients with terminal heart failure. Circ. Research. **73:** 379–385.
13. Pak, P.H., H.B. Nuss, R.S. Tunin, *et al.* 1997. Repolarization abnormalities, arrhythmia and sudden death in canine tachycardia-induced cardiomyopathy. J. Am. Coll. Cardiol. **30:** 576–584.
14. LI, H.G., Jones, D.L., Yee, R. & G.J. Klein. 1992. Electrophysiologic substrate associated with pacing-induced heart failure in dogs: potential value of programmed stimulation in predicting sudden death. J Am Coll Cardiol. **19:** 444–449.
15. VOLDERS, P.G., K.R. Sipido, M.A. Vos, *et al.* 1999. Downregulation of delayed rectifier K(+) currents in dogs with chronic complete atrioventricular block and acquired torsades de pointes. Circulation **100:** 2455–2461.
16. YAN, G.X., S.J. Rials, Y. Wu, *et al.* 2001. Ventricular hypertrophy amplifies transmural repolarization dispersion and induces early afterdepolarization. Am. J. Physiol. Heart Circ. Physiol. **281:** H1968–H1975.
17. POGWIZD, S.M., R.H. Hoyt, J.E. Saffitz, *et al.* 1992. Reentrant and focal mechanisms underlying ventricular tachycardia in the human heart. Circulation **86:** 1872–1887.
18. Anderson, K.P., R. Walker, P. Urie, *et al.* 1993. Myocardial electrical propagation in patients with idiopathic dilated cardiomyopathy. J. Clin. Invest. **92:** 122–140.

19. AKAR, F.G. & D.S. ROSENBAUM. 2003. Transmural electrophysiological heterogeneities underlying arrhythmogenesis in heart failure. Circ. Res. **93:** 638–645.
20. POGWIZD, S.M. & D.M. BERS. 2004. Cellular basis of triggered arrhythmias in heart failure. Trends Cardiovasc. Med. **14:** 61–66.
21. POGWIZD, S.M. 1995. Nonreentrant mechanisms underlying spontaneous ventricular arrhythmias in a model of nonischemic heart failure in rabbits. Circulation **92:** 1034–1048.
22. POGWIZD, S.M., M. Qi, W. Yuan, *et al.* 1999. Upregulation of Na^+/Ca^{2+} exchanger expression and function in an arrhythmogenic rabbit model of heart failure. Circ. Res. **85:** 1009–1019.
23. POGWIZD, S.M., J.P. MCKENZIE & M.E. CAIN. 1998. Mechanisms underlying spontaneous and induced ventricular arrhythmias in patients with idiopathic dilated cardiomyopathy. Circulation **98:** 2404–2414.
24. SIPIDO, K.R., P.G. Volders, S.H. de Groot, *et al.* 2000. Enhanced Ca^{2+} release and Na/Ca exchange activity in hypertrophied canine ventricular myocytes: potential link between contractile adaptation and arrhythmogenesis. Circulation **102:** 2137–2144.
25. SCHLOTTHAUER, K. & D.M. BERS. 2000. Sarcoplasmic reticulum Ca^{2+} release causes myocyte depolarization. Underlying mechanism and threshold for triggered action potentials. Circ. Res. **87:** 774–780.
26. WEHRENS, X.H.T., S.E. Lehnart, F. Huang, *et al.* 2003. FKBP12.6 deficiency and defective calcium release channel (ryanodine receptor) function linked to exercise-induced sudden cardiac death. Cell **113:** 829–840.
27. WEHRENS, X.H.T., S.E. LEHNART & A.R. MARKS. 2005. Intracellular calcium release channels and cardiac disease. Annu. Rev. Physiol. **67:** 69–98.
28. BERS, D.M., D.A. EISNER & H.H. VALDIVIA. 2003. Sarcoplasmic reticulum Ca2+ and heart failure: roles of diastolic leak and Ca2+ transport. Circ. Res. **93:** 487–490.
29. SHANNON, T.R., S.M. POGWIZD & D.M. BERS. 2003. Elevated sarcoplasmic reticulum Ca^{2+} leak in intact ventricular myocytes from rabbits in heart failure. Circ. Res. **93:** 592–594.
30. MARX, S.O., S. Reiken, Y. Hisamatsu, *et al.* 2000. PKA phosphorylation dissociates FKBP12.6 from the calcium release channel (ryanodine receptor): defective regulation in failing hearts. Cell **101:** 365–376.
31. YANO, M., K. Ono, T. Ohkusa, *et al.* 2000. Altered stoichiometry of FKBP12.6 versus ryanodine receptor as a cause of abnormal Ca^{2+} leak through ryanodine receptor in heart failure. Circulation **102:** 2131–2136.
32. O'ROURKE, B., D.A. Kass, G.F. Tomaselli, *et al.* 1999. Mechanisms of altered excitation-contraction coupling in canine tachycardia-induced heart failure, I: experimental studies. Circ. Res. **84:** 562–570.
33. BAARTSCHEER, A., C.A. Schumacher, C.N. Belterman, *et al.* 2003. SR calcium handling and calcium after-transients in a rabbit model of heart failure. Cardiovasc. Res. **58:** 99–108.
34. VERKERK, A.O., M.W. Veldkamp, A. Baartscheer, *et al.* 2001. Ionic mechanism of delayed afterdepolarizations in ventricular cells isolated from human end-stage failing hearts. Circulation **104:** 2728–2733.
35. VERKERK, A.O., M.W. Veldkamp, L.N. Bouman, *et al.* 2000. Calcium-activated Cl(-) current contributes to delayed afterdepolarizations in single Purkinje and ventricular myocytes. Circulation **101:** 2639–2644.
36. VERMEULEN, J.T., M.A. McGuire, T. Opthof, *et al.* 1994. Triggered activity and automaticity in ventricular trabeculae of failing human and rabbit hearts. Cardiovasc. Res. **28:** 1547–1554.
37. POGWIZD, S.M., K.L. Schotthauer, L. Li, *et al.* 2001. Arrhythmogenesis and contractile dysfunction in heart failure: roles of sodium-calcium exchange, inward rectifier potassium current, and residual beta-adrenergic responsiveness. Circ. Res. **88:** 1159–1167.
38. POGWIZD, S.M. & D.M. BERS. 2002. Calcium cycling in heart failure: the arrhythmia connection. J. Cardiovasc. Electrophysiol. **13:** 88–91.
39. COHN, J.N., T.B. Levine, M.T. Olivari, *et al.* 1984. Plasma norepinephrine as a guide to prognosis in patients with chronic congestive heart failure. N. Engl. J. Med. **311:** 819–823.
40. NUSS, H.B., D.C. Johns, S. Kaab, *et al.* 1996. Reversal of potassium channel deficiency

in cells from failing hearts by adenoviral gene transfer: a prototype for gene therapy for disorders of cardiac excitability and contractility. Gene Ther. **3:** 900–912.
41. FILL, M. & J.A. COPELLO. 2002. Ryanodine receptor calcium release channels. Physiol. Rev. **82:** 893–922.
42. MARX, S.O., S. Reiken, Y. Hisamatsu, *et al.* 2001. Phosphorylation-dependent regulation of ryanodine receptors. A novel role for leucine/isoleucine zippers. J. Cell Biol. **153:** 699–708.
43. BRILLANTES, A.B., K. Ondrias, A. Scott, *et al.* 1994. Stabilization of calcium release channel (ryanodine receptor) function by FK506-binding protein. Cell **77:** 513–523.
44. HAIN, J., H. Onoue, M. Mayrleitner, *et al.* 1995. Phosphorylation modulates the function of the calcium release channel of sarcoplasmic reticulum from cardiac muscle. J. Biol. Chem. **270:** 2074–2081.
45. DAAKA, Y., L.M. LUTTRELL & R.J. LEFKOWITZ. 1997. Switching of the coupling of the beta2-adrenergic receptor to different G proteins by protein kinase A. Nature **390:** 88–91.
46. MA, G., Brady, W.J., Pollack, M. & T.C. Chan. 2001. Electrocardiographic manifestations: digitalis toxicity. J. Emerg. Med. **20:** 145–152.
47. NAKAJIMA, T., Y. Kaneko, Y. Taniguchi, *et al.* 1997. The mechanism of catecholaminergic polymorphic ventricular tachycardia may be triggered activity due to delayed afterdepolarization. Eur. Heart J. **18:** 530–531.
48. SANTANA, L.F., H. Cheng, A.M. Gomez, *et al.* 1996. Relation between the sarcolemmal Ca^{2+} current and Ca^{2+} sparks and local control theories for cardiac excitation-contraction coupling. Circ. Res. **78:** 166–171.
49. LEENHARDT, A., V. Lucet, I. Denjoy, *et al.* 1995. Catecholaminergic polymorphic ventricular tachycardia in children. A 7-year follow-up of 21 patients. Circulation **91:** 1512–1519.
50. FISHER, J.D., D. KRIKLER & K.A. HALLIDIE-SMITH. 1999. Familial polymorphic ventricular arrhythmias: a quarter century of successful medical treatment based on serial exercise-pharmacologic testing. J. Am. Coll. Cardiol. **34:** 2015–2022.
51. SWAN, H., K. Piippo, M. Viitasalo, *et al.* 1999. Arrhythmic disorder mapped to chromosome 1q42-q43 causes malignant polymorphic ventricular tachycardia in structurally normal hearts. J. Am. Coll. Cardiol. **34:** 2035–2042.
52. LEHNART, S.E., X.H. Wehrens, P.J. Laitinen, *et al.* 2004. Sudden death in familial polymorphic ventricular tachycardia associated with calcium release channel (ryanodine receptor) leak. Circulation **109:** 3208–3214.
53. LAHAT, H., E. Pras, T. Olender, *et al.* 2001. A missense mutation in a highly conserved region of CASQ2 is associated with autosomal recessive catecholamine-induced polymorphic ventricular tachycardia in Bedouin families from Israel. Am. J. Hum. Genet. **69:** 1378–1384.
54. GEORGE, C.H., G.V. HIGGS & F.A. LAI. 2003. Ryanodine receptor mutations associated with stress-induced ventricular tachycardia mediate increased calcium release in stimulated cardiomyocytes. Circ. Res. **93:** 531–540.
55. MOST, P. & W.J. KOCH. 2003. Sealing the leak, healing the heart. Nat. Med. **9:** 993–994.
56. KOHNO, M., M. Yano, S. Kobayashi, *et al.* 2003. A new cardioprotective agent, JTV519, improves defective channel gating of ryanodine receptor in heart failure. Am. J. Physiol. Heart Circ. Physiol. **284:** H1035–H1042.
57. WEHRENS, X.H., S.E. Lenhart, S.R. Reiken, *et al.* 2004. Protection from cardiac arrhythmia through ryanodine receptor-stabilizing protein calstabin2. Science **304:** 292–296.
58. YANO, M., S. Kobayashi, M. Kohno, *et al.* 2003. FKBP12.6-mediated stabilization of calcium-release channel (ryanodine receptor) as a novel therapeutic strategy against heart failure. Circulation **107:** 477–484.
59. WEHRENS, X.H. & A.R. MARKS. 2004. Novel therapeutic approaches for heart failure by normalising calcium cycling. Nat. Rev. Drug Discov. **3:** 565–573.

Stem Cells for Cardiomyocyte Regeneration: State of the Art

CESARE PESCHLE[a,c] AND GIANLUIGI CONDORELLI[b,d]

[a] *Department of Hematology, Oncology and Molecular Medicine, Istituto Superiore di Sanita, Rome, Italy*

[b] *Laboratory of Molecular Cardiology, San Raffaele Biomedical Science Park and Multimedica Hospital, Rome, Italy*

[c] *Kimmel Cancer Institute, Thomas Jefferson University, Philadelphia, Pennsylvania, USA*

[d] *Institute of Molecular Medicine, University of California San Diego, La Jolla, California 92093, USA*

ABSTRACT: A wide range of stem/progenitor cell types have been used for the regeneration of the infarcted heart. This review details the current status of progress concerning different strategies that have been used to manipulate cardiomyocyte cell growth *in vitro*, for their use in heart failure. The current status of this field involves different types of regenerating cells. Embryonic stem (ES) cells, hematopoietic stem and progenitor cells, mesenchymal stem and progenitor cells, and resident cardiac "stem" cells are discussed here.

KEYWORDS: bone marrow; cardiomyocytes; embryonic stem cells; hematopoietic progenitor; human/murine embryonic stem cells; myocardial infarction; stem cells

INTRODUCTION

Cardiomyocytes: Proliferating or Terminally Differentiated Cells?

Following myocardial infarction, the contractile parenchyma is substituted by fibrotic tissue, and, consequently, the proliferating and self-healing capacity of cardiomyocytes (CMCs) is minimal. The persistence of a Q (death) wave on the electrocardiogram throughout a lifetime suggests that the contractile elements are permanently lost in most of the area affected by a large infarction. *In vitro* and *in vivo* application of bromodeoxyuridine as a tracer suggests that murine CMCs can only proliferate during the first few days after birth,[1] after which their capacity to replicate is lost. However, postnatal CMCs that are bi-, tri-, or tetra-nucleated are also present.[1] Under specific *in vitro* conditions, it has been demonstrated that CMC nuclei can undergo DNA synthesis and thus overcome the G1-S checkpoint. Incubation of CMC nuclei with the cytoplasm of proliferating cells has been shown to induce

Address for correspondence: A Prof. Gianluigi Condorelli, M.D., Ph.D., Institute of Molecular Medicine, University of California San Diego, 9500 Gilman Drive, BSB 5022, La Jolla, CA 92093-0641, USA. Voice: (+1) 858-534-3347; fax: (+1) 858-534-1626.
gcondorelli@ucsd.edu

Ann. N.Y. Acad. Sci. 1047: 376–385 (2005). © 2005 New York Academy of Sciences.
doi: 10.1196/annals.1341.033

DNA synthesis,[2] whereas *in vitro* induction of DNA synthesis by oncogene overexpression in differentiated CMCs induced apoptosis. Furthermore, *in vitro* stimulation of CMCs only with growth factors failed to induce DNA synthesis.[3] In transgenic mice, alpha-myosin heavy-chain (α-MHC) promoter driven overexpression of cyclin D2, but not D1 or D3, induced DNA synthesis and cell proliferation in CMCs *in vivo*, resulting in infarct regression.[4] This would suggest that a small subset of differentiated CMCs can continue to proliferate, if a proliferation gene is expressed during the perinatal age. It has been shown that cardiac-selective overexpression of large T-antigen, driven by the atrial natriuretic factor (ANF) or α-MHC promoter, generated cell lines of differentiated beating *and* proliferating cardiomyoblasts.[5,6] In mouse CMCs, introduction of Cre recombinase resulted in the elimination of large T-antigen expression, inducing a switch from proliferation to differentiation.[7]

Piero Anversa's group[8–11] demonstrated proliferating CMCs in the myocardial tissue of adult human hearts. CMC proliferation was also observed in the myocardium of end-stage heart failure.[8] Further evidence has also shown that human CMCs are capable of division, not only DNA synthesis, following myocardial infarction.[9] *In vivo*, cardiac-specific overexpression of bcl-2,[10] an anti-apoptotic protein, or insulin-like growth factor-1,[11] promoted CMC proliferation.

EMBRYONIC CELLS

Three types of pluripotent stem cell lines have been established from mammalian embryos: embryonic germ (EG), embryonic carcinoma, and embryonic stem (ES) cells. ES cells are derived from the preimplantation embryo, and like EG cells, can be cultivated *in vitro* in their undifferentiated state on feeder layers (mouse embryonic fibroblasts or SIM [Sandoz inbred Swiss mouse] thioguanine-resistant ouabain-resistant cells), or addition of a differentiation inhibitor factor, that is, leukemia inhibitory factor (LIF). They can self-renew without limit, maintaining a relatively stable karyotype. Upon reimplantation into a host blastocyst, they are capable of pluripotent differentiation, including gonad tissue, producing chimeras from which transgenic lineages with homologous recombination of specific genes can be selected.

Murine ES cells (mES) are the most widely studied ES cells, as there are a wide variety of mouse homologous recombination techniques. In the early nineties, Wobus *et al.*, demonstrated that pluripotent mES cells are capable of differentiation into beating CMCs. This group also observed that CMCs are in the initial aggregate called embryoid bodies (EBs), which include a wide range of other specialized cell types.[12] The placement of small drops of ES cells into plastic dishes without any fibronectin resulted in the formation of EBs,[13] which are then dissociated and replated in fibronectin precoated dishes. Soon after plating, CMCs are easily identified as they undergo spontaneous contraction. The functional and ultrastructural characteristics of the terminally differentiated CMCs derived from ES cells are similar to those of neonatal rodent myocytes. Furthermore, the CMCs are capable of cell-to-cell communication due to their functionally coupled gap junctions. Their electrophysiological characteristics are typical of postnatal CMCs, and they possess pharmacological and physiological properties of specialized myocardial cells, including atrial, ventricular, Purkinje, and pacemaker cells.[14] The CMCs have normal contractile sensitivity to calcium, and many features of excitation–contraction (E-C) coupling have been found in isolated fetal or neonatal CMCs.[14] In contrast to early stage mES cell-derived CMCs, terminal CMCs were found to be responsive to β-adrenergic stimulation.[15]

In vitro, the yield of CMCs can be increased in comparison to other cell lineages by transfecting ES cells with an α-MHC promoter-driven aminoglycoside phosphotransferase, and further selection with G418.[16] G418 selected CMCs were capable of functional integration into the myocardium of adult mice.[16] Additional studies demonstrated the generation of gain-of-function phenotypes by electroporetic manipulation of mES cells, with different cardiac-restricted promoters.[14]

Zimmermann's and Eschenhagen's group (Zimmermann *et al.*)[17] has formed reconstituted engineered heart tissue (EHT) *in vitro* using ES cells from a mixture of cardiac myocytes from neonatal rats with liquid collagen type I, matrigel, and serum-containing culture medium. The contractile function of the EHT was normal.[17] These results suggest that ES cells have the capacity to produce both myocytes and myocardial tissue. *In vivo*, the EHT grafts could be used as a patch and function was maintained. The effort of this group is focused on obtaining similar results using ES cells.

Under appropriate *in vitro* conditions, ES-like cells obtained from neonatal mouse testis were induced to differentiate into a variety of somatic cell types, including CMCs. Furthermore, these ES-like cells formed teratomas when inoculated into mice.[18]

Human ES (hES) cell lines were derived from human blastocysts[19] and human EG cell lines[20] from primordial germ cells. Embryos frozen from *in vitro* fertilization procedures were a source of ES cells, and only a few cell lines were recently available. As an alternative, Doug Melton's group (Cowan *et al.*)[21] were able to generate 17 ES cell lines from frozen human blastocysts and share a manual for their use throughout the scientific community.

Like mES cells, hES cells also grow in culture in an undifferentiated state on a feeder layer of mouse embryonic fibroblast (MEF) cells. However, unlike mES cells, hES cells grow in the absence of LIF. Following removal of the cells from the MEF feeder layer and cultivation in suspension, the cells differentiate to form EBs. Spontaneously beating CMCs were generated from hES cells that were dissociated into small clumps of 3 to 20 cells and grown in suspension for 7 to 10 days, to allow the formation of EBs. Rhythmically contracting areas appeared in 8.1% of the EBs 4 to 22 days after plating them onto gelatin-coated culture dishes.[22] The hES cells, therefore, differentiate into CMCs *in vitro* at a slower rate compared to mES cells.[22]

The electrical activity, calcium transients, and chronotropic response to adrenergic agents of beating CMCs from contracting areas expressed cardiac-specific markers, and were similar to early-stage human CMCs. The cells formed a functional syncytium with synchronous action, with potential proliferation and pacemaker activity. CMCs derived from hES cells could form structural and electromechanical connections with cultured rat CMCs and were able to pace swine hearts after transplantation into a swine model of atrioventricular block.[23]

Several problems must be addressed for feasible clinical application of hES cells in cardiac regenerative medicine. First, the CMC yield needs radical improvement by optimizing culture conditions including the use of appropriate cytokines that favor CMC differentiation. Second, the expression of the major histocompatibility complex class I (MHC I) needs to be decreased so as to prevent ES cell rejection after transplantation, as differentiated ES cells express molecules of MHC I, with MHC II expression levels being low or absent.[24] In addition, hES cells grown on mouse feeder layer cells express an immunogenic nonhuman sialic acid Neu5Gc, complete elimination of which would likely require the use of human serum with human feeder layers, ideally starting with fresh hES cells that have never been exposed to animal products.[25] Furthermore, an increase in the efficiency of the differentiation from hES cells is required in order to verify the extent to which CMCs can be considered mature in terms of E-C coupling.

HEMATOPOIETIC STEM AND PROGENITOR CELLS

It has been demonstrated[26] that hematopoietic stem cells (HSCs) can significantly regenerate the infarcted myocardium. The stem cells were isolated from the mononuclear component of rat bone marrow (BM) using negative (immunodepleting cells expressing markers of differentiated hematopoietic lineages, including CD34, CD45, CD20, CD45RO, and CD8) and positive (*c-kit*, the receptor for stem cell factor, SCF) selection.[26] The cells came from male rats and thus could be identified for the Y chromosome (Y-Chr).

Orlic *et al.*[26] injected *c-kit*$^+$/Lin$^-$ bone marrow stem cells (BMSCs) into the myocardium of rats, which were able to repair 60% to 70% of the ischemia damaged tissue, by generating smooth muscle (SM), endothelial, and cardiomyocytic cells. Furthermore, injection of SCF and granulocyte colony stimulating factor (G-CSF) mobilized BMSCs to restore cardiac function following myocardial infarction (MI).[27] Another study conducted by Quaini *et al.*[28] confirmed the potential of BMSCs in cardiac repair. Heart biopsies of female organ donors in male recipients were analyzed a few weeks following heart transplantation. Tissue-specific markers were used to determine the type of stem cell differentiation, and the Y-Chr was used as a marker of cell origin.[28] More than 20% of cells containing the Y-Chr expressed markers of SM, endothelium, or CMCs, which would suggest that BMSCs have the ability to differentiate into all types of cardiac cells. In light of these results, several groups have attempted to obtain similar data with varying degrees of success, with figures for newly generated CMCs originating from BMSCs being much lower than 1%. Laflamme *et al.*[29] observed an approximate mean percentage of CMCs originating from the host of 0.04% with a median of 0.016%, and Y-Chr–positive CMCs were located in regions of acute rejection, indicating an injury event is involved in such chimerism.[29] The discrepancy between the two studies, of more than two log differences in Y-Chr–positive CMCs, was due to differences in tissue histology detection techniques. In a different study,[30] the mean percentage of Y-Chr–positive CMCs in female recipients of sex-mismatched bone marrow transplantation (BMT) was 0.23 ± 0.06%. The control group consisted of four female gender-matched BMT subjects. Fluorescence *in situ* hybridization (FISH) for the Y-Chr was used to identify cells of BM origin and was combined with immunofluorescent labeling for α-sarcomeric actin to identify CMCs. A further study examined autopsy samples taken from male control recipients of female donor hearts, with samples from patients who developed MI following transplantation. The data indicate MI enhanced the invasion of extracardiac progenitor cells that regenerated endothelial cells. However, no significant differentiation into CMCs was observed post ischemia in patients with sex-mismatched heart transplantation.[31] A more recent study revealed G-CSF activates the Jak-Stat pathway in CMCs to promote their survival following MI. It would therefore appear that G-CSF activation of the Jak-Stat pathway is responsible for its therapeutic potential after MI, as opposed to mobilization of BMSCs.[32]

A genetic approach with transgenic animals was used by Jackson *et al.*[33] to demonstrate the ability of *in vivo* BMSC differentiation into CMCs. A highly enriched population of HSCs termed side population (SP) cells were injected into immunocompromised mice that had been lethally irradiated, and subsequently subjected to coronary artery occlusion for 60 min rendering them ischemic, followed by reperfusion. BM cells from Rosa26 transgenic mice expressing lacZ were injected into the peri-infarctual area of the ischemic mice. Approximately 0.02% donor-derived CMCs, identified by expression of lacZ, α-actinin, and lack of expression of CD45 were located in the peri-infarct region. Endothelial engraftment of donor-derived endothelial cells of approximately 3.3% was mainly located in small vessels close to the infarct. The donor-derived cells were identified by expression of lacZ and Flt-1, an endothelial marker absent on SP cells. These results suggest BMSCs have a limited cardiomyogenic potential.

Our group has reported[34] that, *in vitro*, neonatal rat CMCs have the ability to induce freshly isolated embryonic endothelial cells of the aorta-gonad-mesonephros (AGM) region, or those established as homogeneous cells in culture, to transdifferentiate into CMCs and express cardiac markers. The concept of transdifferentiation of BM cells into CMCs has been challenged based on results using the Cre-lox recombination system, which suggested that fusion between cells as opposed to differentiation could account for the difficulty in interpretation of data using BMSCs for myocardium regeneration following MI. *Cre* is a DNA recombinase that cleaves the DNA at specific palindromic sequences, called *lox* sites, with high efficiency. Mice possessing the *cre* gene driven by the ubiquitous cytomegalovirus promoter were crossed with mice having an allele in which the β-galactosidase (β-gal) gene is arranged in a way that once it is cleaved at the lox site by *cre*, it becomes active and the cells turn blue. BM cells from mice carrying the *cre* gene were injected into mice possessing *cre*-activatable, lox-containing β-gal gene. Only a few CMCs, neurons and hepatocytes turned blue, indicating β-gal activation due to fusion between cells of different genetic origin, as opposed to transdifferentiation.[35]

Two other groups used similar methods to address the concept of BM cell transdifferentiation into CMCs. Murry *et al.*[36] used mice carrying the β-gal gene under the α-MHC promoter, where cells turn blue only when the differentiation lineage is activated. No blue cells were observed following MI in α-MHC β-gal mice, suggesting HSCs were unable to trans-differentiate into CMCs. Balsam *et al.*[37] used a model of parabiosis between a mouse expressing green fluorescent protein (GFP) under a universal promoter and another with a wild-type (WT) genome. An infarct was induced in the WT mouse to determine whether GFP^+ BM cells from the donor mouse migrated to the peri-infarct area of the WT mouse, as ischemic tissues produce stem cell chemoattractants.[37] No GFP CMCs were observed, and the only GFP^+ cells were of hematopoietic lineage. The results of these two groups contradict the theory of BMSC trans-differentiation into CMCs.

However, BM cells also contain endothelial progenitor cells. In addition, it has been observed that the $CD34^+$ fraction of BM mononuclear cells (BM-MNCs) can be induced, both *in vivo* and *in vitro*, to differentiate into endothelial cells. Furthermore, infusion of BM-MNCs or expanded endothelial cells improved myocardial perfusion and viability through angiogenesis in murine models of acute ischemic damage.[38] Also, human cord blood $CD34^+$ progenitor cells had the capacity to induce angiogenesis in the ischemic myocardium of nonobese diabetic/severe combined immunodeficiency (NOD/SCID) mice, with only a few CMCs detected staining positive for human nuclei.[39] These cells also produce growth factors and have been shown to have an anti-apoptotic effect on CMCs *in vivo*. The cardiac therapeutic potential of BM cells could therefore be dependent on endothelial cell generation and growth factor synthesis.

MESENCHYMAL STEM AND PROGENITOR CELLS

Located within the BM are mesenchymal stem cells (MSCs) and progenitor cells, which are clonogenic nonhematopoietic stem/progenitor cells that have the capacity to differentiate into multiple mesoderm-type cell lineages (e.g., chondrocytes, osteoblasts, and endothelial cells. *In vitro*, they adhere to plastic and expand in tissue culture with a finite lifespan of 15 to 50 cell doublings.

Multipotent adult progenitor cells (MAPCs) are a relatively rare adherent stem-cell population that have been isolated from BM cells grown for several months in medium containing specific growth factors, such as platelet-derived growth factor and epidermal growth factor. *In vitro*, MAPCs have the capacity to form endodermally, mesodermally, and ectodermally derived cell types such as hepatocytes, endothelial cells, and neurons. Makino *et al.*[40] generated CMCs from murine BM stromal cells, *in vitro*. The stromal cells were immortalized and treated with 5-azacytidine, which induced the generation of spontaneously beating cells. In addition, human mesenchymal stem cells (hMSCs) from adult BM were able to differentiate into cardiomyocytes, when transplanted into the adult murine heart.[41]

Rat mesenchymal stem cells, genetically modified to overexpress the prosurvival gene Akt1, prevented remodeling and restored performance of the infarcted heart.[42] Nonhematopoietic MSCs, cardiomyogenic (CMG) cells, expressing enhanced green fluorescent protein (EGFP), were transplanted into the BM of lethally irradiated mice, MI was induced, and they were treated with G-CSF. The presence of EGFP(+) actinin(+) cells in the ischemic myocardium would indicate that CMG cells were mobilized and differentiated into CMCs. These results suggest that most of the BM-derived CMCs originate from MSCs.[43] Clonally expanded novel human BM-derived multipotent stem cells (hBMSCs), not belonging to any previously described BM-derived stem cell population, were expanded *in vitro* and generated CMCs and cells of all three germ layers in co-culture conditions.[44] Following MI, the transplantation of hBMSCs into the myocardium resulted in engraftment of transplanted cells, which exhibited co-localization with markers of CMC, SM cell, and endothelial cell (EC). Therefore, the hBMSCs differentiated into multiple lineages. Moreover, the hBMSC-transplanted hearts demonstrated upregulation of paracrine factors including angiogenic cytokines, anti-apoptotic factors, and proliferation of host ECs and CMCs.[44]

LOCAL CARDIAC STEM CELLS

The existence of resident cardiac "stem" cells that have the capacity to replicate and differentiate into CMCs has been demonstrated by several groups. There exists a population of cells in BM, muscle, and skin that exclude Hoechst dye, giving a characteristic appearance in fluorescence-activated cell sorting (FACS),[45] known as the side population (SP). These cells have a restricted ability to differentiate into striated skeletal myoblasts. Although rare, similar cardiac SP cells have been identified.[46]

Schneider's group (Oh *et al.*)[47] has identified a population of cells with SP characteristics that express stem cell antigen-1 ([sca-1]$^+$ but *c-kit*$^-$).[47] Initially, these cells do not express cardiac specific genes, nor Nkx2.5. However, *in vitro* they differentiate in response to 5′-azacytidine, a process that is partly dependent on Bmpr1a, a receptor for bone morphogenetic proteins. The administration of these cells following MI resulted in the generation of CMCs from an equal contribution of differentiation and fusion as demonstrated using the Cre-Lox technology. The cells were shown to "home" spontaneously to the injured myocardium, even when injected in the tail vein.[47]

Another resident population of cardiac cells exist that are similar to BMSCs.[48] They are negative for blood lineage markers CD34, CD45, CD20, CD45RO, and CD8 (Lin$^-$) and positive for *c-kit* (*c-kit*$^+$), the receptor for SCF. These cells are multipotent, propagate indefinitely *in vitro*, and can be induced to differentiate into endothelium, SM, and CMCs. In addition, *in vivo* they were shown to support myocardial regeneration following MI.

Undifferentiated adult cardiac stem cells have also been isolated from the human and murine heart, by the group of Giulio Cossu (Messina *et al.*).[49] The cells grow as self-adherent clusters termed "cardiospheres," they are clonogenic and express stem and endothelial progenitor cell antigens/markers.[49] The cells have demonstrated long-term self-renewal and could differentiate *in vitro*, following ectopic or orthotopic transplantation in SCID beige mice, into myocytes and vascular cells.

Furthermore, a subpopulation of cardiac progenitor cells have been isolated from the anterior pharynx, which express the homeobox gene *islet-1* (*isl1*), that is lost during CMC differentiation.[50] These postnatal *isl1*$^+$ cells have been isolated in an undifferentiated state, from the outflow tract, the atria, and the right ventricle, which develop from the "secondary heart field," of mature hearts of newborn rodents and humans.[51] These cells express Nkx2-5 and GATA4, but not Sca-1, CD31, or *c-kit*. The isolated progenitor cells demonstrate both self-renewal and maintain the ability to differentiate into CMCs when cultivated on a cardiac mesenchymal feeder layer. In addition, co-culture with neonatal myocytes showed the *isl1*$^+$ cells differentiated into a mature cardiac phenotype expressing myocytic markers (25%), with no evidence of cell fusion, demonstration of intact Ca^{2+} cycling, and the generation of action potentials. This would indicate that the cells are cardiomyoblasts, and using the Cre-Lox system it has been shown that these cells differentiate *in vitro* and *in vivo*.[51] To date, it is unknown if these cells exist in adults.

CONCLUSIONS

This article reviews the state of the art in stem cell research and their potential in cardiac therapeutics. As one might have noticed, the field is in turmoil, with some apparently contradictory reports. However, the excitement is high because new types of cells are being discovered every year. Obviously, the review is not complete. Although the genome has been sequenced, our current knowledge on the types of cells composing our body is still limited. Also, a number of aspects concerning clinical applications of stem cells have not been included here. Generally speaking, we can state that the road for a rational use of stem cells as a cure for arrhythmias and heart failure is long and difficult, but, as the Latin motto says, *ad astra per aspra*!

ACKNOWLEDGMENT

We thank Dr S. Hourshid for editorial assistance.

REFERENCES

1. Pasumarthi, K.B. & L.J. Field. 2002. Cardiomyocyte enrichment in differentiating ES cell cultures: strategies and applications. Methods Mol. Biol. **185:** 157–168.
2. Engel, F.B., L. Hauck, M. Boehm, *et al.* 2003. p21(CIP1) Controls proliferating cell nuclear antigen level in adult cardiomyocytes. Mol. Cell. Biol. **23:** 555–565.
3. von Harsdorf, R., L. Hauck, F. Mehrhof, *et al.* 1999. E2F-1 overexpression in cardiomyocytes induces downregulation of p21CIP1 and p27KIP1 and release of active cyclin-dependent kinases in the presence of insulin-like growth factor I. Circ. Res. **85:** 128–136.
4. Pasumarthi, K.B., H. Nakajima, H.O. Nakajima, *et al.* 2005. Targeted expression of cyclin D2 results in cardiomyocyte DNA synthesis and infarct regression in transgenic mice. Circ. Res. **96:** 110–118.
5. Steinhelper, M.E., N.A. Lanson, Jr., K.P. Dresdner, *et al.* 1990. Proliferation in vivo and in culture of differentiated adult atrial cardiomyocytes from transgenic mice. Am. J. Physiol. **259:** H1826–H1834.
6. Claycomb, W.C., N.A. Lanson, Jr., B.S. Stallworth, *et al.* 1998. HL-1 cells: a cardiac muscle cell line that contracts and retains phenotypic characteristics of the adult cardiomyocyte. Proc. Natl. Acad. Sci. U.S.A. **95:** 2979–2984.
7. Rybkin, I.I., D.W. Markham, Z. Yan, *et al.* 2003. Conditional expression of SV40 T-antigen in mouse cardiomyocytes facilitates an inducible switch from proliferation to differentiation. J. Biol. Chem. **278:** 15927–15934.
8. Kajstura, J., A. Leri, N. Finato, *et al.* 1998. Myocyte proliferation in end-stage cardiac failure in humans. Proc. Natl. Acad. Sci. U.S.A. **95:** 8801–8805.
9. Beltrami, A.P., K. Urbanek, J. Kajstura, *et al.* 2001. Evidence that human cardiac myocytes divide after myocardial infarction. New Engl. J. Med. **344:** 1750–1757.
10. Limana, F., K. Urbanek, S. Chimenti, *et al.* 2002. bcl-2 overexpression promotes myocyte proliferation. Proc. Natl. Acad. Sci. U.S.A. **99:** 6257–6262.
11. Reiss, K., W. Cheng, A. Ferber, *et al.* 1996. Overexpression of insulin-like growth factor-1 in the heart is coupled with myocyte proliferation in transgenic mice. Proc. Natl. Acad. Sci. U.S.A. **93:** 8630–8635.
12. Wobus, A.M., G. Wallukat & J. Hescheler. 1991. Pluripotent mouse embryonic stem cells are able to differentiate into cardiomyocytes expressing chronotropic responses to adrenergic and cholinergic agents and Ca^{2+} channel blockers. Differentiation **48:** 173–182.
13. Gepstein, L. 2002. Derivation and potential applications of human embryonic stem cells. Circ. Res. 91: 866–876.
14. Boheler, K.R., J. Czyz, D. Tweedie, *et al.* 2002. Differentiation of pluripotent embryonic stem cells into cardiomyocytes. Circ. Res. **91:** 189–201.
15. Maltsev, V.A., G.J. Ji, A.M. Wobus, *et al.* 1999. Establishment of beta-adrenergic modulation of L-type Ca^{2+} current in the early stages of cardiomyocyte development. Circ. Res. **84:** 136–145.

16. KLUG, M.G., M.H. SOONPAA, G.Y. KOH & L.J. FIELD. 1996. Genetically selected cardiomyocytes from differentiating embronic stem cells form stable intracardiac grafts. J Clin Invest. **98:** 216–224.
17. ZIMMERMANN, W.H., K. SCHNEIDERBANGER, P. SCHUBERT, *et al.* 2002. Tissue engineering of a differentiated cardiac muscle construct. Circ. Res. **90:** 223–230.
18. KANATSU-SHINOHARA, M., K. INOUE, J. LEE, *et al.* 2004. Generation of pluripotent stem cells from neonatal mouse testis. Cell **119:** 1001–1012.
19. THOMSON, J.A., J. ITSKOVITZ-ELDOR, S.S. SHAPIRO, *et al.* 1998. Embryonic stem cell lines derived from human blastocysts. Science **282:** 1145–1147.
20. SHAMBLOTT, M.J., J. AXELMAN, S. WANG, *et al.* 1998. Derivation of pluripotent stem cells from cultured human primordial germ cells. Proc. Natl. Acad. Sci. U.S.A. **95:** 13726–13731.
21. COWAN, C.A., I. KLIMANSKAYA, J. MCMAHON, *et al.* 2004. Derivation of embryonic stem-cell lines from human blastocysts. New Engl. J. Med. **350:** 1353–1356.
22. KEHAT, I., D. KENYAGIN-KARSENTI, M. SNIR, *et al.* 2001. Human embryonic stem cells can differentiate into myocytes with structural and functional properties of cardiomyocytes. J. Clin. Invest. **108:** 407–414.
23. KEHAT, I., L. KHIMOVICH, O. CASPI, *et al.* 2004. Electromechanical integration of cardiomyocytes derived from human embryonic stem cells. Nat. Biotechnol. **22:** 1282–1289.
24. DRUKKER, M., G. KATZ, A. URBACH, *et al.* 2002. Characterization of the expression of MHC proteins in human embryonic stem cells. Proc. Natl. Acad. Sci. U.S.A. **99:** 9864–9869.
25. MARTIN, M.J., A. MUOTRI, F. GAGE & A. VARKI. 2005. Human embryonic stem cells express an immunogenic nonhuman sialic acid. Nat. Med. **11:** 228–232.
26. ORLIC, D., J. KAJSTURA, S. CHIMENTI, *et al.* 2001. Bone marrow cells regenerate infarcted myocardium. Nature **410:** 701–705.
27. ORLIC, D., J. KAJSTURA, S. CHIMENTI, *et al.* 2001. Mobilized bone marrow cells repair the infarcted heart, improving function and survival. Proc. Natl. Acad. Sci. U.S.A. **98:** 10344–10349.
28. QUAINI, F., K. URBANEK, A.P. BELTRAMI, *et al.* 2002. Chimerism of the transplanted heart. New Engl. J. Med. **346:** 5–15.
29. LAFLAMME, M.A., D. MYERSON, J.E. SAFFITZ & C.E. MURRY. 2002. Evidence for cardiomyocyte repopulation by extracardiac progenitors in transplanted human hearts. Circ. Res. **90:** 634–640.
30. DEB, A., S. WANG, K.A. SKELDING, *et al.* 2003. Bone marrow-derived cardiomyocytes are present in adult human heart: a study of gender-mismatched bone marrow transplantation patients. Circulation **107:** 1247–1249.
31. HOCHT-ZEISBERG, E., H. KAHNERT, K. GUAN, *et al.* 2004. Cellular repopulation of myocardial infarction in patients with sex-mismatched heart transplantation. Eur. Heart J. **25:** 749–758.
32. HARADA, M., Y. QIN, H. TAKANO, *et al.* 2005. G-CSF prevents cardiac remodeling after myocardial infarction by activating the Jak-Stat pathway in cardiomyocytes. Nat. Med. **11:** 305–311.
33. JACKSON, K.A., S.M. MAJKA, H. WANG, *et al.* 2001. Regeneration of ischemic cardiac muscle and vascular endothelium by adult stem cells. J. Clin. Invest. **107:** 1395–1402.
34. CONDORELLI, G., U. BORELLO, L. DE ANGELIS, *et al.* 2001. Cardiomyocytes induce endothelial cells to trans-differentiate into cardiac muscle: implications for myocardium regeneration. Proc. Natl. Acad. Sci. U.S.A. **98:** 10733–10738.
35. ALVAREZ-DOLADO, M., R. PARDAL, J.M. GARCIA-VERDUGO, *et al.* 2003. Fusion of bone-marrow-derived cells with Purkinje neurons, cardiomyocytes and hepatocytes. Nature **425:** 968–973.
36. MURRY, C.E., M.H. SOONPAA, H. REINECKE, *et al.* 2004. Haematopoietic stem cells do not transdifferentiate into cardiac myocytes in myocardial infarcts. Nature **428:** 664–668.
37. BALSAM, L.B., A.J. WAGERS, J.L. CHRISTENSEN, *et al.* 2004. Haematopoietic stem cells adopt mature haematopoietic fates in ischaemic myocardium. Nature **428:** 668–673.

38. Isner, J.M. & T. Asahara. 1999. Angiogenesis and vasculogenesis as therapeutic strategies for postnatal neovascularization. J. Clin. Invest. **103:** 1231–1236.
39. Botta, R., E. Gao, G. Stassi, *et al*. 2004. Heart infarct in NOD-SCID mice: therapeutic vasculogenesis by transplantation of human CD34+ cells and low dose CD34+ KDR+ cells. FASEB J. **18:** 1392–1394.
40. Makino, S., K. Fukuda, S. Miyoshi, *et al*. 1999. Cardiomyocytes can be generated from marrow stromal cells in vitro. J. Clin. Invest. **103:** 697–705.
41. Toma, C., M.F. Pittenger, K.S. Cahill, *et al*. 2002. Human mesenchymal stem cells differentiate to a cardiomyocyte phenotype in the adult murine heart. Circulation **105:** 93–98.
42. Mangi, A.A., N. Noiseux, D. Kong, *et al*. 2003. Mesenchymal stem cells modified with Akt prevent remodeling and restore performance of infarcted hearts. Nat. Med. **9:** 1195–1201.
43. Kawada, H., J. Fujita, K. Kinjo, *et al*. 2004. Nonhematopoietic mesenchymal stem cells can be mobilized and differentiate into cardiomyocytes after myocardial infarction. Blood **104:** 3581–3587.
44. Yoon, Y.S., A. Wecker, L. Heyd, *et al*. 2005. Clonally expanded novel multipotent stem cells from human bone marrow regenerate myocardium after myocardial infarction. J. Clin. Invest. **115:** 326–338.
45. Gussoni, E., Y. Soneoka, C.D. Strickland, *et al*. 1999. Dystrophin expression in the mdx mouse restored by stem cell transplantation. Nature **401:** 390–394.
46. Martin, C.M., A.P. Meeson, S.M. Robertson, *et al*. 2004. Persistent expression of the ATP-binding cassette transporter, Abcg2, identifies cardiac SP cells in the developing and adult heart. Dev. Biol. **265:** 262–275.
47. Oh, H., S.B. Bradfute, T.D. Gallardo, *et al*. 2003. Cardiac progenitor cells from adult myocardium: homing, differentiation, and fusion after infarction. Proc. Natl. Acad. Sci. U.S.A. **100:** 12313–12318.
48. Beltrami, A.P., L. Barlucchi, D. Torella, *et al*. 2003. Adult cardiac stem cells are multipotent and support myocardial regeneration. Cell **114:** 763–776.
49. Messina, E., L. De Angelis, G. Frati, *et al*. 2004. Isolation and expansion of adult cardiac stem cells from human and murine heart. Circ. Res. **95:** 911–921.
50. Cai, C.L., X. Liang, Y. Shi, *et al*. 2003. Isl1 identifies a cardiac progenitor population that proliferates prior to differentiation and contributes a majority of cells to the heart. Dev. Cell **5:** 877–889.
51. Laugwitz, K.L., A. Moretti, J. Lam, *et al*. 2005. Postnatal isl1+ cardioblasts enter fully differentiated cardiomyocyte lineages. Nature **433:** 647–653.

Extracellular Stimulation in Tissue Engineering

DROR SELIKTAR

Faculty of Biomedical Engineering, Technion, Israel Institute of Technology, Haifa, Israel

ABSTRACT: The field of tissue engineering has created a need for biomaterials that are capable of providing biofunctional and structural support for living cells outside of the body. Most of the commonly used biomaterials in tissue engineering are designed based on their physicochemical properties, thus achieving precise control over mechanical strength, compliance, porosity, and degradation kinetics. Biofunctional signals are added to the scaffold by tethering, immobilizing, or supplementing biofunctional macromolecules, such as growth factors, directly to the scaffold material. The challenge in tissue engineering remains to find the correct balance between the biofunctional and the physical properties of the scaffold material for each application. Moreover, the ability to modulate communication between cells and the extracellular environment using the engineered scaffold as the actuator can provide a significant advantage in tissue engineering. In this study, a unique scaffold material is presented. The material interchangeably combines biofunctional and structural molecules by fusing the two into a single backbone macromolecule. This integration provides the basis for practical, effective, and high-resolution control of both the biofunctional and the physical properties of the scaffold material. This new scaffold material has proven effective with smooth muscle, cardiac, cartilage, and human embryonic stem cell cultures. The advantages of this approach as well as the potential applications of this unique scaffold material are discussed.

KEYWORDS: biofunctional; fibrin; polyethylene glycol; scaffold; structure

INTRODUCTION

The allure of creating living tissues in the laboratory for organ replacement has spawned countless attempts to use laboratory-grown cells to develop 3-D tissue analogs.[1] The establishment of material technologies designed specifically for this task is a true hallmark in this field.[1] The approaches to engineering tissues *in vitro* are now varied and plentiful; selecting the right biomaterial, or scaffold, is as important as choosing which type of cells to use. In fact, choosing a cell type seems easy in comparison to choosing the proper biomaterial from the vast inventory of scaffolds in use today. This choice is particularly complicated by the fact that most biomaterials are

Address for correspondence: Dror Seliktar, Ph.D., Faculty of Biomedical Engineering, Technion, Israel Institute of Technology, Haifa 32000, Israel. Voice: (+972) 4-829-4805; fax: (+972) 4-829-4599. dror@bm.technion.ac.il

Ann. N.Y. Acad. Sci. 1047: 386–394 (2005). © 2005 New York Academy of Sciences.
doi: 10.1196/annals.1341.034

capable of communicating with cells in ways that are yet unknown, and may be the primary reason for so many unpredictable final outcomes in tissue engineering.

Virtually all scaffold materials in tissue engineering belong to one of three categories: natural or biological materials, synthetic materials, and biosynthetic composite or hybrid materials.[2] Natural materials include proteins (e.g., collagen and fibrin) that provide abundant biological cell signaling domains built into the protein backbone. Some of these domains are characterized, but biofunctionality of most domains in these proteins remain elusive and may introduce uncertainty in the regulated communication between the scaffold and the cells. Moreover, the inherent structural characteristics of biological materials are not easily altered, thus limiting control over physical cues received by the cells within. Synthetic scaffolds are created mostly from biologically inactive materials that are easily manipulated, making them ideal for regulating the physical interactions between cells and biomaterial at the biomaterial–tissue interphase.[3] Synthetic scaffolds may inadvertently participate in random biological signaling caused by adsorption of serum proteins, for example, and subsequent activation of cells through contact with these proteins. Biosynthetic composite materials are a complementary blend of natural and synthetic materials designed to regulate between the physical and biological cell signaling of the scaffold. The challenge in tissue engineering is to identify the correct balance between the biofunctional and the physical properties of the material for each application.

A BIOSYNTHETIC HYBRID SCAFFOLD

A new era in scaffold design has emerged with the advent of biosynthetic hybrid materials for tissue regeneration. These sophisticated materials incorporate biologically functional macromolecules directly onto the backbone of synthetic polymers so as to balance between structure and function. High-resolution control over structural properties, including porosity, compliance, and density are directed through the synthetic polymer network.[4,5] Biological cell signaling is controlled through the incorporated cell signaling domains, which may include protein fragments,[6] growth factors,[7–9] or even biologically active peptide sequences.[10,11] When resident cells encounter these biological factors, they initiate important remodeling events, including cell migration, proliferation, and guided differentiation. The structural properties of the scaffold must not impede upon the cells' remodeling ability.

In tissue engineering, the objective of using biosynthetic hybrid materials is to reproduce the intricate extracellular stimulation of the native cellular environment. Already, there are numerous examples of biosynthetic hybrid materials successfully implemented as tissue engineering scaffolds that selectively regulate cellular behavior. For example, Luo and Shoichet recently created agarose hydrogels and modified them with the adhesive fibronectin peptide fragment Gly-Arg-Gly-Asp-Ser (GRGDS) for guided cell migration and neurite outgrowth.[12] Kapur and Shoichet biochemically engineered synthetic polymer scaffolds with immobilized concentration gradients of nerve growth factor to facilitate directional neurite outgrowth.[13] The distinguishing feature in both examples is the ability to coordinate the spatial features of the scaffold with the desired cellular response (i.e., neurite extension). Hubbell and co-workers use a polyethylene glycol (PEG) hydrogel backbone that is cross-linked with proteolytically sensitive oligopeptides,[11,14] providing cells the freedom to move about in the highly dense synthetic PEG network with the aid of their own naturally secreted proteases. Here, one sees how inert synthetic polymers can be transformed into responsive scaffolds for cellular remodeling. There are, of course, many other relevant examples that demonstrate the refined strategies of creating "communicating" scaffolds for tissue engineering. Many of these approaches utilize the design concept of taking a biologically inactive material, adding one or more bioactive elements, and transforming it into a biomimetic extracellular environment for tissue engineering.

A HYBRID PROTEIN-BASED SCAFFOLD

Programming synthetic materials with multiple bioactive domains can be an involved process.[2] An alternative approach to creating hybrid materials is to use whole proteins and modify them with "structural" domains made from either synthetic polymers[15,16] or synthetic cross-linkers.[17] The protein already contains a set of biofunctional cell signaling domains that are suitable for a specific task (i.e., tissue remodeling). In comparison with other hybrid materials, the modified protein materials contain a complete set of bioactive domains, with no way of turning off any undesired extracellular stimulation. Nevertheless, if the choice of the protein is appropriate for the application, one can achieve excellent results. In our laboratory, we use a protein molecule called fibrinogen to create a hybrid hydrogel for tissue regeneration.[16,18] Fibrinogen is the precursor to the commonly known blood clot protein, fibrin. The fibrinogen molecule contains a set of biological cell-signaling cues specific to cellular remodeling, including the cell-adhesive sequence Arg-Gly-Asp (RGD) and a protease degradation substrate.[19] Before the fibrinogen molecules are transformed into a hydrogel scaffold, we enhance the structural versatility of fibrinogen using a common biomedical polymer called PEG. The PEG molecules are covalently attached to the 29 cysteine residues present in the sequence of the fibrinogen protein, and the PEG-modified fibrinogen (PEGylated fibrinogen) is then formed into a solid polymer network by cross-linking between the free functional groups on the PEG (FIG. 1). The result is a structurally versatile hydrogel network with pre-programmed biofunctionality (based on the fibrinogen backbone) and malleable structural characteristics (based on the PEG molecules).

When creating a hybrid scaffold for tissue engineering, the advantages of using PEG and fibrinogen are evident; the fibrinogen is a natural provisional matrix for wound healing and tissue remodeling, and the highly hydrophilic PEG is transparent to cells.[20] One can therefore alter the physical properties of the scaffold material by changing the characteristics of the PEG molecules, without affecting the overall biofunctionality of the fibrin-like scaffold. Using this material for cell and tissue culture reveals its remarkable versatility. Cellular outgrowth experiments demonstrate the invasion ability of different cell types as they penetrate into the surrounding dense matrix using secreted proteases[18] (FIG. 2).

To date, we have used the PEGylated fibrinogen scaffold for culturing smooth muscle cells (SMCs),[16] cartilage cells, dorsal root ganglion cells, cardiac myocytes, and human embryonic stem (unpublished data). We observed cell migration, proliferation, and differentiation in the cultures we tested. However, considering the success of reconstituted fibrin as a tissue-engineering scaffold, it should be no surprise

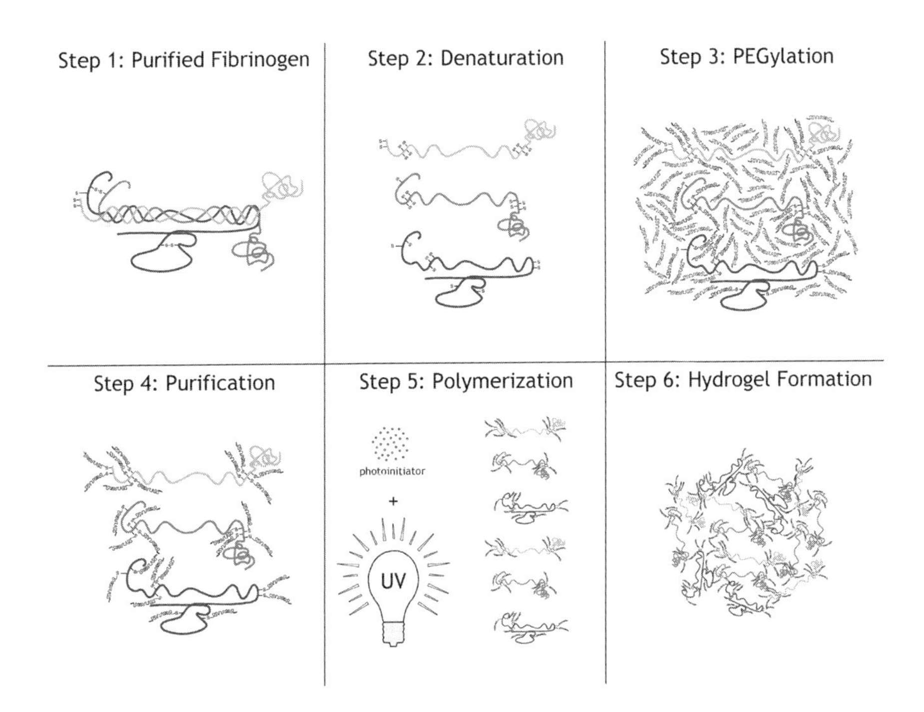

FIGURE 1. Creating a hybrid PEG-protein hydrogel scaffold. (**Step 1**) Purified fibrinogen is dissolved in buffered saline solution containing urea. (**Step 2**) The disulfide bonds in the fibrinogen are reduced using a reducing agent; the fibrinogen is dissociated into the alpha, beta, and gamma fragments of the molecule, leaving reactive thiols (SH) exposed. (**Step 3**) The fibrinogen fragments are reacted in a high molar excess of di-functional polyethylene glycol (PEG-di-acrylate); the PEG-di-acrylate self-selectively reacts with free thiols on the fibrinogen molecule by Michael-type addition. (**Step 4**) The PEGylated fibrinogen molecules are purified from the excess PEG-di-acrylate by acetone precipitation and dialysis. (**Step 5**) A solution of PEGylated fibrinogen is exposed to UV light and photoinitiator to initiate a free-radical polymerization between the unreacted acrylates on the PEG. (**Step 6**) The photopolymerization process results in a solid hydrogel network [in color in Annals Online].

that these cells are capable of thriving within a fibrin-like culture environment. In fact, the true advantage of using these hybrid materials is not their biofunctional versatility but their structural versatility.

CONTROLLING THE PHYSICAL ENVIRONMENT

In the native extracellular environment, the physical cues received by cells via the extracellular matrix (ECM) are vital to proper cellular function. Despite this, scaffold structure has generally been viewed as being passively involved in the regeneration process in many tissue-engineering applications. The major focus in scaffold design has been on creating a material that provides temporary structural support (at the tissue level) during the earliest stages of the development process and degrades

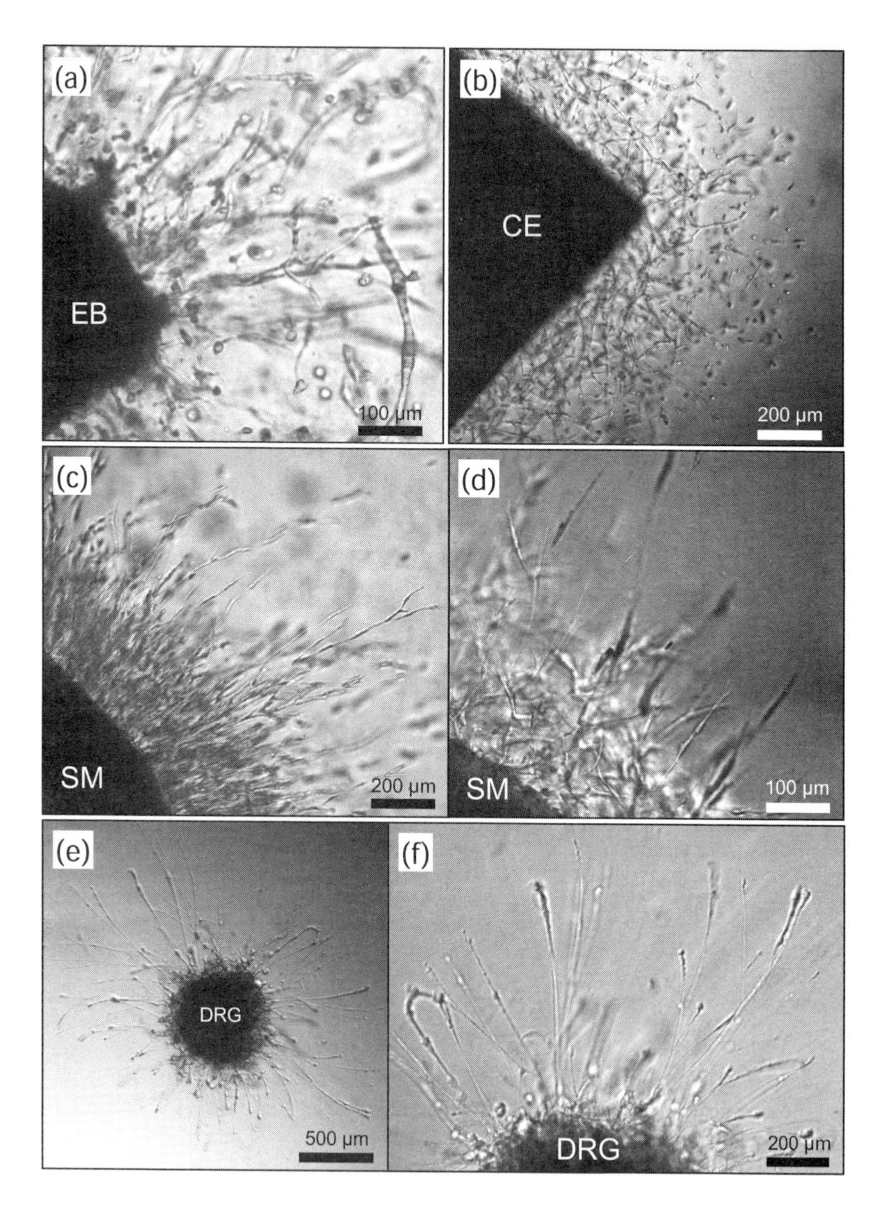

FIGURE 2. Cellular invasion into PEGylated fibrinogen hydrogels. Each image shows an *in vitro* culture of a three-dimensionally entrapped living, natural tissue explant within a dense PEGylated fibrinogen hydrogel matrix. (**a**) Spontaneously beating human embryoid bodies (EB) cultured for 8 days in the hydrogels continue beating and demonstrate 3-D embryonic stem cell sprouting from the EB into the dense gel matrix. (**b**) Cartilage explants (CE) cultured for 14 days exhibit massive 3-D cell migration from the native tissue into the encapsulating gel matrix. (**c**,**d**) Smooth muscle cells migrate out of the dense smooth muscle tissue (SM) into the gel matrix. (**e**,**f**) Dorsal root ganglion (DRG) inside the gel matrix after 24 hours in culture exhibit 3-D neurite outgrowth.

as it is replaced with natural ECM proteins. With the advent of biosynthetic hybrid scaffolds for tissue engineering, the task of manipulating the physical properties of the scaffold in order to regulate cellular behavior without compromising biofunctionality is considerably simplified. Now, scaffold structure can be actively involved in guiding cellular events associated with tissue remodeling, including cell migration,[21,22] cell morphology,[23] and even cellular phenotypic expression.[24]

The ability to independently change physical extracellular cues and biofunctionality in a hybrid material provides much greater control over how cells behave in a tissue-engineered construct. In the context of a PEG-fibrinogen scaffold, the structural features of the material are easily altered by changing the PEG constituent.[16,25] We have found that by decreasing the molecular size of the PEG, for example, from 10 kDa to 6 kDa, the PEG-fibrinogen scaffold is less susceptible to proteolytic degradation.[18] Moreover, PEGylated fibrinogen hydrogels made with excess PEG were structurally denser and more resistant to proteolytic degradation. The experimental results with SMCs evince the profound impact of hydrogel structure on the ability of cells to proteolytically penetrate the scaffold and form cellular processes (FIG. 3). Evidently, the structural changes that results from the use of excess PEG in combi-

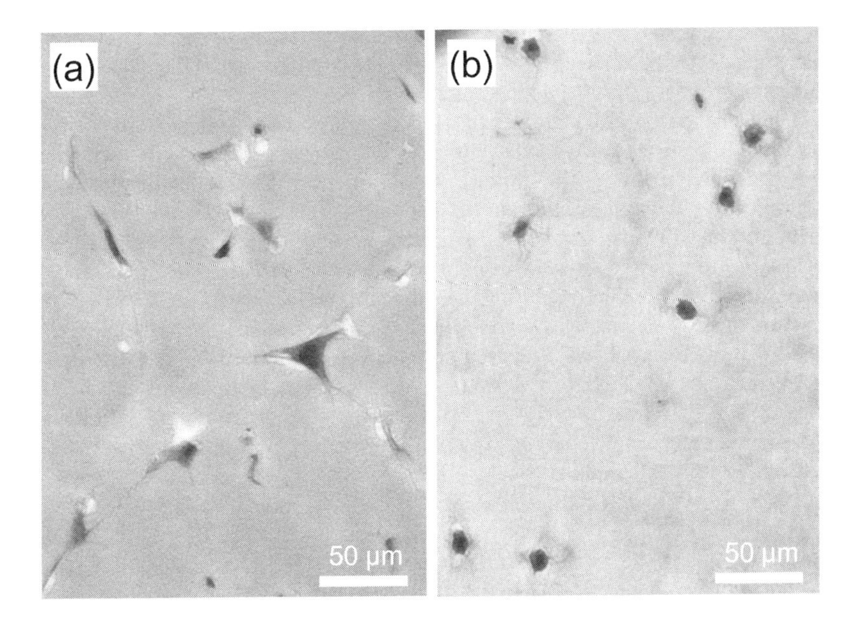

FIGURE 3. Cell spreading within modified PEGylated fibrinogen hydrogels. Cross-sections of smooth muscle cell-seeded hydrogels stained with hematoxylin and eosin. (**a**) Unmodified cell-seeded hydrogels show spindled cell morphology and cellular extensions protruding three-dimensionally into the gel network. (**b**) Structural modification rendered by the addition of 1.5% free PEG-di-acrylate to the PEGylated fibrinogen matrix increases the resistance of the matrix to proteolytic degradation and limits the cellular protrusions. The additional PEG-di-acrylate, therefore, has a profound effect on cell morphology, despite the presence of the biofunctional fibrinogen backbone of the hydrogel network [in color in Annals Online].

nation with the inherently biofunctional fibrinogen backbone, alters the morphological cues for cell spreading (FIG. 3b).

HYBRID MATERIAL FOR OTHER APPLICATIONS

Hybrid biomaterials have not only revolutionized the design of tissue-engineered scaffolds but have also provided alternative approaches to regenerating damaged tissues in the body without cell therapy. For example, damaged organs that have undergone massive tissue loss can be treated with an artificial matrix that promotes tissue regeneration.[2] This matrix can be created from a hybrid biomaterial for direct implantation into the defect site. The hybrid biomaterial functions as a provisional healing substrate that can recruit cells and initiate tissue remodeling.[26] As the body's cells are recruited, they invade the provisional hybrid matrix and replace it with new healthy tissue. This approach has been demonstrated very effectively by several studies,[27,28] including work in our laboratory.[29] We use the hybrid PEGylated fibrinogen hydrogels as a provisional matrix having the biological characteristics of a native blood clot (i.e., fibrin properties) with controllable ultrastructural properties based on the PEG composition. The hydrogel functions as a biomimetic fibrin clot with the added feature of being able to regulate the rate of invasion of cells from the defect site periphery. In the absence of cellular invasion, the hybrid hydrogels remain stable for several months inside the body.

We have tested the PEGylated fibrinogen materials in rat tibias with a critical size defect caused by osteotomy, 7-mm in length.[29] The defect site was filled with a precast hydrogel having the optimum ratio of fibrinogen and PEG to facilitate new bone formation in this type of injury. We have found that the rate of cellular invasion is very important to the healing characteristics: hydrogels that promote rapid cellular invasion are replaced with scar tissue, whereas hydrogels that delay the invasion using excess PEG remain intact for longer durations and consequently promote the formation of new bone tissue. Using an optimal concentration of PEG and fibrinogen in the hydrogel network, we observed complete restoration of the defect with new bone formation after 5 weeks.[29] The results were verified by histology and X-ray of the defect site. Interestingly, hydrogels containing an abundance of PEG constituent remain in the defected site and are only partially invaded by a thin layer of new bone surrounding the implanted hydrogel.

CONCLUSIONS

Interactive or communicating scaffolds made from hybrid biomaterials are fast becoming a standard in tissue engineering and regenerative medicine. These materials present the opportunity to communicate with cells through interactions between cell surface receptors and biological signaling domains or through structural features of the matrix. The choice of synthetic and biological precursors that make up the matrix, as well as the delicate balance between the two, allows custom tailoring of the extracellular stimulation for each clinical or scientific application. Our work with an interactive hybrid material made from fibrinogen and PEG shows evidence that this material can perform both as a cell-signaling scaffold for tissue engineering, as well as a biomimetic-signaling matrix for *in vivo* tissue regeneration. The regulated physical signals received by resident cells provide the means for controlling cell morphology, migration, and differentiation. We anticipate that further refinement of the PEGylated fibrinogen platform technology will enable us to successfully implement this hybrid material for various challenging tissue-engineering applications.

ACKNOWLEDGMENTS

The author gratefully acknowledges the financial support of the Israel Science Foundation, the Israel Ministry of Health, and the Technion V.P.R. research fund; as well as the help of Prof. Lior Gepstein and Dr. Amira Gepstein with embryonic stem cell experiments.

REFERENCES

1. HUBBELL, J.A. 1995. Biomaterials in tissue engineering. Bio/Technology **13:** 565–576.
2. LUTOLF, M.P. & J.A. HUBBELL. 2005. Synthetic biomaterials as instructive extracellular microenvironments for morphogenesis in tissue engineering. Nat. Biotechnol. **23:** 47–55.
3. PEPPAS, N.A. & R. LANGER. 1994. New challenges in biomaterials. Science **263:** 1715–1720.
4. PEPPAS, N.A., Y. HUANG, M. TORRES-LUGO, *et al.* 2000. Physicochemical foundations and structural design of hydrogels in medicine and biology. Annu. Rev. Biomed. Eng. **2:** 9–29.
5. TSANG, V.L. & S.N. BHATIA. 2004. Three-dimensional tissue fabrication. Adv. Drug Deliv. Rev. **56:** 1635–1647.
6. CUTLER, S.M. & A.J. GARCIA. 2003. Engineering cell adhesive surfaces that direct integrin alpha5beta1 binding using a recombinant fragment of fibronectin. Biomaterials **24:** 1759–1770.
7. SELIKTAR, D., A.H. ZISCH, M.P. LUTOLF, *et al.* 2004. MMP-2 sensitive, VEGF-bearing bioactive hydrogels for promotion of vascular healing. J. Biomed. Mater. Res. **68A:** 704–716.
8. ZISCH, A.H., M.P. LUTOLF, M. Ehrbar, *et al.* 2003. Cell-demanded release of VEGF from synthetic, biointeractive cell ingrowth matrices for vascularized tissue growth. FASEB J. **17:** 2260–2262.
9. DELONG, S.A., J.J. MOON & J.L. WEST. 2005. Covalently immobilized gradients of bFGF on hydrogel scaffolds for directed cell migration. Biomaterials **26:** 3227–3234.
10. MANN, B.K., A.S. GOBIN, A.T. TSAI, *et al.* 2001. Smooth muscle cell growth in photopolymerized hydrogels with cell adhesive and proteolytically degradable domains: synthetic ECM analogs for tissue engineering. Biomaterials **22:** 3045–3051.
11. LUTOLF, M.P., J.L. LAUER-FIELDS, H.G. SCHMOEKEL, *et al.* 2003. Synthetic matrix metalloproteinase-sensitive hydrogels for the conduction of tissue regeneration: engineering cell-invasion characteristics. Proc. Natl. Acad. Sci. U.S.A. **100:** 5413–5418.
12. LUO, Y. & M.S. SHOICHET. 2004. A photolabile hydrogel for guided three-dimensional cell growth and migration. Nat. Mater. **3:** 249–253.
13. KAPUR, T.A. & M.S. SHOICHET. 2004. Immobilized concentration gradients of nerve growth factor guide neurite outgrowth. J. Biomed. Mater. Res. A. **68:** 235–243.
14. PRATT, A.B., F.E. WEBER, H.G. SCHMOEKEL, *et al.* 2004. Synthetic extracellular matrices for in situ tissue engineering. Biotechnol. Bioeng. **86:** 27–36.
15. HALSTENBERG, S., A. PANITCH, S. RIZZI, *et al.* 2002. Biologically engineered protein-graft-poly(ethylene glycol) hydrogels: a cell adhesive and plasmin-degradable biosynthetic material for tissue repair. Biomacromolecules **3:** 710–723.
16. ALMANY, L. & D. SELIKTAR. 2005. Biosynthetic hydrogel scaffolds made from fibrinogen and polyethylene glycol for 3D cell cultures. Biomaterials **26:** 2467–2477.

17. BRINKMAN, W.T., K. NAGAPUDI, B.S. THOMAS & E.L. CHAIKOF. 2003. Photo-cross-linking of type I collagen gels in the presence of smooth muscle cells: mechanical properties, cell viability, and function. Biomacromolecules **4:** 890–895.
18. SELIKTAR, D. 2005. Nano-structured 3-D biomaterials for cellular stimulation in tissue engineering. Proceedings of the 4th Congress of the Federation of the Israeli Societies for Experimental Biology–FISEB, February 7–10, 2005, Eilat, Israel.
19. HERRICK, S., O. BLANC-BRUDE, A. GRAY & G. LAURENT. 1999. Fibrinogen. Int. J. Biochem. Cell Biol. **31:** 741–746.
20. MERRILL, E.A. & E.W. SALZMAN. 1983. Polyethylene oxide as a biomaterial. ASAIO J. **6:** 60–64.
21. DUBEY, N., P.C. LETOURNEAU & R.T. TRANQUILLO. 1999. Guided neurite elongation and schwann cell invasion into magnetically aligned collagen in simulated peripheral nerve regeneration. Exp. Neurol. **158:** 338–350.
22. CEBALLOS, D., X. NAVARRO, N. DUBEY, *et al.* 1999. Magnetically aligned collagen gel filling a collagen nerve guide improves peripheral nerve regeneration. Exp. Neurol. **158:** 290–300.
23. TRANQUILLO, R.T., T.S. GIRTON, B.A. BROMBEREK, *et al.* 1996. Magnetically orientated tissue-equivalent tubes: application to a circumferentially orientated media-equivalent. Biomaterials **17:** 349–357.
24. KOFIDIS, T., L. BALSAM, J. DE BRUIN & R.C. ROBBINS. 2004. Distinct cell-to-fiber junctions are critical for the establishment of cardiotypical phenotype in a 3D bioartificial environment. Med. Eng. Phys. **26:** 157–163.
25. TEMENOFF, J.S., K.A. ATHANASIOU, R.G. LEBARON & A.G. MIKOS. 2002. Effect of poly(ethylene glycol) molecular weight on tensile and swelling properties of oligo(poly(ethylene glycol) fumarate) hydrogels for cartilage tissue engineering. J. Biomed. Mater. Res. **59:** 429–437.
26. HUBBELL, J.A. 2003. Materials as morphogenetic guides in tissue engineering. Curr. Opin. Biotechnol. **14:** 551–558.
27. LUTOLF, M.P., F.E. WEBER, H.G. SCHMOEKEL, *et al.* 2003. Repair of bone defects using synthetic mimetics of collagenous extracellular matrices. Nat. Biotechnol. **21:** 513–518.
28. SCHMOEKEL, H., J.C. SCHENSE, F.E. WEBER, *et al.* 2004. Bone healing in the rat and dog with nonglycosylated BMP-2 demonstrating low solubility in fibrin matrices. J. Orthop. Res. **22:** 376–381.
29. PELED, E., M. LIVNAT, C. ZINMAN, *et al.* 2005. A biosynthetic matrix alone promotes bone regeneration in a critical size tibia defect. Proceedings of Regenerate 2005, June 1–3, 2005, Atlanta, GA.

Multiscale Modeling of Cardiac Cellular Energetics

JAMES B. BASSINGTHWAIGHTE, HOWARD J. CHIZECK,
LES E. ATLAS, AND HONG QIAN

*Department of Bioengineering, University of Washington,
Seattle, Washington 98195, USA*

ABSTRACT: Multiscale modeling is essential to integrating knowledge of human physiology starting from genomics, molecular biology, and the environment through the levels of cells, tissues, and organs all the way to integrated systems behavior. The lowest levels concern biophysical and biochemical events. The higher levels of organization in tissues, organs, and organism are complex, representing the dynamically varying behavior of billions of cells interacting together. Models integrating cellular events into tissue and organ behavior are forced to resort to simplifications to minimize computational complexity, thus reducing the model's ability to respond correctly to dynamic changes in external conditions. Adjustments at protein and gene regulatory levels shortchange the simplified higher-level representations. Our cell primitive is composed of a set of subcellular modules, each defining an intracellular function (action potential, tricarboxylic acid cycle, oxidative phosphorylation, glycolysis, calcium cycling, contraction, etc.), composing what we call the "eternal cell," which assumes that there is neither proteolysis nor protein synthesis. Within the modules are elements describing each particular component (i.e., enzymatic reactions of assorted types, transporters, ionic channels, binding sites, etc.). Cell subregions are stirred tanks, linked by diffusional or transporter-mediated exchange. The modeling uses ordinary differential equations rather than stochastic or partial differential equations. This basic model is regarded as a primitive upon which to build models encompassing gene regulation, signaling, and long-term adaptations in structure and function. During simulation, simpler forms of the model are used, when possible, to reduce computation. However, when this results in error, the more complex and detailed modules and elements need to be employed to improve model realism. The processes of error recognition and of mapping between different levels of model form complexity are challenging but are essential for successful modeling of large-scale systems in reasonable time. Currently there is to this end no established methodology from computational sciences.

KEYWORDS: cardiac metabolic systems modeling; constraint-based analysis; energetics; multicellular tissues; oxidative phosphorylation

Address for correspondence: Prof. James B. Bassingthwaighte, M.D., Ph.D., Bioengineering Department, University of Washington, Box 357962, Seattle, WA 98195-7962, USA. Voice: 206-685-2012; fax: 206-685-3300.
jbb@bioeng.washington.edu

**Ann. N.Y. Acad. Sci. 1047: 395–426 (2005). © 2005 New York Academy of Sciences.
doi: 10.1196/annals.1341.035**

INTRODUCTION

Multiscale modeling of human physiology serves to integrate knowledge from genomics and molecular biology, through the levels of cell tissue and organ modeling to integrated systems behavior. Biochemical system models need to be expressed by fully detailed dynamic equations, rather than in matrix forms for steady states. They must account for the thermodynamics, meaning that every reaction is reversible and obeys Haldane constraints, and they must account for appropriate balances of mass, energy, constituent fluxes, including the solute mass bound to enzymes. Tissue models are composed of these cellular models with appropriate anatomic measures of volumes and distances. Reduced forms are in use for steady-state tracer analysis of images from positron emission tomography and magnetic resonance imaging. Physiological modeling, however, requires handling transients in response to external changes, and reductions to steady-state kinetics would not be satisfactory in general.

Models of cell-blood solute exchanges and reactions were necessarily simplified to be computed fast enough to aid in devising hypotheses and designing critical experiments. Reduced-form models must be parameterized to reproduce the response functions of the full model for any given element or module—an optimization that results in a descriptor that is correct over only a limited range of operation. Each successive reduction degrades adaptability. For long-term time-dependent solutions to changes in state or external conditions, the reduced models err. Error correction requires resorting to the more complex and detailed forms of the modules and elements. The process of recognition—going down the hierarchy of model forms and back up—is difficult, prone to error, and is time consuming, but it is essential for modeling large-scale systems in reasonable time. Algorithms for this are still primitive.

MODELING THE RESPONSE TO EXERCISE: AN EXAMPLE

One useful example is the physiological response to exercise, showing rather tight regulation of blood pressure in spite of large changes in body metabolism, cardiac output, and vascular resistance. In general, in the adult, there is a nested hierarchical control, the lowest level being the regulation of transcription and the generation and degradation of proteins. The next levels concern the biophysics and biochemistry of cells and tissues, which is where we start our cell-to-organ-to-cardiorespiratory system modeling, defined in FIGURE 1. The higher levels of organization, tissues, organs, and organism levels are increasingly complex, representing the dynamically varying behavior of the 2 billion cells of the body operating together.

Currently, computational models integrating cellular events into tissue and organ behavior utilize simplifications of the basic biophysical and biochemical models in order to reduce the complexity to practical computable levels. This compromises the ability of the models to respond correctly to dynamic changes in external inputs, for these normally require adjustments in rates at the biophysical, biochemical, and gene regulatory levels.

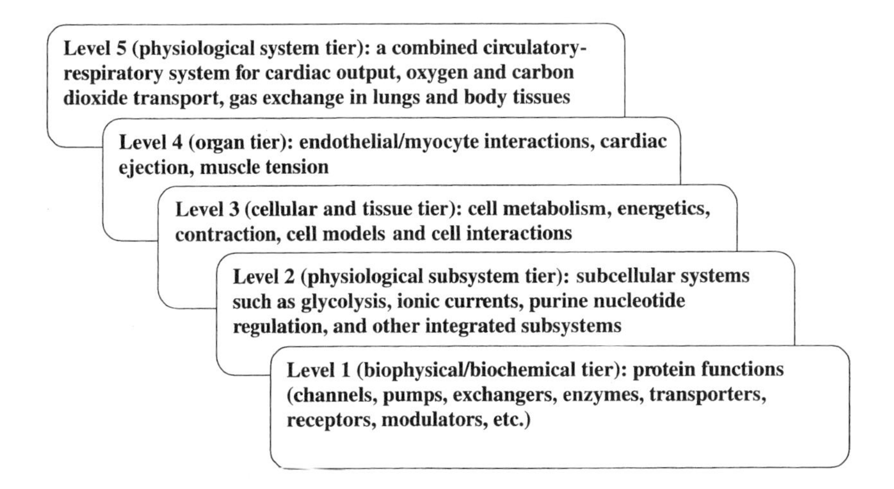

FIGURE 1. Multiscale cardiovascular/respiratory system description.

CONTROLLING COMPUTATIONAL PROCESSES: MODELING BASED ON CONTINUOUS MONITORING AND DYNAMIC CONTROL CONCEPTS

This is a novel approach to carrying out simulations, with continuous monitoring and dynamic control of the computing carried out at subsidiary modules in a multiscale model system. The idea is to consider a computational simulation process no different from other complex engineering processes in Boeing or DuPont. The procedure is based on established methodologies from systems and control engineering, but even so is not definable as a recipe. This is a work in progress, and many others will contribute to advancing this technology. A key difference between simulations in physical sciences and in physiology is that the physiological processes are gradually being discovered and remain incompletely known, while a physical process, such as building an airplane, is composed of exactly defined parts. However, in both cases a common principal goal is to determine integrative behavior. The risk of error in physiological modeling is great, simply because each hypothesized descriptor for a biological function serves not only as the "output" of one computational module but is the "input" of another. Error propagates. The adaptive approach to be described strives to control the complex computational processes in a simulation, but it is not yet clear that any one completely general strategy can be prescribed.

CELLULAR TO SYSTEMS MODELING

This integrative system approach to multiscale modeling extends from four complementary areas of research: cardiovascular physiology and medicine, systems optimization and parameterization, signal analysis and decision control, and molecular and cellular biophysics and mathematical biology. These can be applied together to formulate computationally feasible multiscale system descriptions, allowing adaptation to interventions or to changing external driving forces. The approach encompasses four principles of modern, "open" science: (1) it uses new computational methods spanning multiple scales (regulation of transcription, cellular energetic metabolism, cell-to-cell interactions, integrated organ contractile function); (2) it bridges current levels of modeling from the level of cardiovascular and respiratory system responses to exercise, trauma, and activity with the cellular levels of receptor signaling, cellular metabolic dynamics, and the thermodynamics of energy transformation and transduction, and provides a framework for guiding research into the regulation of transcription in muscular tissues (e.g., cardiac remodeling, training in athletes); (3) it results in highly integrated models of cardiac and skeletal muscle systems showing adaptive responses to lowered or raised oxygen and substrate supply, changed demand for contractile work, sympathetic and parasympathetic neuronal input levels, and encompasses the remarkable heterogeneity of normal regional flows and metabolic function that exists in both heart and skeletal muscles; (4) it will produce models for research and education when operated under an open source philosophy. The results *should* be models with manuals and tutorials, source code, a modeling system for their operation, a Web site for their dissemination, and truly full publication, allowing complete reproducibility of the models by others. This is entirely in tune with the developing emphasis on full cooperativity in scientific research.

An *adaptive simulation methodology* appears to be more or less essential to furthering multiscale modeling of complex physiological systems during the foreseeable future, simply because computing resources are not infinite. It is vital to compute "at the speed of thought," that is, to display model solutions so fast that they are useful as "mind-expanders," projecting the results of "what if?", and the queries exploring the behavior of the system. Consequently, models must be composed of modules of reduced complexity: one does not build a body out of atoms or a truck out of quarks. The term "adaptive" has two implications: In the biological context, it means that the physiological responses must be adaptive to the changing environment of a cell or an organ, and therefore it means that any computational model should be able to deal with this challenge. In the engineering context, our approach is to use adaptive simulation technology, wherein the multiscale simulation is treated as a complex computing process—not very different from a chemical factory or an aircraft plant with an engineering control approach applied to it.

The overall strategy summarized in TABLE 1 can be organized into three sets of tasks:

(1) Steps 1, 2, and 3 are to define a fully detailed multiscale model of the cardiomyocyte and its interactions with smooth muscle cells and endothelial cells, including the signaling pathways, and composition of this overall model out of nested elements and modules. This includes describing the models mathematically, writing and verifying the computer code, and then validating the modules and the system model against high-quality data for all components at each level of this five-tiered model, using the full set of equations for each component at each level.

(2) Steps 4, 5, and 6 are developing sets of hierarchically related reduced-form models, capturing the varied essences of the fully developed models and parameterizing them for specific conditions by optimizing their behavior

TABLE 1. Strategy for developing a practical multiscale model

1. Define the detailed model over all of the levels.
2. Build it of modular elements for each of the levels.
3. Validate the model against multiple, diverse data sets.
4. Build reduced-form modules that are computationally faster.
5. Optimize their parameters to mimic the full-form modules.
6. Insert reduced-form models into the system model for a chosen set of conditions, recheck validity.
7. Define ranges of conditions for validity of the system model when composed of different combinations of reduced-form models.
8. When running systems models, detect violations of valid conditions, select and replace an inadequate module with either an alternative reduced-form module or with the full-form version.

against the fully developed equivalent modules or elements, and revalidating the models against multiple sets of experimental data obtained under diverse but defined experimental conditions, thereby defining the behavior of the fully developed system.

(3) Steps 7 and 8 focus on determining how to identify when automated model reduction is needed and automating the module replacements, thus adapting the systems model to changing states and using efficient computation in a variety of states. One can devise algorithms for detecting when the reduced model forms begin to lose accuracy because of changes in conditions (external or internal) and can implement mechanisms for either resorting to the use of the full subsidiary model or shifting to an alternative form of the reduced model. The goal is automated reconfiguration of the model when changes in conditions occur.

The identification relationships between subcellular module parameters and overall systems behavior—linking, for example, cardiac cellular energetics with contractile force development, beat-to-beat ejection of blood from the cardiac chambers with the regulation of blood pressure, and cardiac output in response to perturbations—will prove useful in defining conditions under which specific subsets of lower level modules might best be replaced simultaneously, each by an alternative reduced-form module. Defining such associations will provide computational strategies and technologies and their application to cardiac and muscle physiology.

The central product of such an effort provides examples of all of the techniques in a form that can be applied to different multiscale modeling applications, and it can be used by other investigators. The specific systems model of the cardiorespiratory system serves as vehicle for research and teaching and for the training of physicians and other medical personnel. The current version is limited to a higher level representation giving appropriate pressures and flows in the circulatory and respiratory systems, gas exchange including oxygen and carbon dioxide binding to hemoglobin, bicarbonate buffering in the blood, H+ production in the tissues, control of the respiratory quotient (RQ), baroreceptor and chemoreceptor feedback control of heart rate (HR), respiratory rate, and peripheral resistance. This is a large-scale model that summarizes the information from key studies done over the past century and adapts to changing conditions in a limited way. It is available now for public use as the model CVRESP (at http://nsr.bioeng.washington.edu/software/models/) and it serves as a basis not merely for the understanding of physiology, but for pathophysiological processes and their consequences. It is not yet a good framework for pharmacokinetics and pharmacodynamics, as the tissue processes are not described broadly enough to handle blood-tissue transcapillary exchange and cellular metabolism.

The next stage is the linking of a representative cell model with the overall circulatory system model in order to characterize the relationships between tissue events (varying requirements for blood flow, generation of metabolites, substrate exchanges). While this can be done at a fairly simple level for handling a few solutes at a time, as we have done using axially distributed capillary-tissue exchange models accounting for gradients along capillaries,[1,2] these models are themselves computationally demanding when used in their fully expanded forms, accounting for intra-organ flow heterogeneity and for exchange, metabolism, and binding in red blood cells, plasma, endothelial cells, interstitium, and the organ's parenchymal cells, requiring up to the equivalent of 80,000 differential equations.

CELL LEVEL MODELING BEGINNING WITH THE "ETERNAL CELL"

To characterize cellular support for a high-level, adaptive model[3] of the contracting heart, one must relate metabolic processes to contractile performance. To do this requires a basic cell unit accounting for metabolism, ionic regulation, and excitation–contraction (E-C) coupling, as well as contractile force development. It thus must be a more comprehensive model than the "eternal cell" model[4] that accounts only for energetics and ionic balance. While a useful cell-level model can lack elements on regulation of transcription and translation to form protein, and lack inflammatory and immune response systems, it will be a basic fundamental structure upon which such features may be added.

A basic cell model adequate to serve as a component of a contracting heart (level 3) contains cellular metabolic networks, oxidative phosphorylation and the redox system, response systems for adrenergic stimuli and insulin, stretch-activated calcium fluxes and their modulation of metabolic fluxes, contractile force development in accord with heterogeneities of electrical activation times and sarcomere lengths. Control of contractile performance is defined by influences external to the cardiomyocyte, namely levels of sympathetic and parasympathetic stimulation and circulating hormones, and cardiac filling pressures (preload), the arterial pressure against which the heart ejects blood (after load), the HR, and the peripheral vascular resistances. Integration of these events into a whole organ model accounting for regional heterogeneities in flow[5,6] and contractile stress[7,8] gives rise to level 4 organ models with patterns of regional flows and oxygen consumption rates that are fractals in space but are stable over time. When put into the context of a fully developed three-dimensional finite element heart with all the correct fiber and sheet directions, one should be able to discern whether the spatial heterogeneity of local flows is governed by the local needs for energy for contractile force development. Because all organs demonstrate marked flow heterogeneity, what is learned from cardiac muscle can be applied to other organs.

Several hundred published models relate to our efforts in composing a five-level model with full sets of equations—some in texts[9–12] and most as articles. At the cardiorespiratory system level, the classic model of Guyton, Coleman, and Granger[13] was the pioneer. Of the many recent models at that level, there are several articles by Clark's group (e.g., Lu *et al.*)[14] that illustrate a coherent and extensive body of work. Exemplary models—meaning carefully documented, published with full sets of equations, parameters, and initial conditions, and reproducible—are remarkably few, but include Hodgkin and Huxley,[15] Beeler and Reuter,[16] Winslow *et al.*,[17–19] Noble,[20] Rideout,[12] and Sherman,[21] some of which can be found at our Web site (http://nsr.bioeng.washington.edu) along with many subcellular level, transport, and electrophysiological models.

Cellular tissue and organ level models will serve as the vehicles for systems integration[22] of knowledge coming from molecular, cellular, and organ level observations of many sorts, including mRNA arrays or array sequences, gas chromatography/mass spectrometry (GCMS), metabolomic arrays, nuclear magnetic resonance (NMR), proteomic arrays,[23] gene regulation, electrophysiology, NMR phosphoenergetics, membrane potential, ion regulation, and so on. While the "eternal cell" lacks the key ingredients for gene regulation, it is precisely positioned to accept recent developments in the synthesis of signaling networks and to carry the developments in specific genetic variants, as in the studies by Saucerman *et al.*[24,25] The subsequent phase is the joining of signaling and genetic regulatory networks.[26] Beyond the metabolic biochemical network, the signaling network (i.e., ligand-receptor binding, guanosine triphosphatase (GTPase) activation, intracellular phosphorylations) and genetic regulatory networks (gene transcription initiation, activations, repression, etc.) have been extensively studied in recent years. In recent work, Palsson and his colleagues have developed interesting approaches to signaling and regulatory networks from the constraint-based approach that integrates these models.[23,27,28] In our own work we have an ideal preparation for studying the regulation of transcription of contractile proteins, namely the chronically paced heart of ambulatory dogs, in which one gets hypertrophy in predictable regions of increased work and atrophy in regions of diminished work.[29,30] The "eternal cell" model, by providing a basic energetically and metabolically based picture, is in our view a key step to augmenting the powers of genomic and molecular biology. It is also a key to determining how genetic and cellular interventions can affect the physiology of the whole body (which is a long way beyond this project).

MULTISCALE MODELING OF CARDIAC PERFORMANCE

Constructing a multiscale model is currently being done by brute force, putting all the pieces into a full "reference" model (Step 2 of Table 1). This is doomed to be computationally slow. The models are to be used both as "mind-expanders" in exploring systems behavior and providing insight and prediction, and as data analysis tools, so it is vital "to compute at the speed of thought." Thus, in aggregating large sets of models at a lower tier to form the next higher tier model, one must reduce the complexity of the individual modules in order to minimize the increase in computation time.

To meet the objectives of Step 3 of Table 1, traditional methods of optimization and parameter identification are being used for the fitting of computationally simpler, reduced forms of the mathematical models, for components at each level are fitted to the fully developed forms. For each component subsystem at each of the five levels of modeling, one or several reduced models are being developed (the "several" indicating that differing levels of approximation are used). Over a five-level hierarchical system, one must use successive levels of reduction, coalescing each time a few lower-level modules into a computationally more efficient composite model. The big problem with this approach is that each up-the-scale reduction limits the range of operational validity of the composite model.

How to determine when one should modify a reduced submodel or resort to the full submodel is important, is difficult, and is the part of any multiscale code that one would like to minimize, since it does not contribute directly to obtaining a good model solution. The tricky aspect of this is that when one is using external driving conditions or changes in internal conditions over time to govern the model solution, then decisions concerning the adequacy of the algorithms need to be made as the solution progresses. New research into devising algorithms is needed for detecting when the reduced model forms begin to lose accuracy in representing the physiology because of changes in external conditions or the shifting of the internal conditions, causing a reduced-form module to move out of its local range of validity. An adjustment requires either making changes in the parameters of the reduced-form module or resorting to an alternative reduced-form module or, as a last resort, reincorporating the full-form module into the computation. The basic decision control here is experiential, as it uses prior determination of the ranges of validity for detection of the need to use an alternative computation.

An alternative approach to avoiding errors due to model reduction is to use signal analysis to detect changes in the character of solution variable values that indicate lessening accuracy or deviation from a range of validity. "Lessening accuracy" implies failure in numerical methods for obtaining the solution. "Deviation from validity" implies that the solution variables show a change in a quantifiable characteristic that is not the variable value itself, but a change in the frequency or power content of a variable or set of variables that is associated with the overall systems model pushing the limits of validity. A detection method might also be based on system stability (e.g., in terms of Lyapunov exponents for subsystems).

It is essential to develop strategies for efficient computation of adaptive multiscale models in order to apply them to solving a major problem in the simulation of multilevel simulations in biology, while retaining adaptability during the computing of whole system responses to changes in activity and environment. Extending this concept within the context of "open source science," it would be reasonable to project combining of all of these algorithms into a well-documented software package to provide to the user community. This has been a central aspect of our Simulation Resource facility, but even there we have found it difficult to bring all the research models up to a level fit for public distribution. More than 90% of the effort put into cleaning up, protecting against incorrect parameterization, validating against data, archiving, and documenting follows after proving out an adequate research-level model. This is an expensive and time-consuming process that is difficult to support financially.

MULTISCALE MODEL DEVELOPMENT AND COMPUTATIONAL CONTROL

Development of Biological Models

The multiscale cardiac and cardiovascular-respiratory system model brings together many transport models[1,9,31–50] with work by us on cardiac cell modeling[3,4,51–62] and by many others,[7,8,17,18,25,63–68] in models incorporating both cardiac and endothelial cell uptake and metabolism;[69–75] and for accounting for regional flow heterogeneities.[5,6,37,76–80] Many of these are available at our Web site.

Transport models have been developed using both analytic and numerical methods of solution. The more detailed ones account for both passive diffusional transport and carrier-facilitated transport across capillary endothelial and parenchymal cell membranes, intracellular and extracellular binding, the different velocities of erythrocytes and plasma, concentration gradients between inflow and outflow, and enzymatically facilitated reactions in all cell types and extracellular regions.[1,2,31,36,47,59,81,82] One general blood-tissue exchange model, GENTEX, accounts for most of these processes (for up to seven solute species simultaneously), and it is written in finite element form. Its code has the virtue that all processes not used by a specific solute are automatically removed from the computation, with the capability of being used as a one-compartment model (single exponential impulse response) or as a fully distributed model that accounts for flow heterogeneity for all species using up to a maximum of 80,000 differential equations.[83] This model is used for analyzing image sequence data from PET (positron emission tomography), obtaining parametric images (e.g., regional blood flows, regional oxygen consumption, regional receptor densities). For the parameter optimization, under our simulation system XSIM we have done the computation on a supercomputer and displayed the results on a laptop or desktop in one's office; this is being implemented within our simulation system, JSim.

Our "eternal cell" model,[4,57] a virtual or computational cell model, is "eternal" in that it assumes there is no proteolysis and no protein synthesis. The initial configuration centers on energetics, fatty acid and glucose metabolism, the tricarboxylic acid cycle, the electrophysiology of time-dependent and voltage-dependent ionic currents, the ionic pumps and exchangers, the calcium cycling and the processes of calcium release and reuptake by the sarcoplasmic reticulum during E-C coupling, purine nucleoside and nucleotide metabolism in myocytes and endothelial cells, oxygen and CO_2 metabolism, and mitochondrial oxidative phosphorylation.

Reaction schemes for biochemical reactions in energy metabolism, biosynthesis, and transport, etc., are well-developed components of the biochemical networks (see TABLE 2). The constraints on the system are physicochemical: balances of mass, charge, osmolarity and volume, oxidoreductive state, and energy balance. The cell system is computed as microcompartmentalized sets of (usually ordinary) differential and algebraic equations. The modules are running under JSim. Some components are incomplete (e.g., the ion exchangers and the Ca pump). For many reactions the equilibrium is known, providing exact thermodynamic (Haldane) constraints.[84] A completely configured model should be ready for distribution in 2005 and will then become a component of the multiscale cardiovascular respiratory systems model. While most of the enzymes involved in these reactions are well characterized by reduced-form Michaelis-Menten-type equations, these account only

TABLE 2. Stages of complexity and realism in modeling metabolic networks

- Stoichiometric matrices → flux balance
- Thermodynamically-constrained steady-state fluxes
- Mass-conservation as a further constraint
- Substrate buffering and dynamics of transients
- Redox: reducing group accounting
- pH and buffering
- Specific mass balances (e.g., P, purine, O_2, N)
- Charge balance, Donnan equation, and membrane potential
- Osmotic and water balance

for fluxes and do not give an exact accounting for mass balance, since they omit the amount of substrate bound to enzyme and must therefore (in most cases) be augmented to account for the additional mass. This is essential for analyzing tracer experiments.

Model Integration from Biophysical Chemistry to Cellular Functions

At the biochemical network level, we use a *thermodynamic based flux-balance analysis of steady-state metabolic networks*. This approach is capable of predicting metabolic flux of large networks with more than hundreds of reactions and biochemical species. On the whole-cell level, steady-state metabolic network models have been extensively characterized via flux balance analysis and constraint-based optimization.[85] Recently this approach was augmented by adding thermodynamic constraints, which establishes a conduit between standard metabolic control analysis (MCA) and biochemical kinetic modeling,[86] not yet taken up by others.[87] More specifically, the constraint-based modeling with thermodynamics is able to provide estimates of the steady-state concentrations of metabolites, as well as rate constants in terms of biochemical conductances, and it provides the flux estimates with much narrower constraints than does simple stoichiometric flux balance analysis. The difference in the equations used is not great, the essential feature being to account for the reversibility of each reaction that is implicit in having an equilibrium condition: the Haldane constraint (FIGURE 2).

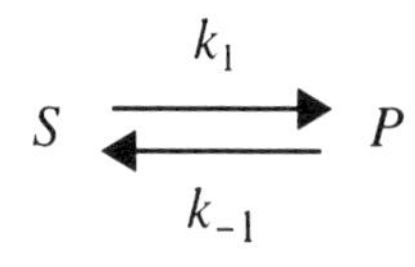

FIGURE 2. The equilibrium condition for the conversion of Substrate (S) to Product (P) is independent of the conditions for the conversion. The ratio of the forward to the backward rate is the Haldane constraint or equilibrium condition, determined by the free energies of S and P.

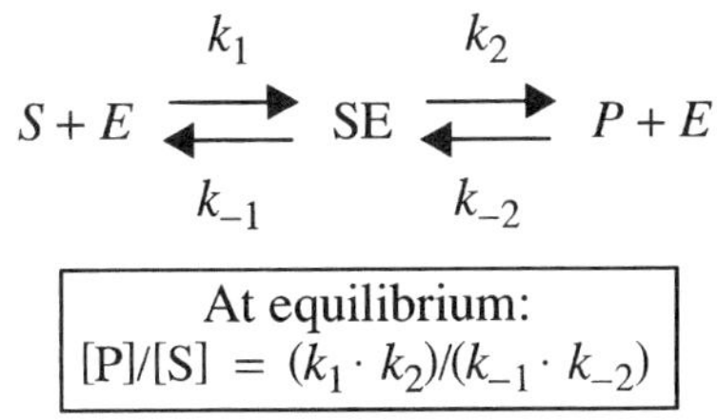

$$S + E \underset{k_{-1}}{\overset{k_1}{\rightleftarrows}} \mathrm{SE} \underset{k_{-2}}{\overset{k_2}{\rightleftarrows}} P + E$$

At equilibrium:
$[\mathrm{P}]/[\mathrm{S}] = (k_1 \cdot k_2)/(k_{-1} \cdot k_{-2})$

FIGURE 3. Haldane conditions with catalysis. As in FIGURE 2, thermodynamic balance is defined at equilibrium. The ratio $(k_1 \cdot k_2)/(k_{-1} \cdot k_{-2}) = K_P/K_S = [\mathrm{P}]/[\mathrm{S}]$.

The existence of a catalyst for the reaction has the life-providing benefit of lowering the energy of activation for the reaction, increasing the probability of reaction often by a million to a billion times, but it does not affect the free energies of either substrate or product (as is shown in FIGURE 3.)

Given that all reactions are reversible, our view of the equations for fluxes needs to be matched to this concept. No longer can we think of an enzyme-facilitated reaction as being only forward $S \rightarrow P$, but the backward reaction must be incorporated into the calculation. Thus, instead of $J_{S\rightarrow P} = (V_{max} \cdot S)/(K_m + S)$, where $J_{S\rightarrow P}$ is the forward flux in moles/s, we calculate a net forward flux. The general expression is

$$J_{S\rightarrow P} = 1 - \frac{[\mathrm{P}]/[\mathrm{S}]}{K_P/K_S} \cdot V_{\mathrm{forward}} \cdot \frac{[\mathrm{S}]/K_S}{1 + [\mathrm{S}]/K_S + [\mathrm{P}]/K_P},$$

where V_{forward} is the maximal forward reaction flux at high [S] and zero [P], K_S is k_{-1}/k_1, and K_P is k_2/k_{-2}. The term in brackets accounts for the reverse reaction rate in accord with the Haldane conditions and therefore accounts for the free energies and the reversibility of the reaction.

While the flux in a metabolic network depends on the network structure, the input and output, as well as the level of gene expression (enzyme activity), the biochemical conductance through a given reaction is related to the rate constants for the reaction, which include the forward flux term, V_{forward}. This is the product of two inseparable terms: the total enzyme concentration times the forward reaction rate to form product. Thus, the flux is dependent on the rate of expression of the protein minus its rate of proteolysis, hence the level of gene expression. The proteolytic rate may be variable: many enzymes are degraded more slowly when substrate is bound, protecting them.[88]

Considerations of *kinetic models for cellular signaling and subsequent phosphorylation-dephosphorylation cycles with GTPase activation* has implications not only for energy utilization but also for the accuracy, efficiency, and robustness of the cell's biological functions through its switches and signal relays.[84,89,90] These studies, based on kinetic models of a protein network, revealed cooperativity different from the traditional allosterism and revealed that sharp responses (high cooperativity) can be achieved through monomeric protein-protein interaction using cellular energy derived from the hydrolysis of adenosine triphosphate (ATP) or GTP.

Cellular biochemical reactions often involve species with low copy numbers. Hence, a realistic model for intracellular signaling network requires stochastic treatment of the biochemical reactions. Stochastic kinetics has become increasingly important in systems biology.[91,92] Qian's recent work[84,93,94] revealed that chemical oscillation is also intimately related to the energy input to the reactions. With the utilization of energy, stochasticity of the biochemical reactions in cell can be suppressed, and more precise temporal behavior is achieved. Different computational methods for carrying out stochastic simulations to improve computational speed have also been studied.

Sigg, Qian, and Bezanilla[95] developed a biophysical model of the thermodynamics and kinetics of voltage-dependent channels based on insight from the field of protein folding. The model is able to fit a wide range of experimental data from K- and Na-channels: in particular, the model fits both the fast and slow kinetics across several orders of time scale and may introduce a new approach to describing channel kinetics to contrast with the traditional Markovian models, just as Qian's stochastic model for heterogeneous flow distribution in vascular networks contrasts with continuous flow spatial forms.[80]

These examples serve to show the level of detail that is required to develop models that are mathematically and scientifically valid. The devil is in the details, and the consequence of simplification is that it is difficult to maintain either mathematical accuracy or scientific validity through successive reductions. This argument would oppose using model reduction to gain speed, but the other side of the coin is that models will not, almost cannot, be used unless the results can be seen rapidly. The trade-offs are summarized in TABLE 3.

Building the Fully Detailed Multiscale Model

The first stage is to define the structure and content of a cardio-respiratory-skeletal muscle system for oxygen exchange and substrate delivery model at five levels of scale, defining components from the biophysical/biochemical level through to the fifth, the system level, using modular elements in a hierarchical fashion. Modules at each level need to be verified (for correct computation), validated (against experimental data), completely defined and documented (for reproducibility), and made publicly available for download on a simulation platform that runs under Linux, Unix, Windows, and Macintosh operating systems. The models and subsidiary mod-

TABLE 3. Multiscale modeling: up the down staircase and down the up staircase

- Successive simplifications of a module are used at successively higher hierarchical levels. Reduced forms must be parameterized and optimized to match the behavior of the original more realistic module.
- Lower level adaptive features are gradually lost in successive reductions for speed.
- The range of conditions for validity inevitably narrows with reduction.
- When conditions change too much, the model is no longer valid.
- Automated screening for invalidity is required in order to go back to the more complex forms, or to select another version of the simplified form that is valid under the conditions.
- Tests for violation of valid conditions must run continuously, and be individualized for many component modules.
- The cost of evaluation compromises the computational speed, a trade-off.

ules provide solutions that give the time courses of the variables and the responses to perturbations such as changes in blood pressure, HR, respiratory minute volume, cardiac contractility, cellular substrate usage, and mitochondrial membrane potential, in response to a demand for increased blood flow to the body such as during exercise. Our most recent models are available at http://nsr.bioeng.washington.edu/PLN/models/.

Many lower-level modules have been developed and validated. For example, we are using, in a cellular level model not yet available on our Web site, the NaK ATPase model of Chapman *et al.*,[96] for which Kootsey provided his code in ScOP (an ordinary differential equation [ODE] based language), translated into the simple equation-based mathematical modeling language of JSim, adding the membrane potential influences. This was embedded in a composite model of cardiac cell electrophysiology built upon that of Rudy's group,[97] Winslow *et al.*,[17–19] and Michailova and McCulloch.[66] The NaK ATPase model was validated against the data provided by Skou[98,99] and compared to the data provided by Lauger[11] and to the solutions of the Post-Albers model, an alternative NaK ATPase model that Lauger described fully. On incorporation into the cell, the parameters for numbers of transporters were adjusted to give appropriate balances for Na and K in the whole cardiac cell beating at 1/s.

An example of a desired multilevel model is one for oxygen transport and consumption in the heart. In papers over the past twenty years there has been debate concerning the length and cause of a delay in the venous oxygen consumption response to an increase in metabolic rate, as occurs following an increase in HR or mitochondrial decoupling. The delay has been attributed to transport time through the blood, myoglobin buffering of O_2, slow adenosine diphosphate (ADP) generation, or slow mitochondrial cytochrome oxidase activation. These are all probable causes, but their relative influences should be apportioned via a "complete" mathematical model incorporating the following elements going from the whole organ down to the biochemical/biophysical level:

(1) A whole-organ, multicapillary, blood-tissue exchange model with RBC, plasma (and plasmatic gaps between RBC), endothelial cells, interstitial fluid space, parenchymal cells and mitochondria, accounting for gradients in concentrations of O_2, CO_2, pH and substrate concentrations along the length of the capillary. Diffusional resistances at the membranes are known to be small for H_2O, O_2, and CO_2, as determined using ^{15}O-labelled compounds.[78,100,101] In fact, these resistances are so small that the exchange of these compounds is essentially flow-limited at even high physiological flows. This puts the onus on the other mechanisms to account for delay greater than the transit time through the system, namely their tissue volumes of distribution divided by their flows of solute-containing fluid.

(2) Oxygen binding to hemoglobin (Hb) and the influences of CO_2, pH, temperature, and oxygen partial pressure on the Hb saturation,[81] and for intracytosolic myoglobin in cardiomyocytes that provides facilitated diffusion as well as buffering storage.[101]

(3) Intracytosolic and intramitochondrial purine nucleosides and nucleotides in endothelial cell and cardiomyocytes using the open system configuration demonstrated by Kroll *et al.*[102] and the transmitochondrial membrane system for ADP/ATP and nicotinamide adenine dinucleotide (NAD) and its

ratio to NADH (NADH is the reduced form of NAD with a hydrogen donator), and the proton-driven ATP formation.[4]

(4) The ubiquinone-cytochrome system for the transfer of reducing groups and reduction of O_2 to form water. Here we account for the observations[103] that the apparent affinity of cytochrome oxidase for O_2 depends upon the energy state (i.e., lowering the phosphorylation potential lowers the dissociation constant at cytochrome c oxidase),[104] while NADH levels are affected by not only the energy state but also by the activation of mitochondrial dehydrogenases and the level of cytochrome c oxidation (the ratio c2+/c3+).[105]

This four-level model becomes a five-level model at the whole contracting heart level when intramural gradients in intratissue pressures and subendocardial to subepicardial flows are considered, especially at reduced perfusion pressures where subendocardial ischemia comes into play. The current contracting heart models of the investigators account for this compression and flow reduction but do not extend it to the compromising of cellular metabolism preferentially in the subendocardium.

The cellular level model, the "eternal cell," a useful element in itself, covers three levels, from biophysical/biochemical models to cellular models. Major pieces already exist, due to the work of the past decades by many investigators. Cell models (myocyte, smooth muscle cell, endothelial cell) all interact in governing and in being influenced by interstitial fluid (ISF) concentrations. For example, in hypoxia where ATP is released from myocytes, the extracellular hydrolysis of ATP to adenosine 5′-monophosphate (AMP) is due to the ISF alkaline phosphatase, which is abundant at the arteriolar end of capillaries, not at the venous end. This localization requires using axially distributed models[2,31,35–37,39,42,71,72,74,83,100] for representation of the convection-diffusion-reaction events. Next, the enzyme ecto-5′-nuclotidase, like its intracellular analog, hydrolyzes AMP to adenosine, which in turn relaxes smooth muscle by binding to a receptor or is rapidly taken up by endothelial and cardiac cells or degraded to inosine.[2,59] While we hypothesized and accounted for the ecto-5′-nucleotidase for our modeling analysis,[2,59] only recently has the importance of the enzyme been thoroughly demonstrated.[106,107]

The *verification* of such models is difficult because there are no analytical solutions with which to make comparisons. Partial approaches are to check numerical stability at different time step lengths and under different conditions of stiffness, to check conservation (of mass, charge, etc.), verifying steady-state behavior in limiting cases and where possible determining agreement with known analytical results or simplified solutions. Such procedures do not guarantee but help to ensure the correctness of the mathematical model (irrespective of its validity regarding experimental data), computational algorithms, and computer code.

The *validation*, by comparison of model solutions with data, that the model is a reasonable representation of the experimental data with its physicochemical anatomical and mass constraints, is the scientific core of the project, without which the modeling would be a trivial exercise in defining a concept. True "validation" requires testing the model, component-by-component and level-by-level (all the way to the integrated systems level), against high-quality, carefully chosen data. Many of these data sets have already been selected implicitly, for each of the basic biophysical models has been built on experimental data. At this point we do not have all of those data sets in our computer system in digital form. Obtaining them is difficult and tedious, as it often requires going back to the original authors to ask them for their data. This often fails, for many investigators do not store their original data, feeling they have summarized them adequately in their papers; the papers, however, seldom provide enough of the raw observations and leave us with estimated means and variances. This is a big problem and is inspiring the National Institutes of Health (NIH) to demand that authors preserve data. (At the same time, however, the NIH is failing the research community by failing to provide repositories for physiological, pharmacological, immunological, or higher-level clinical information, except that which comes out of clinical trials.)

The digitized data that are available are mostly of time course responses to stimuli such as changes in concentrations, pH, tracer addition, oxygen deprivation, and so on. Other data are observations of enzyme levels, metabolite concentrations, anatomic measures, and composition. Fully detailed metabolic profiles that provide data during transients are needed and await technical developments in the field. We have shown that the use of constrained mass, density, and volume considerations powerfully resolves multiple observations into a self-consistent mass-conservative context for biochemical systems analysis.[108]

For validation we fit dynamical model solutions to the data using *a priori* constraints wherever possible from the anatomy and the thermodynamics. A model is considered a viable "working hypothesis" when the model fit is good for a variegated set of observations, preferably from more than one laboratory. (Examples of validated models from our own lab are found in Refs. 36, 49, and 59.)

REDUCED MODEL FORMS AND THEIR PARAMETERIZATION

Stochastic vs. Markovian, Lumped vs. Distributed Models

Given that one must reduce model complexity, the questions are, "Where are you starting in your thinking about the system?" and "What are the possible approaches?" At the molecular level we know that everything is stochastic: molecules diffuse by jumps with collisions and react when they collide with an appropriate molecule. In 1977, Gillespie[109] gave a clear approach to the mathematics of simulating the system at this level; despite later improvements, the computation is orders of magnitude slower than using the equivalent continuous expressions, assuming stationary or Markovian rate constants. The choice depends on copy numbers, the numbers of molecules within the volume being considered, and whether averaging over many such regions puts one in a situation where the continuous methods are valid.

One approach to capturing the behavior of systems of this type is to represent them as switching between different models, due to an underlying Markov process. Stochastic control theory for this is well developed and can be applied to systems that abruptly "jump" between different models.[110–118] An alternative is to apply methods of adaptive control to automatically adjust models to reflect underlying changes. For example, this approach has been applied in the control of functional electrical stimulation for neural prostheses.[119–131]

For the mathematical modeling, system identification and adaptive control of large-scale biomedical systems, it is necessary to define mathematically, write computer code, and verify computational accuracy for computationally efficient reduced-complexity models for the components at each of the levels of this five-level model. These reduced form models must (1) accurately match the behavior of the full models (for the components of interest, at that level) over a prescribed range of conditions, (2) be computationally more efficient than the full model, (3) interact correctly with other components at the same level, and (4) allow for correct incorporation into the next higher-level models, with the goal being to attain accurate simulations of the desired behavior while minimizing computation.

Cellular models are chosen as the central target for testing model reductions, since they currently exist, are detailed, mechanistic, and extensively studied in computational biology. The reduced model forms use physically meaningful quantities under the constraint that the reduced model forms are descriptive of the behavior of the full models within a defined margin of error over a prescribed range of conditions. Each is an approximate representation that improves computational efficiency but sacrifices some accuracy in being applicable only over a restricted set of conditions.

Reducing a complex chemical reaction system into a simplified kinetic model has had a long history in chemical dynamics, a subfield of chemical physics.[132] One standard approach is to reduce a kinetic model by its time scales. With a given time scale as the objective, there will be fast time scales on which all the kinetics are treated as quasi-steady-state (QSS), and slow time scales on which the dynamic variables are considered constant. This is formally known as singular perturbation in applied mathematics.[133,134] Well-known examples of such are the studies on Belousov-Zhabotinsky reaction (oregonator) and Hodgkin-Huxley theory (FitzHugh-Nagumo).[135] The advantage of this approach is that the reduced model usually preserves a high level of connection with the original mechanistic model; hence, the model parameters are physically meaningful. The drawback is that this is a task for trained specialists.

The general approaches to reduced model derivation that will be pursued involve kinetic model reductions. Essentially all the computational cellular biology models that are based on nonlinear chemical kinetics involve enzymes. There is a large literature (that is still growing) on model reduction in enzyme kinetics. This reflects the difficulty of multiscale model reduction, as well as the central role of enzyme kinetics in computational biology.[136] A careful biochemical analysis of enzyme catalysis or channel or ion pump function usually leads to a detailed mechanistic scheme with multiple steps and multiple intermediate conformational states (species). While molecular biochemists strive for a more complete description of the details, not all of these details are important to the description of the dynamics and physiological functions.[91,92] Because it is impractical to compute large networks of protein reactions at the level of detailed molecular motions, model reduction is useful to move from biophysical to cellular-level integration. The key to faithfully modeling physiological function through model reduction is to know which aspects of the model require preservation at higher levels, and which aspects can be relatively crudely approximated without compromising the robustness of the system.[137]

Discussing kinetics of just one enzyme serves as an illustration. The full model is based on the law of mass action: the full equations force mass conservation and are constrained also by the thermodynamics governing the reversibility. However, a widely used approximation in the biochemical community[138] is the QSS approximation that neglects the rapid loading of the enzyme by substrate,[139] assuming that the enzyme concentration is negligible compared to that of substrate. However, when enzyme and substrate concentrations are similar, then one must account for the binding. Likewise, if the enzyme substrate complex influences another reaction, then the QSS approximation is incorrect or inappropriate as a reduced model.[140] One can often, however, reduce the model, assuming that substrate levels change slowly relative to the fast-binding kinetics (something known as "inner solution" rather than "outer solution" to applied mathematicians). A further reduction is to assume rapid binding and augment the QSS Michaelis-Menten expression with a term accounting exactly for single-site binding. The proposed approach to integrate the QSS and the full equation models with constraint-based flux balance and energy balance analysis is to apply the idea from MCA, in which we determine the sensitivities of the steady-state concentrations as functions of enzyme activities, which are the "control coefficients" of MCA.[85] Since some of the control coefficients are experimentally measured, this gives us the possibility of integrating the Michaelis-Menten flux equations and the MCA approaches with the experimental data as constraints.

For each component subsystem, at each of the levels of the five-level model, there may be several reduced-order models fit from observed inputs and outputs to the component subsystem. These families of identified models will generally be nested and of increasingly complexity (like a series of Russian *matryoshka* nesting dolls). How to determine when or if resorting to lower-level models is important, and deciding which reduced-complexity model is the best choice is quite difficult.

Parameter Identification

Parameter identification for the reduced models will be individual to each reduced form. All require the fitting of the reduced form to the original full model. Reduction in biological systems analysis has many variants. Standard pharmacokinetic analysis usually assumes lumped (instantaneously mixed) volumes of distribution of agents, an oversimplified approach that yields parameters having statistical meaning but having little physiological meaning except to determine associations.[141] These can be improved by accounting for continuously changing volumes where that is appropriate.[142–145] Parameter estimation in nonlinear systems such as muscle and joint dynamics[146,147] is often trickier or requires more data. Generalized kernel approaches, as described by Marmarelis,[148] tend to be descriptive only and use a minimal number of parameters, and therefore lose sight of the biophysical level complexity. Discrete time representation and system identification of electrically stimulated muscle has been used in real-time adaptive control and detection of changes in muscle properties such as fatigue.[120,121,149,150,151] At the molecular level, dynamic algebraic system representation allows for the prediction of local variations of DNA helical parameters, with the base pair sequence (as one goes from one end of the helix to the other) considered as the system input signal, and geometric features of the DNA molecule (at each location along the helix) considered as the output.[152] Real-time gait event detection in spinal cord injury patients during paraplegic walking, using machine intelligence methods,[153,154] bases control decisions upon observations. The simultaneous real-time system identification of nonlinear recruitment properties in the quadriceps muscle of spinal cord injury subjects[117] involved the use of an identification algorithm that imposed equality constraints on the identified parameters, which is a key aspect of the work proposed here in mapping between different levels of model complexity.

Optimization of the fits of models to data for parameter identification should be viewed as a combination of art and science. One thinks automatically of least squares minimization as the method of choice, but this is not a defined choice. The art, or personal bias, comes into the choices of domain (time, frequency, or some other space of representation of data and model), of distance function (linear or log spaces, evenly weighted or points individually weighted to account for variations in experimental error), the choices of which experimental data are most reliable or most consistent with the state to be represented by the model, and on and on. For any given situation it is worthwhile to develop guidelines to assist in weighting the decisions, but there are no guaranteed effective algorithms. Realization of the nature of the art is particularly important in large systems where the model is necessarily a compromise developed to reconcile differences in observations and to render a view that is self-consistent. Choosing amongst variants of weighted recursive least square algorithms is a part of the art. The parameter identification algorithms can be used to fit models that are linear in parameters, which includes linear times series models, linear differential or difference equation models, and linear systems combined with memory-less nonlinearities (which can be expressed polynomials or other functions of the model parameters). These algorithms can be extended to situations that are not linear in parameters, through use of methods such as the extended least squares and generalized least squares algorithms.[155]

Constraints in parameter estimation in large models, when substantial data are available, are often so severe that one would have to consider the parameters to be overdetermined. This is not always the case, as many published models have not been subjected to extensive validity testing against carefully obtained quantitative data. But large models tend to take years to put together and are put together for the purpose of integrating diverse data sets and for reconciling disparities amongst conflicting data sets and mutually contradictory hypotheses. In science, alternative models must be considered, in line with Platt's remonstrance[156] to develop an alternative to one's favorite hypothesis and to design the experiment that distinguishes between two hypotheses: the experiment must disprove one of them and advance science. Thus, the working hypothesis defined by the large model is usually a parsimonious but accurate portrayal of the system compatible with the best data available, and with the contradictions reconciled and removed. Removal of contradictions means that judgment was used; publications should report on how such decisions were made. (This view assumes that the definer of the working hypothesis is determined to provide the most accurate and predictive representation of the system under study.)

The result of this kind of effort is that usually the numbers of data sets available for validation and parameter estimation by far exceeds the numbers of free parameters. In complex interconnected systems like the circulatory and respiratory systems or large biochemical networks, there are many fewer degrees of freedom than there are equations and parameters. This is because in interactive networks the behavior of the system is most heavily influenced by the combined behavior of the set of equations, rather than by individual equations. One can make number counts to push this argument: count the number of data points in the multiple sets of experiments, versus the numbers of free parameters in the model equations. This is not a properly refined way of estimating degrees of freedom—something statisticians would like us to do—but it is in the right direction. Determining formal degrees of freedom is truly difficult, because in both the models and in the data sets, neither the individual equations nor the individual points are truly independent of each other. For example, in analysis of a time series (e.g., a blood pressure response to hemorrhage or a change in glycolytic rate with exercise), the points at each time are correlated with the near neighbors. Likewise, the equations for blood pressure in neighboring segments of the circulation are necessarily related to each other, both in parameter values and in variable values. Extending this view, one can see that the actual degrees of freedom in a closed loop integrated system are many fewer than in an open loop modeling of a part of a system.

The optimization routines used to adjust model parameters to data are useful in getting an idea of the accuracy of parameter estimates. Most provide an estimate of the covariance matrix that gives numerical measures of the correlation among the different free parameters, and they provide confidence limits on parameters when fitting a given data set. How one arrives at these estimates always involves judgment calls; the confidence limits for a parameter are narrowed by imposing constraints upon, or fixing the values of, other parameters. These parameters can be provided by prior information. An example is that the use of anatomic or mass or energy balance constraints on a system reduces the degrees of freedom in the fitting of data and narrows confidence limits on the individual parameters.

Validation of Reduced-Form Models

The validation of reduced-form models has to take into account that there may be a few reduced-form models each an appropriate analog to the full-complexity parent model over a limited range of conditions and parameter settings and therefore, that each version has to be validated. These different forms can either be completely different mathematical equations or different sets of parameter settings. The validation therefore requires several steps; namely validation against the full-complexity model, defining the range of conditions over which the approximation is valid, and reaffirming the validation against experimental data, seeking data over wide ranges of experimental or physiological conditions.

CONTROL OF SYSTEMS LEVEL SIMULATION

When using the full-form systems model, one could presumably simply let it run, changing parameters or variables on the fly to explore the effects of particular interventions or events. This implies that the models have been built to accommodate unsteady states and are not tightly constrained in their validity.

However, with the inclusion of reducing-form modules or elements, adaptation to changing conditions necessitates consideration of the limited ranges of validity of the modules, and one would like to minimize substitutions in the interest of computational efficiency. How to do this is an undeveloped technology, a new art form. One would like to substitute one module for another when the first reaches the edge of its range of validity. This raises the question of the smoothness of the transition from one reduced model form to another. Taking again enzyme kinetics as an example, after finding the two reduced forms for fast time scale and slow time scale (inner and outer solutions) to the full mass-action based kinetics, one needs to establish the "time zone" in which these two solutions join smoothly. In applied mathematics this is known as "matching asymptotics."[132]

Methods used for system identification and adaptive control in pharmacological regulation of physiological processes are useful in thinking about model control. For example, self-tuning adaptive control of arterial pressure in dogs under anesthesia can be as good as equivalent with that of an anesthesiologist,[157] using only the simultaneous regulation of mean arterial pressure and cardiac output with two drugs (sodium nitroprusside and dopamine/dobutamine).[158–160] The method was similar to a detection and control algorithm for a pharmacological stress test.[161] Similar control strategies are built into humans (e.g., adaptive buffering of breath-by-breath variations of end-tidal CO_2)[162] using physiological information to prevent instabilities.[163] Refinements in algorithms for adaptive control help where the identified parameters are subject to inequality constraints,[164] or more generally subject to statistical constraints.[165]

Detecting Lost Accuracy

Detecting when the reduced model forms begin to lose accuracy because of changes in conditions (external or internal), and revising and implementing mechanisms for either resorting to the use of the full subsidiary model or shifting to an alternative form of the reduced model, are a part of the required new art form. Changes in the character of the behavior of the variables (signals) provide information about potential submodel inadequacy.

Time-frequency density functions have been applied in acoustics, radar, machine monitoring, and biomedical signal analysis as a framework for multiscale combinations of density-type functions, such as concentrations and energies, for example, in examining molecular signal processing and detection in T-cells.[166] Recent work in multiscale modulation spectral decompositions[167,168] offers a new approach to signal decomposition, and has been applied to audio coding and detection of acoustic signal change. This modulation spectra approach offers a new approach for multiscale decompositions of potential-like signals, where phase details and shift-invariance properties need to be maintained. In this approach, a signal's components are treated as a product of a slowly varying modulator and a higher-frequency carrier.[167] This modulation spectral approach allows recurrent decompositions into modulators and carriers, thus providing an approach to representation of signal change that integrates across multiple scales.[169]

These methods for recognizing change are almost surely stronger than more conventional techniques (such as correlation dimension; Kolmogorov entropy; approximate entropy; auto and cross correlation; autoregressive, autoregressive moving average, autoregressive integrated moving average, or fractional autoregressive integrated moving average; fractal dimension; or wavelet analysis[170]), but have not yet been proven out in the setting of adaptive model modification.

The needs for biological systems models are quite different from engineering uses of change detection. For example, system variables are usually potentials or energy levels; these require different representations and mathematics. A pressure or voltage potential can be either positive or negative, relative to some reference level. Sums of potentials are well behaved and the linear combinations needed for calculations of usual representations for change detection are thus acceptable mathematical operations. However, energy levels, concentrations, and other observed quantities such as volumes do not have similar mathematical combination properties. For example, energies and concentrations are never legitimately negative. By carefully considering these quantities as densities, which are non-negative and normalized to range between zero and one, estimators for change detection are possible. But the underlying mathematics used for these estimators needs to change to be consistent with density-type quantities.[171] An analogous problem exists in combinations of time-frequency density functions, which have been extensively studied (see, for example, Ref. 172). Multiscale time-frequency density functions have been previously proposed.[173] These techniques then identify variance of model solutions from expected behavior and hand over the task of revising the model by automated means.

AUTOMATED MODEL ADAPTATION TO DETECTED CHANGES IN MODEL VALIDITY

Modulating the Model Form

On detecting a change in "texture" or signal characteristic amongst the variables, the next question is what to do. Methods for either resorting to the use of the full subsidiary model or shifting to an alternative form of the reduced model are complicated if there is more than one possible improved form. Decisions are balanced between demands for speed versus validity. Also, one must keep in mind that the systems-level models demonstrate emergent behavior characteristic of nonlinear dynamical systems, although this is not commonly found to be characterizable as a low-order system with a measurable Lyapunov exponent.[77] For example, if the changed behavior is due to a bifurcation, the representation given by the reduced model forms might still be entirely correct, in which case one should *not* switch to a more complex model. The goal is automated reconfiguration of the model when changes in conditions occur. Which model version to choose might also depend on the resolution and accuracy of the data against which the solution is being matched, but in principle this dependency ought to be secondary.

Three different approaches can be applied to adaptive model modification:

(1) Reparameterizing a reduced-complexity model to make it more accurate under current operating conditions. This approach is often problematic. In "black box" models, the identified parameters of the model may not have intuitive physical meaning. Often "obvious" facts about the system, which can be expressed as constraints on the allowable values of identified parameters, may even be contradicted by the identified parameters of the model.

    This problem can be reduced when, under certain conditions, *a priori* information about the system can be used in the system identification process to obtain parameter estimates that are more realistic and meaningful than if this knowledge is ignored. This is done through the imposition of inequality or equality constraints on the identified parameters. For example, a constraint might be imposed to ensure that the parameters estimated result in a subsystem gain that has the right sign, or that its magnitude is within a known-to-be-true range. The constraints will ensure that the models interact correctly with other components at the same level.

(2) Substituting one reduced-complexity model for another (different Matryoshka dolls) when conditions change or inadequacies are detected. This

approach is useful when prior exploration has defined ranges within which different model variants are applicable. Then it is a simple decision to switch to a different model variant when the system variables being computed cross a defined limit. An important requirement for smoothness in the transition is that the two variants should have very similar behavior in the neighborhood of the defined limit.

(3) Using a higher complexity model at a lower level to recalibrate the reduced form model. This is the last resort. The price is the loss of computational speed. The implication is that when one has been forced to this extreme, none of the reduced forms are valid; perhaps another reduced form model variant can be developed if the circumstances warrant the effort in developing a new model, and coding, verifying, validating, archiving, and placing it in the decision tree.

For all of these parameter identification algorithms, a "weighting factor" can be chosen that allows the model parameters to track slowly varying system changes (by discounting past input and output data, relative to newer information). This use of weightings in the identification algorithms allows the identified reduced-complexity models to be adaptively fit to changing conditions.

SUMMARY

Multiscale multilevel systems modeling is in its infancy, but formulating and operating such models is critical to future success in integrative systems biology. The problems associated with using reduced-form components within large systems models stem primarily from the limitations to their validity. Developing high-level models with such components thus encourages the formulation of multiple subsidiary or component modules that can be substituted for one another when conditions demand it. Such adaptive modeling is presaged by substantial developments in engineering practices, and though it is not at all straightforward, it can be anticipated to apply to the engineering analysis and design of biological systems.

ACKNOWLEDGMENTS

This research was supported by grants from NIH EB 1973-24 and HL 073598-1, and DARPA W81XWH-04. We appreciate the assistance of J.E. Lawson in the preparation of the manuscript.

REFERENCES

1. Bassingthwaighte, J.B., C.Y. Wang & I.S. Chan. 1989. Blood-tissue exchange via transport and transformation by endothelial cells. Circ. Res. **65:** 997–1020.
2. Schwartz, L.M., T.M. Bukowski, J.H. Revkin & J.B. Bassingthwaighte. 1999. Cardiac endothelial transport and metabolism of adenosine and inosine. Am. J. Physiol. Heart Circ. Physiol. **277:** H1241–H1251.
3. Bassingthwaighte, J.B., H. Qian & Z. Li. 1999. The Cardiome Project: An integrated

view of cardiac metabolism and regional mechanical function. Adv. Exp. Med. Biol. **471:** 541–553.
4. BASSINGTHWAIGHTE, J.B. 2001. The modelling of a primitive "sustainable" conservative cell. Philos. Trans. R. Soc. Lond. A **359:** 1055–1072.
5. KING, R.B., J.B. BASSINGTHWAIGHTE, J.R.S. HALES & L.B. ROWELL. 1985. Stability of heterogeneity of myocardial blood flow in normal awake baboons. Circ. Res. **57:** 285–295.
6. KING, R.B., G.M. RAYMOND & J.B. BASSINGTHWAIGHTE. 1996. Modeling blood flow heterogeneity. Ann. Biomed. Eng. **24:** 352–372.
7. VETTER, F.J. & A.D. MCCULLOCH. 1998. Three-dimensional analysis of regional cardiac function: a model of rabbit ventricular anatomy. Prog. Biophys. Mol. Biol. **69:** 157–183.
8. VETTER, F.J. & A.D. MCCULLOCH. 2001. Mechanoelectric feedback in a model of the passively inflated left ventricle. Ann. Biomed. Eng. **29:** 414–426.
9. BASSINGTHWAIGHTE, J.B., C.A. GORESKY & J.H. LINEHAN. 1998. Modeling in the analysis of the processes of uptake and metabolism in the whole organ. *In* Whole Organ Approaches to Cellular Metabolism. J.B. Bassingthwaighte, C.A. Goresky & J.H. Linehan, Eds.: 3–27. Springer Verlag. New York.
10. KEENER, J. & J. SNEYD. 1998. Mathematical Physiology. Springer-Verlag. New York.
11. LAUGER, P. 1991. Kinetic basis of voltage dependence of the Na,K-pump. Soc. Gen. Physiol. Ser. **46:** 303–315.
12. RIDEOUT, V.C. 1991. Mathematical Computer Modeling of Physiological Systems. Prentice Hall. Englewood Cliffs, NJ.
13. GUYTON, A.C., T.G. COLEMAN & H.J. GRANGER. 1972. Circulation: overall regulation. Ann. Rev. Physiol. **34:** 13–46.
14. LU, K., J.W. CLARK, JR., F.H. GHORBEL, *et al.* 2001. A human cardiopulmonary system model applied to the analysis of the Valsalva maneuver. Am. J. Physiol. Heart Circ. Physiol. **281:** H2661–H2679.
15. HODGKIN, A.L. & A.F. HUXLEY. 1952. A quantitative description of membrane current and its application to conduction and excitation in nerve. J. Physiol. **117:** 500–544.
16. BEELER, G.W., JR., & H. REUTER. 1977. Reconstruction of the action potential of ventricular myocardial fibres. J. Physiol. (Lond.) **268:** 177–210.
17. GREENSTEIN, J.L., R. WU, S. PO, *et al.* 2000. Role of the calcium-independent transient outward current I_{to1} in shaping action potential morphology and duration. Circ. Res. **87:** 1026–1033.
18. MAZHARI, R., J.L. GREENSTEIN, R.L. WINSLOW, *et al.* 2001. Molecular interactions between two long-QT syndrome gene products, HERG and KCNE2, rationalized by *in vitro* and *in silico* analysis. Circ. Res. **89:** 33–38.
19. WINSLOW, R.L., J.L. GREENSTEIN, G.F. TOMASELLI & B. O'ROURKE. 2001. Computational models of the failing myocyte: relating altered gene expression to cellular function. Philos. Trans. R. Soc. Lond. A **359:** 1187–1200.
20. NOBLE, D. 2002. Modeling the heart—from genes to cells to the whole organ. Science **295:** 1678–1682.
21. SEDAGHAT, A.R., A. SHERMAN & M.J. QUON. 2002. A mathematical model of metabolic insulin signaling pathways. Am. J. Physiol. Endocrinol. Metab. **283:** E1084–E1101.
22. WESTERHOFF, H.V. & B.O. PALSSON. 2004. The evolution of molecular biology into systems biology. Nat. Biotechnol. **22:** 1249–1252.
23. VO, T.D., H.J. GREENBERG & B.O. PALSSON. 2004. Reconstruction and functional characterization of the human mitochondrial metabolic network based on proteomic and biochemical data. J. Biol. Chem. **279:** 39532–39540.
24. SAUCERMAN, J.J., L.L. BRUNTON, A.P. MICHAILOVA & A.D. MCCULLOCH. 2003. Modeling beta-adrenergic control of cardiac myocyte contractility in silico. J Biol. Chem. **278:** 47997–48003.
25. SAUCERMAN, J.J., S.N. HEALY, M.E. BELIK, *et al.* 2004. Proarrhythmic consequences of a KCNQ1 AKAP-binding domain mutation: computational models of whole cells and heterogeneous tissue. Circ. Res. **95:** 1216–1224.
26. HUANG, C.Y., D.L. BUCHANAN, R.L. GORDON, JR., *et al.* 2003. Increased insulin-like

growth factor-I gene expression precedes ventricular cardiomyocyte hypertrophy in a rapidly-hypertrophying rat model system. Cell. Biochem. Funct. **21:** 355–361.
27. Herrgard, M.J., M.W. Covert & B.O. Palsson. 2004. Reconstruction of microbial transcriptional regulatory networks. Curr. Opin. Biotechnol. **15:** 70–77.
28. Papin, J.A. & B.O. Palsson. 2004. The JAK-STAT signaling network in the human B-cell: an extreme signaling pathway analysis. Biophys. J. **87:** 37–46.
29. Van Oosterhout, M.F.M., T. Arts, J.B. Bassingthwaighte, *et al.* 2002. Relation between local myocardial growth and blood flow during chronic ventricular pacing. Cardiovasc. Res. **53:** 831–840.
30. Wakatsuki, T., J. Schlessinger & E.L. Elson. 2004. The biochemical response of the heart to hypertension and exercise. Trends Biochem. Sci. **29:** 609–617.
31. Bassingthwaighte, J.B. 1974. A concurrent flow model for extraction during transcapillary passage. Circ. Res. **35:** 483–503.
32. Bassingthwaighte, J.B., T. Yipintsoi & R.B. Harvey. 1974. Microvasculature of the dog left ventricular myocardium. Microvasc. Res. **7:** 229–249.
33. Bassingthwaighte, J.B. & B. Winkler. 1982. Kinetics of blood to cell uptake of radiotracers. *In* Biological Transport of Radiotracers. L.G. Columbetti, Ed.: 97–146. CRC Press. Baton Rouge, FL.
34. Bassingthwaighte, J.B., T. Yipintsoi & T.J. Knopp. 1984. Diffusional arteriovenous shunting in the heart. Microvasc. Res. **28:** 233–253.
35. Bassingthwaighte, J.B., H.V. Sparks, Jr., I.S. Chan, *et al.* 1985. Modeling of transendothelial transport. Fed. Proc. **44:** 2623–2626.
36. Bassingthwaighte, J.B., I.S. Chan & C.Y. Wang. 1992. Computationally efficient algorithms for capillary convection-permeation-diffusion models for blood-tissue exchange. Ann. Biomed. Eng. **20:** 687–725.
37. Beard, D.A. & J.B. Bassingthwaighte. 2000. The fractal nature of myocardial blood flow emerges from a whole-organ model of arterial network. J. Vasc. Res. **37:** 282–296.
38. Beyer, R.P., J.B. Bassingthwaighte & A.J. Deussen. 2002. A computational model of oxygen transport from red blood cells to mitochondria. Comput. Methods Programs Biomed. **67:** 39–54.
39. Gorman, M.W., J.B. Bassingthwaighte, R.A. Olsson & H.V. Sparks. 1986. Endothelial cell uptake of adenosine in canine skeletal muscle. Am. J. Physiol. Heart Circ. Physiol. **250:** H482–H489.
40. Kellen, M.R. & J.B. Bassingthwaighte. 2003. An integrative model of coupled water and solute exchange in the heart. Am. J. Physiol. Heart Circ. Physiol. **285:** H1303–H1316.
41. Kellen, M.R. & J.B. Bassingthwaighte. 2003. Transient transcapillary exchange of water driven by osmotic forces in the heart. Am. J. Physiol. Heart Circ. Physiol. **285:** H1317–H1331.
42. King, R.B., A. Deussen, G.R. Raymond & J.B. Bassingthwaighte. 1993. A vascular transport operator. Am. J. Physiol. Heart Circ. Physiol. **265:** H2196–H2208.
43. Kroll, K., T.R. Bukowski, L.M. Schwartz, *et al.* 1992. Capillary endothelial transport of uric acid in the guinea pig heart. Am. J. Physiol. Heart Circ. Physiol. **262:** H420–H431.
44. Kroll, K., N. Wilke, M. Jerosch-Herold, *et al.* 1996. Modeling regional myocardial flows from residue functions of an intravascular indicator. Am. J. Physiol. Heart Circ. Physiol. **271:** H1643–H1655.
45. Kuikka, J., M. Levin & J.B. Bassingthwaighte. 1986. Multiple tracer dilution estimates of D- and 2-deoxy-D-glucose uptake by the heart. Am. J. Physiol. Heart Circ. Physiol. **250:** H29–H42.
46. Levin, M., J. Kuikka & J.B. Bassingthwaighte. 1980. Sensitivity analysis in optimization of time-distributed parameters for a coronary circulation model. Med. Prog. Technol. **7:** 119–124.
47. Li, Z., T. Yipintsoi & J.B. Bassingthwaighte. 1997. Nonlinear model for capillary-tissue oxygen transport and metabolism. Ann. Biomed. Eng. **25:** 604–619.
48. Lightfoot, E.N., J.B. Bassingthwaighte & E.F. Grabowski. 1976. Hydrodynamic models for diffusion in microporous membranes. Ann. Biomed. Eng. **4:** 78–90.
49. Safford, R.E., E.A. Bassingthwaighte & J.B. Bassingthwaighte. 1978. Diffusion of water in cat ventricular myocardium. J. Gen. Physiol. **72:** 513–538.

50. WILKE, N., K. KROLL, H. MERKLE, *et al.* 1995. Regional myocardial blood volume and flow: first-pass MR imaging with polylysine-Gd-DTPA. J. Magn. Reson. Imaging **5:** 227–237.
51. BARTA, E., S. SIDEMAN & J.B. BASSINGTHWAIGHTE. 2000. Facilitated diffusion and membrane permeation of fatty acid in albumin solutions. Ann. Biomed. Eng. **28:** 331–345.
52. BASSINGTHWAIGHTE, J.B.& H. REUTER. 1972. Calcium movements and excitation-contraction coupling in cardiac cells. *In* Electrical Phenomena in the Heart. W.C. DeMello, Ed.: 353–395. Academic Press. New York.
53. BASSINGTHWAIGHTE, J.B., C.H. FRY & J.A.S. MCGUIGAN. 1976. Relationship between internal calcium and outward current in mammalian ventricular muscle; a mechanism for the control of the action potential duration? J. Physiol. **262:** 15–37.
54. BASSINGTHWAIGHTE, J.B. The myocardial cell. 1991. *In* Cardiology: Fundamentals and Practice. 2nd Ed. E.R. Giuliani, V. Fuster, B.J. Gersh, *et al.*, Eds.: 113–149. Mosby-Year Book, Inc. St. Louis, MO.
55. BASSINGTHWAIGHTE, J.B., B. WINKLER & R.B. KING. 1997. Potassium and thallium uptake in dog myocardium. J. Nucl. Med. **38:** 264–274.
56. BASSINGTHWAIGHTE, J.B., Z. LI & H. QIAN. 1998. Blood flows and metabolic components of the cardiome. Prog. Biophys. Mol. Biol. **69:** 445–461.
57. BASSINGTHWAIGHTE, J.B. & K.C. VINNAKOTA. 2004. The computational integrated myocyte: a view into the virtual heart. Ann. N.Y. Acad. Sci. **1015:** 391–404.
58. KUIKKA, J., E. BOUSKELA & J. BASSINGTHWAIGHTE. 1979. D-, L-, and 2-deoxy-D-glucose uptakes in the isolated blood perfused dog hearts. Bibl. Anat. **18:** 239–242.
59. SCHWARTZ, L.M., T.R. BUKOWSKI, J.D. PLOGER & J.B. BASSINGTHWAIGHTE. 2000. Endothelial adenosine transporter characterization in perfused guinea pig hearts. Am. J. Physiol. Heart Circ. Physiol. **279:** H1502–H1511.
60. VAN DER VUSSE, G.J., J.F.C. GLATZ, F.A. VAN NIEUWENHOVEN, *et al.* 1998. Transport of long-chain fatty acids across the muscular endothelium. *In* Skeletal Muscle Metabolism in Exercise and Diabetes. E.A. Richter, H. Galbo, B. Kiens & B. Saltin, Eds. Adv. Exp. Med. Biol. **441:** 181–191.
61. WONG, A.Y.K. & J.B. BASSINGTHWAIGHTE. 1981. The kinetics of Ca-Na exchange in excitable tissue. Math. Biosci. **52:** 275–310.
62. WONG, A.Y.K., A. FABIATO & J.B. BASSINGTHWAIGHTE. 1992. Model of calcium-induced calcium release mechanism in cardiac cells. Bull. Math. Biol. **54:** 95–116.
63. BERS, D.M. 2001. Excitation-Contraction Coupling and Cardiac Contractile Force. 2nd Ed. Kluwer Academic Publishers. Dordrecht, Netherlands.
64. CABRERA, M.E., G.M. SAIDEL & S.C. KALHAN. 1998. Role of O_2 in regulation of lactate dynamics during hypoxia: mathematical model and analysis. Ann. Biomed. Eng. **26:** 1–27.
65. JAFRI, M.S., J.J. RICE & R.L. WINSLOW. 1998. Cardiac Ca^{2+} dynamics: the roles of ryanodine receptor adaptation and sarcoplasmic reticulum load. Biophys. J. **74:** 1149–1168.
66. MICHAILOVA, A. & A. MCCULLOCH. 2001. Model study of ATP and ADP buffering, transport of Ca^{2+} and Mg^{2+}, and regulation of ion pumps in ventricular myocyte. Biophys. J. **81:** 614–629.
67. PUGLISI, J.L. & D.M. BERS. 2001. LabHEART: an interactive computer model of rabbit ventricular myocyte ion channels and Ca transport. Am. J. Physiol. Cell Physiol. **281:** C2049–C2060.
68. SALEM, J.E., G.M. SAIDEL, W.C. STANLEY & M.E. CABRERA. 2002. Mechanistic model of myocardial energy metabolism under normal and ischemic conditions. Ann. Biomed. Eng. **30:** 202–216.
69. BASSINGTHWAIGHTE, J.B. & H.V. SPARKS. 1986. Indicator dilution estimation of capillary endothelial transport. Annu. Rev. Physiol. **48:** 321–334.
70. BASSINGTHWAIGHTE, J.B., L. NOODLEMAN, G.J. VAN DER VUSSE & J.F.C. GLATZ. 1989. Modeling of palmitate transport in the heart. Mol. Cell. Biochem. **88:** 51–58.
71. CALDWELL, J.H., G.V. MARTIN, J.M. LINK, *et al.* 1990. Iodophenylpentadecanoic acid-myocardial blood flow relationship during maximal exercise with coronary occlusion. J. Nucl. Med. **30:** 99–105.

72. CALDWELL, J.H., G.V. MARTIN, G.M. RAYMOND & J.B. BASSINGTHWAIGHTE. 1994. Regional myocardial flow and capillary permeability-surface area products are nearly proportional. Am. J. Physiol. Heart Circ. Physiol. **267:** H654–H666.
73. DEUSSEN, A. & J.B. BASSINGTHWAIGHTE. 1996. Modeling [^{15}O] oxygen tracer data for estimating oxygen consumption. Am. J. Physiol. Heart Circ. Physiol. **270:** H1115–H1130.
74. GORMAN, M.W., R.D. WANGLER, J.B. BASSINGTHWAIGHTE, *et al.* 1991. Interstitial adenosine concentration during norepinephrine infusion in isolated guinea pig hearts. Am. J. Physiol. Heart Circ. Physiol. **261:** H901–H909.
75. WANGLER, R.D., M.W. GORMAN, C.Y. WANG, *et al.* 1989. Transcapillary adenosine transport and interstitial adenosine concentration in guinea pig hearts. Am. J. Physiol. Heart Circ. Physiol. **257:** H89–H106.
76. BASSINGTHWAIGHTE, J.B., R.B. KING, & S.A. ROGER. 1989. Fractal nature of regional myocardial blood flow heterogeneity. Circ. Res. **65:** 578–590.
77. BASSINGTHWAIGHTE, J.B., L.S. LIEBOVITCH & B.J. WEST. 1994. Fractal Physiology. Oxford University Press. New York, London.
78. BASSINGTHWAIGHTE, J.B. & D.A. BEARD. 1995. Fractal ^{15}O-water washout from the heart. Circ. Res. **77:** 1212–1221.
79. BASSINGTHWAIGHTE, J.B., D.A. BEARD & Z. LI. 2001. The mechanical and metabolic basis of myocardial blood flow heterogeneity. Basic Res. Cardiol. **96:** 582–594.
80. QIAN, H. & J.B. BASSINGTHWAIGHTE. 2000. A class of flow bifurcation models with lognormal distribution and fractal dispersion. J. Theor. Biol. **205:** 261–268.
81. DASH, R.K. & J.B. BASSINGTHWAIGHTE. 2004. Blood HbO_2 and $HbCO_2$ dissociation curves at varied O_2, CO_2, pH, 2,3-DPG and temperature levels. Ann. Biomed. Eng. **32:** 1676–1693.
82. DASH, R.K., Z. LI & J.B. BASSINGTHWAIGHTE. 2005. Simultaneous blood-tissue exchange of oxygen, carbon dioxide, bicarbonate, and hydrogen ion. Ann. Biomed. Eng. In press.
83. BASSINGTHWAIGHTE, J.B. & G.A. RAYMOND. 2005. A multi-metabolite model for blood-tissue exchange. Philos. Trans. R. Soc. Lond. A. In press.
84. QIAN, H. 2003. Thermodynamic and kinetic analysis of sensitivity amplification in biological signal transduction. Biophys. Chem. **105:** 585–593.
85. FELL, D.A. 1997. Understanding the Control of Metabolism. Portland Press. London.
86. BEARD, D.A., H. QIAN & J.B. BASSINGTHWAIGHTE. 2004. Stoichiometric foundation of large-scale biochemical system analysis. *In* Modeling in Molecular Biology. G. Ciobanu & G. Rozenberg, Eds.: 1–19. Springer-Verlag Berlin Heidelberg. New York.
87. RAMAKRISHNA, R., J.S. EDWARDS, A. MCCULLOCH & B.O. PALSSON. 2001. Flux-balance analysis of mitochondrial energy metabolism: consequences of systemic stoichiometric constraints. Am. J. Physiol. Regul. Integr. Comp. Physiol. **280:** R695–R704.
88. GRISOLIA, S. 1964. The catalytic environment and its biological implications. Physiol. Rev. **44:** 657–714.
89. KRAUSS, G. 2003. Biochemistry of Signal Transduction and Regulation: Third, Completely Revised Edition. Wiley-VCH. Darmstadt, Germany.
90. LI, G. & H. QIAN. 2002. Kinetic timing: a novel mechanism for improving the accuracy of GTPase timers in endosome fusion and other biological processes. Traffic **3:** 249–255.
91. QIAN, H., G.M. RAYMOND & J.B. BASSINGTHWAIGHTE. 1998. On two-dimensional fractional Brownian motion and fractional Brownian random field. J. Phys. A: Math. Gen. **31:** L527–L535.
92. QIAN, H., G.M. RAYMOND & J.B. BASSINGTHWAIGHTE. 1999. Stochastic fractal behavior in concentration fluctuation and fluorescence correlation spectroscopy. Biophys. Chem. **80:** 1–5.
93. QIAN, H., S. SAFFARIAN & E.L. ELSON. 2002. Concentration fluctuations in a mesoscopic oscillating chemical reaction system. Proc. Natl. Acad. Sci. U.S.A. **99:** 10376–10381.
94. BEARD, D.A. & H. QIAN. 2004. Thermodynamic-based computational profiling of cellular regulatory control in hepatocyte metabolism. Am. J. Physiol. Endocrinol. Metab. **288:** E633–E644.

95. SIGG, D., H. QIAN & F. BEZANILLA. 1999. Kramer's diffusion theory applied to gating kinetics of voltage-dependent ion channels. Biophys. J. **76:** 782–803.
96. CHAPMAN, J.B., J.M. KOOTSEY & E.A. JOHNSON. 1979. A kinetic model for determining the consequences of electrogenic active transport in cardiac muscle. J. Theor. Biol. **80:** 405–425.
97. LUO, C.-H. & Y. RUDY. 1994. A dynamic model of the cardiac ventricular action potential I. Simulations of ionic currents and concentration changes. Circ. Res. **74:** 1071–1096.
98. KLODOS, I. & J.C. SKOU. 1975. The effect of Mg2+ and chelating agents on intermediary steps of the reaction of Na+,K+-activated ATPase. Biochim. Biophys. Acta **391:** 474–485.
99. SKOU, J.C. 1974. The (Na^+ + K^+) activated enzyme system and its relationship to transport of sodium and potassium. Q. Rev. Biophys. **7:** 401–434.
100. BEARD, D.A. & J.B. BASSINGTHWAIGHTE. 2000. Advection and diffusion of substances in biological tissues with complex vascular networks. Ann. Biomed. Eng. **28:** 253–268.
101. BEARD, D.A. & J.B. BASSINGTHWAIGHTE. 2001. Modeling advection and diffusion of oxygen in complex vascular networks. Ann. Biomed. Eng. **29:** 298–310.
102. KROLL, K., D.J. KINZIE & L.A. GUSTAFSON. 1997. Open system kinetics of myocardial phosphoenergetics during coronary underperfusion. Am. J. Physiol. Heart Circ. Physiol. **272:** H2563–H2576.
103. TAKAHASHI, E. & K. ASANO. 2002. Mitochondrial respiratory control can compensate for intracellular O_2 gradients in cardiomyocytes at low P_{O_2}. Am. J. Physiol. Heart Circ. Physiol. **283:** H871–H878.
104. WILSON, D.F., C.S. OWEN & A. HOLIAN. 1977. Control of mitochondrial respiration: a quantitative evaluation of the roles of cytochrome c and oxygen. Arch. Biochem. Biophys. **182:** 749–762.
105. WILSON, D.F. 1994. Factors affecting the rate and energetics of mitochondrial oxidative phosphorylation. Med. Sci. Sports Exerc. **26:** 37–43.
106. KOSZALKA, P., B. OZUYAMAN, Y. HUO, *et al.* 2004. Targeted disruption of cd73/ecto-5′-nucleotidase alters thromboregulation and augments vascular inflammatory response. Circ. Res. **95:** 814–821.
107. OLSSON, R.A. 2004. Cardiovascular ecto-5′-nucleotidase: an end to 40 years in the wilderness? Circ. Res. **95:** 752–753.
108. VINNAKOTA, K. & J.B. BASSINGTHWAIGHTE. 2004. Myocardial density and composition: a basis for calculating intracellular metabolite concentrations. Am. J. Physiol. Heart Circ. Physiol. **286:** H1742–H1749.
109. GILLESPIE, D.T. 1977. Exact stochastic simulation of coupled chemical reactions. J. Phys. Chem. **81:** 2340–2361.
110. CHIZECK, H.J., A.S. WILLSKY & D. CASTANON. 1986. Discrete-time Markovian-jump linear quadratic optimal control. Int. J. Control. **43:** 213–231.
111. FENG, X., K.A. LOPARO, Y. JI & H.J. CHIZECK. 1992. Stochastic stability properties of jump linear systems. IEEE Trans. Automatic Control. **37:** 38–53.
112. JI, Y. & H.J. CHIZECK. 1988. Controllability, observability and discrete-time Markovian jump linear quadratic control. Int. J. Control **48:** 481–498.
113. JI, Y. & H.J. CHIZECK. 1989. Optimal quadratic control of jump linear systems with separately controlled transition probabilities. Int. J. Control **49:** 481–491.
114. JI, Y. & H.J. CHIZECK. 1989. Bounded sample path control of discrete time jump linear systems. IEEE Trans. Syst. Man Cybern. **19:** 277–284.
115. JI, Y. & H.J. CHIZECK. 1990. Jump linear quadratic Gaussian control. Steady-state solution and testable conditions. Control Theory Adv. Technol. **6:** 289–319.
116. JI, Y. & H.J. CHIZECK. 1990. Controllability, stabilizability, and continuous-time Markovian jump linear quadratic control. IEEE Trans. Autom. Control **37:** 777–788.
117. JI, Y., H.J. CHIZECK, X. FENG & K.A. LOPARO. 1991. Stability and control of discrete-time jump linear systems. Control Theory Adv. Technol. **7:** 247–270.
118. JI, Y. & H.J. CHIZECK. 1992. Jump linear quadratic Gaussian control in continuous time. IEEE Trans. Autom. Control **37:** 1884–1892.
119. ABBAS, J.J. & H.J. CHIZECK. 1991. Feedback control of coronal plane hip angle in

paraplegic subjects using functional neuromuscular stimulation. IEEE Trans. Biomed. Eng. **38:** 687–698.

120. ABBAS, J.J. & H.J. CHIZECK. 1995. Neural network control of functional neuromuscular stimulation systems: computer simulation studies. IEEE Trans. Biomed. Eng. **42:** 1117–1127.
121. BERNOTAS, L.A., P.E. CRAGO & H.J. CHIZECK. 1986. A discrete-time model of electrically stimulated muscle. IEEE Trans. Biomed. Eng. **33:** 829–838.
122. BERNOTAS, L.A., P.E. CRAGO & H.J. CHIZECK. 1987. Adaptive control of electrically stimulated muscle. IEEE Trans. Biomed. Eng. **34:** 140–147.
123. CHIZECK, H.J., P.E. CRAGO & L.S. KOFMAN. 1988. Robust closed-loop control of isometric muscle force using pulsewidth modulation. IEEE Trans. Biomed. Eng. **35:** 510–517.
124. CHIZECK, H.J., N. LAN, L.S. PALMIERI & P.E. CRAGO. 1991. Feedback control of electrically stimulated muscle using simultaneous pulse width and stimulus period modulation. IEEE Trans. Biomed. Eng. **38:** 1224–1234.
125. CRAGO, P.E, R.J. NAKAI & H.J. CHIZECK. 1991. Feedback regulation of hand grasp opening and contact force during stimulation of paralyzed muscle. IEEE Trans. Biomed. Eng. **38:** 17–28.
126. KOLACINSKI, R.M., W. LIN & H.J. CHIZECK. 2000. Control of an antagonistic biomimetic actuator system. Int. J. Control **73:** 804–818.
127. LAN, N., P.E. CRAGO & H.J. CHIZECK. 1991. Control of end-point forces of a multijoint limb by functional neuromuscular stimulation. IEEE Trans. Biomed. Eng. **38:** 953–965.
128. LAN, N., P.E. CRAGO & H.J. CHIZECK. 1991. Feedback control methods for task regulation by electrical stimulation of muscles. IEEE Trans. Biomed. Eng. **38:** 1218–1223.
129. PETERSON, D.K. & H.J. CHIZECK. 1987. Linear quadratic control of a loaded agonist-antagonist muscle pair. IEEE Trans. Biomed. Eng. **34:** 790–796.
130. VELTINK, P.H., H.J. CHIZECK, P.E. CRAGO & A. EL-BIALY. 1992. Nonlinear joint angle control for artificially stimulated muscle. IEEE Trans. Biomed. Eng. **39:** 368–380.
131. WILHERE, G.F., P.E. CRAGO & H.J. CHIZECK. 1985. Design and evaluation of a digital closed-loop controller for the regulation of muscle force by recruitment modulation. IEEE Trans. Biomed. Eng. **32:** 668–676.
132. EPSTEIN, I.R. & J.A. POJMAN. 1998. An Introduction to Nonlinear Chemical Dynamics. Oxford University Press. London.
133. KEVORKIAN, J. & J.D. COLE. 1996. Multiple Scale and Singular Perturbation Methods. Springer-Verlag. Berlin.
134. MURRAY, J.D. 1990. Mathematical Biology. Springer-Verlag. New York.
135. MURRAY, J.D. 1977. Lectures on Nonlinear-Differential-Equation Models in Biology. Clarendon Press. Oxford.
136. DE ATAURI, P., D. ORRELL, S. RAMSEY & H. BOLOURI. 2004. Evolution of 'design' principles in biochemical networks. Syst. Biol. **1:** 28–40.
137. KITANO, H. Biological robustness. 2004. Nat. Rev. Gen. **5:** 826–837.
138. SCHNELL, S. & P.K. MAINI. 2003. A century of enzyme kinetics: reliability of the K_M and v_{max} estimates. Comm. Theoret. Biol. **8:** 169–187.
139. PALSSON, B.O. 1987. On the dynamics of the irreversible Michaelis-Menten reaction mechanism. Chem. Eng. Sci. **42:** 447–458.
140. SCHNELL, S. & P.K. MAINI. 2000. Enzyme kinetics at high enzyme concentration. Bull. Math. Biol. **62:** 483–499.
141. CABRERA, M.E. & H.J. CHIZECK. 1996. On the existence of a lactate threshold during incremental exercise: a systems analysis. J. Appl. Physiol. **80:** 1819–1828.
142. OLIVERO, W.C., H.L. REKATE, H.J. CHIZECK, *et al.* 1988. Relationship between intracranial and sagittal sinus pressure in normal and hydrocephalic dogs. Pediatr. Neurosci. **14:** 196–201.
143. REKATE, H.L., S. ERWOOD, J.A. BRODKEY, *et al.* 1985–1986. Etiology of ventriculomegaly in choroid plexus papilloma. Pediatr. Neurosci. **12:** 196–201.
144. REKATE, H.L., J.A. BRODKEY, H.J. CHIZECK, *et al.* 1988. Ventricular volume regulation: a mathematical model and computer simulation. Pediatr. Neurosci. **14:** 77–84.

145. REKATE, H.L., F.C. WILLIAMS, JR., J.A. BRODKEY, *et al.* 1988. Resistance of the foramen of Monro. Pediatr. Neurosci. **14:** 85–89.
146. SHUE, G., P.E. CRAGO & H.J. CHIZECK. 1995. Muscle-joint models incorporating activation dynamics, moment-angle, and moment-velocity properties. IEEE Trans. Biomed. Eng. **42:** 212–223.
147. STEIN, R.B., E.P. ZEHR, M.K. LEBIEDOWSKA, *et al.* 1996. Estimating mechanical parameters of leg segments in individuals with and without physical disabilities. IEEE Trans. Rehabil. Eng. **4:** 201–211.
148. MARMARELIS, V.Z. 2004. Nonlinear Dynamic Modeling of Physiological Systems. Wiley-Interscience. Hoboken, NJ.
149. CHIA, T.L., P.C. CHOW & H.J. CHIZECK. 1991. Recursive parameter identification of constrained systems: an application to electrically stimulated muscle. IEEE Trans. Biomed. Eng. **38:** 429–442.
150. CHIZECK, H.J., S. CHANG, R.B. STEIN, *et al.* 1999. Identification of electrically stimulated quadriceps muscles in paraplegic subjects. IEEE Trans. Biomed. Eng. **46:** 51–61.
151. ERFANIAN, A., H.J. CHIZECK & R.M. HASHEMI. 1998. Using evoked EMG as a synthetic force sensor of isometric electrically stimulated muscle. IEEE Trans. Biomed. Eng. **45:** 188–202.
152. HAWLEY, S. & H.J. CHIZECK. 2004. Parameter estimation on algebraic systems: application to sequence dependent structure of DNA. IASTED Conference on Biomedical Engineering (BioMED 2004). Paper No. 417-022. February 2004.
153. NG, S.K. & H.J. CHIZECK. 1997. Fuzzy model identification for classification of gait events in paraplegics. IEEE Trans. Fuzzy Systems **5:** 536–543.
154. SKELLY, M.M. & H.J. CHIZECK. 2001. Real-time gait event detection for paraplegic FES walking. IEEE Trans. Neural Syst. Rehabil. Eng. **9:** 59–68.
155. LJUNG, L. 1999. System Identification: Theory for the User. 2nd Ed. PTR Prentice-Hall. Upper Saddle River, NJ.
156. PLATT, J.R. 1964. Strong inference. Science **146:** 347–353.
157. STERN, K.S., H.J. CHIZECK, B.K. WALKER, *et al.* 1985. The self-tuning controller: comparison with human performance in the control of arterial pressure. Ann. Biomed. Eng. **13:** 341–357.
158. VOSS, G.I., H.J. CHIZECK & P.G. KATONA. 1987. Regarding self-tuning controllers for nonminimum phase plants. Automatica **23:** 405–408.
159. VOSS, G.I., P.G. KATONA & H.J. CHIZECK. 1987. Adaptive multivariable drug delivery: control of arterial pressure and cardiac output in anesthetized dogs. IEEE Trans. Biomed. Eng. **34:** 617–623.
160. VOSS, G.I., H.J. CHIZECK & P.G. KATONA. 1988. Self-tuning controller for drug delivery systems. Int. J. Control **47:** 1507–1520.
161. VALCKE, C.P. & H.J. CHIZECK. 1997. Closed-loop drug infusion for control of heart-rate trajectory in pharmacological stress tests. IEEE Trans. Biomed. Eng. **44:** 185–195.
162. MODARRESZADEH, M., K.S. KUMP, H.J. CHIZECK, *et al.* 1993. Adaptive buffering of breath-by-breath variations of end-tidal CO2 of humans. J. Appl. Physiol. **75:** 2003–2012.
163. TIMMONS, W.D., H.J. CHIZECK & P.G. KATONA. 1991. Adaptive control is enhanced by background estimation. IEEE Trans. Biomed. Eng. **38:** 273–279.
164. TIMMONS, W.D., H.J. CHIZECK, F. CASAS, *et al.* 1997. Parameter-constrained adaptive control. Ind. Eng. Chem. Res. **36:** 4894–4905.
165. CHIA, T.L., D. SIMON & H.J. CHIZECK. 2005. Kalman filtering with statistical state constraints. Int. J. Control. Intelligent Systems. In press.
166. KEANE, J.F. & L.E. ATLAS. 2002. Molecular signal processing and detection in T-cells. Proc. IEEE Int. Conf. Acoustics Speech Signal Processing. **2:** 1597–1600.
167. ATLAS, L. & S.A. SHAMMA. 2003. Joint acoustic and modulation frequency. EURASIP J. Appl. Sign. Process **2003:** 668–675.
168. LI, Q. & L. ATLAS. 2005. Properties for modulation spectral filtering. Proc. IEEE Int. Conf. Acoustics Speech Signal Processing, March, Philadelphia, PA.

169. Sukittanon, S., L.E. Atlas & J.W. Pitton. 2004. Modulation-scale analysis for content identification. IEEE Trans. Signal Process. **52:** 3023–3035.
170. Kingsbury, N. 1999. Image processing with complex wavelets. Philos. Trans. R. Soc. Lond. A **357:** 2543–2560.
171. Goyal, V.K., J. Kovacevic & J.A. Kelner. 2001. Quantized frame expansions with erasures. Appl. Comput. Harmon. Anal. **10:** 203–233.
172. Loughlin, P., J. Pitton & B. Hannaford. 1994. Approximating time-frequency density functions via optimal combinations of spectrograms. IEEE Signal Proc. Lett. **1:** 199–202.
173. Umesh, S., L. Cohen, N. Marinovic & D.J. Nelson. 1999. Scale transform in speech analysis. IEEE Trans. Speech Audio Process. **7:** 40–45.

Index of Contributors

Ann. N.Y. Acad. Sci. 1047: 425–428 (2005).